Mechanical Waves: Free, Forced, and Transition

Leonid I. Slepyan

Mechanical Waves: Free, Forced, and Transition

Springer

Leonid I. Slepyan
Faculty of Engineering
Tel Aviv University
Tel Aviv, Israel

ISBN 978-3-032-15863-5 ISBN 978-3-032-15864-2 (eBook)
https://doi.org/10.1007/978-3-032-15864-2

This Springer imprint is published by the registered company Springer Nature Switzerland AG
The registered company address is: Gewerbestrasse 11, 6330 Cham, Switzerland

Acknowledgements I would like to take this opportunity to express my deep gratitude to Profs. Gennady Mishuris and Alexander Movchan for their constant support and advice, without which this book would hardly have seen the light of day.

Competing Interests The author has no competing interests to declare that are relevant to the content of this manuscript.

Introduction

Waves are the universal manifestation of matter—from surface waves on water and seismic waves, to light and gravity, and from waves in media and structures to the microscopic behavior of particles. First of all, the waves differ by their physical nature. Gravitational waves, as a gravitation field dynamics, are fundamental. They propagate through (and interact with) matter, and their speed is detected uniquely by observes independent of their own speeds. Electromagnetic waves propagate with the same fundamental speed (as they say, *in an empty space*). This speed is usually called the *light speed*), and it is a physical constant.

Mechanical waves are different: they depend on the properties of the medium and the structure through which they propagate. Therefore, it is not surprising that there exist many types of mechanical waves.

The diversity of waveguides leads to the diversity of mechanical waves.

In this connection, the helical structure presents a nontrivial example. Under an axial tension, it should rotate according its structure; however, no angular momentum can arise, since there is no such an action. The contradiction is resolved by the emergence of a *binary wave*, Sect. 3.2.4

As a simplest example of a wave, we present a line of uniformly distributed rigid beads, that provides a clear representation a wave, Sect. 1.1.1.

The waveguide can be represented, in particular, by a string or a beam 1D models, by liquid or elastic media, and by their types: longitudinal and transverse, tension-compression and shear waves in space, and the surface waves. Although, they are characterized by their form, amplitude and the *wave velocity* (we use this term for the velocity of the wave front/quasifront). Note that there are waves consisting of some parts moving with different speeds, as, for example, a fracture wave.

In this book, we consider general characteristics of mechanical waves and explore their specific properties. In this way, we study waves in various media and structures—both continuous and discrete—including waveguides resting on dynamic foundation equipped with a dynamic microstructure. Note that waves accompanies a quasi-static crack grows, Slepyan (1981), Bitzek et al. (2015).

The "In brief" part provides a first-level theory of the free, forced, and transition waves. It introduces the definition of mechanical waves as disturbances propagating *through* a medium, and presents the mechanical waves in comparison with the gravitational and electromagnetic fundamental waves.

In the complex-valued representation valid for linear waves, the wave frequency as a function of the wavenumber (7), that is a function of the wave length, defines the wave dispersion.

In this context, the phase velocity represents the *visible* motion of the wave and is defined as the ratio of the frequency to the wave number. The group velocity, which corresponds to the propagation of wave energy and the advance of the wave front or quasifront, is defined as the derivative of frequency with respect to wave number.[1] The phase and group velocities can differ not only in magnitude but also in direction, as it is demonstrated by waves in a compressed beam rested on an elastic foundation, Sect. 1.2.8.

If a wave requires energy to propagate (and to grow in the resonant case, Sect. 1.3.5), this energy can generally be supplied either by an external action or directly from the waveguide itself, provided the latter possesses a distributed energy source. The corresponding mechanisms are discussed in the section Forced Waves, Sect. 1.3, where the energy flux and the phase shift between the applied load and the wave phase velocity are determined, Sect. 1.3.7.

Along with such a *feeding energy*, the transient wave with a front/qusifront can spend energy to overcome a resistant to its spread. The simplest example is a 1D pressure wave reflecting at an obstacle moving under its pressure, Sect. 1.2.6.

Note that the energy partition (as the relationship between kinetic and potential energy) remained unrecognized for a long time (Lord Rayleigh noted this uncertainty in 1877), until it was clarified in 2015 for linear (where there is equipartition) and some classes of nonlinear waves, Sect. 1.3.6, Sect. 2.2.

In this book, the wave mass and momentum carried by waves in a uniform waveguide is considered in general, based on the conservation laws and a kind of the causality principle, Slepyan (2016). Specifically, we examine (without resorting to constitutive data) steady-state waves of an arbitrary shape, periodic waves which speed (the front/quasifront speed) differs from the speed of its form and a binary wave carrying self-equilibrated momentum, Sect. 3.2.4. The approach allows us to represent momentum as a product of the *wave mass* and the wave speed, where the propagating wave mass, positive or negative, is defined as the excess of that in the wave over its initial value of the waveguide. This general representation is valid for mechanical waves of arbitrary nature and intensity.

Then, we study the basic wave properties. We show that the orthogonality of waves of different lengths follows directly from Betti's theorem and a symmetry condition, without resorting to constitutive data. This three-line statement of the Orthogonality Theorem is a good example of the fact that **generality leads to simplicity**. We

[1] Note that the term group velocity reflects the conventional manner of its definition, although in fact it corresponds to single waves.

further determine the conditions for wave localization along its propagation line and illustrate both general statements with specific examples.

Concerning the wave excitation, it can be seen in the $k-\omega$ plane with the dispersion relation, $\omega(k)$, is plotted. There are three possibilities for the uniformly moving-oscillation load with the load line on that plane

$$\omega = \omega_0 + kv, \tag{(1)}$$

where ω_0 is the unmoving load frequency, and v is the speed of the load:

(a) The load line does not cross the dispersion relation graph. In this case, no propagating wave is excited.
(b) The load line crosses (but not tangent to) the dispersion relation graph. Now the load excites waves hose frequencies and wavenumbers correspond to those of the cross points.
(c) The load excites a resonant wave growing in time—for the load line tangent to the dispersion relation graph.

In the integral representation of a forced wave, the above possibilities correspond to a regular expression (a), to the first-order pole (b), and to a higher singularity (c).

In this book, we obtain expressions for the energy flux from the load to the wave, and determine the phase shift between the oscillating load and the energy flux-related the body displacement rate. For example, in a regular (non-resonant) periodic excitation of an oscillator, the phase shift $\varphi = \pi/2$, since the energy flux averaged is at zero. However, in the resonant excitation, $\varphi = 0$.

In the Brief part, we also consider several types of transition waves—those that transform the waveguide including waves in a bistable structure.

Along with the general properties of mechanical waves, waves in elastic space and half-space are consider in detail in the Dynamic Elasticity part, Part IV.

Together with the linear waves, we consider some types of nonlinear waves. As already noted, there exist general statement valid for both the linear and nonlinear waves.

In conclusion, we discuss here the general and specific properties of mechanical waves, without attempting to give an exhaustive account of wave theory, since the diversity of waves is immense.

Contents

Part IV Some Nonlinear Waves

Part VI Some General Relations

Chapter 1
On Waves in Brief

1.1 Preliminaries

1.1.1 What Is Wave

Among many types of mechanical waves, the wave on water is likely the first one that comes to mind when you hear this term. At the same time, surface water waves allow us to detect directly some fundamental properties of mechanical waves.

Looking at a wavy water surface, one can see that while the wave propagates, any small floating body (as the water particles) moves up and down, fore and back, remaining almost in the same place. So, the water surface wave, in contrast to a water flux, travels *through* the water. That is the main distinguishing feature of mechanical waves.

The *wave is a disturbance propagating through the medium*, although a (usually relatively slow) drift of the latter, the Stokes drift, may accompany the wave.

Note that some *dynamic flows* with a moving front are also often called waves, as in the case of a tsunami; it is not a wave in our meaning.

Throw a pebble on calm water. You may see another remarkable event: the wave crests propagate faster than the wavefront (in fact, it is not a sharp front but a slowly expanding area, a *quasifront*). The crests disappear reaching the latter.

This observation reflects another general wave property:

The wave *phase velocity* can differ from the front (or quasifront) velocity. The latter corresponds to the *energy flux* (in the linear formulation, it is the *group velocity*).

So, the three values of the velocity characterize the steady-state wave propagation: the front/quasifront velocity–the wave velocity, the velocity of its form, and the particle velocity. For linear waves the first two are the *group velocity and the* phase velocity, respectively.

L. I. Slepyan, *Mechanical Waves: Free, Forced, and Transition*,
https://doi.org/10.1007/978-3-032-15864-2_1

Concerning the velocity, we note that, in general, any movement is relative, because there is no absolute frame in nature. The only electromagnetic and gravitational wave speed is absolute in the sense that it is the same for any moving observer. There is no such uncertainty for mechanical waves, whose velocity is reasonably considered as that relative to the disturbed medium, or a structure (the initial state ahead of the wave), through which the wave propagates.

We start with the simplest example of a wave, which can be demonstrated at home.

1.1.1.1 Collision Wave

Consider a finite number of identical rigid beads periodically placed along a straight line.

Let the beads be initially at rest, and let, under an impact, the left beads move to the right, to the next one, at speed v.

It follows from the energy and momentum conservation laws that, under the perfect collision, the first bead transfers its speed to the second one, and so on. Thus, the collision wave propagates at speed $V = v(a + d)/a$, where a is the initial between-beads distance, and d is the bead diameter. Finally, the last bead starts to move at speed v, while the others displaced by the distance a are at rest.

In this model, $V \to \infty$ ($a \to 0$), and at $a = 0$ after the impact, the last bead rebounds, while the others remain at the initial places.

1.1.2 Ggravitational, Electromagnetic, and Mechanical Waves

The gravitational field, generally dynamic, is the most fundamental field in nature. It defines the natural reference frame, determining the trajectories of freely moving bodies, as well as those of light and other waves. The gravitational wave predicted by Albert Einstein in 1916 was first detected in 2015 (Abbott et al., 2016). It was an awe-inspiring event occurred at a distance of one and a half billion light-years (one and a half billion years ago) when two black holes lost in collision 5% of their mass radiated by the gravitational wave.

Although light is fundamental in the sense that, under a given gravitational field, different observers detect the same its speed.

Christian Huygens proposed the wave theory of light In 1678. Light and other electromagnetic waves, different by frequency ranges, propagate at the same speed as the gravitational waves. The speed $c = 3.00 \times 10^8$ m/s (more precisely 299792458 m/s) relative any observer. It is the limiting speed, which nobody of a nonzero rest mass can be close but cannot reach. This speed is a fundamental physical constant.[1]

These original values for regions of different gravity levels, measured at a *fixed region*, turn out to be dependent on the gravity level (the phenomenon known as a *gravitational time delation*).

Light possess energy and the corresponding mass (the *mass of movement*), as follows from Einstein's fundamental mass-energy relation

$$E = mc^2 , \tag{1.2}$$

where E is the energy, and m is the mass (relativistic mass). The light has no *rest mass* but only the mass of movement: $m = E/c^2$.

For a particle (or a body) of the rest mass $m_0 > 0$, moving at speed v, the above relation (1.2) is still valid, but with the mass $m = \Lambda m_0$, where the Lorenz factor

$$\Lambda = \frac{1}{\sqrt{1 - v^2/c^2}} . \tag{1.3}$$

Thus, the other masses attract light. This conclusion was confirmed by measuring the deflection of the stellar beam caused by the attraction of the Moon. The results of the measurements coincided with those predicted by Albert Einstein's theory.

Mechanical waves, as already noted, propagate due to the motion of a medium, and its speed is much lower. The highest speeds of mechanical waves are of the order of $10^{-5}c$. For example, the *Rayleigh waves* localized at the Earth's surface propagate at speed of the order of 3.000m/s.

The speed of sound in air is 340m/s, and impulses in a nerve propagate not faster than 100m/s. The speed of *capillary waves* on the water is only 1/4 m/s. So, as a rule, mechanical waves are the subject of classical physics.

1.1.3 Linear and Nonlinear Waves

The linear formulation assumes that the considered body under external actions remains physically and geometrically unchanged despite strains and displacements. In particular, the boundary conditions correspond to the fixed counter (surface). With this formulation, the *superposition principle* is valid.

The nonlinearity can have a geometrical or physical nature. The former usually becomes essential when the rotation of a material element is not small enough. The

---

[1] There are three related original physical quantities, *speed*: the speed of light, *time: 1 second is defined as the duration of 9,192,631,770 periods of the radiation corresponding to the transition between two hyperfine energy levels of the ground state of the cesium-133 atom, and* mass: m_{Planck} is defined as $2.176 * 10^{-8}$ kg, where the Planck mass $m_{Planck} = \sqrt{hc/(2\pi G)}$ ($h = 6.62607015 * 10^{-34} Js$ is the Planck constant, c is the light speed, and $G = 6.67430 * 10^{-11}\,\hat{A}\, m^3/(kgs^2)$ is the gravitational constant).

physical nonlinearity corresponds to nonlinear stress-strains (or stress-displacement) relations. The nonlinear formulation can lead to some effects that disappear under linearization.

For example, a hardening nonlinearity leads to a shock wave with an energy release in its front The solitary waves represent another example. The class of very stable *solitary waves* called *solitons*, as if they were solids, originated from the water solitary wave in a canal discovered by John Scott Russell in 1844 (Russell, 1845a). A theory of this wave was developed 60 years later by Korteweg and de Vries (1895), who introduced the governing equation (now well-known as the KdV equation). Enrico Fermi, John Pasta, and Stanislaw Ulam (1955a) found such a solitary wave, the FPU wave, in a nonlinear chain.

Slepyan et al. (1995a) and Slepyan et al. (1998) found solitary waves in an inextensible and extensible helical fibers, respectively. Nonlinear waves theory is discussed below, Part V.

1.2 Free Waves

1.2.1 Wave Equation

The (nonuniform) dynamic equation is

$$\varrho \frac{\partial^2 \mathbf{u}}{\partial t^2} = \nabla S + \mathbf{Q} \qquad ((\nabla \mathbf{S})_\mathrm{i} = @œ_\mathrm{ij}/@\mathbf{x}_\mathrm{j})\,, \tag{1.4}$$

where ϱ, $\mathbf{u}$, S, and $\mathbf{Q}$ are density, displacement, stress tensor, and external force, respectively.

We call a steady-state wave free if no external forces act on the wave in the considered area. Formally, this means that the wave satisfies the uniform governing equations. In this case, external forces existed in *the past* or acting outside the domain under consideration. Also, the wave may receive energy initially accumulated in the waveguide.

We start with the harmonic waves. Such waves are typical, like sound, water gravity waves, and the Rayleigh waves in solids. Besides, a sum or the Fourier integral of harmonic waves can represent a general linear wave.

1.2.2 Complex Wave

We use the complex representation of a linear harmonic wave

$$\mathbf{u} = \mathbf{A}e^{\mathrm{i}(\omega t - \mathbf{kx})}\,, \tag{1.5}$$

where $\mathbf{x}$ is the space vector, $\omega = \omega(\mathbf{k})$ is the frequency, and $\mathbf{k}$ is the wavevector. The wave period and length are

$$T = \frac{2\pi}{|\omega|}, \quad \lambda = \frac{2\pi}{|k|}. \tag{1.6}$$

In particular, when the x-axis is directed as the wave,

$$\mathbf{u} = \mathbf{A}e^{i(\omega t - kx)}, \tag{1.7}$$

where k is the wavenumber.

The time derivative and the kinetic energy density are

$$\dot{u} = i\omega u, \quad K = \frac{1}{2}\dot{u}\bar{\dot{u}} = \frac{1}{2}|A|^2\omega^2. \tag{1.8}$$

The frequency ω is that in an unmoving framework. As directly follows from the complex wave representation (1.7), the frequency, ω_v, detected in coordinates moving at speed v, $x = vt + \xi$, $\xi = \text{const}$, is

$$e^{i(\omega t - k(vt+\xi))} = e^{i(\omega_v t - k\xi))}, \quad \omega_v = \omega - kv. \tag{1.9}$$

The latter relation represents the well-known Doppler effect.

In a complex wave, at the coordinate fixed (x = const), the particle velocity is

$$\frac{\partial u}{\partial t} = A_w e^{i\omega t}, \quad A_w = i\omega e^{-ikx} = \text{const}. \tag{1.10}$$

So, as it follows from the linear formulation, the particles oscillate, remaining in the same place. However, strictly speaking, we have oscillations at a fixed coordinate, not a fixed particle. As already mentioned, the difference, the *particle drift* (the Stokes drift), which is an effect of nonlinearity, is usually much lower the wave speed.

As an example, when the particle velocity may not be small, consider a longitudinal pressure step wave, $vt < x < Dt$, with the constant front and positive particle velocities, D and v, respectively, defined by the applied pressure and the corresponding material stiffness. The relation v/D corresponds to strain, $0 < v/D = -\varepsilon$. It is crucial that $v/D < 1$.

1.2.3 Wave Dispersion. The Phase and Group Velocities

The main characteristic of the sinusoidal wave is the *dispersion relation*

$$\omega = \omega(k). \tag{1.11}$$

In the model of a 3D isotropic medium, the frequency is independent of the direction of the wavenumber vector, depending only on its magnitude. Additionally, we assume the symmetry of time here. Under these conditions, the dispersion relation in three dimensions is given by

$$\omega^2 = \omega^2(k^2)\,, \quad k^2 = k_x^2 + k_y^2 + k_z^2\,. \tag{1.12}$$

This equality defines the sphere (circle in the 2D case). Now we return to 1D case, $k_y = k_z = 0$.

Generally speaking, there is no one-to-one correspondence between the wavenumber and the frequency. The relation can have a set of branches, each uniquely defining $\omega(k)$. We here consider the wave corresponding to a single branch. The *phase velocity* of a wave, v_p (also called the *wave velocity*), follows directly from the definition (1.7)

$$u = A\mathrm{e}^{\mathrm{i}(\omega(k)t - kx)} \quad \Longrightarrow \quad v_p = \frac{\omega(k)}{k}\,. \tag{1.13}$$

In a common case, where the wave corresponds to a differential equation, it directly follows that the dispersion relation can be obtained by substituting the wave expression in the equation. More details can be seen in Sect. 3.4.3.

There is no wave dispersion if $\omega(k) = vk$ with $v = \text{const}$. Indeed, in this case, the phase speed is independent of the wavenumber, and the sum (finite or infinite) of waves of different k propagates as a solid, with no change of its shape. Otherwise, the wave shape varies while propagate.

The phase velocity, as the velocity of the wave phase, does not relate to the matter or energy propagation. Therefore the phase velocity has no upper limit.

The dispersion relation also defines the *group velocity*, the velocity of the envelope of a group of waves of close wavenumbers, and the corresponding frequencies. So, the group velocity is the *energy flux velocity* (the energy flux ratio to the energy density).

The energy flux velocity, the *group velocity*, is defined by the derivative

$$v_g = \frac{\mathrm{d}\omega(k)}{\mathrm{d}k}\,. \tag{1.14}$$

W.R. Hamilton first proposed a group velocity distinct from the wave's phase velocity in 1839, Hamilton (1839). Lord Rayleigh gave the treatment, including the connection with the energy flux in 1877 in his "Theory of Sound.", Rayleigh (1877). This nontrivial and general fact attracted attention repeatedly (e.g., Brillouin, 1960).

For a linear dispersion relation, $\omega = vk$ with $v = \text{const}$, the group and phase velocities coincide (it is true for 1D case only, see Sect. 1.2.4). Generally, the phase and group velocities are different; one can be greater or lower than the other, and they can be different in sign, too (see Sect. 1.2.8).

The appearance of the two speeds is demonstrated below by considering a package of two converging waves:

$$
\begin{gathered}
U = e^{i(\omega(k_1)t - k_1 x)} + e^{i(\omega(k_2)t - k_2 x)} \\
= 2\exp\left(i\left(\frac{\omega(k_1)+\omega(k_2)}{2}t - \frac{k_1+k_2}{2}x\right)\right)\cos\left(\frac{\omega(k_1)-\omega(k_2)}{2}t - \frac{k_1-k_2}{2}x\right) \\
\sim 2e^{i(\omega(k)t - kx)}\cos\left(\frac{1}{2}(\delta\omega(k)\,t - \delta k\,x)\right), \\
k_1 \to k_2 \to k\,, \quad k_1 - k_2 = \delta k\,, \quad \omega(k_1) - \omega(k_2) = \delta\,\omega(k)\,, \quad \delta k \to 0\,.
\end{gathered}
\tag{1.15}
$$

The above expression contains the complex wave, the carrier, propagating with the phase speed $v_c \to v_p = \omega(k)/k$, and the envelope propagating with the group speed, $v_e \to v_g = \mathrm{d}\omega(k)/\,\mathrm{d}k$.

The energy cannot flow through the zero points of the envelope. Therefore, its value in a single package is invariable. It follows that the energy propagates with the group speed v_g.

The configuration of the two-wave group presented in Fig. 1.1 is based on the (non-dimensional) dispersion relation $\omega = \sqrt{1+k^2}$ corresponding to the Klein–Gordon equation considered below as an example.

Note that the statement follows not only from the limit of two converging waves. We come to the same result considering a single wave forced by a load moving with the group speed. Indeed, in the integral representation of such a wave, there exist a higher order of the singularity corresponding to a resonant wave, where the load speed coincides with that of the wave energy.

Thus, in fact, the group velocity is the energy flux velocity in a single wave. This statement is also valid for transient waves.

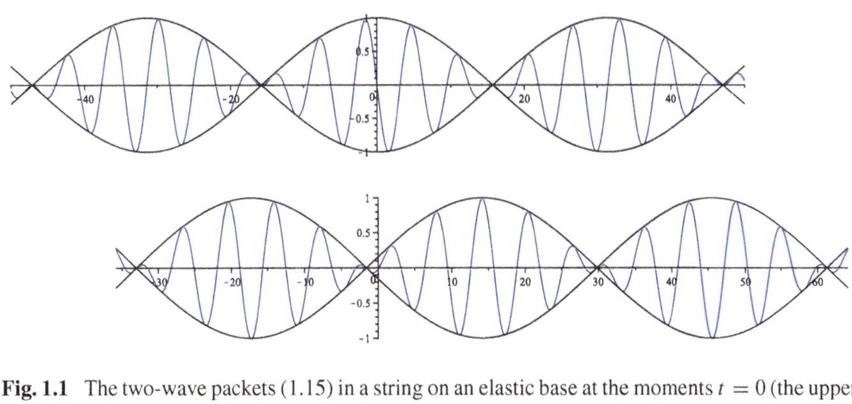

Fig. 1.1 The two-wave packets (1.15) in a string on an elastic base at the moments $t = 0$ (the upper plot) and $t = 20$ (the lower one). The sinusoidal waves obey the (non-dimensional) dispersion relation $\omega = \sqrt{1+k^2}$ with $k_1 = 1.05$, $k_2 = 0.95$. The carrier and envelope speeds are $v_c = 1.3473$ and $v_e = 0.7069$; $v_c \to v = \sqrt{2}$, $v_e \to v_g = \sqrt{2}/2$ $(k_1 \to k_2 \to 1)$

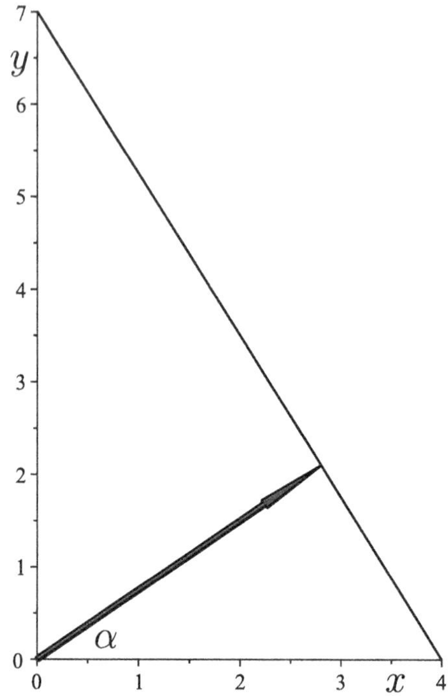

Fig. 1.2 Wave in x, y-plane. The arrow show the wave phase velocity, v, while $v_x > v$, $v_y > v$ correspond to the cross points of the wave crest with the axes

1.2.4 Phase and Group Velocities in 2D Isotropic Medium

Let the wave propagate in the x, y-plane at angle α to the x-axis, Fig. 1.2, where the arrow corresponds to the phase velocity, v. In the dispersion free case, $\omega^2 = v^2(k_x^2 + k_y^2)$, the phase velocities are

$$v = \frac{\omega}{k}\,, \quad v_x = \frac{\omega}{k_x} = \frac{v}{\cos\alpha}\,, \quad v_y = \frac{\omega}{k_y} = \frac{v}{\sin\alpha}\,, \tag{1.16}$$

while the group velocities are

$$v_g = \frac{\mathrm{d}\omega}{\mathrm{d}k}\,, \quad (v_g)_x = \frac{\partial\omega}{\partial k_x} = \frac{vk_x}{k} = v\cos\alpha\,, \quad (v_g)_y = \frac{\partial\omega}{\partial k_y} = \frac{vk_y}{k} = v\sin\alpha\,. \tag{1.17}$$

So, the phase and group velocities on the coordinate axes oppositely depend on the wave direction.

Now consider surface waves on water.

1.2.5 Surface Waves on Water

The sound speed in water, 1400–1500 m/s (depending on temperature), is much higher than the speed of surface waves, which allows us to neglect the water compressibility.

Also, ignoring the water viscosity we express the particle velocity, $\boldsymbol{w}$, in terms of the scalar potential, φ. With respect to the surface wave, we write

$$\varphi(x, z, t) = A(z)\mathrm{e}^{\mathrm{i}(\omega t - kx)}, \quad \boldsymbol{w} = \nabla\varphi. \tag{1.18}$$

From here and the incompressibility condition

$$\Delta\varphi = \frac{\mathrm{d}^2 A(z)}{\mathrm{d}z^2} - k^2 A(z) = 0, \tag{1.19}$$

it follows that for a uniform depth of the water area, $-h < z < 0$,

$$A(z) = A_0\mathrm{e}^{kz}\left(1 + \mathrm{e}^{-2k(h+z)}\right), \quad A_0 = \mathrm{const}. \tag{1.20}$$

Next, the dynamic equation and the pressure at $z = 0$ are

$$\begin{aligned} \varrho\frac{\partial \boldsymbol{w}}{\partial t} &= -\nabla p \quad (-h < z \leq 0), \\ p = \varrho g u_z - T\frac{\partial^2 u_z}{\partial x^2} &= (\varrho g + Tk^2)u_z \quad (z = 0), \end{aligned} \tag{1.21}$$

where g, p, T, and u_z are the gravity acceleration (on the Earth surface), pressure at the surface, surface tension and the surface elevation in the wave, respectively, ($\partial u_z/\partial t = w_z$ at $z = 0$). Note that $\varrho g \approx 100N/m^3$, while $T \approx 72mN/m$. So, the surface tension becomes essential for tiny wavelengths, $\lambda \approx 0.2$ m or less, where the term ripples are adequate. In the following, we consider 'long' waves, and neglect the surface tension.

So, for $z = 0$, where u_z is defined, the horizontal component of the dynamic equation (for $k \neq 0$) is

$$\frac{\partial^2\varphi}{\partial t^2} + g\frac{\partial\varphi}{\partial z} = 0 \quad (z = 0). \tag{1.22}$$

The dispersion relation follows as

$$\begin{aligned} &\omega = \sqrt{gk\tanh(kh)}, \quad \omega \sim k\sqrt{gh} \ (kh \to 0), \quad \omega \sim \sqrt{gk} \ (kh \to \infty), \\ &v_p \sim v_g \sim \sqrt{gh} \ (kh \to 0), \quad v_g \sim \frac{1}{2}v_p \sim \sqrt{g/k} \ (kh \to \infty). \end{aligned} \tag{1.23}$$

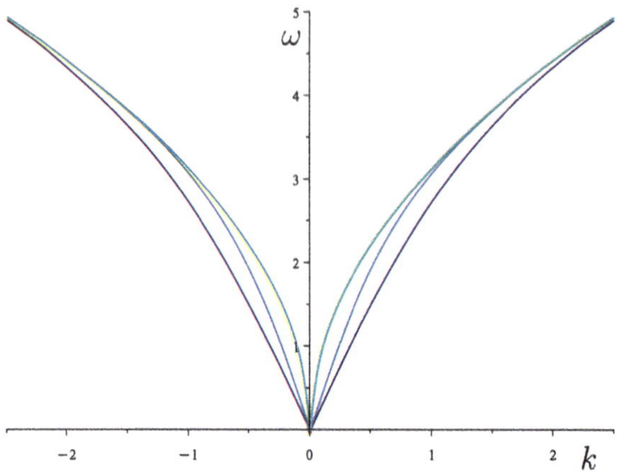

Fig. 1.3 Water surface waves. Dispersion relations (plotted for $\omega > 0$) for the water depths (in meters) 1, 2, and ∞; $\omega_1 < \omega_2 < \omega_\infty$, respectively ($g = 9.81\, m/sec^2$)

See the dispersion relation plot in Fig. 1.3. The group-to-phase speeds ratio

$$\frac{v_g}{v_p} = \frac{1}{2} + \frac{kh}{\sinh(2kh)} \tag{1.24}$$

is plotted in Fig. 1.4.

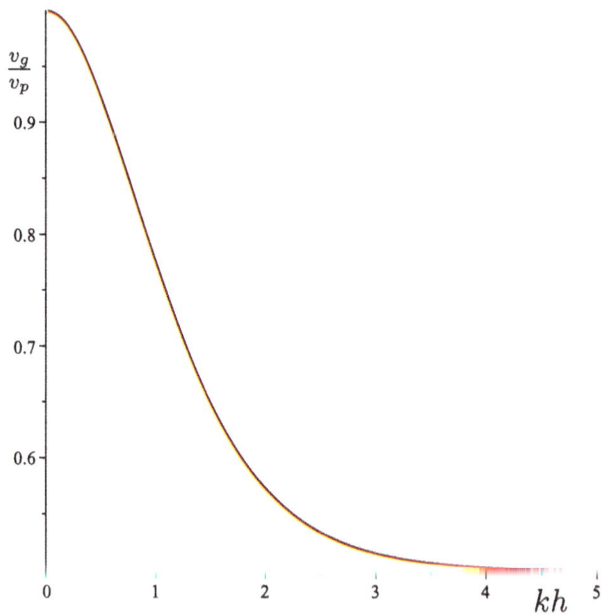

Fig. 1.4 The group to phase speeds ratio for the water surface waves as a function of the depth-to-wavelength ratio, $kh = 2\pi h/\lambda$

1.2.6 Longitudinal Waves: Wavefront Conditions and the Energy Release

In a (generally nonlinear) *longitudinal wave*, the particle velocities direct as the wave propagates. The question is whether the wave has a front as a moving surface of discontinuity. To this end, we consider the mass and momentum *conservation laws* which restrict mass and momentum fluxes in or out the wave front (we here do not consider cases where there is an exchange of mass and momentum between the waveguide and surrounding medium). Also, we discuss the energy relation.

In a *step wave*, strains and particle velocities at the wavefront abruptly reach some values, which then remain constant. The mass and momentum conservation relations are

$$\varrho(c - v) = \varrho_0 c\,, \quad \sigma = -\varrho_0 v c\,, \tag{1.25}$$

where ϱ_0 and ϱ are the densities ahead of and behind the wave front, respectively, c, v and σ are the wave speed, particle velocity and stresses in the wave, respectively. From the first relation we have

$$\varepsilon = \frac{\sigma}{E} = \frac{\varrho_0}{\varrho} - 1 = -\frac{v}{c} \quad \Longrightarrow \quad v = -\varepsilon c\,, \tag{1.26}$$

and the momentum conservation relation now gives us

$$c = \sqrt{\frac{E}{\varrho_0}}\,, \tag{1.27}$$

where $E = E(\sigma)$ is the secant modulus.

The work of the stresses per unit time and the corresponding rate of the kinetic energy are[2]:

$$N = -\sigma v\,, \quad \mathcal{K} = \frac{1}{2}\varrho_0 c v^2\,, \tag{1.28}$$

and per unit length (unit volume)

$$N_0 = \frac{N}{c} = \sigma\varepsilon\,, \quad \mathcal{K}_0 = \frac{\mathcal{K}}{c} = \frac{1}{2}\varrho_0 c^2\varepsilon^2 = \frac{1}{2}\sigma\varepsilon\,. \tag{1.29}$$

[2] The relations between kinetic and potential energies in linear and nonlinear waves (Slepyan, 2015a), are considered in Sect. 2.2.

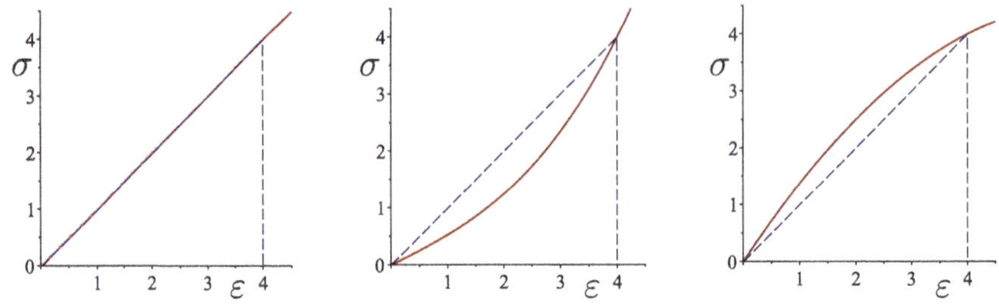

Fig. 1.5 The three types of the stress-strain diagrams (left to right): linear, hardening and softening. Stresses as functions of strains (solid curves) and the linear dependence corresponding to the energy conservation (dashed lines)

The strain energy is (see Fig. 1.5)

$$\begin{aligned} \mathcal{P}_0 &= \mathcal{K}_0 \quad \text{(the linear case)}\,, \\ \mathcal{P}_0 &< \mathcal{K}_0 \quad \text{(hardening nonlinearity)}\,, \\ \mathcal{P}_0 &> \mathcal{K}_0 \quad \text{(softening nonlinearity)}\,. \end{aligned} \tag{1.30}$$

We now see that the energy conservation law holds only in the linear elasticity. For the hardening stress-strain relation, a part of the energy disappears, and the softening one corresponds to an *energy flux from the step wave front*.

The step wave propagating in the hardening nonlinear material is called the *shock wave*. The mechanical energy disappearing in its front transverses to heat. In the case of the softening diagram, the step wave could only be possible if there exists an energy flux from the material, as, for example, takes place in a detonation wave. Otherwise, the step front cannot exist, and stresses reaches the final level gradually.

Note that the energy release occurs not only in shock waves. We also meet it in transition waves, and in the moving contact problem of linear elasticity, Chap. 8.

Physically, the energy release means that it transfers to the micro-level, which does not present in the classical continuous model. It is its fundamental weakness (or the strength, from another point of view). We repeatedly discuss this issue, see in particular, Sects. 1.2.12 and 6.2.

We accept the following *energy condition*:

A singular point (line or surface) can be an energy sink but not a source (if there is no natural energy source at this singularity). This condition should be satisfied not only in the problem under consideration but in general (see, e.g., Sect. 1.4.3).

As commonly accepted, the released energy goes to oscillations on the micro-level, which creates the *heat pressure*. In dependence on the thermal expansion coefficient, the latter elevates the isothermal stress-strain diagram (the *cold adiabate*) to the so-called *Hugoniot adiabate* (the *hot adiabate*).[3] Note, however, that in a regular elastic lattice, the 'heat' oscillations can be radiated away from the source. The radiation power and hence the thermal expansion can depend on the lattice structure and the wave velocity.

The limiting case of the hardening nonlinearity, Fig. 1.5, corresponds to a set of the uniformly distributed rigid masses with spaces between them (like a line of billiard balls or a 1D model of a gas, as discussed above, Sect. 1.1.1.1).

1.2.7 Waves in a String Placed on the Winkler Foundation

The linear dynamic equation (the 1D *Klein–Gordon equation*) for the transverse displacement u is

$$\varrho\ddot{u}(x,t) - Tu''(x,t) + \varkappa u(x,t) = 0\,, \tag{1.31}$$

where ϱ, T and $\varkappa$ are the density as the mass per unit length of the string, the tensile force and Winkler's constant (the foundation stiffness), respectively. The time- and x-derivatives are denoted as usual by a dot and a prime, respectively. *We here assume* that $T > 0$, $\varkappa \geq 0$.

Substituting the wave expression (1.7) we obtain the *dispersion relation*

$$\omega = \omega(k) = \pm c\sqrt{k^2 + \frac{\varkappa}{T}}\,, \quad c = \sqrt{\frac{T}{\rho}}\,, \tag{1.32}$$

where c is the wave speed in the string. To express the relation in a non-dimensional form we introduce a length unit a. We have

$$\hat{\omega} = \sqrt{\hat{k}^2 + \hat{\varkappa}}\,, \quad \hat{\omega} = \frac{a\omega}{c}\,, \quad \hat{k} = ak\,, \quad \hat{\varkappa} = \frac{a^2\varkappa}{T}\,. \tag{1.33}$$

The dispersion relation for the above non-dimensional expression with $\hat{\varkappa} = 1$ is plotted in Fig. 1.6.

[3] The shock waves and related topics are considered in detail in the book by Zel'dovich and Raizer (2002).

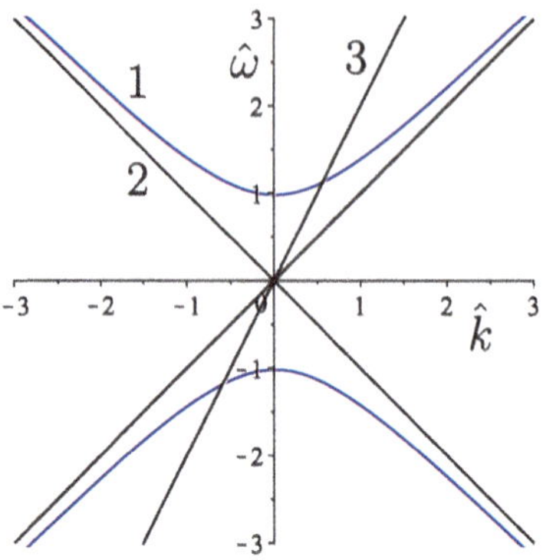

Fig. 1.6 Dispersion relations for the sinusoidal wave in the string on an elastic foundation (1) and for a free string (2). The line (3) corresponds to $\hat{\omega} = 2\hat{k}$. It can be seen that the phase speed has only the lower bound, $1 < \hat{v}_p < \infty$, whereas the group speed, $\hat{v}_g$, is always finite, $0 \leq \hat{v}_g < 1$. As the wave length grows ($\hat{k} \to 0$), $\hat{v}_p \to \infty$ while $V_g \to 0$. Otherwise, as $\hat{k} \to \infty$, these speeds asymptotically coincide, $\hat{v}_g \to \hat{v}_p \to 1$, and we have the case of a free string

1.2.8 Waves in a Compressed Beam, and the Negative Group Speed

Consider the dispersion relation for flexural waves in an axially compressed bending beam rested on the Winkler foundation. The homogeneous Bernoulli-Euler dynamic equation respective the transverse displacement $u(x, t)$ is

$$ESr^2 u^{IV} - Tu'' + \varkappa u + \varrho \ddot{u} = 0\,, \tag{1.34}$$

where E, T, S, r, ϱ, and $\varkappa$ are the elastic modulus, compressive force, cross-section area, the inertia radius of the latter, the material density, and the foundation stiffness, respectively(usually the bending stiffness, ESr^2, is denoted by D). *Here we assume* that $\varkappa > 0, T \leq 0$.

Dividing the expression (1.34) by $\varkappa$ we rewrite it in a non-dimensional form

$$\hat{\omega} = \sqrt{\hat{k}^4 + \hat{T}\hat{k}^2 + 1}\,, \quad \hat{k} = \left(\frac{ESr^2}{\varkappa}\right)^{1/4} k\,, \quad \hat{T} = \frac{T}{\sqrt{ES\varkappa r^2}}\,, \quad \hat{\omega} = \sqrt{\frac{\varrho}{\varkappa}}\,\omega\,. \tag{1.35}$$

Dispersion relations (1.35) for some values of $\hat{T}$ are presented in Fig. 1.7. It can be seen that for $-2 < \hat{T} \leq 0$ the frequency $\omega(k)$ has a minimum, and the group speed is negative at the left of it, whereas the phase speed is everywhere positive. The beam loses stability under the critical force $\hat{T} = -2$.

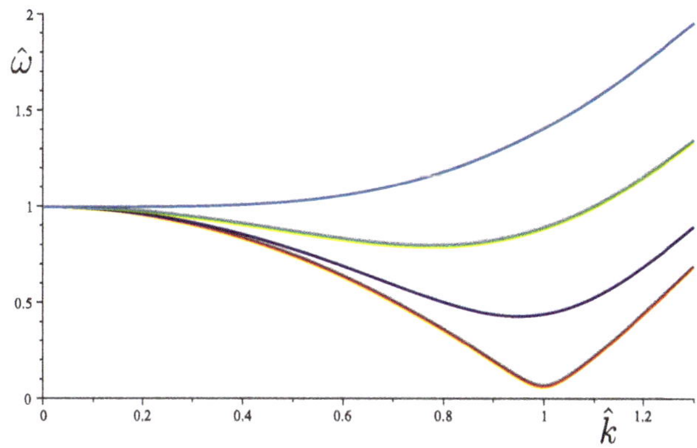

Fig. 1.7 Dispersion curves for the sinusoidal flexural wave in the elastic beam on the Winkler foundation. The corresponding non-dimensional force (top-down) $\hat{T} = 0, -1.2, -1.8, -1.995$

1.2.9 Wave in a Discrete Chain

Consider a *mass-spring chain*. In the non-dimensional variables where the spring stiffness and particle mass are taken as the units, the dynamic equation is

$$\ddot{u}_n = u_{n-1} + u_{n+1} - 2u_n = 0\,, \tag{1.36}$$

where n is the particle number. A wave in this periodic discrete system, in terms of the particle displacements, can be described the same as in a continuous system, namely, as the sinusoidal wave

$$u = A\sin(\omega t - kx)\,, \tag{1.37}$$

where only integer values of x have a real sense (intermediate values unrelated to the movement of spring points!). Note that such a representation can be treated as corresponding to a *quasi-continuum* (Kunin, 1982, 1983). The dispersion relation

$$\omega = \pm 2\sin\frac{k}{2}\,. \tag{1.38}$$

is a 2π-periodic function; however, it can be replaced by a 2π-segment, say, $-\pi < k < \pi$. Indeed, since only integer values of x matter, replacement $k \to k \pm 2\pi$ does not change the particle displacement. Nevertheless, as can be seen below, in the case of forced waves Part VI, it is useful to preserve the whole domain of k, $-\infty < k < \infty$.

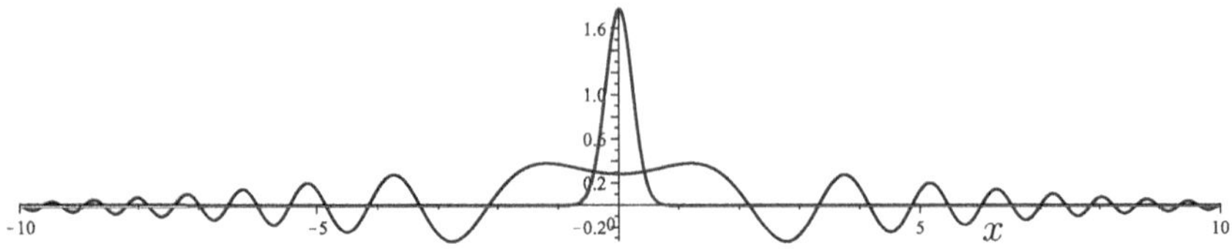

Fig. 1.8 Symmetric stretching of the localized flexural wave packet: the localized $t = 0$ and stretched (t = 0.5) shapes of the beam

1.2.10 Evolution of a Localized Wave Packet

Consider a wave packet comprised of sinusoidal flexural waves (we use non-dimensional variables)

$$u(x,t) = \frac{1}{2\pi}\int_{-\infty}^{\infty} \exp\left(-\frac{\alpha k^2}{4}\right)\cos(k^2 t - kx)\,\mathrm{d}k \quad (\omega = k^2\,,\ \alpha > 0)\,. \tag{1.39}$$

Initially it looks like a solitary disturbance

$$u(0,x) = \frac{1}{\sqrt{\pi\alpha}}\exp\left(-\frac{x^2}{\alpha}\right) \quad \left(\int_{-\infty}^{\infty} u(0,x)\,\mathrm{d}x = 1\right), \tag{1.40}$$

which localizes and tends to the delta function at $x = 0$, Sect. 17.1.3, as α tends to zero.

The shape of the initially unmoving beam, $\dot{u} = 0$ $(t = 0)$, at $t = 0$ and $t = 0.5$ is shown in Fig. 1.8. In this case, the packet stretches symmetrically. Changing the representation (1.39) to

$$u(x,t) = \frac{1}{\pi}\int_{0}^{\infty} \exp\left(-\frac{\alpha k^2}{4}\right)\cos(k^2 t - kx)\,\mathrm{d}k \tag{1.41}$$

we obtain the same initial shape but with the waves propagating only to the right, Fig. 1.9. The corresponding initial particle velocities are distributed as in Fig. 1.10. Note that such an analysis of linear waves in a uniform waveguide is usually carried out using the Fourier integral transform, Sect. 18.2

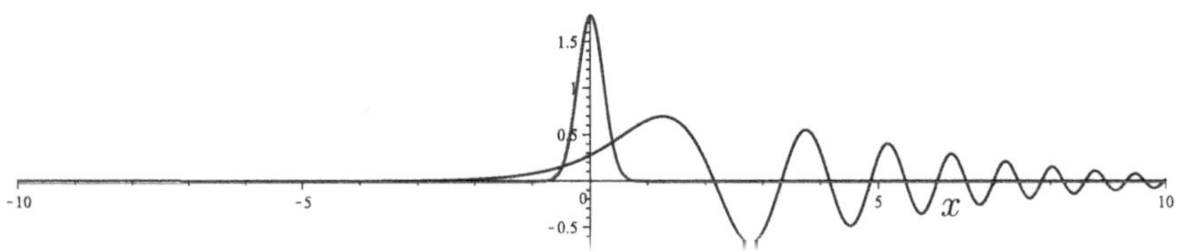

Fig. 1.9 Right-directional stretching of the localized flexural wave packet: the localized $t = 0$ and stretched (t = 0.5) shapes of the beam

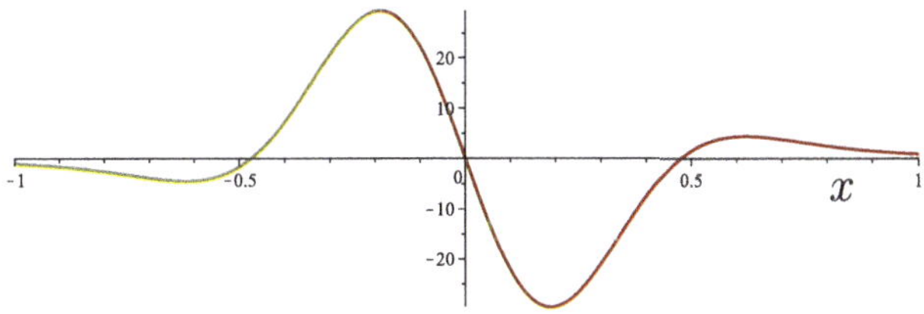

Fig. 1.10 Initial particle velocities resulting in the right-directional stretching of the localized flexural wave packet

1.2.11 Waveguide with a Dynamic Structure

1.2.11.1 Structure Versus Microstructure

A structured waveguide is usually called the one with a *microstructure*. Such terms, the microstructure and nanostructure, reflect the structure scale, micrometer and nanometer. The question is whether a dynamic process on such a level can be modelled on the macro level.

It is so, and both macro and micro structure has no specific size and time units, and the microstructure differs only by the size. In this case, the size by itself doesn't matter, but only the non-dimensional ratios.

Otherwise, if such a unit exists on one or both levels, the specific properties of the structure(s) be taken into account, for example, gravity forces on the macrolevel and electro-mechanical forces on the micro level.

Note, however, that the discreteness can be modelled on the macro level. A mass-spring lattice is a natural model of such a microstructure. Its behavior can be modeled in a similar lattice of an arbitrary scale. The continuous, nonstructural medium model is only a long-wave approximation (compared with the lattice cell size).

In some cases, the macrostructure model can be developed, as for example in the macro-model of fracture, where the surface energy reflecting the microstructure is taken into account.

Slepyan (1967) considered the evolution of a wave in a cylindrical elastic shell with the attached dynamic *macro*-structure (modeled by uniformly distributed mass-spring oscillators). The same wave (with the corresponding time scale) can propagate in a continuous waveguide with the microstructure. Also, in this case, the wave evolution is scale-insensitive.

A theory of a dynamic elastic medium with a dynamic microstructure was introduced by Milton and Willis (2007), where the complex system is considered as the nonstructural elastic medium, for which a modified Newton's second law is valid.

Waves in various waveguides were the subject of many works. In particular, in Bigoni (2002, 2013) and Mishuris et al. (2019, 2020), where the corresponding references are listed. Waves in a complex waveguide are topical in various applications. For example, Engelbrecht (2020) and Peets et al. (2021) studied nerve signals propagation in a natural micro-structured waveguide. Eun Bok et al. (2018) discussed the role of a nontrivial structural dynamic layer as a water-air interface.

1.2.12 Attached Microstructure Rearranges Dispersion Pattern

Consider a free string equipped by distributed oscillators as a uniform continuous mass-spring medium. The latter can also be envisioned as another string with a vanishing tensile force.[4] Instead of equation (1.31) we have a system

$$\varrho\ddot{u}(x,t) - Tu''(x,t) + \varkappa(u(x,t) - w(x,t)), \quad \varrho_0\ddot{w}(x,t) - \varkappa(u(x,t) - w(x,t)) = 0\,. \tag{1.42}$$

Dispersion relations for the sinusoidal wave follows from (1.42) consisting of two branches

$$\omega_{1,2}^2 = \frac{1}{2}\left(\frac{\varkappa(\varrho+\varrho_0)}{\varrho\varrho_0} + c^2k^2 \pm \sqrt{\left(\frac{\varkappa(\varrho+\varrho_0)}{\varrho\varrho_0} + c^2k^2\right)^2 - \frac{4\varkappa}{\varrho_0}c^2k^2}\right), \tag{1.43}$$

where only positive part of the dependence is presented (the negative one differs by the sign only), $c = \sqrt{T/\varrho}$ is the wave speed in the string.

It follows:

$$\begin{gathered}\omega_1 = \sqrt{\frac{\varkappa}{\varrho_0}\left(1+\frac{\varrho_0}{\varrho}\right)}, \quad \omega_2 = 0 \quad (k=0)\,,\\ \omega_1 \sim \pm ck\,, \quad \omega_2 \sim \sqrt{\frac{\varkappa}{\varrho_0}} \quad (k\to\pm\infty)\,.\end{gathered} \tag{1.44}$$

Thus, there is the bandgap

$$\sqrt{\frac{\varkappa}{\varrho_0}} < \omega < \sqrt{\frac{\varkappa}{\varrho_0}\left(1+\frac{\varrho_0}{\varrho}\right)}, \tag{1.45}$$

where no wave can propagate. The dispersion dependencies are plotted in Fig. 1.11.

Note that for the waveguide with the distributed oscillators, the inertia term, as the "modified Newton's second law", becomes

$$\omega^2\varrho \implies \omega^2\varrho\left(1+\frac{\varkappa}{\varkappa-\omega^2\varrho}\right). \tag{1.46}$$

[4] This model was introduced by Slepyan (1967) in the study of how a set of elastically attached masses affects a step wave in an elastic cylindrical shell; also see Mishuris et al. (2019).

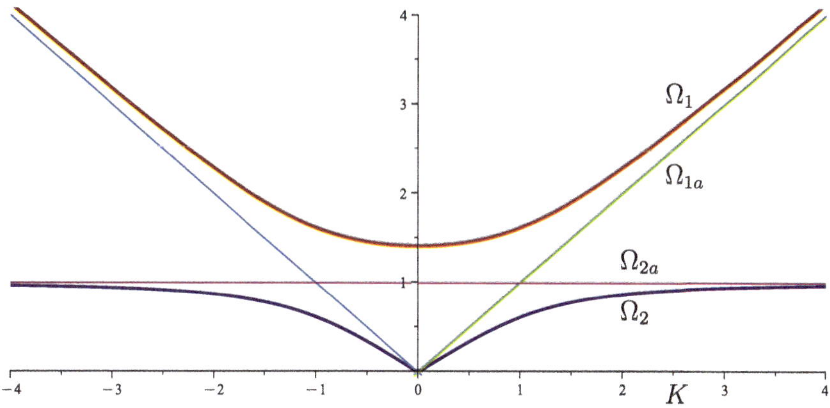

Fig. 1.11 The positive part of the dispersion branches, Ω_1, Ω_2, for a free string equipped by distributed oscillators, $c = \varkappa = \varrho_0 = \varrho = 1$. The dependence for the nonstructural string, $\varrho_0 = 0$, is also shown, Ω_{1a}. Also the asymptote of ω_2 ($k \to \pm\infty$) is presented, Ω_{2a}. The negative part differs only by the sign, (1.43)

This modification does not concern the potential-energy-related terms of the dynamic equation, and remains valid for various waveguides. In particular, in the case of the bending beam, (1.34), the dispersion relation follows from that for the string, (1.43), where c^2k^2 is replaced by Dk^4.

1.2.12.1 Dynamic Elastic Foundation

As another structure, consider a dynamic elastic foundation, a uniformly distributed supports without interconnections (as usually adopted in statics but with the account of inertia forces). Let E_0, ϱ_0, $c_0 = \sqrt{E_0/\varrho_0}$ be the support material parameters (in usual notations), and l_0, b_0 and y be its length and width and the (normal to x) coordinate along it. The equation and boundary conditions in terms of the displacement $w(t, y)$ for an arbitrary x and the boundary conditions are

$$\varrho_0 b \frac{\mathrm{d}^2 w(t, y)}{\mathrm{d}t^2} - E_0 \frac{\mathrm{d}^2 w(t, y)}{\mathrm{d}y^2} = 0\,, \quad w(t, 0) = 0\,, \quad w(t, l) = u(t, x)\,, \tag{1.47}$$

where $u(t, x)$ is the transverse displacement the waveguide.

In the following, we consider the latter as an elastic bending beam, the the uniform equation of each is

$$\varrho S \frac{\mathrm{d}^2 u(t, x)}{\mathrm{d}t^2} + D \frac{\mathrm{d}^4 u(t, x)}{\mathrm{d}x^4} - P(t) = 0\,, \tag{1.48}$$

where S and D are the cross-section area and bending stiffness, respectively, and $P(t)$ is the (x-distributed force by which the support acts on the beam.

For the complex wave, $u(t, x) = \exp(\mathrm{i}(\omega t + kx))$, the the above relations result in

$$\begin{aligned} w(t, y) &= \frac{\sin(\omega y/c_0)}{\sin(\omega l/c_0)} \mathrm{e}^{\mathrm{i}\omega t}, \\ P = -\frac{\mathrm{d}w(t, y)}{\mathrm{d}y} E_0 b_0 &= \frac{E_0 b_0 \omega}{c_0} \cot\left(\frac{\omega l}{c_0}\right) \mathrm{e}^{\mathrm{i}\omega t}. \end{aligned} \tag{1.49}$$

Finally, we come to the (inverse) dispersion relation

$$K = \pm(\Omega(\Omega - \lambda\cot(\Omega))^{1/4}, \tag{1.50}$$

where only real values should be accepted, and the (non-dimensional) values are

$$\Omega = \frac{\omega l}{c_0}, \quad K = k\left(\frac{Dl^2}{\varrho S c_0^2}\right)^{1/4}, \quad \lambda = \frac{\varrho_0 b l}{\varrho S}. \tag{1.51}$$

The dispersion relation $\Omega(K)$ for $\lambda = 1$ is plotted in Fig. 1.12.

Note that $K = 0$ at the points where $\Omega = \lambda\cot(\Omega)$, and

$$K \to \pm\infty \quad (\Omega \to n\pi - 0, n = 0, \pm 1, \pm 2, \ldots). \tag{1.52}$$

Physically, the first case corresponds to the oscillation of the support with the (straight-line) beam. The second one appears at the resonant oscillations of the support. In these connections, note that the bending beam model is valid only for rela-

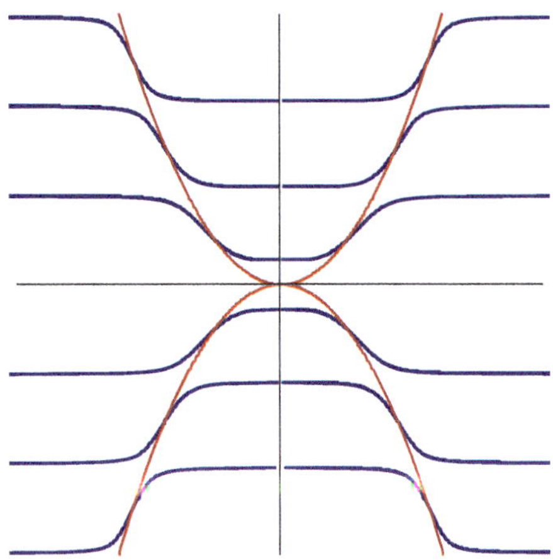

Fig. 1.12 Dispersion dependence $\Omega(K)$ for a bending beam on the dynamic elastic foundation, (1.50). The brown curves, $\Omega = \pm K^2$, correspond to the free beam ($b_0 = 0$). The horizontal asymptotes of the blue lines, $K \to \pm\infty$, reflect the resonant oscillations

tively long waves, that is, not too large wavenumbers. So, the latter may not increase unboundedly.

1.2.13 2D Wave Propagating Along a Layered Structure

Consider an x-uniform, y-periodic composite structure, Fig. 1.13, where each layer is of an isotropic material. Let the phase velocities be different, $v_1 < v_2$, that is, $k_2 < k_1$. Doe to isotropy, for 2D waves

$$k_{x1}^2 + k_{y1}^2 = k_1^2 \,, \quad k_{x2}^2 + k_{y2}^2 = k_2^2 \,. \tag{1.53}$$

So,

$$k_{x1}^2 = k_1^2 - k_{y1}^2 < k_1^2 \ (k_{y1}^2 > 0) \,, \quad k_{x2}^2 = k_2^2 - k_{y2}^2 > k_2^2 \ (k_{y2}^2 < 0) \,. \tag{1.54}$$

For an x-propagating wave the phase velocities in the layers are equal two each other. We have

$$k_1^2 - k_{y1}^2 = k_2^2 - k_{y2}^2 = k_x^2 \ \ (k_{y2}^2 < 0) \,, \tag{1.55}$$

where k_x is an unknown. We suppose that

$$k_2 \leq k_x \leq k_1 \,. \tag{1.56}$$

Since the middle x-line of any layer is the line of symmetry, the wave can be represented as

$$\begin{aligned} u_1 &= A_1 \cos(k_{y1} y) e^{i(\omega t - k_x x)} \ \ (-a < y < a) \,, \\ u_2 &= A_2 \cosh(|k_{y2}| y) e^{i(\omega t - k_x x)} \ \ (-b < y < b) \,, \end{aligned} \tag{1.57}$$

where $2a$ and $2b$ are the widths of the first and the second layers, respectively, and y is the layer-related coordinates started at the middle line of the layer.

In a simplest case, where the layers obey the second order wave equation, as for a membrane, we have two continuity relations

$$A_1 \cos(k_{y1} a) = A_2 \cosh(|k_{y2}| b) \,, \quad A_1 k_{y1} \sin(k_{y1} a) = A_2 |k_{y2}| \sinh(|k_{y2}| b) \,. \tag{1.58}$$

These two equations define, possibly nonuniquesly, the two unknowns, one of the constants $A_{1,2}$ (the other can be taken arbitrary), and the wavenumber k_x (for higher order equations there are more conditions and more related functions). The wavenumber k_x as a function of the first layer density, ϱ_1 (multiple graphs of the owned speed $v_1 = \sqrt{T/\omega_1}$, $T =$ const) is plotted in Fig. 1.14. The wave localization at a finite waveguides of some structure is considered in a similar way as above in Chap. 5.

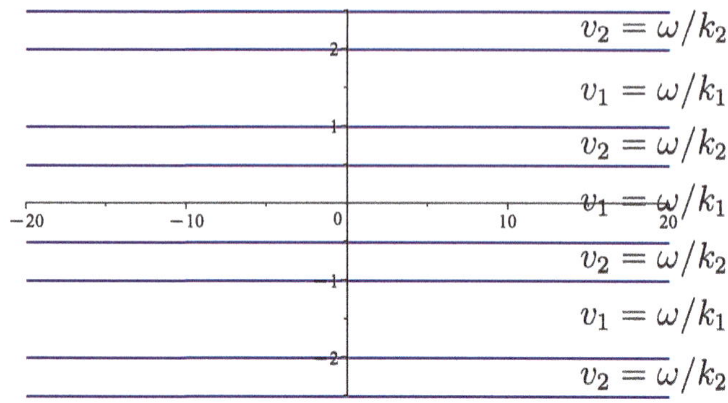

Fig. 1.13 Layered structure, the phase velocities, and wavenumbers under the same frequency

Fig. 1.14 The wavenumber k_x as (multiple) functions of the first layer density, ϱ_1 (plotted by the "implicitplot" command of the Maple)

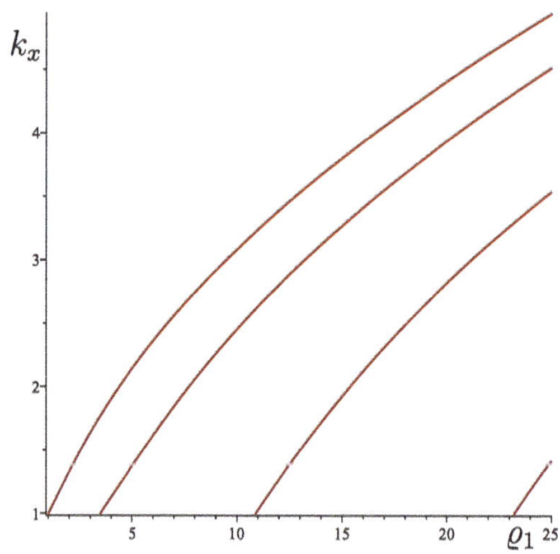

1.2.14 *Reflection and Refraction*

We consider waves in two isotropic half-planes of non-equal wave speeds. In this problem, in difference from those discussed in Sect. 1.2.13 and Chap. 5, there is no symmetry.

The incident wave, u_0, and the reflecting one, u_1, propagate in the upper half-plane, $y > 0$, while the refracting wave, u_2, propagates in the lower one, $y < 0$,

$$u_0 = A_0 e^{i(\omega t - k_{0x}x + k_{0y}y)}, \quad u_1 = A_1 e^{i(\omega t - k_{0x}x - k_{0y}y)}, \quad u_2 = A_2 e^{i(\omega t - k_{0x}x + k_{2y}y)}. \quad (1.59)$$

Due to isotropy,

$$k_{0x}^2 + k_{0y}^2 = k_1^2\,, \quad k_{0x}^2 + k_{2y}^2 = k_2^2\,, \tag{1.60}$$

where the frequency and wavenumbers, ω, k_1^2, k_2^2, and the incident wave x, y-wavenumbers, k_{0x}, k_{0y} (satisfying the relation in (1.60)) are given.

In the simplest case, there are three unknowns, $A_{1,2}$ and k_{2y}, and three equations, the continuity relations for the x-wavenumber, the wave and its y-derivative at the separate line, $y = 0$,

$$k_{2x} = k_{0x}\,, \quad A_0 + A_1 = A_2\,, \quad k_{0y}(A_0 - A_1) = k_{2y}A_2\,. \tag{1.61}$$

It follows that

$$A_1 = \frac{k_{0y} - k_{2y}}{k_{0y} + k_{2y}} A_0\,, \quad A_2 = \frac{2k_{0y}}{k_{0y} + k_{2y}} A_0\,,$$
$$k_{2y} = \sqrt{k_2^2 - k_{0x}^2} > 0 \;\; (k_{0x}^2 < k_2^2)\,, \quad k_{2y} = -\mathrm{i}\sqrt{k_{0x}^2 - k_2^2} \;\; (k_2^2 < k_{0x}^2;\, y < 0)\,. \tag{1.62}$$

1.3 Forced Waves

We consider waves in a uniform waveguide excited by the moving-oscillating load. The main question is *What waves are generated*? Together with the wave determination, the *wave resistance* (to the load action) is determined as the wave energy.

As in the case of free waves, the gravity surface wave, generated by a moving ship (or any body moving on water) represents us a visible example considered in Sect. 1.3.4 after some 1D waves.

1.3.1 What Waves are Generated

Consider a 1D waveguide excited by a moving-oscillation point load

$$Q = Q_0 \mathrm{e}^{\mathrm{i}\omega_0 t}\delta(x - v_0 t)\,. \tag{1.63}$$

This load can excite a propagating wave (or several waves). We assume that the wave propagation is supported by a nonzero energy flux from the load.

The first statement we are based on is that **the wave frequency, as detected at the moving load, coincides with the load frequency**, ω_0. Indeed, otherwise there is no energy flux which could get the wave.

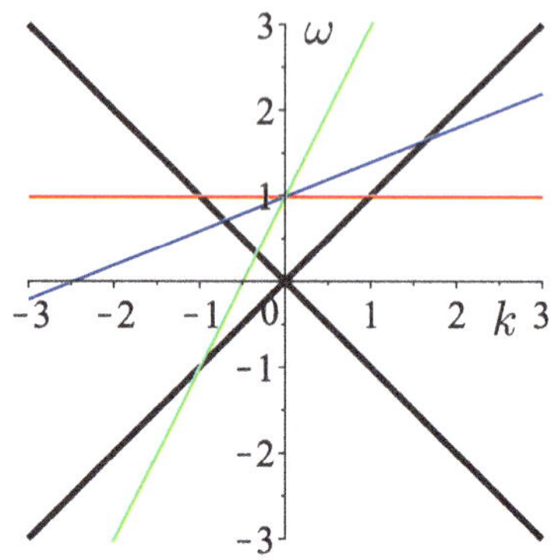

Fig. 1.15 Wave excitation in a non-dispersion 1D waveguide, $\omega(k) = \pm k$ (bold black lines). The frequency of the load $\omega_0 = 1$ $(k = 0)$. The lines correspond to the unmoving oscillating force (red line, $v_0 = 0$), and to the subsonic (blue line, $0 < v_0 < 1$) and supersonic loads (green line, $v_0 > 1$). The cross-points with the $\omega(k) = \pm k$ lines at $k > 0$ correspond to the waves propagating in front of the load and vice versa. There is no right-propagating wave in the supersonic load's speed, and only one cross-point exists corresponding to the left-propagating wave

Let the wave frequency detected at an unmoving and moving points be $\omega(k)$ and $\omega_0(k)$, respectively. The connection between these values, the Doppler effect, (1.15), follows directly from the wave expression, (1.7), as already mentioned in (1.9).

Graphically, the wavenumber and frequency of the propagating wave correspond to a cross-point of the dispersion curve, $\omega(k)$, with the *load line* $\omega = \omega_0 + v_0 k$. Generally, there can be zero, two, or more numbers of the waves excited by a given couple (ω_0, v_0), see Figs. 1.15 and 1.16.

Regular waves correspond to load lines *crossing* the dispersion curve, but not the tangent

$$v_0 \neq v_g(k) \quad (\omega(k) = \omega_0 + k v_0)\,. \tag{1.64}$$

The equality, $v_0 = v_g(k)$, matches to a resonant wave, Sect. 1.3.5.

In particular, if the load does not move then the frequencies coincide

$$\omega(k) = \omega_0 \;\; (v_0 = 0)\,, \tag{1.65}$$

and if the moving force does not oscillate, then the phase speed of the wave coincides with the speed of the force

$$\omega(k) = k v_0 \quad (\omega_0 = 0)\,. \tag{1.66}$$

In the latter case, the non-oscillating moving load excites sinusoidal waves moving with load velocity, $v_p = v_0$, and both frequencies, detected by a moving observer, are at zero.

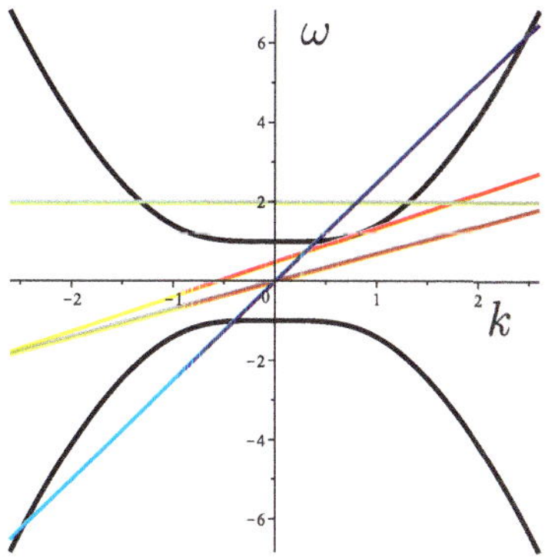

Fig. 1.16 Wave excitation in a beam on an elastic foundation, $\omega(k) = \pm\sqrt{1+k^4}$ (bold black lines). The load lines: 1. $\omega = 0.7k$, $\omega_0 = 0$ (brown), no propagating waves are excited. 2. $\omega = 2$, $v_0 = 0$ (khaki), two waves are excited, the right and left ones; 3. $\omega = 0.5 + 0.86k$ (red), the resonant wave is excited growing and moving together with the load (the load line is tangent to the dispersion curve). 4. $\omega = 2.5k$, $\omega_0 = 0$ (navy). Four waves are excited propagating to the right, one couple (as the cosinusoidal wave) propagates in front of the load, and another one—behind it

1.3.2 The String Under a Moving Force

A string under transverse force P moving uniformly with speed v_0 can serve as an example of the special case $\omega_0 = 0$. The linear equation is

$$\varrho\ddot{u}(x,t) - Tu''(x,t) = P\delta(x - v_0 t)\,, \tag{1.67}$$

where ϱ is the string mass per unit length.

For the subsonic load speed, $0 \le v_0 < c$, the solution can be represented as

$$u = u_+(x - ct)\ \ (x > v_0 t)\,,\quad u = u_-(x + ct)\ \ (x < v_0 t)\qquad c = \sqrt{T/\varrho}\,, \tag{1.68}$$

where $u_\pm$ are waves propagating to the right and left, respectively, c is the wave speed in the string.

The solution under zero initial conditions is

$$u_+ = \frac{P}{2\varrho c(c - v_0)}(ct - x)\,,\qquad u_- = \frac{P}{2\varrho c(c + v_0)}(v_0 t + x)\,. \tag{1.69}$$

For the supersonic load speed, $c \le v_0$

$$\begin{gathered} u_+ = 0\,, u_- = u_{-1} + u_{-2}\,, \\ u_{-1} = \frac{P}{\rho(v_0^2 - c^2)}(v_0 t - x)(ct < x < v_0 t)\,,\qquad u_{-2} = \frac{P}{2\rho c(c + v_0)}(ct + x)\qquad (-ct < x < ct)\,, \end{gathered} \tag{1.70}$$

The string shape for these cases is shown in Figs. 1.17 and 1.18.

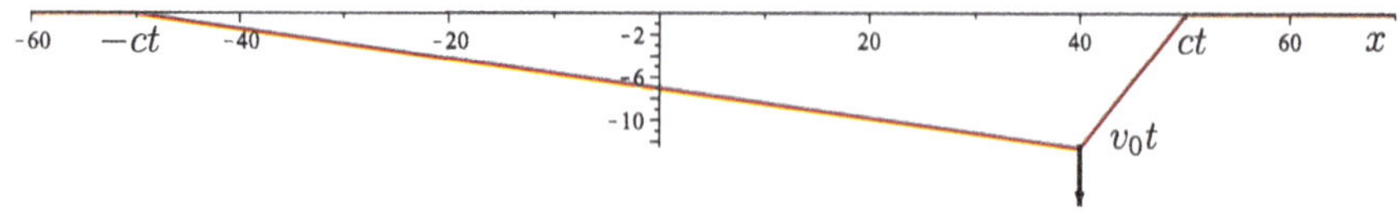

Fig. 1.17 A string under a moving force. The subsonic regime. The force movement speed v_0 is set as 4/5 of the string's sound speed, c. The plot corresponds to the solution (1.69) (the scaling is not constrained)

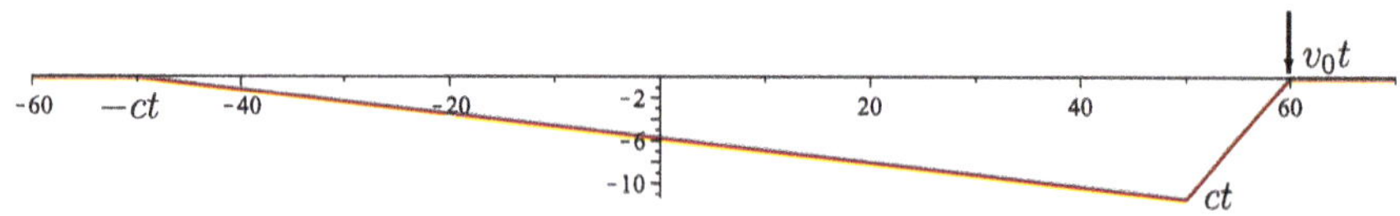

Fig. 1.18 A string under a moving force. The supersonic regime. The force movement speed v is set as 6/5 of the string's sound speed, c. The plot corresponds to the solution (1.70) (the scaling is not constrained)

Note that under the load's critical speed, $v_0 = c$, the established regime breaks down, and a resonant wave develops (resonant waves excited by moving and oscillating loads are considered in detail in Chap. 16.

1.3.3 Elastic Beam on the Winkler Foundation

We now return to Eq.(1.34) for uncompressed beam, $T = 0$, and plot three rays corresponding to the (non-dimensional) force speed, $v_0 = (1, \sqrt{2}, 2)$. There are two cross-points of the ray ($v_0 = 2$) with the dispersion curve, for the lower one $v_g < v$, and the radiated wave places behind the external force. The opposite inequality corresponds to the upper point, and the wave propagates ahead of the force. The ray ($v_0 = \sqrt{2}$) corresponds to the resonant regime, and the lower ray ($v_0 = 1$) corresponds to an exponentially localized wave (Fig. 1.19).

The wave in a beam excited by an unmoving oscillating force is derived in Sect. 1.3.7.

1.3.4 Kelvin Wedge Behind a Moving Ship

Consider a 2D problem for the configuration of steady-state gravity waves induced by a moving ship. The scheme is shown In Fig. 1.20, where this connection applies to a wave propagating normal to line 3 (as the thin arrow directs). Let v_p be its phase velocity (its value corresponds to a in the sectors the length of the normal from the origin to line 3). Then the phase velocity of its trace on the horizontal line is

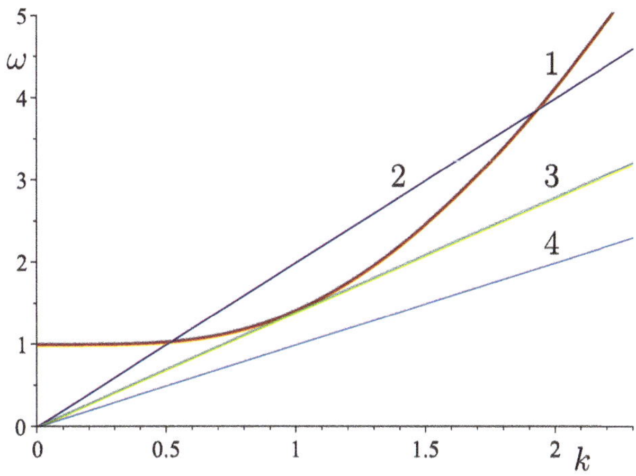

Fig. 1.19 The dispersion relation for the sinusoidal flexural wave in the elastic beam on the Winkler foundation: 1. $\omega(k)$, 2. $v_0 = 2$ (the regular excitation), 3. $v_0 = \sqrt{2}$ (the resonant case), and 4. $v_0 = 1$ (no sinusoidal wave is excited)

$$v = \frac{v_p}{\sin\alpha}, \tag{1.71}$$

where α is the angle between these lines. In a steady-state regime, the trace phase velocity is equal to the ship velocity v_0 shown by the horizontal arrow. So

$$v_p = v_0 \sin\alpha\,. \tag{1.72}$$

It follows that in the steady-state regime, the waves can exist which correspond to (1.72) with α in the sectors $0 < \sigma < \pi/2$ and $-\pi < \sigma < -\pi/2$. However, the wavefront propagates twice as low as shown in Fig. 1.20 by line 2, which makes angle β with the horizontal 'ship line'; its speed corresponds to the thick arrow normal to line 1. These considerations are valid for any wave in the sector, the upper bound of which corresponds to the admissible waves' maximal group velocity.

The relation between the angles

$$\cot\beta = \frac{4}{\sin 2\alpha} - \tan\alpha \tag{1.73}$$

follows from the scheme in Fig. 1.20, where the value of the arrow 5 is half the distance 4 to the (prospective) phase wavefront 3.

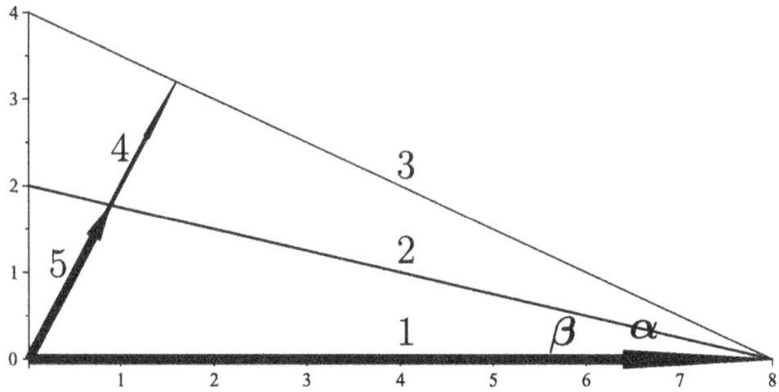

Fig. 1.20 Wave excitation by a ship moving uniformly on deep water. In this scheme, the horizontal arrow 1 presents the ship velocity, v_0. The line 2 (inclined at angle β to the ship velocity) is the wedge boundary defined by the wave group velocity, v_g. The line 3 (inclined at angle α to the ship velocity) is the prospective wave phase velocity front (not realized because the group velocity, $v_g = v_p/2$, defines the wavefront). The normal to the phase velocity front thin arrow 4 starts from the origin, and the thick arrow 5 directs the same as 4 but to the wedge boundary

For the maximal angle β, at which the line 2 in Fig. 1.20 is the boundary of all steady-state waves in the *Kelvin wedge* excited by the ship, we have $\mathrm{d}\beta/\mathrm{d}\alpha = 0$. This equality yields

$$\alpha = \arctan(\sqrt{2}) \approx 0.9553\,, \quad \beta = \arcsin(1/3) \approx 0.3398\,. \tag{1.74}$$

This value of α corresponds to the angle between the ship velocity and the wave crests at the Kelvin wedge boundary.[5]

The Kelvin wedge is similar to the Mach angle for a shock wave induced by a body supersonically moving in a medium. Also, note that the above-used way of consideration is applicable for any relation between phase and group velocities under the condition $v_g < v_p$.

As to the *wave resistance*, we here only can say that it depends, in particular, on the *Froude number*

$$Fr = \frac{v}{\sqrt{gL}}\,, \tag{1.75}$$

where v, L and g are the speed, length and gravity acceleration.

1.3.5 Resonant Waves

In a finite body, the resonance arises when the load frequency coincides with that of a free oscillation. For the resonant wave, the frequency and speed of the load must coincide with the frequency (detected at the moving load) and group speed of the

[5] The results in (1.74) was first found by William Thomson (Lord Kelvin) in 1887. Also, see Lighthill (1978). Rabaud and Moisy (2013) presented a detailed analysis and many illustrations.

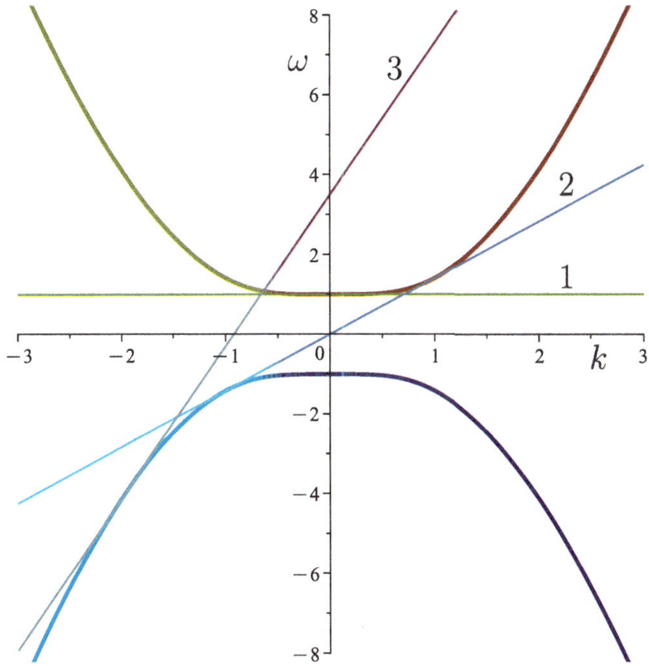

Fig. 1.21 The three of possible load lines for the resonant excitation of flexural waves in the elastic beam on the Winkler foundation, $\omega = \pm\sqrt{1+k^4}$: $\omega_0 = 1$, $v_0 = 0$ (1), $\omega_0 = 0$, $v_0 = \sqrt{2}$ (2), $\omega_0 = 3.5$, $v_0 = 15^{3/4}/2$ (3). The corresponding resonant points, $(k, \omega(k))$, are (0,1), (1, $\sqrt{2}$), $(-15^{1/4}, -4)$, respectively

wave (the energy condition). Graphically, the load line must be tangent at the related point of the dispersion curve. At last, mathematically, that corresponds to a stronger singularity of the forced wave equation.

$$\omega_0 + kv_0 = \omega(k) \quad \text{and} \quad \mathrm{v}_0 = \mathrm{v}_\mathrm{g}(\mathrm{k})\,, \tag{1.76}$$

where v_0 is the load speed, and ω_0 is the wave frequency detected at the load moving support. So, *any* point of the dispersion curve becomes resonant for the load which line is tangent to $\omega(k)$ at this point. The three typical resonant load acting on a beam on the elastic foundation are shown in Fig. 1.21. For a non-dispersion waveguide, the resonant load line coincides with one of the $\omega(k)$ straight line.

The growing resonant wave shape is determined using the integral transforms in Chap. 16, Part VII.

1.3.6 The Wave Energy

As follows from the string dynamic equation (1.67), its potential and kinetic energy densities are

$$\mathcal{P} = \frac{1}{2}T(u')^2 = \frac{1}{2}\varrho c^2(u')^2\,, \quad \mathcal{K} = \frac{1}{2}\varrho(\dot{u})^2\,. \tag{1.77}$$

Clearly, for free waves, $u(x \pm ct)$, these energies are equal.[6]

Hear the potential energy is not a strain one since, in the considered model, the string is flexible and inextensible. It is the work against the tension force, T. Indeed, the horizontal distance δL corresponds to the inclined element length $\sqrt{1+(u')^2}\delta L$. So, the work of a unit string element displacement is

$$A = (\sqrt{1+(u')^2} - 1)T \sim \frac{1}{2}(u')^2 T\,. \tag{1.78}$$

Strictly speaking, a longitudinal wave must accompany the nonuniform string displacement. However, its amplitude is of the second order of magnitude, and the linear formulation neglects the longitudinal motion.

Energy aspects of the wave superposition are considered in Sect. 3.4.8.

As the speed of the force crosses the wave speed (while force accelerates),[7] a wave pulse arises as strong as the acceleration is low (Chap. 16). The phenomenon is similar to a sound boom when a plane breaks through the sound barrier.

1.3.7 The Phase Shift

The phase shift phenomenon was first met in 1827 when George Simon Ohm considered the relationship between electrical voltage and current in a conductor. In mechanical waves, the phase shift is the shift between the oscillating load and the energy flux-related the body displacement rate. It plays the crucial role in the energy flux.

In a regular (non-resonant) periodic excitation of an oscillator, the pase shift $\varphi = \pi/2$, since the energy flux averaged is at zero. However, in the resonant excitation, $\varphi = 0$.

As an example, where there is no this phenomenon, consider a string under a moving-oscillating point force. The equation is

[6] General statements concerning the energy partition in oscillations and waves (Slepyan, 2015a), are discussed in Sect. 2.2.

[7] This problem for a string on an elastic foundation was considered by Kaplunov and Muravskii (1986).

$$\varrho \frac{\partial^2 u(x,t)}{\partial^2 t^2} - T \frac{\partial^2 u(x,t)}{\partial^2 x^2} = Q = Q_0 \delta(\eta) e^{i\omega_0 t} , \quad \eta = x - v_0 t , \tag{1.79}$$

where ϱ and T are the density as the mass per unit length of the string, and the tensile force, respectively. The displacement is

$$u(x,t) = A(\exp(i(\omega(k_+)t - k_+ x))H(\eta) + \exp(i(\omega(k_-)t - k_- x))H(-\eta) , \tag{1.80}$$

with

$$\omega(k_\pm) = \omega_0 + k_\pm v . \tag{1.81}$$

In these expressions, for $\omega_0 > 0, 0 < v < c = \sqrt{T/\varrho}$, the wavenumbers $k_+ > 0, k_- < 0$ (see Fig. 1.15).

Directly from the dynamic equation, (1.79), (defining a discontinuity of u' at $\eta = 0$) we find

$$A = -i \frac{Q_0}{T(k_+ - k_-)} ; \tag{1.82}$$

hence

$$\frac{\partial u(x,t)}{\partial t} = \frac{Q_0 \omega_0}{T(k_+ - k_-)} > 0 \implies \varphi = 0 . \tag{1.83}$$

This conclusion remains valid for the resonant wave, where

$$Q = Q_0 \delta(x - ct) , \quad u(x,t) = \frac{Q_0}{\varrho c^2}(x + ct)H(x + ct)H(ct - x) . \tag{1.84}$$

Now consider a similar problem for the bending beam under an oscillating force, where the phase shift in the wave excitation is nontrivial. The equation is

$$\varrho \frac{d^2 u(t,x)}{dt^2} + D \frac{d^4 u(t,x)}{dx^4} = Q(t,x) = Q_0 \delta(x) e^{i\omega_0 t} , \tag{1.85}$$

where ϱ is the beam density per unit length, D is the bending stiffness, and we assume that Q_0 and ω_0 to be real. The energy flux-related displacement and the corresponding equation are

$$u = U(x) e^{i\omega_0 t} , \quad \frac{d^4 U(x)}{dx^4} - \Omega^2 U(x) = \frac{Q_0}{D} \delta(x) , \tag{1.86}$$

with

$$\Omega = \sqrt{\varrho/D}\, \omega , \quad U(-x) = U(x) , \quad U'(0) = 0 , \quad U'''(+0) = Q_0/(2D) . \tag{1.87}$$

A non-growing (as $x \to \infty$) solution is

$$u(t, x) = -\frac{Q_0 \mathrm{i}}{4D\Omega^{3/2}} \mathrm{e}^{\mathrm{i}\omega_0 t} \left(\mathrm{e}^{-\mathrm{i}\sqrt{\Omega} x} - \mathrm{i}\mathrm{e}^{-\sqrt{\Omega} x} \right) \quad (x > 0)\,, \tag{1.88}$$

and

$$\frac{\mathrm{d}u(t, 0)}{\mathrm{d}t} = \frac{Q_0}{4\sqrt{\varrho D\Omega}} \mathrm{e}^{\mathrm{i}\omega_0 t} (1 - \mathrm{i})\,. \tag{1.89}$$

Thus, the phase shift $\varphi = -\pi/4$, and it owes its existence to the presence of the localized partial solution.

1.3.8 The Load-to-Wave Energy Flux

In general, for the moving-oscillating point force, the rate of the energy flux to the wave is

$$N = \Re Q \,\Re \dot{u} + \Im Q \,\Im \dot{u} \tag{1.90}$$

with, at the moving-oscillating load

$$Q = Q_0 \mathrm{e}^{\mathrm{i}\omega_0 t}\,, \quad \dot{u} = W_0 \mathrm{e}^{\mathrm{i}\omega_0 t + \varphi}\,, \tag{1.91}$$

where Q is the external force (moment), and $\dot{u}$ is the energy-corresponding linear (angular) velocity; Q_0 and W_0 are assumed to be real constants with $Q_0 W_0 = N_0 > 0$.

We have

$$N = N_0(\cos(\omega_0 t)\cos(\omega_0 t + \varphi) + \sin(\omega_0 t)\sin(\omega_0 t + \varphi)) = N_0 \cos\varphi\,. \tag{1.92}$$

So, the phase shift, φ, leads to the energy flux decrease, as reflected by the factor $\cos\varphi$.

In the case of the beam considered

$$N = N_0 \cos\pi/4 = \frac{\sqrt{2}}{2} N_0\,. \tag{1.93}$$

Thus, in this case, the phase shift significantly reduces the energy flux rate.

In Fig. 1.22, the energy flux rate is shown corresponding the real parts of the load and wave respond, $\cos(\omega_0 t)\cos(\omega_0 t + \varphi)$.

One can see that the energy flux rate oscillates from - 0.05 to 0.3. The negative flux arises due to a phase shift between the sinusoids of the load and the beam, Fig. 1.23.

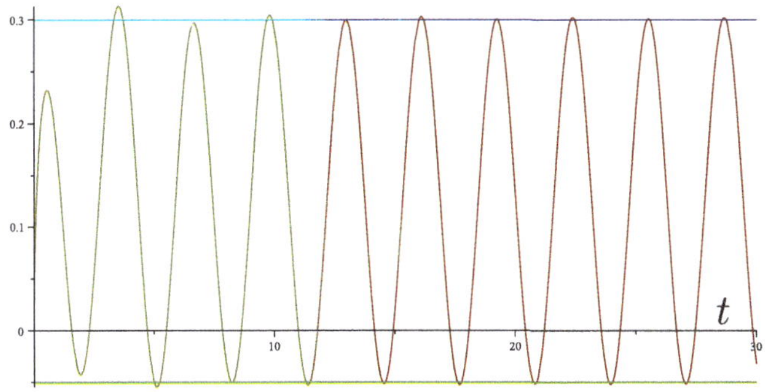

Fig. 1.22 The rate of the energy flux from the concentrated force to the beam. It is negative in part, that reflects the phase shift between the load and beam oscillations

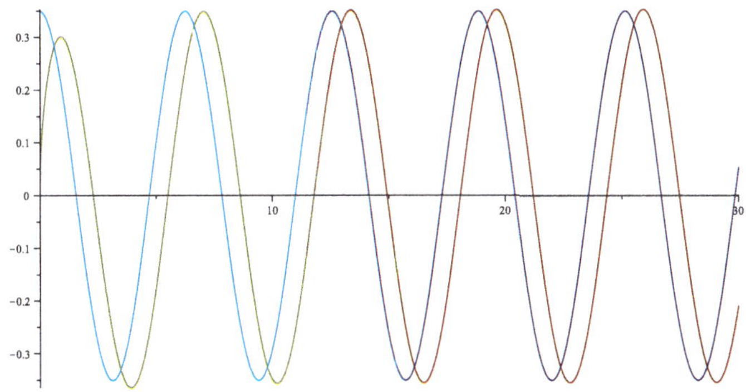

Fig. 1.23 The phase shift between the sinusoids of the load, $\sin(t)$ (navy), and the beam oscillation speed (brown)

1.4 Transition Waves

We here consider waves transforming the waveguide.

A good (more correct, bad) example is the collapse of the long San Saba railway bridge in Central Texas collapsed in May 2013 as a result of initial damage caused by fire (see Brun et al., 2014). The event occurred as a transition (destruction) wave. The latter is similar to the *domino wave*, Sect. 1.4.2

A crushing wave in a brittle bar excited by the longitudinal impact (Slepyan & Troyankina, 1969) is a transition wave of a different kind. A glass bar was crushing while impacting a steel plate. The propagating wavefront separated the intact part from the crushing area. In the experiment, the latter widened like a flaming tail of a rocket taking off, while the wavefront looked plane.

Below, we shortly discuss this topic.

1.4.1 On the Transition Wave Formulation

The (dynamic) equilibrium of a system (or any its part) corresponds to the extremum of the Lagrangian. In Mechanics, the latter is the difference between kinetic and potential energies

$$L = \mathcal{K}(\dot{\boldsymbol{u}}) - \mathcal{P}(\boldsymbol{u}). \tag{1.94}$$

In the case of the system parameters fixed, the external (Newtonian) forces follow as the derivatives of the energy by the corresponding displacement components and vice versa. However, if the system parameters vary, the energy is no longer defined by the Newtonian forces.

To preserve the one-to-one energy-forces relation, non-Newtonian 'forces' are introduced, also as the derivatives of the energy by displacements, but the displacements related to the system configuration or (and) any other its properties. In particular, it can be a displacement of an immaterial object, e.g. a singular point (or a line) or an interface between two different phases.

Such 'forces' have different names: the *configuration forces* as related to a transformation of a body configuration, as it called by Morton Gurtin, see Gurtin (2000), the *Eshelbian forces*, John Eshelby introduced the forces in elasticity, Eshelby (1951; 1956), the *material forces*, because they act in the *material space*, as it called by Maugin (2015), or the *driving force* as in thermodynamics, see e.g. Dill and Bromberg (2003).

Some words on the Eshelbian forces:

In contrast to Newtonian forces, these forces do not participate in a body's equilibrium, but only its configuration.

In the case of the *force on singularity*, the Eshelbian force is only another name for the through-the-singularity energy release. (The singularity moves if the 'force' value is enough to overcome the resistance to the movement.)

The Eshelbian forces, Sect. 17.2.2.3, as well as singularities, can exist in a continuous model. However, that model is only valid for a long-wave field (at the microstructure scale). Thus, although the continuous model with a singularity is internally consistent, the microstructure is advised to be involved.

A moving phase transition surface or a line is the main characteristic of a transition wave. In this sense, a dynamic crack separating an elastic material also behaves as a transition wave, Slepyan (1981, 2002). Here we outline the related theory considering some other illustrative examples.

1.4.2 Transition Wave Taking Energy Stored in a Waveguide

Let the dominos be placed vertically on a line with a distance between neighbors. As is typical for such transition waves, the domino line is initially stable, and some energy input is required to overcome the energy barrier for the dominos to fall. For the rotation of a single domino around its angle, it is the energy of the maximal elevation of its center

$$E_0 = \frac{1}{2}(\sqrt{h^2 + a^2} - h)gm\,, \tag{1.95}$$

where h, a, m and G are the domino height, thickness and mass, and the gravity acceleration. But after that, the initial potential energy of the first domino is released (as it falls becoming almost horizontal). And the released energy is sufficient for the second domino to repeat, and so on. So, after the initial pulse, the falling wave propagates.[8]

An interesting, but not so simple example of such a wave propagating under the stored energy release, the *Cobra wave*, is discussed in Boucher et al. (2017).[9]

1.4.3 Reflection at a Moving Obstacle

A characteristic feature of a transition wave is an *energy release*. This phenomenon can be demonstrated on a simple model. Let a constant-amplitude tension wave propagate in an elastic rod from the left to its right end, $x = 0$. It causes a reflected one propagating to the left. In each of these two cases, where the end is free or fixed, there is no energy release. The latter, however, arises if the wave reflects at a cross-section *moving through the rod*. Let v be its speed and $0 < v < c$, where c is the wave speed.

The displacements of the incident and reflected waves are

$$u_i = A_i(x - ct)\,, \quad u_r = A_r(x + ct)\,, \tag{1.96}$$

respectively, and the displacement rates at the moving interface

$$w = A_i(v - c) + A_r(v + c)\ \ (x = vt - 0)\,, \quad w = 0\ \ (x = vt + 0)\,. \tag{1.97}$$

Due to the mass conservation, $w = 0$ at the left too, $(v - c)A_i + (v + c)A_r = 0$. Thus,

$$A_r = \frac{c - v}{c + v}A_i\,. \tag{1.98}$$

---

[8] See https://video.search.yahoo.com/search/video?fr=tightropetb&p=domino+line+video#id=2&vid=ab0a07402219b47687eae080e98ed3dc&action=click.

[9] See https://www.bing.com/search?q=cobra+wave&cvid=194e10901b5d4b348e39c6f6f4be8a99\&aqs=edge.0.0l9.13431j0j1\&pglt=41\&FORM=ANSPA1\&PC=LCTS.

The left-directed force acting on the moving cross-section from the left and the energy release rates is

$$Q = ES(A_i + A_r) = ES\frac{2c}{c+v}A_i\,, \tag{1.99}$$

where $E = \varrho c^2$ and S are the elastic modulus and the cross-section area, and ϱ is density. The force Q work per the crack unit development is

$$\mathcal{E}_Q = Q\varepsilon = \frac{Q^2}{ES}\,, \tag{1.100}$$

while, as follows from (1.98) and (1.99), the additional energy of the road is

$$\mathcal{E}_R = \frac{1}{2}ES(A_i + A_r)^2 + \frac{1}{2}S\varrho c^2(A_r - A_i)^2 = \frac{Q^2}{2ES}\left(1 + \frac{v^2}{c^2}\right)\,. \tag{1.101}$$

Thus, the energy release (also per unit lengs of the crack growth) is

$$G = \mathcal{E}_Q - \mathcal{E}_R = \frac{Q^2}{2ES}\left(1 - \frac{v^2}{c^2}\right)\,. \tag{1.102}$$

1.4.4 Pulling the Rod Out from the Ground

Now let the rod, such as a pile, be partially but deeply embedded in the ground, and let a cylindrical crack, $x < vt$, grow at the rod boundary under the external tension, separating it from the surrounded medium. The fracture energy, $G_c = \gamma\ell$, resists the crack propagation, where ℓ is the length of the rod cross-section boundary. We here assume that the crack resistance γ is independent of the crack speed, and that there is any other lost, that is, the energy release, (1.102), $G = G_c$. We find that in the steady-state regime,

$$v = c\sqrt{1 - \frac{2ESG_c}{Q^2}} \quad (Q > \sqrt{2ESG_c}), \quad Q = \sqrt{2ESG_c} \ \text{(quasi-static pulling)}. \tag{1.103}$$

1.4.5 Wave in a Bistable Structure

Consider a two-phase material, the stress-strain diagram for which is shown in Fig. 1.24.

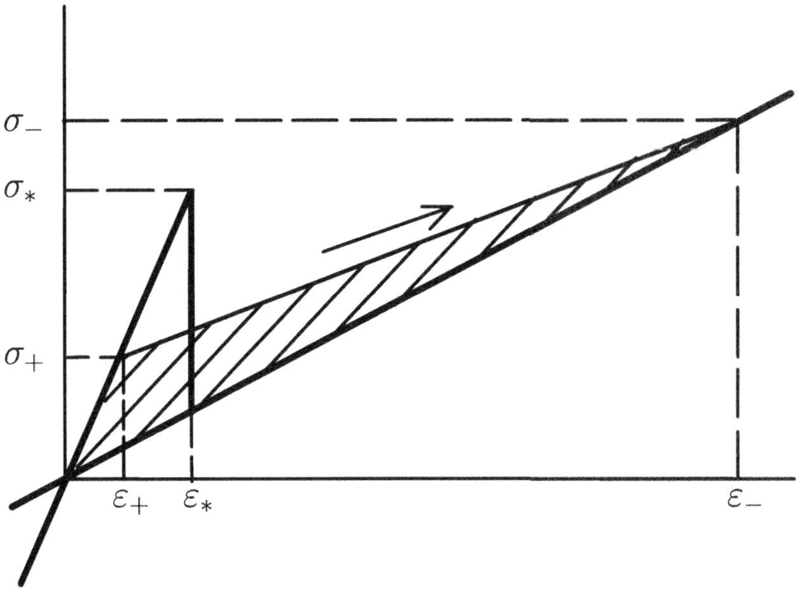

Fig. 1.24 The bi-stable stress-strain diagram. The shaded area corresponds to the energy release at the transition front, where the stresses change by a stresses jump from σ_+ to σ_-

We assume that a constant-stress wave, propagates to the right and reflects and transmits at the transition front, $x = v_0 t$, where the higher-modulus phase, $\sigma = E_+\varepsilon$ is replaced by the lower-modulus one, $\sigma = E_-\varepsilon, 0 < E_- < E_+$. The waves are

$$u_i = \varepsilon_i(x - c_-t)H(v_0t - x)\,, \quad u_r = \varepsilon_r(x + c_-t)H(v_0t - x)\,, \\ u_t = \varepsilon_t(x - c_+t)H(x - v_0t)\,, \tag{1.104}$$

where $c_\pm = \sqrt{E_\pm/\varrho_0}$ are the wave speeds ahead and behind the transition front $x = v_0 t$, respectively, ϱ_0 is the initial density, and $\varepsilon_i, \varepsilon_r, \varepsilon_t$ are constant strains.

For the three unknowns, v_0, ε_r and ε_t, there are two wavefront conditions: the mass and momentum conservation laws, and, as the third one addition, the phase transition energy (or another transition criterion) should be involved.

The mass conservation under the transition at $x = v_0 t$ is satisfied by zero mass concentration rate at this point

$$\varepsilon_i(v_0 - c_-) + \varepsilon_r(v_0 + c_-) - \varepsilon_t(v_0 - c_+) = 0\,. \tag{1.105}$$

Taking into account that the densities ahead and behind the transition front are

$$(1 + \varepsilon_+)\varrho_+ = (1 + \varepsilon_-)\varrho_- = \varrho_0\,, \quad \varepsilon_+ = \varepsilon_t, \,, \quad \varepsilon_- = \varepsilon_i + \varepsilon_r \tag{1.106}$$

and ϱ_0 is the initial density. The laws can be expressed as follows

$$\begin{gathered}\frac{v_- - v_0}{1+\varepsilon_-} + \frac{v_0 - v_+}{1+\varepsilon_+} = 0\,, \\ \left(\frac{v_-}{1+\varepsilon_-} - \frac{v_+}{1+\varepsilon_+}\right)\varrho_0 v_0 = \sigma_+ - \sigma_-\,, \quad \sigma_- = \sigma_i + \sigma_r\,, \quad \sigma_+ = \sigma_t\,,\end{gathered} \tag{1.107}$$

which for small strains and particle velocities (as assumed in the following) become

$$\begin{aligned} v_- - v_+ &= (\varepsilon_+ - \varepsilon_-)v_0\,, \\ v_- - v_+ &= \frac{\sigma_+ - \sigma_-}{\varrho_0 v_0} \end{aligned} \tag{1.108}$$

with the identity

$$\varrho_0 v_0^2 = \frac{\sigma_- - \sigma_+}{\varepsilon_- - \varepsilon_+}\,. \tag{1.109}$$

The displacement continuation at $x = v_0 t$:

$$\varepsilon_i(v_0 - c_-) + \varepsilon_r(v_0 + c_-) = \varepsilon_t(v_0 - c_+)\,, \tag{1.110}$$

We now note that the first relation in (1.25) is still valid for the right-propagating incident and transmitted waves, and also, but with the opposite sign, for the reflecting wave, namely

$$V_i = -\frac{\sigma_i}{\varrho_0 c_-}, \quad V_+ = -\frac{\sigma_+}{\varrho_0 c_+}, \quad V_r = V_- - V_i = \frac{\sigma_- - \sigma_i}{\varrho_0 c_-}, \quad \Longrightarrow \quad V_- = \frac{\sigma_- - 2\sigma_i}{\varrho_0 c_-}. \tag{1.111}$$

We now can exclude the particle velocities from the conservation laws and to express them as follows

$$\begin{aligned} \left(1+\frac{v_0}{c_-}\right)\sigma_- + \frac{c_-}{c_+}\left(1-\frac{v_0}{c_+}\right)\sigma_+ &= 2\sigma_i\,, \\ (c_- + v_0)\sigma_- - c_-\left(1-\frac{v_0}{c_+}\right)\sigma_+ &= 2v_0\sigma_+\,, \end{aligned} \tag{1.112}$$

which result in expressions for the stresses

$$\sigma_- = \frac{c_-(c_+ + v_0)}{(c_- + c_+)(c_- + v_0)}2\sigma_i\,, \quad \sigma_+ = \frac{c_+^2(c_- - v_0)}{c_-(c_- + c_+)(c_+ - v_0)}2\sigma_i\,. \tag{1.113}$$

Since in each of the considered waves, incident, reflected and transmitted, the kinetic and potential energies are equal, the energy flux to the transition front release per area and time and the energy release rate per unit volume are

$$N = \frac{\sigma_i^2}{E_-}(c_- - v_0) - \frac{\sigma_r^2}{E_-}(c_- + v_0) - \frac{\sigma_+^2}{E_+}(c_+ - v_0)$$
$$= \frac{(c_+ - c_-)(c_+ + v_0)(c_- - v_0)}{(c_+ + c_-)(c_+ - v_0)(c_- + v_0)} \frac{2\sigma_i^2 v_0}{E_-}, \quad G = \frac{N}{v_0}. \tag{1.114}$$

Note that this result is identical to that presented as the shaded area in the stress-strain diagram, Fig. 1.24

$$G = \frac{1}{2}[(\sigma_+ + \sigma_-)(\varepsilon_- - \varepsilon_+) + \sigma_+\varepsilon_+ - \sigma_-\varepsilon_-] = \frac{\sigma_+\sigma_-}{2}\left(\frac{1}{E_-} - \frac{1}{E_+}\right), \tag{1.115}$$

what can be verified using the above expressions for the stresses (1.113).

The normalized energy release rate

$$\tilde{G} = \frac{(c_+ + c_-)E_-G}{(c_+ - c_-)2\sigma_i^2} = \frac{(1 + \lambda\tilde{v}_0)(1 - \tilde{v}_0)}{(1 - \lambda\tilde{v}_0)(1 + \tilde{v}_0)},$$
$$\lambda = \frac{c_-}{c_+}, \quad \tilde{v}_0 = \frac{v_0}{c_-} \tag{1.116}$$

as a function of $\tilde{v}_0$ for some values of λ is presented in Fig. 1.25.

If the critical, probably speed-dependent value of the energy release is known, the above solution allows us to determine the incident wave's critical amplitude. However, in the force criterion where σ_* is given, the homogeneous body model leaves uncertainty: we can only say that σ_+ must be below the critical value σ_* (the equality is inadmissible).

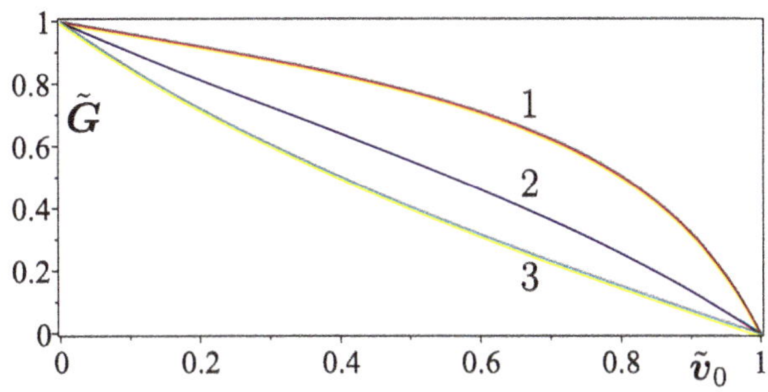

Fig. 1.25 The normalized energy release rate $\tilde{G}$ as a function of the transition front speed $\tilde{v}_0$, (1.116), for three values of $\lambda = c_-/c_+$: $\lambda = 0.8$ (1), 0.5 (2) and 0.2 (3)

These stresses and also σ_- can be much lower. Additional stresses arise due to oscillations caused by the energy release. Thus, to resolve the problem completely, the underlining structure, the 'microstructure', should be taken into account in the formulation. In this regard, bistable mass-spring chains were considered (Slepyan & Troyankina, 1984; Cherkaev et al., 2005; Slepyan et al., 2005), see Sect. 1.2.9.

Different (non-steady-state) established transition wave modes, clustering (Mishuris et al., 2009), and forerunning (Slepyan, 2015a) are considered in relevant section.

In conclusion, note that in dynamic transitions, the role of the structure associated with the continuous medium model can play a considerable and even a decisive role, see relevant section.

Part I
Wave Energy and Momentum

Chapter 2
Wave Energy

2.1 The Energy

As follows directly from the complex wave expression (1.5), the wave kinetic energy of a unit volume is

$$\mathcal{K} = \frac{1}{2}\varrho\omega^2|A|^2\,, \tag{2.1}$$

and it will be seen below that, in the linear wave, the potential energy is the same.

2.2 The Energy Partition

It has often been noticed, in particular cases of progressive waves, that the potential and kinetic energies are equal; but I do not call to mind any general treatment of the question.

Lord Rayleigh, On Progressive Waves. Proceedings of the London Mathematical Society, 1877, IX, 21–26.

This issue was resolved (Slepyan, 2015a), for the linear and some classes of non-linear dynamical systems, with the Lagrangian represented by a sum of homogeneous functions of the displacements and their derivatives. The energy partition is defined solely by the homogeneity orders, and follows directly from the Euler–Lagrange equation and Euler's theorem on homogeneous function. For solitary waves, where the potential energy consists of two functions of different orders, the Derrick–Pohozaev identity is also involved. The general results are illustrated by examples

L. I. Slepyan, *Mechanical Waves: Free, Forced, and Transition*,
https://doi.org/10.1007/978-3-032-15864-2_2

of various oscillations and waves, linear and nonlinear, homogeneous and forced, steady-state and transient, periodic and non-periodic, parametric and resonant, regular and solitary.

For free linear oscillations and sinusoidal waves, it is long recognised that kinetic and potential energies averaged over the period are equal to each other. In some partial cases, this is observed directly, see, e.g., Mask and Jay (1967), Lighthill (1978), Antenucci and Imburger (2001), and Falnes (2007). This fact is also confirmed by Whitham (1974) for a non-specified linear sinusoidal wave.

We consider this topic in terms of classical mechanics and obtain the partition relations for some classes of generally nonlinear oscillations and waves, where the Lagrangian is represented by a sum of homogeneous functions of the displacements and their derivatives, possibly not only the first order. The Lagrangian can also depend on time and spatial coordinates explicitly. Starting from the Euler–Lagrange equation of a general view and taking into account Euler's theorem on homogeneous functions we obtain both the energy partition relation and the conditions defining the regions of averaging (it can be the oscillation period, if exists, and not only it). The partition relation is fully defined by the homogeneity orders regardless of other parameters of the system and the dynamic process. In the case where the potential energy is represented by a single homogeneous function, as well as the kinetic energy, the partition between these energies is defined uniquely.

For a steady-state solitary wave, where the potential energy consists of two functions of different orders, there are three different energy terms, and the single equality is not sufficient to obtain the desired energy partition. At the same time, the wave is exponentially localized, that allows us to use the Derrick–Pohozaev identity as an additional relation (Derrick, 1964; Pohozaev, 1965; Dodd et al., 1984).

The identity is based on the representation of the steady-state solitary wave as a static state. Note that in the steady-state mode, the kinetic energy can be considered as such, as a function of $\dot{u} = \partial u/\partial t$, and also as a function of $u' = \partial u/\partial x$. In the latter case, it can be treated as an additional (negative) part of the potential energy. Different energy relations, one discussed in the present paper and another given by the Derrick–Pohozaev identity, follow from the static representation as a result of different variations. While the former corresponds to the displacement variation $\delta u = \varepsilon u$, the latter is a uniform x-extension/compression. The use of both relations allows us to separate the kinetic and potential energies completely.

First, we consider general relations for finite discrete systems. The relations obtained for the latter are applicable, with minor additions, to a finite continuous-material body and to an infinite continuous or discrete waveguide as discussed below. Then some examples of free and forced oscillations are presented, which evidence that the linearity is neither necessary nor sufficient condition for the equipartition. Next, some examples of linear and nonlinear oscillations are presented, where the energy partition and the regions of averaging are clearly seen. Homogeneous and forced, steady-state and transient, periodic and non-periodic, regular, parametric and resonant oscillations are examined.

Further, the general relations are presented to be applicable for a finite continuous body and for homogeneous, periodic-structure and discrete waveguides. Some examples of linear and nonlinear, periodic, steady-state, transient and solitary waves are presented, where the energy partition is also demonstrated.

2.2.1 The Main Relations

Consider the Euler–Lagrange equations (17.45)

$$\frac{\partial L}{\partial u_i} - \frac{\mathrm{d}}{\mathrm{d}t}\frac{\partial L}{\partial \dot{u}_i} = 0\,, \quad L = L(\boldsymbol{u}, \dot{\boldsymbol{u}}, t)\,, \quad i = 1, 2, \dots\,, \tag{2.2}$$

where $\boldsymbol{u} = (u_1, u_2, \dots)$. Multiplying the equations by u_i and integrating over an arbitrary segment, $t_1 \le t \le t_2$, we obtain (with summation on repeated indices)

$$\int_{t_1}^{t_2} \left(\frac{\partial L}{\partial u_i} u_i + \frac{\partial L}{\partial \dot{u}_i} \dot{u}_i \right) \mathrm{d}t + B_1 - B_2 = 0\,, \tag{2.3}$$

where

$$B_{1,2} = \frac{\partial L}{\partial \dot{u}_i} u_i \quad (t = t_{1,2})\,. \tag{2.4}$$

Thus

$$\int_{t_1}^{t_2} \left(\frac{\partial L}{\partial u_i} u_i + \frac{\partial L}{\partial \dot{u}_i} \dot{u}_i \right) \mathrm{d}t = 0 \tag{2.5}$$

if

$$B_1 = B_2\,. \tag{2.6}$$

Regarding this formulation, we note that the relation (2.5) follows from Hamilton's principle of least action with the variation $\delta\boldsymbol{u}$ replaced by $\boldsymbol{u}$. This, however, entails a change in the additional conditions. Namely, in the variational formulation, the integration limits, $t_{1,2}$, are arbitrary under the condition $\delta\boldsymbol{u} = 0$ at $t = t_{1,2}$, while in the modified formulation they are not arbitrary but must meet the condition (2.6).

We now suppose that the Lagrangian is a sum of homogeneous functions of $\boldsymbol{u}$, $\dot{\boldsymbol{u}}$

$$L(\lambda\boldsymbol{u}, \lambda\dot{\boldsymbol{u}}, t) = \lambda^{\nu_n} L_n(\boldsymbol{u}, \dot{\boldsymbol{u}}, t)\,, \tag{2.7}$$

where ν_n is the homogeneity order. In this expression, ν_n can be not only integer, and it is assumed that only a positive value can have a fractional exponent. For example, in the latter case, $|u_i|^\nu$ can be present in the Lagrangian but not u_i^ν. Accordingly, in the homogeneity definition, we take $\lambda > 0$. It is also assumed that $|u_i|^\nu \geq 0$. Note that the homogeneity condition excludes possible non-uniqueness in the definition of the energies.

With refer to Euler's theorem on homogeneous functions

$$\frac{\partial f(\boldsymbol{x})}{\partial x_i} x_i = \nu f(\boldsymbol{x}) \quad \text{if} \quad f(\lambda \boldsymbol{x}) = \lambda^\nu f(\boldsymbol{x}) \tag{2.8}$$

and the condition (2.6) we can rewrite the relation (2.5) in the form

$$\int_{t_1}^{t_2} \nu_n L_n \, \mathrm{d}t = 0 \,. \tag{2.9}$$

The latter relation can serve for the determination of the energy partition. For example, let L be a difference between the kinetic and potential energies, $L = \mathcal{K} - \mathcal{P}$, where the, $\mathcal{K}$ and $\mathcal{P}$, are homogeneous functions of the orders μ and ν, respectively. In this case, the relation between the averaged energies

$$\langle \mathcal{K} \rangle = \frac{1}{\tau} \int_{t_1}^{t_1+\tau} \mathcal{K} \, \mathrm{d}t \,, \quad \langle \mathcal{P} \rangle = \frac{1}{\tau} \int_{t_1}^{t_1+\tau} \mathcal{P} \, \mathrm{d}t \quad (\tau = t_2 - t_1) \tag{2.10}$$

is

$$\frac{\langle \mathcal{K} \rangle}{\langle \mathcal{P} \rangle} = \frac{\nu}{\mu} \quad \Longrightarrow \quad \langle \mathcal{K} \rangle = \frac{\nu \mathcal{E}}{\mu + \nu} \,, \quad \langle \mathcal{P} \rangle = \frac{\mu \mathcal{E}}{\mu + \nu} \,, \tag{2.11}$$

where $\mathcal{E} = \mathcal{K} + \mathcal{P}$ is the total energy. Note that the above expressions in terms of the total energy as a constant are valid only for conservative systems.

In particular, in the linear case, where $\nu = \mu = 2$, the averaged energies are equal. Cases $\nu > \mu$ and $\nu < \mu$ correspond to hardening and softening nonlinearity, respectively. The potential energy vanishes as $\nu/\mu \to \infty$. In the limit, this case corresponds to the periodic perfect collisions of rigid particles.

It is remarkable that the energy ratio in (2.11) is equal to the homogeneity order ratio regardless of the other parameters of the system and the dynamic process. In turn, the homogeneity orders generally depend on the constitutive properties but not the material constants. In the case where the potential energy is represented by a sum of two or more homogeneous functions of different orders, the relation (2.9) is only one for three or more terms. In this case, it is not sufficient for the determination of ratios between all terms but still allows one to make some conclusions concerning the energy partition.

The condition in (2.6) is satisfied in periodic oscillations

$$\boldsymbol{u}(t + \tau) = \boldsymbol{u}(t) \tag{2.12}$$

and also in the case where $\boldsymbol{u} = 0$ at $t = t_{1,2}$. In addition, if the kinetic energy being a homogeneous function of order μ with respect to u_i, $\dot{u}_i$ is such a function of order 2 with respect to $\dot{u}_i$

$$\mathcal{K}(\lambda\boldsymbol{u}, \lambda\dot{\boldsymbol{u}}) = \lambda^{\mu}\mathcal{K}(\boldsymbol{u}, \dot{\boldsymbol{u}}), \quad \mathcal{K}(\boldsymbol{u}, \lambda\dot{\boldsymbol{u}}) = \lambda^2\mathcal{K}(\boldsymbol{u}, \dot{\boldsymbol{u}}), \tag{2.13}$$

then

$$B_{1,2} = B^0_{1,2}u_i\dot{u}_i \tag{2.14}$$

and $B_1(B_2) = 0$ if $\dot{\boldsymbol{u}} = 0$ at $t = t_1(t_2)$. Therefore the condition (2.6) is also satisfied in the case of non-periodic oscillations, where $\boldsymbol{u}$ or/and $\dot{\boldsymbol{u}}$ vanish at points. This can be seen in the below examples of oscillations and periodic and solitary waves. In the case of regular or irregular oscillations, does not matter, where the condition (2.6) is satisfied for each of the neighboring points, $t_1 < t_2 < \cdots < t_m < \cdots$, the energy partition averaged over any interval between two of such points is the same. So the energy partition appears fixed for a large range of time.

In this section, we have considered systems of a finite number of material points. The main relations presented here are applicable, with minor additions, to a finite continuous-material body and to a infinite continuous or discrete waveguide as discussed in Sects. 2.2.5 and 2.2.6.

While the energy partition is defined solely by the homogeneity orders, below we calculate the energy distribution function, $\mathcal{L}(t)$, as a function of time

$$\mathcal{L}(t) = \int_0^t \nu_n L_n(t)\,\mathrm{d}t \tag{2.15}$$

assuming that $B_1 = 0$ at $t = t_1 = 0$. This function must vanish at $t = t_2$ (2.9) if the condition (2.6) is satisfied at that point. This representation allows us to demonstrate how the partition varies during the (t_1, t_2)-interval and how $\mathcal{L}(t)$ follow the statement (2.9) (also see the expression in (2.14)). As a rule, dimensionless quantities are used since this simplification does not affect the partition.

2.2.2 *Discrete Systems*

2.2.2.1 The Linearity and Equipartition

Three examples are presented below which evidence that the linearity is neither necessary nor sufficient condition for the equipartition. First, consider a piecewise linear equation

$$m\ddot{u}(t) + [\mathcal{P}_+H(u) + \mathcal{P}_-H(-u)]u(t) = 0\,, \tag{2.16}$$

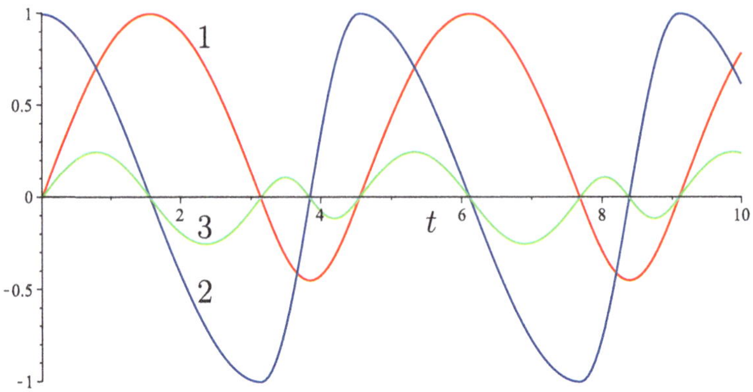

Fig. 2.1 Peace-linear oscillator. Displacements $u(t)$ (1), speed $\dot{u}(t)$ (2) and the energy distribution function $\mathcal{L}(t)/2$ (3) for $\mathcal{P}_+ = 1, \mathcal{P}_- = 5$. It is seen that $\mathcal{L} = 0$ at points where $u = 0$ and where $\dot{u} = 0$

where H is the Heaviside step function and $\mathcal{P}_\pm$ are different positive constants. The corresponding Lagrangian, kinetic and potential energies and homogeneity orders are, respectively

$$L = \mathcal{K} - \mathcal{P}, \quad \mathcal{K}(t) = \frac{m\dot{u}^2(t)}{2}, \quad \mathcal{P}(t) = [\mathcal{P}_+ H(u) + \mathcal{P}_- H(-u)]\frac{u^2(t)}{2}, \quad \mu = \nu = 2. \tag{2.17}$$

Equation (2.16) defines periodic oscillations. The displacement $u(t)$, speed $\dot{u}(t)$ and the energy distribution function $\mathcal{L}(t)$ (2.15) are presented in Fig. 2.1 for $\mathcal{P}_+ = 1, \mathcal{P}_- = 5$ and initial conditions $u(0) = 0, \dot{u}(0) = 1$.

It can be seen that the energies averaged over the period, $\langle\mathcal{K}\rangle$ and $\langle\mathcal{P}\rangle$ (in this case, period $\tau = (1 + 1/\sqrt{5})\pi \approx 4.5465556$), are equal:

$$\mathcal{L}(\tau) = \int_0^\tau 2L(t)\,\mathrm{d}t = \int_t^{t+\tau} 2L(t)\,\mathrm{d}t = 2\tau[\langle\mathcal{K}\rangle - \langle\mathcal{P}\rangle] = 0. \tag{2.18}$$

Next, we consider a nonlinear equation related to centrosymmetric oscillations of a bubble in unbounded, perfect, incompressible liquid, where the added mass is proportional to the bubble radius cubed. Let the (dimensionless) energies be

$$\mathcal{K} = \frac{1}{2}r^3(t)\dot{r}^2(t), \quad \mathcal{P} = \frac{1}{5}r^5(t), \quad \mu = \nu = 5 \quad (\text{the radius } r(t) \geq 0). \tag{2.19}$$

The corresponding dynamic equation is

$$r(t)\ddot{r}(t) + \frac{3}{2}\dot{r}^2(t) + r(t)^2 = 0. \tag{2.20}$$

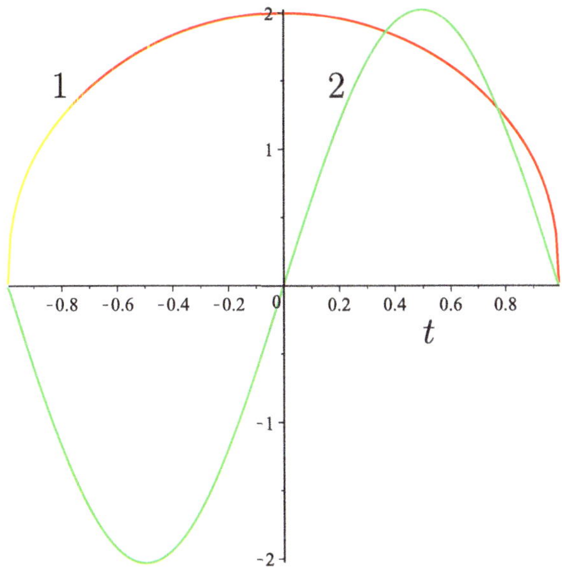

Fig. 2.2 The bubble radius, $r(t)$ (1) and the energy distribution function $\mathcal{L}(t)/2$ (2)

The radius, $r(t)$ and the energy distribution function $\mathcal{L}(t)$ under conditions $r(0) = 2, \dot{r}(0) = 0$ are plotted in Fig. 2.2.

It is seen that the equipartition does take place in this strongly nonlinear situation as well.

Next consider forced resonant oscillations as an inhomogeneous linear problem

$$\begin{gathered}\ddot{u} + u = P \sin t\,, \quad \mathcal{K} = \frac{\dot{u}^2}{2} \quad (\mu = 2)\,, \\ \mathcal{P} = \mathcal{P}_1 + \mathcal{P}_2\,, \quad \mathcal{P}_1 = \frac{u^2}{2} \quad (\nu_1 = 2)\,, \quad \mathcal{P}_2 = -Pu \sin t \quad (\nu_2 = 1)\,.\end{gathered} \tag{2.21}$$

It follows that under zero initial conditions

$$u = \frac{P}{2}(\sin t - t \cos t)\,. \tag{2.22}$$

The displacements and the energy distribution function $\mathcal{L}(t)$ are plotted in Fig. 2.3 for $P = 1$.

Thus, there is no equipartition between the kinetic and potential energies in this inhomogeneous linear problem:

$$\langle \mathcal{K} \rangle = \langle \mathcal{P}_1 \rangle + \frac{1}{2} \langle \mathcal{P}_2 \rangle < \langle \mathcal{P} \rangle = \langle \mathcal{P}_1 \rangle + \langle \mathcal{P}_2 \rangle\,. \tag{2.23}$$

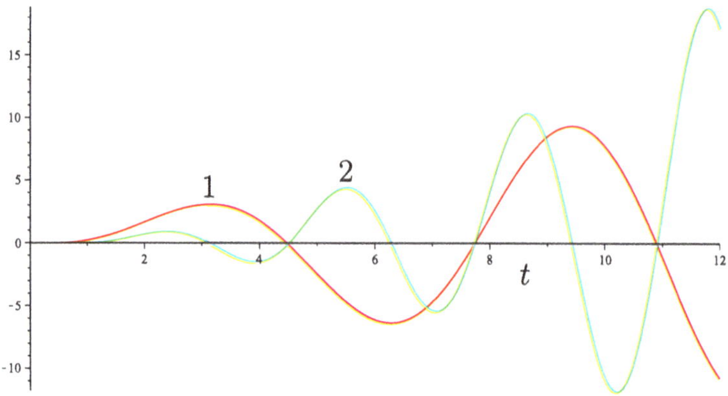

Fig. 2.3 Inhomogeneous problem: The ordinary resonance. The displacement $u(t)$ (1) and the energy distribution function $\mathcal{L}(t)$ (2). It is seen that zeros of the latter correspond to zeros of $u(t)\dot{u}(t)$

2.2.3 *Simplest Example of Unequal Partition*

Consider a bead of mass M thrown upwards with speed $\dot{u}(0) = v_0$. The energies are

$$\mathcal{K} = \frac{M\dot{u}^2}{2}\,, \quad \mathcal{P} = Mgu \quad (\mu = 2\,,\ \nu = 1)\,, \tag{2.24}$$

where g is the gravity acceleration. Here $u = 0$ at $t = t_1 = 0$ and $\dot{u} = 0$ at $t = t_2 = v_0/g$. It follows from (2.9) that the energies averaged over the (t_1, t_2)-segment must satisfy the relation

$$\langle \mathcal{P} \rangle = 2\langle \mathcal{K} \rangle\,, \tag{2.25}$$

which can also be obtained by direct calculations. Indeed,

$$\begin{gathered} \dot{u} = v_0 - gt\,, \quad u = v_0 t - \frac{gt^2}{2}\,, \\ \langle \mathcal{K} \rangle = \frac{1}{2t_2}\int_0^{t_2} M(v_0 - gt)^2\,\mathrm{d}t = \frac{Mv_0^2}{6}\,, \\ \langle \mathcal{P} \rangle = \frac{1}{t_2}\int_0^{t_2} Mg\left(v_0 t - \frac{gt^2}{2}\right)\mathrm{d}t = \frac{Mv_0^2}{3} = 2\langle \mathcal{K} \rangle\,. \end{gathered} \tag{2.26}$$

2.2.4 Linear and Nonlinear Oscillators

Consider oscillators, which kinetic and potential energies are

$$\mathcal{K} = m\frac{\dot{u}^2}{2}, \quad \mathcal{P} = \mathcal{P}_\pm \left(\frac{|u|}{u_0}\right)^\nu, \tag{2.27}$$

where the coefficients $\mathcal{P}_+$ and $\mathcal{P}_-$ correspond to positive and negative values of u, respectively, and can be different, $(m, \mathcal{P}_\pm, u_0, \nu) > 0$. For these systems the ratio of the kinetic-to-potential averaged energies is equal to $\nu/2$. Plots of the displacements and energies for $\nu = 6/5, 2$ and 6 are shown in Figs. 2.4, 2.5, 2.6 and 2.7. To have the same period, 2π, for all these cases we take $\mathcal{P}_+ = \mathcal{P}_- = 0.6152, 1/2, 0.1067$ for $\nu = 6/5, 2$ and 6, respectively, and $m = u_0 = 1$, $\mathcal{E} = 1/2$ for each of them. It is seen that zeros of $\mathcal{L}(t)$ correspond to the condition (2.6) with (2.14). Note that an analytical solution exists corresponding to the above equation (2.27). However here we are not interested in it since the energy partition is defined solely by the homogeneity orders.

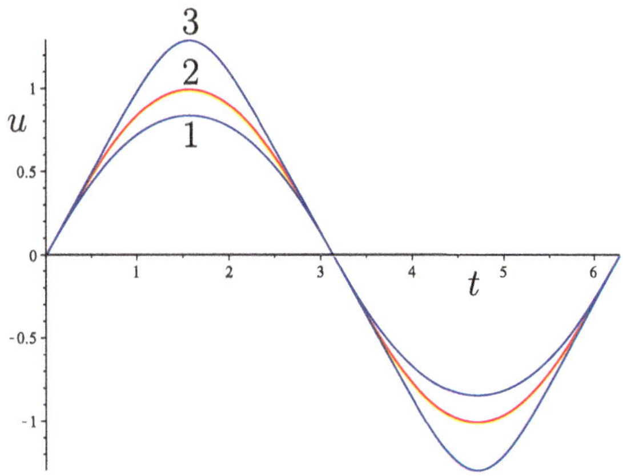

Fig. 2.4 The oscillators: $\nu = 6/5$, $\mathcal{P}_0 = 0.6152$ (1), $\nu = 2$, $\mathcal{P}_0 = 1/2$ (2) and $\nu = 6$, $\mathcal{P}_0 = 0.1067$ (3). Displacements under initial conditions $u = 0$, $\dot{u} = 1$ for each of the oscillators

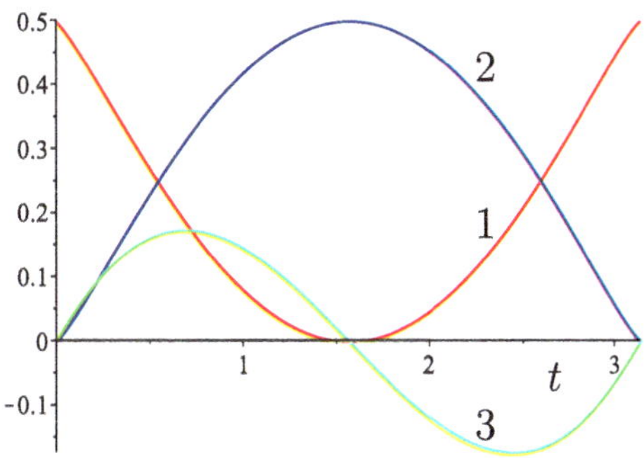

Fig. 2.5 Nonlinear oscillator, $\nu = 6/5$, $\mathcal{P}_0 = 0.6152$. Kinetic energy (1), potential energy (2) and the energy distribution function $\mathcal{L}(t)/2$ (3)

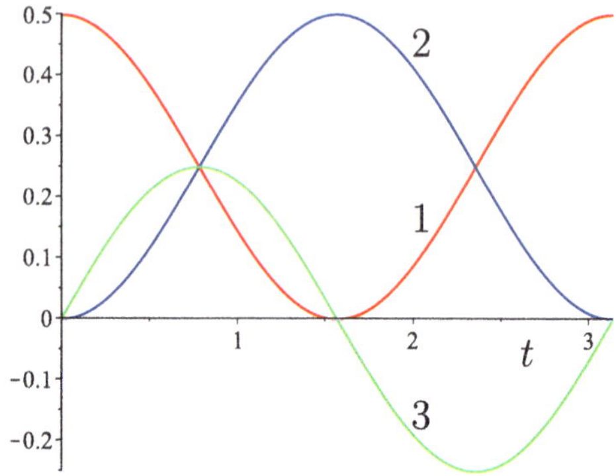

Fig. 2.6 Linear oscillator, $\nu = 2$, $\mathcal{P}_0 = 1/2$. Kinetic energy (1), potential energy (2) and the energy distribution function $\mathcal{L}(t)/2$ (3)

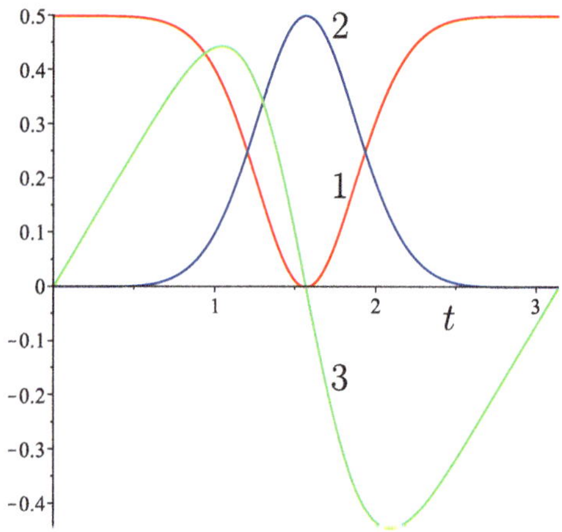

Fig. 2.7 Nonlinear oscillator, $\nu = 6$, $\mathcal{P}_0 = 0.1067$. Kinetic energy (1), potential energy (2) and the energy distribution function $\mathcal{L}(t)/2$ (3)

2.2.4.1 Two-Degree-of-Freedom Nonlinear System

Consider a mass-spring chain, Fig. 2.8, which energies are

$$\mathcal{K} = \frac{m_1 \dot{u}_1^2}{2} + \frac{m_2 \dot{u}_2^2}{2}, \quad \mathcal{P} = \mathcal{P}_{1\pm} \left(\frac{|u_1|}{u_0} \right)^{\nu} + \mathcal{P}_{2\pm} \left(\frac{|u_1 - u_2|}{u_0} \right)^{\nu}. \quad (2.28)$$

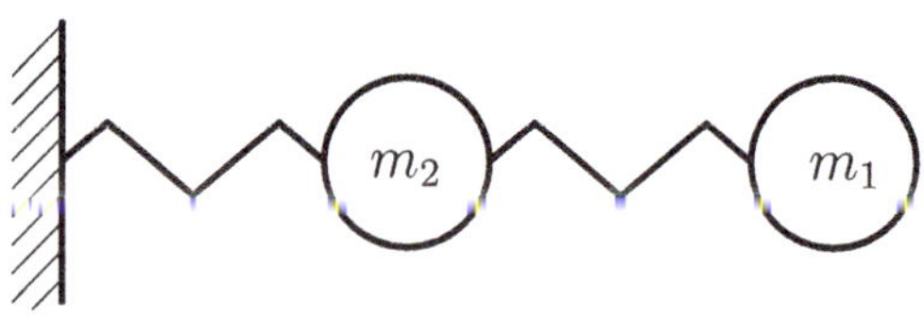

Fig. 2.8 Nonlinear mass-spring system of two degree of freedom

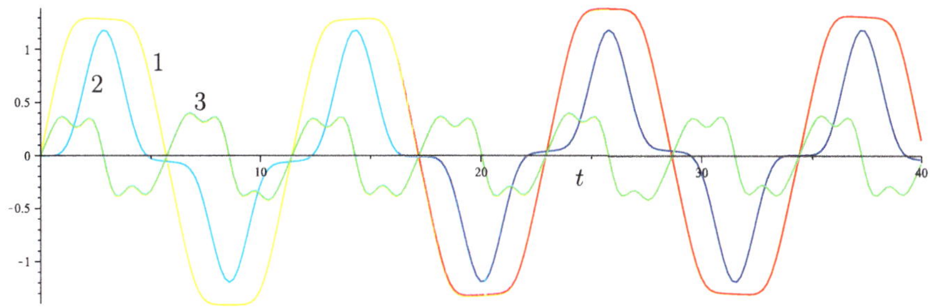

Fig. 2.9 Nonlinear mass-spring system of two degree of freedom. Displacements $u_1(t)$ (1), $u_2(t)$ (2) and $\mathcal{L}(t)/2$ (3); $\nu = 4$. Zeros of $\mathcal{L}(t)$ at the beginning and the end of the visible period, $t = 0$ and $t \approx 34.5$, appear in accordance with condition (2.6) with (2.14). Also, a number of intermediate zeros can be observed

Plots of $u_1(t)$, $u_2(t)$ and $\mathcal{L}(t)$ for $m_1 = m_2 = u_0 = 1, \mathcal{P}_1 = \mathcal{P}_2 = 1/4$ and $\nu = 4$ based on the corresponding dynamic equations

$$\frac{\mathrm{d}^2 u_1}{\mathrm{d}t^2} = -(u_1 - u_2)^3\,, \quad \frac{\mathrm{d}^2 u_2}{\mathrm{d}t^2} = (u_1 - u_2)^3 - u_2^3 \tag{2.29}$$

are shown in Fig. 2.9. An agreement with the relations (2.9) and (2.6) with (2.14) is demonstrated.

2.2.4.2 Oscillators Under Variable Stiffness

Consider oscillators which potential energy depends on t explicitly. Let in the first one the energies be

$$\mathcal{K} = \frac{\dot{u}^2}{2}\,, \quad \mathcal{P} = \frac{tu^2}{2}\,, \quad \nu = \mu = 2\,. \tag{2.30}$$

It follows from (2.5) that, in spite of the aperiodicity of the oscillations, there is the energy equipartition. It concerns the kinetic and potential energies averaged over any interval, t_1, t_2, where $B_1 = B_2$ (2.6) The corresponding dynamic equation is

$$\ddot{u} + tu = 0\,. \tag{2.31}$$

Its solution, $u(t)$, under conditions $u(0) = 0$, $\dot{u}(0) = 1$ and the energy distribution function $\mathcal{L}(t)$ (2.15) are plotted in Fig. 2.10. It is seen that zeros of $\mathcal{L}(t)$ coincide with zeros of $u(t)\dot{u}(t)$, as should be.

Next, we consider parametric resonance based on Mathieu's differential equation. The energies are

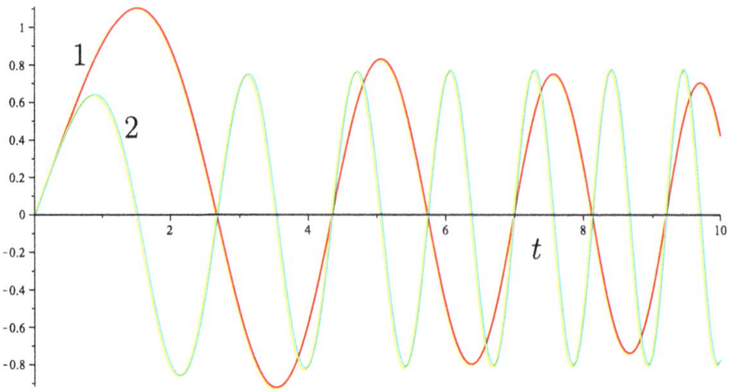

Fig. 2.10 Oscillator with linearly increasing stiffness. The displacement $u(t)$ (1) and the energy distribution function $\mathcal{L}(t)$ (2). It is seen that zeros of the latter correspond to zeros of $u(t)\dot{u}(t)$

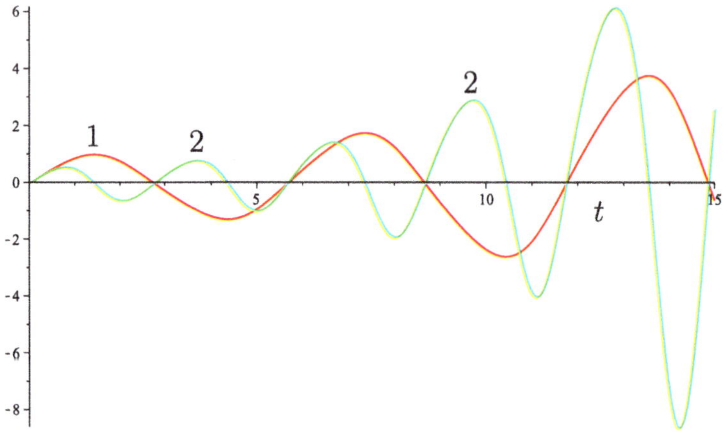

Fig. 2.11 Parametric resonance. The displacement $u(t)$ (1) and the energy distribution function $\mathcal{L}(t)$ (2). It is seen that zeros of the latter correspond to zeros of $u(t)\dot{u}(t)$

$$\mathcal{K} = \frac{\dot{u}^2}{2}, \quad \mathcal{P} = \frac{(1 + 0.5 \sin 2t)u^2}{2}, \quad \nu = \mu = 2. \tag{2.32}$$

The corresponding equation is

$$\ddot{u} + (1 + 0.5 \sin 2t)u = 0. \tag{2.33}$$

Its transient solution, $u(t)$, under conditions $u(0) = 0$, $\dot{u}(0) = 1$ and the energy distribution function $\mathcal{L}(t)$ (2.15) are plotted in Fig. 2.11. It is seen that zeros of $\mathcal{L}(t)$ coincide with zeros of $u(t)\dot{u}(t)$, as should be.

2.2.5 A Finite Continuous Body

In this case, the Lagrangian depends, in addition, on derivatives with respect to the spatial variables. The corresponding Euler–Lagrange equation multiplied by $\boldsymbol{u}(\boldsymbol{x}, t)$ is now integrated by parts over both the time-segment (t_1, t_2) and the body material volume, Ω. As a result we obtain the relation (2.9), where L is the Lagrangian incorporating the energy of the whole body including its boundaries, whereas the condition (2.6) becomes

$$\int_{\Omega} (B_2 - B_1)\, \mathrm{d}\Omega = 0\,. \tag{2.34}$$

Note that the integration limits, in the energy partition relation (2.9), correspond to the boundary terms annihilation condition (2.6) [or (2.34)]. In contrast, no condition is imposed on the boundary terms associated with the integration by parts over Ω. Such terms reflect the energy located on body boundaries.

As an example consider the collision of a linearly elastic rod, $0 < x < 1$, with a rigid obstacle. For this linear problem the (dimensionless) energies are

$$\mathcal{K} = \frac{1}{2}\int_0^1 \dot{u}^2\, \mathrm{d}x\,, \quad \mathcal{P} = \frac{1}{2}\int_0^1 (u')^2\, \mathrm{d}x\,, \tag{2.35}$$

and the homogeneity orders $\mu = \nu = 2$. The corresponding 1D wave equation is

$$\ddot{u}(x,t) - u''(x,t) = 0\,, \quad (0 < x < 1) \tag{2.36}$$

with additional conditions

$$u(x,0) = 0\,, \;\; \dot{u}(x,0) = 1\,, \;\; u(0,t) = 0 \;\; (0 < t < 2)\,, \;\; u'(1,t) = 0\,. \tag{2.37}$$

The collision period: $0 < t < 2$. Dependencies for the energies, $\mathcal{K}$ and $\mathcal{P} = \mathcal{E} - \mathcal{K}$ (the total energy $\mathcal{E} = 1/2$), and the energy distribution function $\mathcal{L}(t)$ are plotted in Fig. 2.12.

Next, we consider oscillations of a nonlinearly elastic rod, $0 < x < l$. Let the energies be

$$\mathcal{K} = \int_0^l \frac{\dot{u}^2}{2}\, \mathrm{d}x\,, \quad \mathcal{P} = \int_0^l \frac{(u')^4}{4}\, \mathrm{d}x\,, \quad \mu = 2\,, \; \nu = 4\,. \tag{2.38}$$

We solve the corresponding dynamic equation

$$\ddot{u} - 3(u')^2 u'' = 0 \tag{2.39}$$

by separation of variables, $u(x,t) = X(x)T(t)$, that results in equations

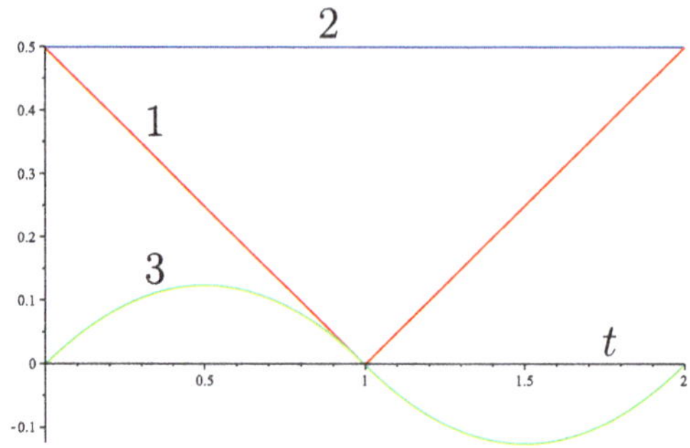

Fig. 2.12 Collision of a rod with a rigid obstacle. The kinetic energy (1), the total energy (2) and the energy distribution function $\mathcal{L}(t)$ (3). Zeros of the latter are at $t = 0$ ($u(x, 0) = 0$), $t = 1$ ($\dot{u}(x, 1) = 0$) and $t = 2$ ($u(x, 2) = 0$)

$$(X')^2 X'' + \frac{\lambda}{3} X = 0\,, \quad \ddot{T} + \lambda T^3 = 0\,, \tag{2.40}$$

where λ is an arbitrary constant. Note that these equations are solvable analytically. To simplify our considerations we take values of λ and additional conditions and obtain the corresponding length of the rod and the oscillation amplitude. Namely, we take

$$\lambda = 1\,, \quad X(0) = X'(l) = T(0) = 0\,, \quad X'(0) = \dot{T}(0) = 1 \tag{2.41}$$

and obtain the rod length and the oscillation period as

$$l = 1.4674161382\,, \quad \iota = 3.118169\,. \tag{2.42}$$

The dependencies of the displacements and energies on the coordinate and time and the energy distribution function $\mathcal{L}(t)$ are presented in Figs. 2.13, 2.14 and 2.15.

2.2.6 *Energy Partition in Waves*

In the case where the total energy of a wave is infinite, we have to choose a finite segment of the waveguide (x_1, x_2), similar to (t_1, t_2) in time, to avoid boundary terms in integration by parts over this segment. Namely, in the conversions

$$\begin{aligned} \int_{x_1}^{x_2} u_i \frac{\mathrm{d}}{\mathrm{d}x}\frac{\partial L}{\partial u_i'}\,\mathrm{d}x &= -\int_{x_1}^{x_2} \frac{\partial L}{\partial u_i'} u_i'\,\mathrm{d}x + \frac{\partial L}{\partial u_i'} u_i\Big|_{x_1}^{x_2} \\ \int_{x_1}^{x_2} u_i \frac{\mathrm{d}^2}{\mathrm{d}x^2}\frac{\partial L}{\partial u_i''}\,\mathrm{d}x &= \int_{x_1}^{x_2} \frac{\partial L}{\partial u_i''} u_i''\,\mathrm{d}x + \frac{\mathrm{d}}{\mathrm{d}x}\frac{\partial L}{\partial u_i''} u_i\Big|_{x_1}^{x_2} - \frac{\partial L}{\partial u_i''} u_i'\Big|_{x_1}^{x_2} \end{aligned} \tag{2.43}$$

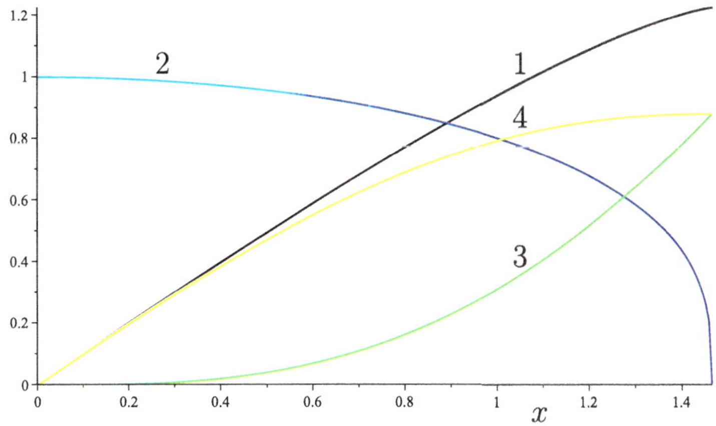

Fig. 2.13 Oscillations of a nonlinearly elastic rod. Functions $X(x)$ (1), $X'(x)$ (2), $\int_0^x X^2(x)\,\mathrm{d}x$ (3) and $\int_0^x (X')^4(x)\,\mathrm{d}x$ (4)

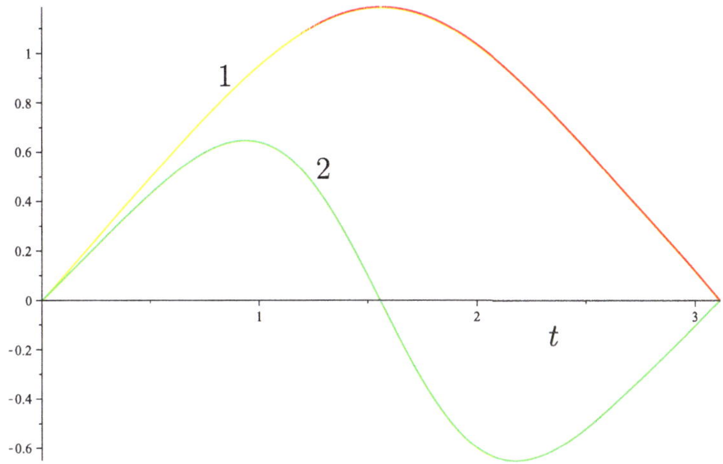

Fig. 2.14 Oscillations of a nonlinearly elastic rod. Functions $T(t)$ (1) and $\mathcal{L}(t)$ (2)

and so on, the segment should be chosen such that the boundary terms vanish. In this case, the energy partition relation (2.9) remains valid with respect to the energies averaged over the space-time region ($x_1 < x < x_2, t_1 < t < t_2$).

However, in some classes of waves, which are considered below, the averaging over one variable, t or x, appears to be sufficient.

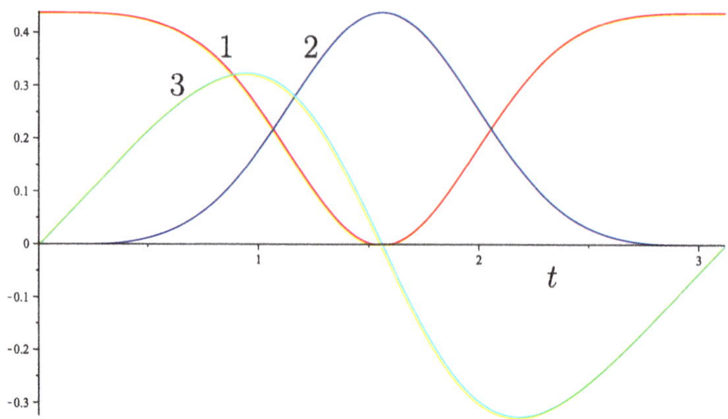

Fig. 2.15 Oscillations of a nonlinearly elastic rod. The kinetic energy (1), potential energy (2) and $\mathcal{L}(t)$ (3)

2.2.6.1 Waves in a Homogeneous Waveguide

For waves depending only on one variable, $\eta = x - vt$, propagating in such a waveguide, the averaging over t- or x-period, that is the averaging over η-period, is sufficient. So, in this case, the above relation (2.9) is valid with respect to any cross-section of the waveguide.

2.2.6.2 Nonlinearly Elastic Beam

Consider a wave in a beam where the energies are

$$\mathcal{K} = \frac{\dot{u}^2(\eta)}{2}, \quad \mathcal{P} = \frac{(u''(\eta))^4}{4}, \quad \eta = x - vt\,. \tag{2.44}$$

For a periodic wave the dynamic equation

$$v^2 u''(\eta) + \left((u''(\eta))^3\right)'' = 0 \tag{2.45}$$

can be reduced to

$$u''(\eta) + [v^2 u(\eta)]^{1/3} = 0 \tag{2.46}$$

The kinetic and potential energies and the energy distribution function $L(\eta)$ calculated based on this equation are presented in Fig. 2.16.

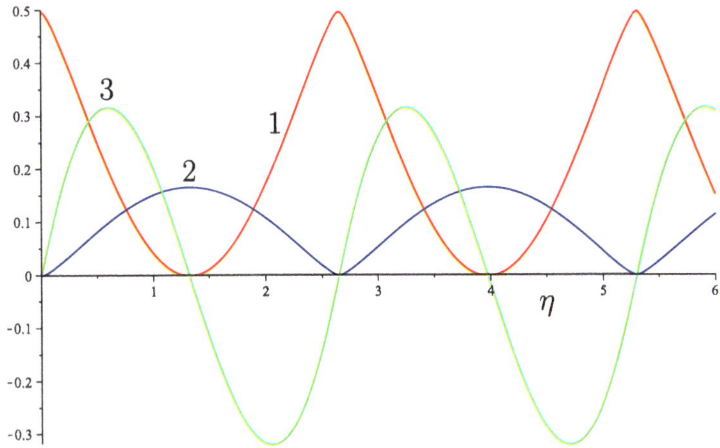

Fig. 2.16 Periodic wave in a nonlinearly elastic beam. Kinetic energy (1), potential energy (2) and the energy distribution function $\mathcal{L}(t)$ (3) are calculated for $\nu = 4$, $v = 1$, $u(0) = 0$, $\dot{u}(0) = 1$

2.2.6.3 Linear and Nonlinear Klein–Gordon Equations

Consider periodic waves, $u(\eta)$, $\eta = x - vt$, with the energies

$$\mathcal{K} = \frac{\dot{u}^2}{2} = \frac{v^2(u')^2}{2}, \quad \mathcal{P} = \frac{|u'|^\nu}{\nu} + \frac{|u|^\nu}{\nu}. \tag{2.47}$$

The corresponding equation is

$$\left(v^2 - (\nu - 1)|u'|^{\nu-2}\right) u'' + |u|^{\nu-2}u = 0. \tag{2.48}$$

The energies and the energy distribution function $\mathcal{L}(\eta)$ are plotted for $\nu = 2$ (the linear wave), 3 and 6 in Fig. 2.17, 2.18 and 2.19, respectively. Note that, in the case of waves, the total energy varies during the period.

2.2.7 *Wave in Periodic Structures*

For a periodic structure we consider Floquet-Bloch wave, where the displacements at discrete points $x = an + x_0$, $0 \le x_0 < a$, depend on x_0 and $\eta = an - vt$, and possibly other variables, but not on integer n and time separately (here a is the spatial period). This means that each of the cross-sections with the same x_0 describes the same periodic trajectory different only by a shift in time. We assume that (generalised) displacements of the cross-sections corresponding to a fixed x_0, say, $x_0 = 0$, defined those for the other cross-sections.

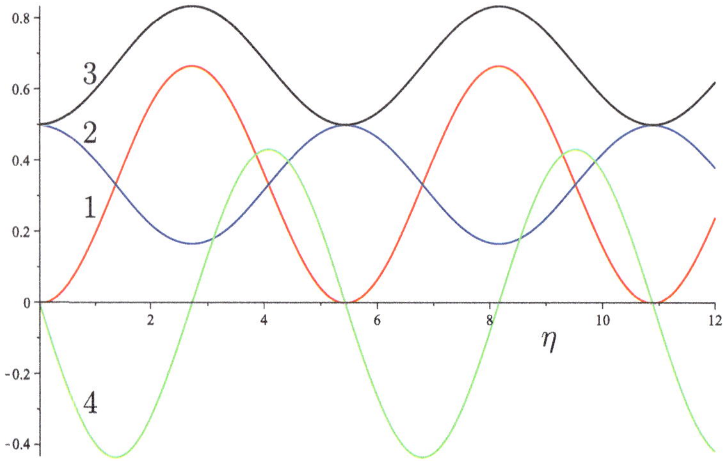

Fig. 2.17 Klein-Gordon linear equation, $\nu = 2$. Kinetic energy (1), potential energy (2), total energy (3) and the energy distribution function $\mathcal{L}(t)$ (4)

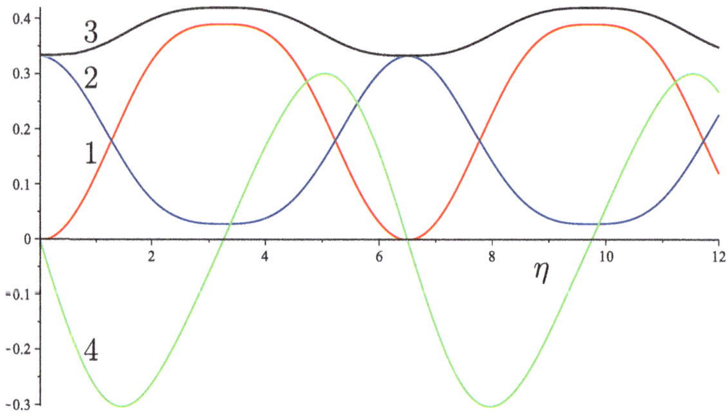

Fig. 2.18 Klein-Gordon nonlinear equation, $\nu = 3$. Kinetic energy (1), potential energy (2), total energy (3) and the energy distribution function $\mathcal{L}(t)$ (4)

In the considered case of an infinite waveguide, the energy of a periodic wave is also infinite. However, in the case of a finite radius of the interaction, where the nth cross-section is directly connected with a finite number of the others (with the same x_0), the displacement $\boldsymbol{u}_n$ appears only in a finite number of the infinite series representing the Lagrangian. Accordingly, the averaging (2.9) should correspond to the structure period

$$\int_{t_1}^{t_2} \int_a \nu_m L_m \, \mathrm{d}x \, \mathrm{d}t = 0 , \qquad (2.49)$$

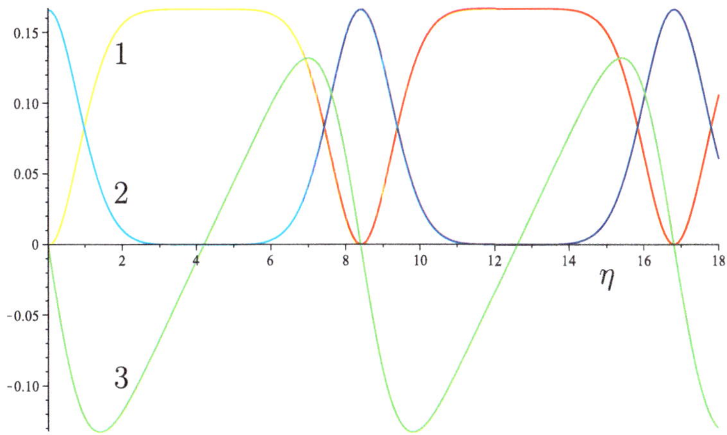

Fig. 2.19 Klein-Gordon nonlinear equation, $\nu = 6$. Kinetic energy (1), potential energy (2) and the energy distribution function $\mathcal{L}(t)$ for 8th cell (3)

where a is the spatial period of the structure. Note that the spatial integral becomes a sum in the case of a discrete structure.

As an example consider a linear mass-spring chain. The (non-dimensional) energies are

$$\mathcal{K} = \frac{1}{2}\dot{u}_n^2\,, \quad \mathcal{P} = \frac{1}{4}(u_{n+1} - u_n)^2 + \frac{1}{4}(u_{n-1} - u_n)^2\,, \tag{2.50}$$

In this case, $\mu = \nu = 2$ and there is equipartition. This result also follows from the dynamic equation

$$\ddot{u}_n = u_{n+1} + u_{n-1} - 2u_n\,, \tag{2.51}$$

which is satisfied by the wave with the corresponding dispersion relation

$$u_n = \exp[\mathrm{i}(\omega t - kn)]\,, \quad \omega = \pm 2\sin\frac{|k|}{2}\,. \tag{2.52}$$

From this and (2.50) we find

$$\mathcal{K} = \frac{\omega^2}{2} = \mathcal{P} = 2\sin^2\frac{k}{2}\,. \tag{2.53}$$

Wave in a nonlinear chains are considered in the next section.

2.2.8 *Solitary Waves*

In the case of a solitary wave, the partition of its total energy is considered, and the integration limits become infinite, $t_{1,2} = \mp\infty$.

2.2.8.1 Phi-Four Equation

Consider the so-called Phi-Four equation

$$\ddot{u}(x,t) - u''(x,t) - u(x,t) + u^3(x,t) = 0\,, \tag{2.54}$$

where dots and primes still mean the derivatives with respect to t and x, respectively. This equation is satisfied by a kink

$$u(x,t) = U(\eta) = \pm\tanh\left(\frac{\eta}{\sqrt{2(1-v^2)}} + \varphi_0\right)\,, \quad \eta = x - vt\,, \tag{2.55}$$

where the arbitrary constants, v ($v^2 < 1$) and φ_0, are the kink's speed and the 'initial' phase, respectively. The corresponding energies are

$$\begin{gathered}
\mathcal{K} = \frac{\dot{u}^2(\eta)}{2} = v^2\frac{(u'(\eta))^2}{2}\,, \quad \mathcal{P} = \mathcal{P}_0 + \mathcal{P}_2 + \mathcal{P}_4\,, \\
\mathcal{P}_0 = \frac{1}{4}\,, \quad \mathcal{P}_2 = \frac{(u'(\eta))^2}{2} - \frac{u^2(\eta)}{2}\,, \quad \mathcal{P}_4 = \frac{u^4(\eta)}{4}\,,
\end{gathered} \tag{2.56}$$

where the value of the arbitrary constant $\mathcal{P}_0$ ($v = 0$) is chosen to limit the potential energy.

While the moving kink corresponds to the transition from one zero-energy state to another, the particle velocity

$$\dot{u}(\eta) = \mp\frac{v}{\sqrt{2(1-v^2)}}\left[\cosh^2\left(\frac{\eta}{\sqrt{2(1-v^2)}} + \varphi_0\right)\right]^{-1}\,, \tag{2.57}$$

represents a true solitary wave. As a result, $u\dot{u} \to 0$ ($\eta \to \pm\infty$) and our statement is valid with respect to the total energies of the wave

$$\int_{-\infty}^{\infty} \mathcal{L}(\eta)\,\mathrm{d}\eta = 0\,, \quad \mathcal{L} = 2(\mathcal{K} - \mathcal{P}_2) - 4\mathcal{P}_4\,. \tag{2.58}$$

The kink, the solitary wave and the energies as functions of η are shown in Figs. 2.20 and 2.21. Far away from the transition region there is a static state, where, in accordance with (2.56) and (2.58), $\mathcal{P} = 0$, $\mathcal{P}_2 = -u^2(\eta)/2 = -2\mathcal{P}_4$.

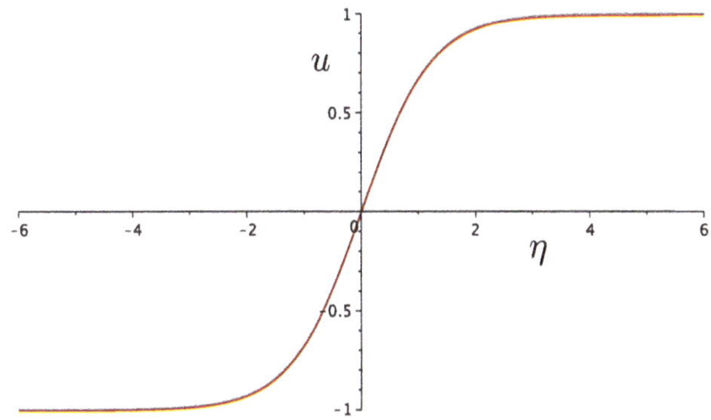

Fig. 2.20 The kink

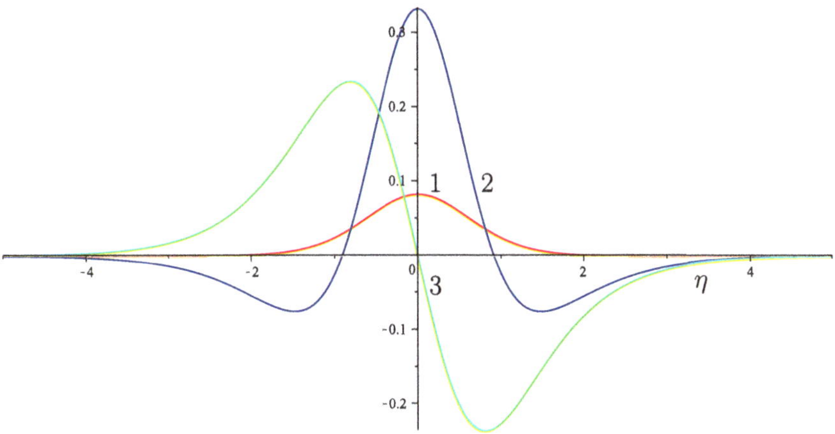

Fig. 2.21 The energies, $\mathcal{K}$ (1), $P_2 + 2P_4$ (2) and the energy distribution function $\mathcal{L}(\eta)$ (3) for the kink, $v = 1/2$

2.2.8.2 Derrick–Pohozaev Identity

In the above problem, there are three unknown energy terms, $\mathcal{K}$, $\mathcal{P}_2$ and $\mathcal{P}_4$. The single equality (2.9) is not sufficient to obtain the $\mathcal{K} - \mathcal{P}$ relation but only $\mathcal{K} - (\mathcal{P}_2 + 2\mathcal{P}_4)$ one. Fortunately, for such a solitary steady-state wave the Derrick–Pohozaev identity can be used as an additional energy relation. The identity is based on the representation of the steady-state solitary wave as a static state. Note that in the steady-state mode, the kinetic energy can be considered as such, as a function of $\dot{u} = \partial u/\partial t$, and also as a function of $u' = \partial u/\partial x$.

$$\mathcal{K} = \frac{\dot{u}^2}{2} \equiv \mathcal{P}_- = \frac{v^2(u')^2}{2} \,. \tag{2.59}$$

In the latter case, it can be treated as an additional (negative) part of the potential energy. Different energy relations, one discussed in the present paper and another given by the Derrick–Pohozaev identity, follow from the static representation as a result of different variations. While the former corresponds to the displacement variation $\delta u = \varepsilon u$, the latter is a uniform x-extension. The use of both relations allows us to separate the kinetic and potential energies completely.

The Derrick–Pohozaev identity is found for the nonlinear Klein–Gordon equation with a general function of u (see [8–9] and [10], Sect. 7.4). Derived for a multidimensional case it reads in 1D as the equality of two parts of the energy, one as a function of u and another as a function of u'. In our terms the identity is

$$\mathcal{E}_1 = \mathcal{E}_2\,, \quad \mathcal{E}_1 = \frac{1-v^2}{2}(u')^2\,, \quad \mathcal{E}_2 = \frac{1}{4}(1-u^2)^2\,. \tag{2.60}$$

The solution (2.55) satisfies this equality. Thus, we now have two independent relations, one in (2.58) and another in (2.60). Using both of them we can completely separate the kinetic and potential energies. The partition between the total energies and the energies themselves are found as

$$\int_{-\infty}^{\infty} \mathcal{K}\,\mathrm{d}\eta = \frac{v^2}{2-v^2}\int_{-\infty}^{\infty} \mathcal{P}\,\mathrm{d}\eta = \frac{v^2}{\sqrt{2(1-v^2)}} \quad (-1 < v < 1)\,. \tag{2.61}$$

The kinetic and total potential energy graphs are shown in Fig. 2.22. The equality between the total energies holds with respect to the energy densities, $\mathcal{K}$ and $\mathcal{P}$. So

$$\mathcal{K}(\eta) = \frac{v^2}{2-v^2}\mathcal{P}(\eta) \quad (-1 < v < 1)\,.. \tag{2.62}$$

2.2.8.3 FPU Problem

Next, we consider a transient problem for a wave excited by a pulse applied to the same but 'semi-infinite' linear chain. No solitary wave can exist in the latter, and we study this problem to compare the result with that for a nonlinear chain first examined by Fermi et al. (1955a). The displacements of the first twenty masses under unit pulse acting on the first left mass and also the corresponding speeds are presented in Figs. 2.23 and 2.24, respectively.

The wave itself, especially the particle velocity wave, and the energy distribution function, $\mathcal{L}(t)$, can be compared with those in the FPU problem [12] for the same but nonlinear chain, Figs. 2.25 and 2.26, where a stable solitary wave arises (compare Fig. 2.24 with Fig. 2.26), and with 'pure nonlinear' one corresponding to the potential energy without linear terms. In a version of the FPU problem, the energies are taken as

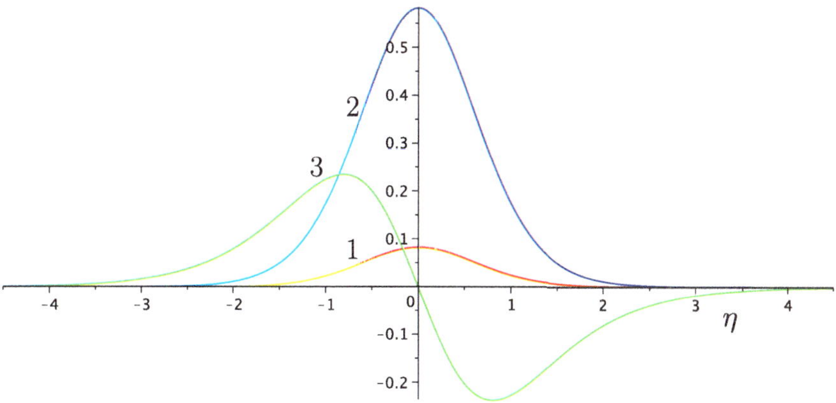

Fig. 2.22 Functions $\mathcal{K}$ (1), $\mathcal{P} = 7\mathcal{K}$ (2) and the energy distribution function $\mathcal{L}(\eta)$ (3) for the kink, $v = 1/2$

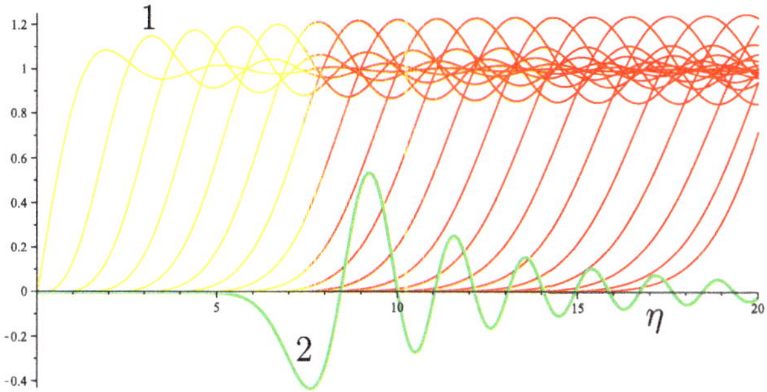

Fig. 2.23 Transient wave in the linear chain under a pulse. The displacements of the masses (1) and $20\mathcal{L}(t)$ for 8th cell (2)

$$\mathcal{K} = \frac{1}{2}\dot{u}_n^2, \quad \mathcal{P} = \frac{1}{4}(u_{n+1} - u_n)^2 + \frac{1}{4}(u_{n-1} - u_n)^2 + \frac{1}{2}(u_{n+1} - u_n)^4 + \frac{1}{2}(u_{n-1} - u_n)^4 . \tag{2.63}$$

Here $\mu = \nu_1 = 2$, $\nu_2 = 4$ and $\int_{-\infty}^{\infty}(\mathcal{K} - \mathcal{P}_1 - 2\mathcal{P}_2)\,\mathrm{d}t = 0$.

A fast-established strongly localized solitary wave appears in the nonlinear chain with quadratic terms in the expression of the potential energy (2.63) removed. In this case, $\mathcal{K} = 2\mathcal{P}$. The calculation results are presented in Figs. 2.27 and 2.28.

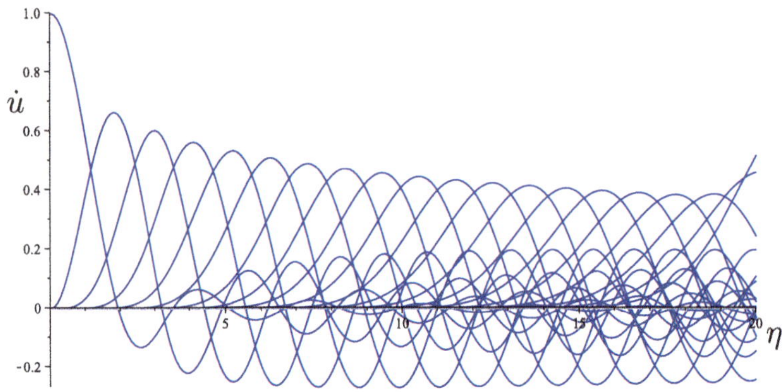

Fig. 2.24 Transient wave in the linear chain under a pulse. The decreasing wave of particle velocities

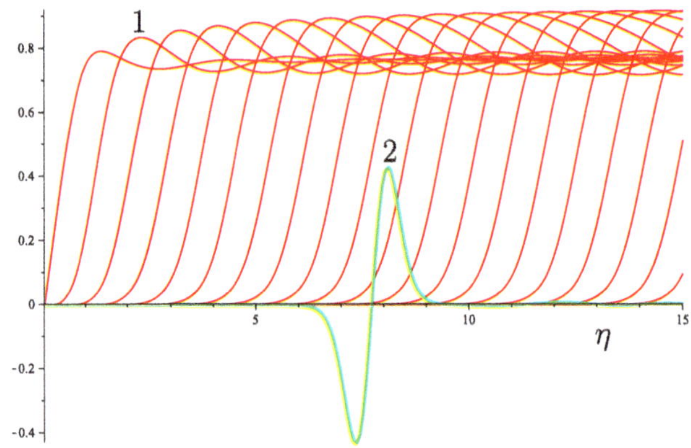

Fig. 2.25 Transient wave in the nonlinear chain under a pulse (FPU problem). The displacements of the masses (1) and $8\mathcal{L}(t)$ for 10th cell (2)

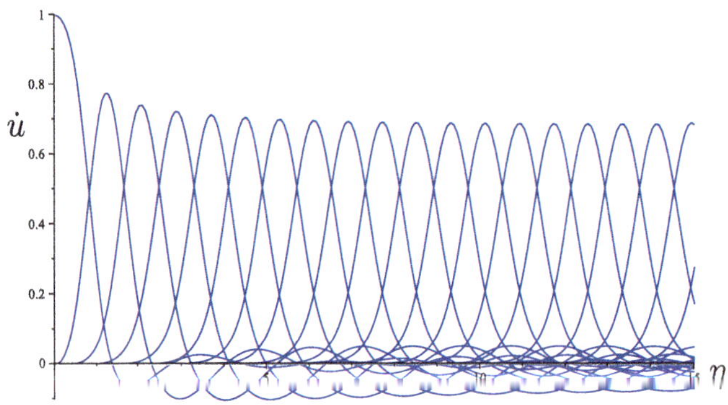

Fig. 2.26 Transient wave in the nonlinear chain under a pulse (FPU problem). The establishing solitary wave of particle velocities.

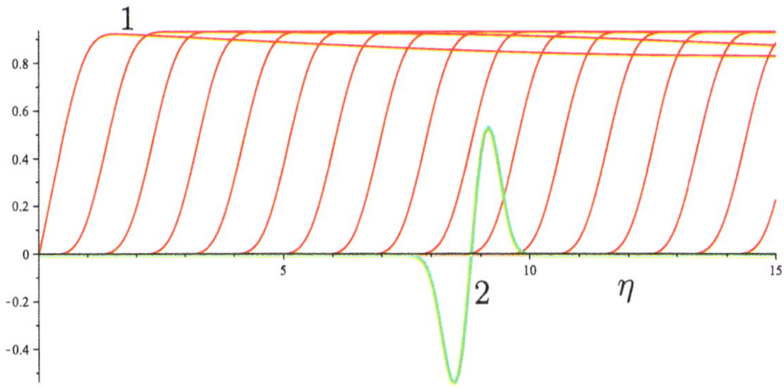

Fig. 2.27 Solitary wave in the nonlinear chain with no quadratic terms in the expression of the potential energy (2.63). The displacements of the masses (1) and $4\mathcal{L}(t)$ for 10th cell (2)

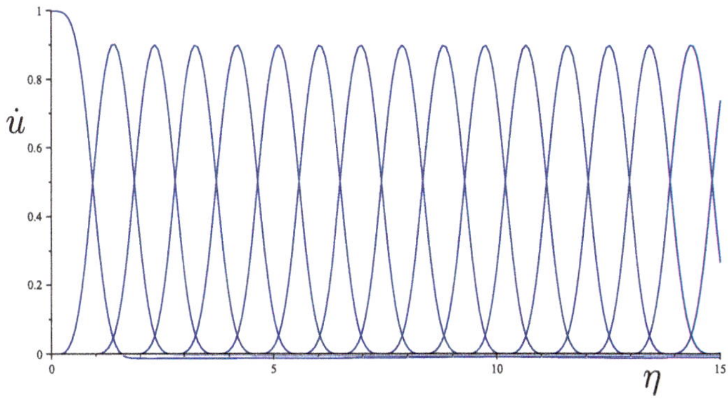

Fig. 2.28 Solitary wave in the nonlinear chain with no quadratic terms in the expression of the potential energy (2.63). The particle velocities

Chapter 3
Wave Momentum

3.1 On the Wave Momentum

Axial momentum carried by waves in a uniform waveguide is considered based on the conservation laws and a kind of the causality principle, Slepyan (2016). Specifically, we examine (without resorting to constitutive data) steady-state waves of an arbitrary shape, periodic waves which speed differs from the speed of its form and binary waves carrying self-equilibrated momentum. The approach allows us to represent momentum as a product of the *wave mass* and the wave speed. The propagating wave mass, positive or negative, is the excess of that in the wave over its initial value. This general representation is valid for mechanical waves of arbitrary nature and intensity. The finite-amplitude longitudinal and periodic transverse waves are examined in more detail. It is shown in particular, that the transverse excitation of a string or an elastic beam results in the binary wave. The closed-form expressions for the drift in these waves functionally reduce to the Stokes' drift in surface water waves (a half the latter by the value). Besides, based on the general representation an energy-momentum relation is discussed and the physical meaning of the so-called "wave momentum" is clarified.

Along with energy, momentum plays a defining role in wave actions, and different related questions were debated since Lord Rayleigh's works on the theory (Rayleigh, 1902, 1905). Brillouin (1925), McIntyre (1981), Ostrovsky and Potapov (1888), Peskin (2010), Falkovich (2011) and Maugin and Rousseau (2015) are among others who considered various aspects of this topic (a comprehensive list of references can be found in the latter book by Maugin and Rousseau).

Note, however, that momentum, in contrast to the energy, generally is not a non-negative value and it is very sensitive to the formulation. In particular, it may be lost with linearization. By definition, momentum is the product of density and the particle velocity, $p = \varrho \boldsymbol{v}$, and (possibly small) variations of both of them should be taken into account. The question is whether it is possible to represent the quantity in

L. I. Slepyan, *Mechanical Waves: Free, Forced, and Transition*,
https://doi.org/10.1007/978-3-032-15864-2_3

a classical manner as the product of a mass and the wave velocity, where only one of the multipliers, the mass can be variable. Below, we introduce such a representation.

Most known results correspond to specific problems and low-intensity waves. The above-discussed representation results in a straightforward way for determining the momentum of high-intensity waves. In dispersive sinusoidal waves, the energy propagates with the group velocity. Does the latter play a similar role for momentum? We show that affirmative reply is correct.

Under a linearized formulation where momentum is lost, could it be possible to extract it based on the simplified description of the wave? We show that such a possibility does exist, and the lost momentum can be obtained considering the mass distribution or even only the waveform (as, for example, in the case of the flexural wave considered below).

In the wave theory, along with the classical definition of momentum, the so-called "wave momentum" $-\varrho_0 u' v$ is considered (ϱ_0 is the initial density, u' and v are the derivative of the displacement and the particle speed). It is "... a much debated notion" (Maugin & Rousseau, 2015, p. 2). "Of course, considering a "definition" of wave momentum ... would be quite reasonable from a strict mathematical viewpoint, but the physical meaning would be doubtful." (ibid, p.14). Below, based on the general representation the physical meaning of the "wave momentum" is given.

We show that the essence of the issue and useful relations immediately follow from the mass and momentum conservation with the assumption that ahead of the wave the waveguide is in a uniform static state until the wave arrives (this is a kind of the *causality principle*). Abstracting from constitutive relations and other particularities a greater clarity can be retrieved, and the results obtained directly from these first principles are valid for mechanical waves of any nature, form and intensity.

We consider a homogeneous straight-line waveguide of an arbitrary cross-section and three types of mechanical waves: a steady-state wave of an arbitrary shape, a periodic wave which speed differs from the speed of its form (for a linear sinusoidal wave these are the group and phase velocities, respectively) and a binary wave carrying a self-equilibrated momentum.

The momentum follows as a product of the *wave mass* and the wave speed. Based on the general representation the wave-induced drift and the connection between the momentum and energy are discussed. Periodic longitudinal and flexural waves are considered as nontrivial examples.

Along with the *reference frame* associated with this initial state, we use a frame moving with the wave. Eulerian (spatial) variables are used, unless otherwise is specified. We denote the longitudinal coordinate by x. The mass and momentum densities per unit x-length and the x-component of the particle velocity are denoted by $m = m(x, t)$, $p = p(x, t)$ and $v = v(x, t)$, respectively. In the initial state of the waveguide, ahead of the wave, $m = m_0$, $v = 0$ (or asymptotically equal to these quantities as $x \to \infty$). As usual, the prime and dot mean the partial derivatives on the coordinate x and time t.

3.2 Steady-State Wave

We are interested in axial momentum of a wave of a general form and intensity propagating along the x-axis with the speed c (note that the wave can possess both the axial and angular momenta, see, e.g., Slepyan et al., 1995a). In the frame moving with the wave, the mass distribution is assumed to be fixed over the wave domain as well as ahead of the wave where the uniform waveguide is in the initial static state (condition $\mathcal{A}$).

3.2.1 *The Wave Mass and Momentum*

As the first state consider the wave in the moving frame, where the waveguide medium flows through the wave domain, and momentum in front of the wave is equal to $-m_0c$. For such a steady movement, it follows from the mass conservation and the above condition $\mathcal{A}$ that

$$\frac{\partial p}{\partial x} = \frac{\partial}{\partial x}\int_S \varrho v \,\mathrm{d}S = -\frac{\partial m}{\partial t} = 0 \quad\Longrightarrow\quad p = p_- = -m_0 c\,, \tag{3.1}$$

where ϱ and v are density and the x-component of the particle velocity, and S is the cross-section area. Now, adding the corresponding right-directed rigid momentum $p_+ = mc$ (to return to the reference frame) we obtain the momentum density

$$p = m_w c\,, \quad m_w = m - m_0\,. \tag{3.2}$$

The excess of the mass in the wave m_w can be called the *wave mass* (per unit length). It does not consist of the same particles but propagates as the wave. Note that both these parameters, the wave mass and momentum, can be positive, zero or negative. In the latter case, the momentum directs opposite to the wave.

A step wave is the simplest example. The mass conservation reads as

$$\varrho(c - v) = \varrho_0 c \quad\Longrightarrow\quad p = \varrho v = (\varrho - \varrho_0)c\,. \tag{3.3}$$

Recall that v is the particle velocity and ϱ and ϱ_0 are the actual and initial mass densities.

The advantage of the representation (3.2) as compared with the initial definition, $p = mv$, is that it contains only one variable m. In certain cases, this allows immediately to indicate the presence of momentum and to determine it based on the linearized expression of the wave, as in the case of a flexural wave considered below.

An arbitrary transverse wave in an inextensible string under a constant tensile force T is a simple example. The above relations are valid for this case, and the wave mass corresponding to an x-segment is $\varrho_0(\ell_s - \ell)$, where ℓ_s is the arc length and ℓ is the segment length.

3.2.2 Wave Momentum, Energy, and Wave Shift

Assuming the flow in the first state to be irrotational consider a stream filament. Let s, v_s and ϱ_s be the curvilinear coordinate, the speed along it and the mass per unit s-length, respectively. For brief, we now take the case where $s(x)$ is a single-valued function. In a similar manner as above, we call ϱ and ϱ_0 the densities of the mass per unit x-length in the actual and initial states of the filament. Now v is the x-projection of the velocity v_s, and the mass conservation reads as $\varrho_s v_s = -c\varrho_0$. It is also convenient to introduce the stretch λ of the distance between (close) filament *material* cross-sections normal to the x-axis, $\lambda = \varrho_0/\varrho$.

The expressions (3.1) and (3.2) are also valid for the filament. Indeed

$$\varrho = \varrho_s\, \mathrm{d}s/\,\mathrm{d}x\,, \quad v_- = v_s\, \mathrm{d}x/\,\mathrm{d}s\,, \quad \Longrightarrow \quad \varrho v_- = \varrho_s v_s = -\varrho_0 c\,, \tag{3.4}$$

where v_- is the particle speed in the first state, and

$$p_- = \varrho v_- = -\varrho_0 c\,, \quad p = p_- + \varrho c = \varrho_w c\,, \quad \varrho_w = \varrho - \varrho_0\,. \tag{3.5}$$

The filament wave mass density (per unit x-length), ϱ_w, in Eulerian (E) and Lagrangian (L) description is

$$\varrho_w = \begin{cases} \varrho_0(1/\lambda - 1) = -\varrho_0 u' & (E) \\ \varrho_0(1-\lambda) = -\varrho_0 u' & (L)\,, \end{cases} \tag{3.6}$$

where u is the longitudinal displacement (recall that here λ is the stretch of the distance but not of the medium element). The value of λ is the same for both expressions, whereas u' is different if $\lambda \neq 1$.

The kinetic energy density is

$$\mathcal{K} = \frac{\varrho}{2}[(c + v_s \cos\alpha)^2 + (v_s \sin\alpha)^2] = \frac{\varrho c^2}{2}\left[1 + \left(\frac{\varrho_0}{\varrho_s}\right)^2 - 2\frac{\varrho_0}{\varrho}\right], \tag{3.7}$$

where α is the angle between the filament and the x-axis. In particular, if $\varrho_s = \varrho_0$ this expression becomes

$$\mathcal{K} = pc = \varrho_w c^2\,. \tag{3.8}$$

Note that the kinetic–potential energy partition is considered in Slepyan (2015a).

The wave passing can be accompanied by a progressive shift. In the stream filament, there is a one-to-one correspondence between the shift and momentum density. Indeed, the shift per unit x-length

$$q = \frac{1-\lambda}{\lambda} = -u' \ (\mathrm{E})\,, \quad q = 1 - \lambda = -u' \ (\mathrm{L}) \ \Longrightarrow \ q = \frac{p}{\varrho_0 c} \ (\text{E and L})\,. \tag{3.9}$$

The wave shift is typical for solitary waves. The well-known Stokes' drift in surface water waves (Stokes, 1847) is also an example.

3.2.3 Physical Meaning of the "Wave Momentum"

The so-called "wave momentum", $-\varrho_0 u'v$, can be treated as the momentum without a zero-mean term. Indeed, represent the momentum density with a linear part separated. With refer to (3.6) we obtain

$$p = \varrho v = p_0 + p_w\,, \quad p_0 = \varrho_0 v\,, \quad p_w = (\varrho - \varrho_0)v = -\varrho_0 u'v\,, \tag{3.10}$$

where $\varrho - \varrho_0$ is the *wave mass density*. The separation makes sense for the case of a periodic wave with a zero mean v. Here the first term vanishes on the averaged, and only the second one remains. Thus, in this case, the "wave momentum" coincides on the average with the original definition of this quantity.

3.2.4 Binary Wave

A binary wave in a helical fiber was described by Krylov and Slepyan (1997). Caused by an axial force with no torque it was a wave consisting of two parts rotating with opposite directed angular momenta of the same value. Similarly, a binary wave with a self-equilibrated axial momentum can be emitted without axial action.

Consider the corresponding piecewise constant two-step wave consisting of a forerunner (the front part of the wave) and the successive part (the wave excited transversely). Let the former be a longitudinal wave propagating with the speed c_+, the mass density m_+ and particle velocity v_+, whereas m and v be the corresponding values for the flexural wave, and let D be the speed of the interface between these parts.

The mass conservation relations are

$$(c_+ - v_+)m_+ = c_+ m_0\,, \quad (D - v)m = (D - v_+)m_+\,. \tag{3.11}$$

The axial momentum per unit time in the flexural wave, P, and in the forerunner, P_+, can be obtained in a similar way as above for the single wave. However, the fact should be taken into account that in front of the former, the mass density and speed are m_+ and v_+, not m_0, c. In this problem, the energy densities $\mathcal{E}$ and CE_+ as well as the energy flux speeds c_g and c_+ (for the flexural wave and forerunner, respectively) are to be taken into account. Note that, in a true steady-state wave, $c_g = c$; however, for greater generality, we denote them as the group and phase speeds, respectively. In a linear longitudinal wave, where the kinetic and potential energies are equal, $\mathcal{E}_+ = m_+ v_+^2$.

With refer to the mass conservation relations (3.11) the momentum densities per unit time can be expressed as

$$
\begin{aligned}
P = mv(D - v) = [(m - m_+)D + (m_+ - m_0)c_+](D - v)\,, \\
P_+ = (m_+ - m_0)c_+(c_+ - D)\,.
\end{aligned}
\tag{3.12}
$$

For the four unknowns, v, v_+, m_+ and D (c_g, c_+, m_0 and $m = m_0 + m_w$ are the input data) there are four equations, two in (3.11) plus the momentum and energy conservation relations

$$
P + P_+ = 0\,, \quad \mathcal{E}(c_g - D) = \mathcal{E}_+(c_+ - D)\,. \tag{3.13}
$$

An explicit solution can be presented for a low-intensity wave where $|m_w/m_0| \ll 1$. The corresponding asymptotic values are

$$
\begin{aligned}
v \sim \frac{c_g c_+}{c_+ + c_g}\frac{m_w}{m_0}\,, \quad v_+ \sim -\frac{c_+ c_g^2}{c_+^2 - c_g^2}\frac{m_w}{m_0}\,, \\
m_+ - m_0 \sim -\frac{c_g^2}{c_+^2 - c_g^2} m_w\,, \quad c_g - D \sim (c_+ - c_g)\frac{m_0 c_+^2 c_g^4}{(c_+^2 - c_g^2)^2 \mathcal{E}}\left(\frac{m_w}{m_0}\right)^2.
\end{aligned}
\tag{3.14}
$$

The momentum densities per unit time follow as

$$
P = -P_+ = mv(D - v) \sim m_0 c_g v \sim \frac{c_+ c_g^2 m_w}{c_+ + c_g}\,. \tag{3.15}
$$

Thus, the densities per unit length asymptotically equal to

$$
p \sim \frac{P}{c_g} \sim \frac{c_+ c_g m_w}{c_+ + c_g}\,, \quad p_+ \sim -\frac{P}{c_+ - c_g} \sim -\frac{c_+ c_g^2 m_w}{c_+^2 - c_g^2}. \tag{3.16}
$$

In addition, we compare the energy densities of these parts of the binary wave. For the linearized formulation the potential and kinetic energies should be equal. It follows that

$$
\mathcal{E} = mv^2 \sim \frac{m_0 c_g^2 c_+^2}{(c_+ + c_g)^2}\frac{m_w^2}{m_0^2}\,, \quad \mathcal{E}_+ = m_+ v_+^2 \sim \frac{m_0 c_g^4 c_+^2}{(c_+^2 - c_g^2)^2}\frac{m_w^2}{m_0^2}\,, \tag{3.17}
$$

and the ratio is

$$
\frac{\mathcal{E}_+}{\mathcal{E}} \sim \frac{c_g^2}{(c_+ - c_g)^2}\,. \tag{3.18}
$$

The binary wave arising under transverse harmonic excitation of an elastic beam is considered in Sect. 3.3.2.

3.3 Periodic Waves

Here we consider a periodic wave which speed c_g differs from the speed c_p of its form. Usually, such a wave is characterized by a quasifront which is considerable in a slowly expanding transition area in a vicinity of $x = c_g t$. In the present study, taking into account the fact that as the wave propagates the quasifront area becomes asymptotically negligible, we consider it as a periodic wave with the front at $x = c_g t$.

We use the c_g-moving frame where the wave does not propagate but the waveform moves with the speed $c - c_g$. In this state, there is not a fixed but a periodic distribution of the mass and particle velocities, and the expressions for the mass conservation and for p_- (3.1) and p_+ (3.2) are still valid being averaged over the period. We have

$$\langle p_- \rangle = -m_0 c_g \,, \quad \langle p_+ \rangle = \langle m \rangle c_g \,, \quad \langle p \rangle = m_w c_g \,, \quad m_w = \langle m - m_0 \rangle \,. \tag{3.19}$$

So, momentum propagates with the same speed c_g as the energy.

The expressions in (3.6) can also be used for the filament

$$\langle p \rangle = -\varrho_0 \langle u' \rangle c_g = \frac{\varrho_0 c_g}{L} [\langle u(x) - u(x+L) \rangle] \,, \tag{3.20}$$

where the value of the period L depends on what the formulation, Eulerian or Lagrangian, is used.

The period-averaged drift speed is

$$v_d = -\langle u' \rangle c_g = \frac{\langle p \rangle}{\varrho_0} \,. \tag{3.21}$$

3.3.1 *Longitudinal and Flexural Waves*

3.3.1.1 Longitudinal Oscillating Wave

Here we consider a one-dimensional, longitudinal, periodic wave of a zero mean particle velocity. The latter condition can be formulated concerning the velocity of fixed particles (as the wave excited by the *Lagrangian emitter*) or regarding the particle velocity detected at a fixed spatial point (the wave excited by the *Eulerian emitter*). The results corresponding to these different formulations could be different (see Falkovich, 2011, pp. 75–76). In the former case considered by McIntyre (1981) momentum is at zero regardless of the wave form and intensity. The latter case was examined by iterations for a small ratio of the particle displacements to the wavelength (Falkovich, 2011). Below, based on the representation presented in (3.5) and (3.6) we express momentum in a closed form valid for the oscillating wave of a finite amplitude. The only condition is assumed that the particle velocity is below the wave speed.

Consider a 1D longitudinal wave where the particle velocity can be represented in the form

$$v = \frac{\mathrm{d}x(t)}{\mathrm{d}t} = v_0 f\left(t - \frac{x(t)}{c}\right) H(ct - x(t)), \tag{3.22}$$

where $-1 \le f \le 1, 0 < v_0 < c$ and for any $x = x_*$ fixed and period T

$$\int_T f(t - x_*/c)\,\mathrm{d}t = 0. \tag{3.23}$$

The displacement follows from this as

$$u = v_0 \int_{x/c}^{t} f\left(t - \frac{x(t)}{c}\right) \mathrm{d}t = v_0 c \int_0^{\zeta} \frac{f(\zeta)\,\mathrm{d}\zeta}{c - v_0 f(\zeta)}, \quad \zeta = t - \frac{x(t)}{c}. \tag{3.24}$$

Thus

$$u' = -\frac{v_0 f(\zeta)}{c - v_0 f(\zeta)} \tag{3.25}$$

and

$$p = \frac{\varrho_0 v_0 c f(\zeta)}{c - v_0 f(\zeta)} \tag{3.26}$$

In particular, if $v \ll c$

$$p \sim \varrho_0 v_0 f(\zeta) + \frac{\varrho_0 v_0^2}{c} f^2(\zeta) + \cdots, \tag{3.27}$$

and if $f(\zeta) = \sin(\omega t - kx)$ the averaged wave drift and momentum are

$$v_d = -cu' \sim \frac{v_0^2}{2c} = \frac{a^2\omega^2}{2c}, \quad \langle p \rangle = \varrho_0 v_d, \tag{3.28}$$

where $a = v_0\omega$ is the wave amplitude. Note that for a dispersive wave the phase speed $c = \omega/k$ must be replaced by group speed c_g. In the case $c = c_g$

$$v_d \sim \frac{1}{2} a^2 \omega k, \quad \langle p \rangle = \varrho_0 v_d. \tag{3.29}$$

Note that the drift speed induced by the 1D low-intensity longitudinal sinusoidal wave, that reflected by the expression in (3.29), is functionally the same but twice as less as the Stokes' drift on the surface of a 2D wave propagating on deep water (see, e.g., Lighthill, 1978). In the next section, we show that the momentum of the flexural sinusoidal wave appears the same as for the longitudinal one.

Also, note that the above results do not follow directly from the linearized formulation, which is incorrect for any value turned out to be zero. So, if the linearized description evidences that the wave does not possess momentum, it does not mean that the latter is absent. It should be added that the linearized mathematical theory can be perfect by itself but does not necessarily reflect the corresponding physical applications correctly. In such a case, further analysis as done above is required.

3.3.2 Flexural—Longitudinal Binary Wave

Consider a semi-infinite string or an elastic beam, $0 < x < \infty$. Let a flexural wave, be excited by a transverse periodic action at $x = 0$, and the transverse displacements in the wave area be

$$w = af(x - ct)\,. \tag{3.30}$$

Here the representation of momentum through the wave mass is especially useful. Indeed, since no axial force is applied, we may conclude that in the established regime, the string axis is not stretched, and the wave mass is

$$m_w = m_0\sqrt{(1 + u')^2 + (w')^2}\,, \tag{3.31}$$

and its value averaged over the period ℓ is

$$\langle m_w \rangle = \frac{1}{\ell}\int_\ell m_w \,\mathrm{d}x = \frac{L}{\ell} - 1\,, \tag{3.32}$$

where L is the arc length corresponding to the hord length ℓ.

Thus, the axial momentum appeared without axial action, which may happen if the flexural wave excites the longitudinal one with the same but opposite directed momentum. That is, the transverse action results in a binary wave as discussed above.

As an example consider a sinusoidal flexural wave in a linearly elastic beam. For a large wave-length-to-amplitude ratio, the classical Bernoulli-Euler dynamic equation is valid, possibly with a term corresponding to the tensile force. The equation is written regarding the transverse displacements, and it may seem that there is no axial momentum; however, it is not so as we already know.

To determine the momentum of the binary wave, using the results obtained in Sect. 3.2.4, we need the input parameters, densities of the initial mass as well as the averaged wave mass and energy for the flexural wave, and also the wave speeds, c_g and c_+.

In terms of the transverse and longitudinal displacements, w and u, respectively, with small derivatives w' and u', the binary wave can be described as follows. Since there is no axial tensile force in the flexural wave region, we based on the equation

$$c_+^2 r^2 w^{IV} + \ddot{w} = 0 , \tag{3.33}$$

where $r = \sqrt{J/A}$ is the radius of inertia moment J of the beam cross-section area A. The flexural wave is

$$w = a\cos(\omega t - kx) , \quad k = \sqrt{\frac{\omega}{c_+ r}} . \tag{3.34}$$

The frequency as a function of k and the group speed c_g are

$$\omega = c_+ r k^2 , \quad c_g = 2c_+ r k = 2c_+ \sqrt{\frac{\omega r}{c_+}} . \tag{3.35}$$

The initial and actual mass densities can be expressed as

$$m_0 = A\varrho_0 , \quad m = \frac{m_0}{\lambda} , \quad \lambda = \frac{1}{1-u'} . \tag{3.36}$$

The derivative u' can be found from the fact that the arc length of a beam material segment remains the same as its corresponding initial length. This results in the equality

$$\sqrt{1+(w')^2} = 1 - u' \quad \Longrightarrow \quad u' = 1 - \sqrt{1+(w')^2} \sim -\frac{(w')^2}{2} . \tag{3.37}$$

Thus, the mass and the wave mass are

$$\begin{gathered} m = m_0(1-u') \sim m_0\left(1 + \frac{(w')^2}{2}\right) , \quad m_w = m - m_0 \sim m_0 \frac{(w')^2}{2} , \\ \langle m_w \rangle = \frac{m_0 a^2 k^2}{4} = \frac{m_0 a^2 \omega}{4c_+ r} . \end{gathered} \tag{3.38}$$

Finally, the flexural wave energy density is

$$\mathcal{E} = \varrho_0(\dot{w})^2 = \varrho_0 a^2 \omega^2 \sin^2(\omega t - kx) , \quad \langle \mathcal{E} \rangle = \frac{1}{2}\varrho_0 a^2 \omega^2 . \tag{3.39}$$

Now all the values presented in (3.14)–(3.16), except the expression for $c_g - D$, can be calculated replacing the variable values by their averaged representations.

As to the $c_g - D$ difference, we note that the original equation (the second relation in (3.11)) contains not negligible products of variable values, whereas the product averaged does not equal to the product of the averaged factors. However, for the low-intensity wave considered here, the difference is very small; it is of the order of $(ak)^4$. So, with the accuracy adopted in the linearized formulation, it only evidences that $D < c_g$ with $D \sim c_g$.

In particular, the ratio of the energy densities defined in (3.18) now is

$$\frac{\langle \mathcal{E}_+ \rangle}{\langle \mathcal{E} \rangle} \sim \frac{4r^2k^2}{(1-2rk)^2} \sim 4r^2k^2 \ll 1 . \tag{3.40}$$

since $|rk| \ll 1$ by the definition. This estimation of the energy ratio is also valid for the total energies. Indeed

$$\frac{\mathcal{E}_+(c_+ - D)}{\mathcal{E}(D-v)} \sim \frac{\mathcal{E}_+ c_+}{\mathcal{E} c_g} \sim 2rk . \tag{3.41}$$

Nevertheless, the longitudinal wave possesses the same momentum, by its value, as the much more powerful flexural wave. However, the momentum is rather small. It asymptotic value coincides with that following directly from the mass conservation (3.2)

$$P = -P_+ = m_+ v_+ (c_+ - D) \sim m_0 c_+ v_+ \sim m_w c_g^2 , \quad p = m_w c_g . \tag{3.42}$$

So, the momentum carrying by the flexural wave can be found using the expression in (3.2), with $c = c_g$, disregarding the forerunner. Surprisingly, it is asymptotically the same as in the longitudinal wave. Namely, as follows from (3.38) with the expression for the group speed in the beam, $c_g = 2\omega/k$, the averaged value of momentum is

$$\langle p \rangle \sim \frac{m_0 a^2 k^2 c_g}{4} = \frac{1}{2} m_0 a^2 \omega k , \tag{3.43}$$

and the drift is also a half of that in the surface of the water waves

$$v_d = \frac{1}{2} a^2 \omega k . \tag{3.44}$$

Thus, in this case, the flexural wave carries a momentum, while the associated longitudinal extensional wave has equal, opposite directed momentum. At the same time, in such a problem but for the infinite beam, due to the symmetry, there is no momentum in the flexural waves, while the associated longitudinal waves possess momenta directed to the origin.

3.4 The Complex Wave

3.4.1 Uniform Waveguide

3.4.1.1 Translation Symmetry

Consider an infinite linearly elastic cylinder, $-\infty < x < \infty$, of an arbitrary cross-section. The material properties and the homogeneous boundary conditions posted on the cylindrical surface are assumed to be (x, t)-independent. The dynamics of such a body obeys partial differential or integral-differential equations with no explicit dependence on the x-coordinate and time, t. In other words, the equation considered, together with additional conditions, admits translation symmetry with respect to these variables, that is substitutions $x = x' + a, t = t' + b$, where a and b are arbitrary real constants, do not change the governing relations. In the following, the boundary is assumed to be unable to accumulate energy; otherwise, the energy consumption structure at the boundary must be included in the bulk of the waveguide. For example, in the case of a beam on a Winkler foundation, the total energy is a sum of the strain and kinetic energies of the beam and the strain energy of the foundation, and this fact is taken into account in the equation for the beam.

The simplest example is a one-dimensional wave equation

$$T\frac{\partial^2 u}{\partial x^2} - \varrho S\frac{\partial^2 u}{\partial t^2} = 0 \tag{3.45}$$

with the tensile force, T, density, ϱ, and the cross-section area, S, as constant parameters. This equation is commonly called the *string equation*. It also corresponds to an elastic rod model. A general linear one-dimensional equation of such a kind is

$$K(x,t) ** u(x,t) = 0\,, \tag{3.46}$$

where the double *convolution* is defined as

$$\begin{aligned} K(x,t) ** u(x,t) &= \int_{-\infty}^{\infty}\int_{-\infty}^{\infty} K(x-\xi, t-\tau)u(\xi,\tau)\,\mathrm{d}\xi\,\mathrm{d}\tau \\ &= \int_{-\infty}^{\infty}\int_{-\infty}^{\infty} K(\xi,\tau)u(x-\xi, t-\tau)\,\mathrm{d}\xi\,\mathrm{d}\tau \end{aligned} \tag{3.47}$$

and the kernel, K, is a generalized function. In particular, if the kernel includes an nth order derivative of the delta function, convolution (3.46) with this term is the corresponding derivative of u:

$$\delta^{(m)}(x)\delta^{(n)}(t) ** u(x,t) = \delta(x)\delta(t) ** \frac{\partial^{m+n}u}{\partial x^m \partial t^n} = \frac{\partial^{m+n}u}{\partial x^m \partial t^n}\,. \tag{3.48}$$

A regular term in K corresponds to a *nonlocal interaction* material (Kunin, 1982; 1983). Similar considerations are applicable in the case of two- and three-dimensional translation-symmetry systems.

3.4.1.2 Translation-Rotation Symmetry

Uniformity, which is required for a steady-state solution to exist, concerns not only the material properties and the waveguide cross-section, but the waveguide axis geometry as well. The most general uniform curve is a helix; its curvature and torsion are invariable. The helix is defined by its radius, r, and the angle, α, between the helix and its straight-line axis, x. A straight line and a circumference are the limiting cases of the helix with $r = 0$ and $\alpha = \pi/2$ $(r \neq 0)$, respectively.

Let the waveguide material properties and geometry be uniform from the point of view of an observer moving and rotating with the triad natural to the helix. Owing to this uniformity, translation-rotation symmetry holds, which leads to the existence of a propagating-rotating steady-state wave. In terms of the cylindrical coordinate system, x, r, θ, associated with the straight-line axis of the helix, this complex wave can be represented as

$$\boldsymbol{u}(x, y, z, t) = \boldsymbol{A}(r, \theta - 2\pi x/l)\mathrm{e}^{\mathrm{i}(\omega t - kx)}\,, \tag{3.49}$$

where $l = 2\pi r \cot\alpha$ is the pitch of the helix. For real k and ω the wave appears invariant under the condition $x = x_0 + \omega t/k, \theta = \theta_0 + 2\pi\omega t/(kl)$ with invariable r, x_0 and θ_0. This wave is a generalization of that defined in (3.50).

Note that the pitch of the helix can remain nonzero for the straight line which can be considered as a limit where $\alpha \to 0$ together with r in such a way that $\lim l > 0$. This remark relates to the fact that a propagating-rotating steady-state wave natural to the helix can exist in the case of a straight-line waveguide with proper symmetry as well. Nonlinear propagating-rotating solitary waves in a helix are considered in Sects. 10.6 and 10.8. In the following, the wave (3.50) for a straight-line uniform waveguide is considered.

3.4.2 Translation-Reflection Symmetry and the Complex Wave

An x, t-translation symmetry linear equation is satisfied by a *complex wave*

$$\boldsymbol{u}(x, y, z, t) = \boldsymbol{A}(y, z)\mathrm{e}^{\mathrm{i}(\omega t - kx)}\,, \tag{3.50}$$

where the *frequency*, ω, and the *wavenumber*, k, are, in general, complex-valued parameters as well as the waves, $\Re\boldsymbol{u}$ and $\Im\boldsymbol{u}$, different only by a shift in the *phase* and separated by the imaginary unit as a multiplier

$$\begin{aligned} \boldsymbol{u} &= \Re\boldsymbol{u} + \mathrm{i}\Im\boldsymbol{u}\,, \\ \Re\boldsymbol{u} &= \boldsymbol{R}\exp\left(-t\Im\omega + x\Im k\right)\,, \\ \Im\boldsymbol{u} &= \boldsymbol{I}\exp\left(-t\Im\omega + x\Im k\right)\,, \\ \boldsymbol{R} &= \Re\boldsymbol{A}\cos(t\Re\omega - x\Re k) - \Im\boldsymbol{A}\sin(t\Re\omega - x\Re k)\,, \\ \boldsymbol{I} &= \Re\boldsymbol{A}\sin(t\Re\omega - x\Re k) + \Im\boldsymbol{A}\cos(t\Re\omega - x\Re k)\,. \end{aligned} \tag{3.51}$$

In the case of equations and boundary conditions with real parameters (here call all these relations *the equations*), the real and imaginary parts of the complex wave (3.50) are separated after substitution into the equations as well as before this and each part with an arbitrary amplitude satisfies the equation. Hence the conjugate wave

$$\boldsymbol{u} = \overline{\boldsymbol{A}(y,z)}\mathrm{e}^{-\mathrm{i}(\overline{\omega}t - \overline{k}x)} \tag{3.52}$$

exists as well.

In representation (3.51), the exponential multiplier is called the *envelope*, while the trigonometric one—the *carrier wave*. The carrier *wavelength and period* are, respectively

$$\lambda = \frac{2\pi}{\Re k}, \quad T_0 = \frac{2\pi}{\Re\omega}\,. \tag{3.53}$$

The use of a complex representation instead of a real one is more convenient because the exponential dependence remains the same under the action of any x, t-homogeneous linear operator. For example

$$\begin{aligned} \frac{\partial^n}{\partial x^n}\mathrm{e}^{\mathrm{i}(\omega t - kx)} &= (-\mathrm{i}k)^n\mathrm{e}^{\mathrm{i}(\omega t - kx)}\,, \\ \frac{\partial^n}{\partial t^n}\mathrm{e}^{\mathrm{i}(\omega t - kx)} &= (\mathrm{i}\omega)^n\mathrm{e}^{\mathrm{i}(\omega t - kx)}\,, \\ \mathcal{C} \equiv K(x,t) ** \mathrm{e}^{\mathrm{i}(\omega t - kx)} &= K^{FF}(k,-\omega)\mathrm{e}^{\mathrm{i}(\omega t - kx)}\,, \\ K^{FF}(k,-\omega) &= \int_{-\infty}^{\infty}\int_{-\infty}^{\infty} K(\xi,\tau)\exp\left[-\mathrm{i}(\omega\tau - k\xi)\right]\,\mathrm{d}\xi\,\,\mathrm{d}\tau\,. \end{aligned} \tag{3.54}$$

In addition, if the waveguide possesses reflection symmetry, that is, if the change x to $-x$ does not change the equation, there exists symmetric waves, where displacements, u, and tractions, $\boldsymbol{\sigma}_x$, possess the following symmetry.

Let a wave exists with displacements $\boldsymbol{u}$ in a cross-section and traction $\boldsymbol{\sigma}_x$ acting on the cross-section from the right

$$\boldsymbol{u}\,(u_x, u_y, u_z) \quad \text{and} \quad \boldsymbol{\sigma}_x(\sigma_{xx}, \sigma_{xy}, \sigma_{xz})\,. \tag{3.55}$$

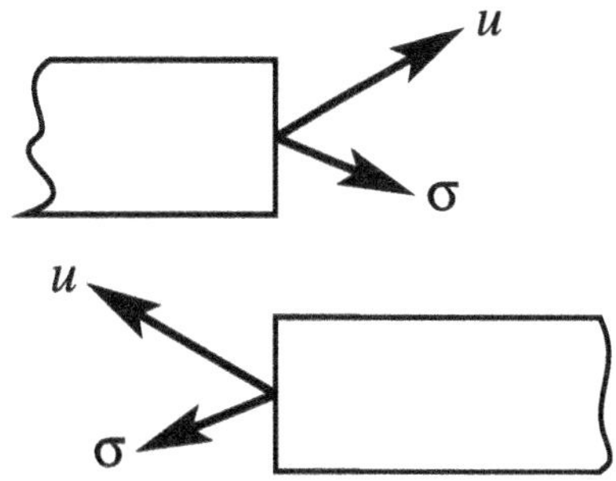

Fig. 3.1 Displacements and tractions in symmetrical waves

Then a symmetric wave exists with symmetric displacements and the symmetric traction, which act on the cross-section from the left (Fig. 3.1), such that the displacement and stress components are

$$(-u_x, u_y, u_z) \quad \text{and} \quad (\sigma_{xx}, -\sigma_{xy}, -\sigma_{xz})\,. \tag{3.56}$$

Note that the reflection symmetry need not exist in the general case of an anisotropic medium.

In addition, note that the change of t to $-t$ may be possible in a perfect system (a system with dissipation does not permit this). Lastly, in the considered linear elasticity, the wave amplitude can be multiplied by any complex number.

Similarly two- and three-dimensional complex waves are introduced as

$$\boldsymbol{u}(\boldsymbol{x}, t) = \boldsymbol{A}\mathrm{e}^{\mathrm{i}(\omega t - \boldsymbol{k}\boldsymbol{x})}\,, \tag{3.57}$$

where vector k defines the wave propagation direction; it is called the *wave vector*.

If the complex wave expression is substituted into the equation of the considered type, an equation defined on the cross-section of the waveguide is obtained with the parameters ω and k. A solution of the latter equation must satisfy the boundary conditions and this yields a *dispersion relation* between ω and k. In the one-dimensional case, the dispersion relation follows immediately from the primary equation.

3.4.3 Dispersion Relation as the Fourier Transform

Now consider a uniform-waveguide linear dynamic equation admitting the complex wave $u(t, x) = \exp(\mathrm{i}(\omega t - kx))$. Due to the uniformity, the equation coefficients are independent of the 'absolute' space-time coordinates but introduced as convolutions. In this connection, the derivatives are convolutions with the delta function derivatives, (17.37).

So, let the equation be as

$$\varrho\omega^2 u(t,x) - f(x) * u(t,x) = 0\,, \tag{3.58}$$

where $f(x)$ is a non-specified integrable function, possibly generalized one, and the convolution (see (17.28)) is

$$f(x) * u(t,x) = \int_{-\infty}^{\infty} f(\xi)u(t, x-\xi)\,\mathrm{d}\xi\,. \tag{3.59}$$

Remaind that for $f(x) = \delta^n(x)$ the convolution becomes the n-derivative
So, the convolution (3.59) is

$$\begin{aligned} f(x) * u(t,x) &= \int_{-\infty}^{\infty} f(\xi)\mathrm{e}^{\mathrm{i}(\omega t - k(x-\xi)} \\ &= \int_{-\infty}^{\infty} f(\xi)\mathrm{e}^{\mathrm{i}k\xi}\,\mathrm{d}\xi\, u(t,x) = f^F(k)u(t,x)\,. \end{aligned} \tag{3.60}$$

Thus, the general dispersion-related expression is obtained using the Fourier transform. In particular, the dispersion relation corresponding the equation in (3.58) is

$$\omega(k) = \pm\sqrt{\frac{1}{\varrho} f^F(k)}\,, \tag{3.61}$$

Such a result for a system of similar equations for $u_m(t,x), f_m(x)).(m = 1, 2, \dots M)$ follows in the same way.

3.4.4 *Examples*

3.4.4.1 Waves in a String

For string equation (3.45) we find

$$\varrho S\omega^2 - Tk^2 = 0 \quad \Longrightarrow \quad \omega = \pm ck\,, \quad c = \sqrt{\frac{T}{\varrho S}}\,. \tag{3.62}$$

3.4.4.2 Waves in an Elastic Beam

The dynamic equation

$$D\frac{\partial^4 w}{\partial x^4} + \varrho S\frac{\partial^2 w}{\partial t^2} = 0 \tag{3.63}$$

yields the dispersion relation

$$\omega = \pm k^2 \sqrt{\frac{D}{\varrho S}} \,. \tag{3.64}$$

Note that $D = EI$ for a beam (E is the elastic modulus and I is the moment of inertia of the cross-section area relative to the neutral axis) and $D = Eh^3/[12(1-\nu^2)]$ for a plate (h is the plate thickness and ν is the Poisson's ratio).

In the dispersion relations for propagating waves, both, ω and k, are real. However, there also exist waveguides, where, along with the propagating waves, there exist complex wavenumbers for a real frequency, as in a beam

$$k = \pm\left(\frac{\varrho S \omega^2}{D}\right)^{1/4}, \quad k = \pm i\left(\frac{\varrho S \omega^2}{D}\right)^{1/4}. \tag{3.65}$$

3.4.4.3 Elastic Beam on an Elastic Foundation

From the equation

$$D\frac{\partial^4 w}{\partial x^4} + \varkappa w + \varrho S \frac{\partial^2 w}{\partial t^2} = 0 \tag{3.66}$$

it follows that

$$\omega = \pm\sqrt{\frac{1}{\varrho S}(\varkappa + Dk^4)} \,. \tag{3.67}$$

Thus, in this case, there is a positive lower bound of the propagating wave frequencies.

3.4.4.4 Two-Dimensional Wave Equation for a Strip

Consider the wave equation

$$\frac{\partial^2 u}{\partial x^2} + \frac{\partial^2 u}{\partial y^2} - \frac{1}{c^2}\frac{\partial^2 u}{\partial t^2} = 0 \tag{3.68}$$

defined in $-\infty < x < \infty$, $-h < y < h$, with the traction-free boundary conditions

$$\frac{\partial u}{\partial y} = 0 \quad (y = \pm h)\,, \tag{3.69}$$

and a complex wave, which can propagate in this *waveguide*, $u = A(y)e^{i(\omega t - kx)}$. It follows that

$$\frac{\partial^2 A}{\partial y^2} + \Omega^2 A = 0\,, \quad \Omega^2 = \frac{\omega^2}{c^2} - k^2\,; \tag{3.70}$$

hence there are symmetric and antisymmetric solutions with

$$A(y) = \cos \Omega y\,, \quad \text{and} \quad A(y) = \sin \Omega y\,. \tag{3.71}$$

Condition (3.69) yields dispersion relation as

$$\omega = \omega_n = \pm \frac{c}{h}\sqrt{h^2 k^2 + \frac{n^2\pi^2}{4}}\,, \quad n = 0, 1, \ldots\,, \tag{3.72}$$

where even n correspond to the symmetric mode and odd n correspond to the antisymmetric ones. Thus, in this case, there are infinite number of the propagating wave modes.

3.4.4.5 Two-Dimensional Wave Equation for an Unbounded Plane

Let Eq. (3.68) be defined in $-\infty < (x, y) < \infty$. In this case, the dispersion relation for for the two-dimensional complex wave (3.57) is

$$\omega = \pm c\sqrt{k_x^2 + k_y^2}\,, \tag{3.73}$$

that is, it corresponds to a right circular conical surface in the (k_x, k_y, ω) space.

3.4.5 Phase and Group Velocities

The carrier *phase velocity*, v, is defined as the speed of a constant phase of the wave

$$Re\varphi \equiv t\Re\omega - x\Re k = \text{const}\,. \tag{3.74}$$

It follows that

$$v = \frac{\Re\omega}{\Re k} = \frac{\mathrm{d}x}{\mathrm{d}t} \quad (t\Re\omega - x\Re k = \text{const})\,. \tag{3.75}$$

The *group velocity*, v_g, is defined, Hayes (1977), in the same way but for the vanishing envelope (which propagates with the energy flux velocity, see (3.93)). The envelop velocity follows from the below expression for the imaginary part of the wave phase

$$\Im\varphi \equiv t\Im\omega - x\Im k = \text{const}, \tag{3.76}$$

as

$$v_e = \frac{\Im\omega}{\Im k}, \quad \Im\omega \sim \frac{\mathrm{d}\omega}{\mathrm{d}k}\Im k \quad (\Im k \to 0). \tag{3.77}$$

It follows that, for the real frequency and wavenumber, the energy as well as the wave front (or quasifront) propagates with the speed

$$v_g = \frac{\mathrm{d}\omega}{\mathrm{d}k}. \tag{3.78}$$

The name "group velocity" originated from the fact that this is the velocity of a group of sinusoidal waves with close frequencies and wavenumbers (Rayleigh, 1877, 1894). In fact, it is the velocity of the non-exponential envelope of these waves. For example,

$$\begin{aligned} &\sin(\omega t - kx) + \sin[(\omega + v_g\varepsilon)t - (k+\varepsilon)x] \\ &= 2\cos\left[\frac{\varepsilon}{2}(v_g t - x)\right]\sin\left[\left(\omega + \frac{v_g\varepsilon}{2}\right)t - \left(k + \frac{\varepsilon}{2}\right)x\right] \\ &\sim 2\cos\left[\frac{\varepsilon}{2}(v_g t - x)\right]\sin(\omega t - kx). \end{aligned} \tag{3.79}$$

In this expression, the cosine represents the envelope, while the carrier is represented by the sine.

If the phase and group velocities are the same, as in the case of the string (3.62), that is, if the frequency is proportional to the wavenumber, the velocity is k-independent and all the sinusoidal waves propagate with the same speed. In this case, there is no *wave dispersion*: the wave packet consisting of waves of different lengths propagates with no change of its size and shape. Otherwise, if the phase and group velocities are different, only a sinusoidal wave or its complex representation propagates intact, but not a wave packet. Note that in the case of the beam (3.64) $v_g = 2v$, whereas for surface waves in a deep sea $v_g = v/2$. These speeds can also differ sign, that is, a sinusoidal wave and a group of sinusoidal waves almost the same length can propagate in opposite directions. For example, from the equation for an elastic beam, axially compressed by force T and supported by an elastic foundation of stiffness $\varkappa$

$$D\frac{\partial^4 w}{\partial x^4} + T\frac{\partial^2 w}{\partial x^2} + \varkappa w + \varrho S\frac{\partial^2 w}{\partial t^2} = 0 \tag{3.80}$$

the dispersion relation is

$$\omega = \pm\sqrt{(\varkappa - Tk^2 + Dk^4)/(\varrho S)}. \tag{3.81}$$

The beam is stable if $T^2 < 4D\varkappa$. In this case, for ω and k positive

$$v = \frac{\omega}{k} > 0\,, \quad v_g = \frac{d\omega}{dk} = \frac{k}{\omega}(-T + 2Dk^2) < 0 \quad \left(0 < k < \sqrt{\frac{T}{2D}}\right). \tag{3.82}$$

Comment 3.1 The phase and group velocities are the most important notions in the wave theory. The phase velocity is the apparent velocity of the sinusoidal wave form; it is a kinematical notion, whereas the group velocity is a dynamic notion; as was shown below it is the energy flux velocity in the wave.

The second derivative, $\omega''(k)$, which can be called the *dispersion number*, is also important. It plays a decisive role in the wave quasifront and the resonant wave formations, Chaps. 13 and 16.

For two- or three-dimensional sinusoidal waves the phase velocity can be defined in the same way as the frequency-to-wavenumber ratio, where the wavenumber is the wave vector modulus $k = |\boldsymbol{k}|$ (or $-k$). In this case, however, the apparent phase speed along an axis is greater than the phase speed itself, if directions of the wave vector and the axis does not coincide with each other. Indeed, for a constant phase

$$\varphi = \omega t - k_1 x_1 - k_2 x_2 - k_3 x_3 = \text{const} \tag{3.83}$$

the phase speed of the wave, v, and the speed of the wave along the axes, v_x, v_y, v_z, are

$$v = \frac{\omega}{k}\,, \quad v_i = \frac{\omega}{k_i} = \frac{v}{\cos\alpha_i}\,, \tag{3.84}$$

where $\cos\alpha_i$ are the directional cosines, $i = 1, 2, 3$. Thus the coordinate axes wave speeds are not the phase speed vector components. In this connection, instead of the wave vector a *wave slowness* vector is introduced with the components as

$$\frac{k_1}{\omega}\,, \quad \frac{k_2}{\omega}\,, \quad \frac{k_3}{\omega}\,. \tag{3.85}$$

Components of the wave slowness vector do represent the coordinate axes wave slowness. In this terms, the phase velocity vector is defined to be directed as the slowness vector, but to have the speed of the wave. More precisely, the phase velocity vector is the slowness vector divided by its squared modulus

$$\boldsymbol{v} = \frac{(k_1/\omega, k_2/\omega, k_3/\omega)}{(k_1/\omega)^2 + (k_2/\omega)^2 + (k_3/\omega)^2} = \frac{\omega}{k^2}(k_1, k_2, k_3)\,. \tag{3.86}$$

The group velocity is defined as the derivative with respect to the wave vector

$$\boldsymbol{v}_g = \left(\frac{\partial}{\partial k_1}, \frac{\partial}{\partial k_2}, \frac{\partial}{\partial k_3}\right)\omega(\boldsymbol{k})\,. \tag{3.87}$$

It is the *group velocity vector*, and the same as above for the one-dimensional case it can be considered as the complex wave with a vanishing envelope.

3.4.6 Energy Flux in a Wave

3.4.6.1 The Concept

Consider a material plane with the unit normal $\boldsymbol{n}$. The energy flux density through the plane in the $\boldsymbol{n}$-direction is defined as

$$N_n = -\boldsymbol{\sigma}_n \frac{\partial \boldsymbol{u}}{\partial t}, \tag{3.88}$$

where $\boldsymbol{\sigma}_n$ is the stress vector which acts on the plane from the positive direction of $\boldsymbol{n}$, that is the projection of the stress tensor onto the normal $\boldsymbol{n}$. Accordingly

$$N_i = -\boldsymbol{\sigma}_i \frac{\partial \boldsymbol{u}}{\partial t} = -\sigma_{ij} \frac{\partial u_j}{\partial t} \tag{3.89}$$

is the energy flux density through the coordinate plane with x_i as the normal, $i = 1, 2, 3$. Recall that summation by repeated indices is assumed. Thus there exists an energy flux density vector $\boldsymbol{N}$ whose projections onto the axes are N_i.

The energy flux relative to an unmoving frame, N_E, is

$$\boldsymbol{N}_E = \boldsymbol{N} + (W + \mathcal{K}) \frac{\partial \boldsymbol{u}}{\partial t}, \tag{3.90}$$

where W and $\mathcal{K}$ are the strain and kinetic energy densities, respectively. The first term, $\boldsymbol{N}$, is called the *inductiveenergy flux inductive* part of the flux, while the other term is the *convective energy flux convective* part.

Note that these expressions are valid for the regular case of *local interactions*, where the interaction between two parts of the domain occurs only at the interface (see Kunin, 1982; 1983).

3.4.7 Energy Flux Associated with the Complex Wave

Now consider the energy flux through a cross-section of an elastic waveguide for the complex wave (3.50). The energy flux density is defined here as that associated with both the real and imaginary parts of the wave. The total density can be represented as

$$N_x = -\Re\boldsymbol{\sigma}_x\Re\frac{\partial \boldsymbol{u}}{\partial t} - \Im\boldsymbol{\sigma}_x\Im\frac{\partial \boldsymbol{u}}{\partial t} \\ = -\Re\left(\mathrm{i}\omega\boldsymbol{u}\overline{\boldsymbol{\sigma}_x}\right) = \Im\left(\omega\boldsymbol{u}\overline{\boldsymbol{\sigma}_x}\right) . \quad (3.91)$$

In a general linear case with (x, t)-uniformity, the traction vector can be expressed similarly to the displacement (3.50), as

$$\boldsymbol{\sigma}_x = \boldsymbol{\sigma}_x^0 \exp\left[\mathrm{i}(\omega t - kx)\right] , \quad (3.92)$$

where the amplitude, $\boldsymbol{\sigma}_x^0$, is independent of x and t. This representation follows from the fact that an arbitrary (x, t)-uniform linear operator preserves the exponential function (3.54). The energy flux is thus

$$N_x = \Im\left(\omega\boldsymbol{A}\overline{\boldsymbol{\sigma}_x^0}\right) \exp\left[-2(\Im\omega\, t - \Im k\, x)\right] , \quad (3.93)$$

while the total energy flux through the cross-section is

$$N = \int_S N_x\, dS = \int_S \Im\left(\omega\boldsymbol{A}\overline{\boldsymbol{\sigma}_x^0}\right)\, \mathrm{d}S \exp\left[-2(\Im\omega\, t - \Im k\, x)\right] . \quad (3.94)$$

3.4.7.1 Energy Density

The energy density in a complex wave is defined as the sum of two terms corresponding to the real and imaginary parts of the wave, as is assumed with respect to the energy flux. For a homogeneous linearly elastic material the energy density $\mathcal{E}_0 = W + \mathcal{K}$ can be calculated as

$$W = \frac{1}{2}\left(\Re\sigma_{ij}\Re\frac{\partial u_j}{\partial x_i} + \Im\sigma_{ij}\Im\frac{\partial u_j}{\partial x_i}\right) , \\ \mathcal{K} = \frac{\varrho}{2}\left(\Re\frac{\partial u_j}{\partial t}\Re\frac{\partial u_j}{\partial t} + \Im\frac{\partial u_j}{\partial t}\Im\frac{\partial u_j}{\partial t}\right) , \\ \mathcal{E}_0 = \frac{1}{2}\Re\left(\overline{\sigma_{ij}}\frac{\partial u_j}{\partial x_i} + \varrho\frac{\partial \overline{u_j}}{\partial t}\frac{\partial u_j}{\partial t}\right) . \quad (3.95)$$

Since in each product one of the two multipliers is a conjugate complex value, it follows that the carrier exponential, $\exp[\mathrm{i}(\Re\omega t - \Re kx)]$, does not appear in this expression as well as in the expression associated with the energy flux (3.93). Note that this conclusion is true for any other model of an elastic waveguide in which stresses and displacements are represented by generalized values with the same carrier exponential. For example, in the case of the model for waves in a beam, the generalized stress is the bending moment, M, and the energy density can be expressed as

$$\mathcal{E}_0 = \frac{1}{2}\Re\left[\overline{M}\frac{\partial^2 w}{\partial x^2} + \varrho S\left(\frac{\partial \overline{w}}{\partial t}\right)^2\right]$$
$$= \mathcal{E}^0 \exp\left[-2(\Im\omega\, t - \Im k\, x)\right] , \tag{3.96}$$

where $\mathcal{E}^0$ is independent of x and t.

3.4.7.2 Energy Flux Velocity

Let $\Im k$ be nonzero. In a moving coordinate system, $x = v_g t + x'$, $v_g = \Im\omega/\Im k$, the energy density is time-independent as well as the energy flux (3.93). In addition, the energy density is integrable $[0 < x' < \infty$ $(\Im k < 0)$ or $-\infty < x' < 0$ $(\Im k > 0)]$ and hence the energy in the corresponding part of the waveguide is invariable. Thus, in this moving coordinate system, the energy flux is zero. This reveals that, in the unmoving coordinate system, *the energy flux velocity is equal to the group velocity of the wave.*

For the case of a *sinusoidal wave* where the frequency and the wavenumber are real, the energy flux can be considered as a limit of that when the imaginary parts tend to zero. Note that the energy flux, by its definition (3.94), depends on the wavenumber continuously. Thus the statement is true for the sinusoidal wave as well. Recall that, in this case, the group velocity is $v_g = \mathrm{d}\omega/\,\mathrm{d}k$ (3.78).

The energy flux velocity is the ratio of the energy flux to the energy density. Thus, the energy per unit length of the waveguide is

$$\mathcal{E} = \int_S \mathcal{E}_0 \,\mathrm{d}S = \frac{N}{v_g}$$
$$= \frac{1}{v_g}\int_S \Im\left(\omega \boldsymbol{A}\overline{\boldsymbol{\sigma}_x^0}\right)\,\mathrm{d}S \exp\left[-2(\Im\omega\, t - \Im k\, x)\right] . \tag{3.97}$$

Note that these results are in agreement with the *energy conservation law*, which for the considered case can be expressed as

$$\frac{\partial \mathcal{E}}{\partial t} = -\frac{\partial N}{\partial x} . \tag{3.98}$$

Indeed,

$$\frac{\partial \mathcal{E}}{\partial t} = -2\Im\omega\, \mathcal{E} , \quad \frac{\partial N}{\partial x} = 2\Im k\, N \tag{3.99}$$

and as follows from (3.98)

$$\frac{N}{\mathcal{E}} = \frac{\Im\omega}{\Im k} = v_g . \tag{3.100}$$

3.4.7.3 Energy and Energy Flux in a Real Wave

The expressions for the energy flux (3.94) and energy per unit length (3.97) represent these quantities as a sum of those for the real and imaginary parts of the complex wave. These parts represent the same waves, but in different phases (the phases differ by $\pi/2$). As a result, the energy and the energy flux defined for the complex wave do not contain the oscillating multiplier. This, however, is not true for each of the two waves separately. For example, the real part of the complex wave (3.50) yields

$$\begin{aligned} N_{x,real} &= \frac{1}{2} N_x + N_{x,oscillation} , \\ N_{x,oscillation} &= \frac{1}{2} \Im\{\omega \boldsymbol{A} \boldsymbol{\sigma}^0 \exp[2\mathrm{i}(\Re\omega\, t - \Re k\, x]\} \mathrm{e}^{-2(\Im\omega\, t - \Im k\, x)} . \end{aligned} \tag{3.101}$$

Usually the oscillating part of the energy (represented by the second term) is not important since it does not affect the long-term energy flux. The imaginary part yields the same, but with a different sign for the second term. Taking into account the contributions of both the real and imaginary parts of the complex wave the *energy flux doubled* is obtained without the oscillating term.

Along with the relation between the energy and the energy flux in complex waves, $N = v_g \mathcal{E}$, such a ratio for real waves is of interest. The latter has a simple expression for real ω and k or for a slow-varying envelope, $\Im\omega \ll \Re\omega$, $\Im k \ll \Re k$. In this case, based on expression (3.101), a similar expression for the energy density (the energy per unit length of the waveguide) and on the energy conservation law (3.98), it can be found that *while the averaged parts are connected by the group speed, the oscillation parts are connected by the phase speed*

$$\langle N_{real} \rangle = v_g \langle \mathcal{E}_{real} \rangle , \quad N_{oscillation} = v \mathcal{E}_{oscillation} , \tag{3.102}$$

where $\langle N_{real} \rangle$ and $\langle \mathcal{E}_{real} \rangle$ are the averaged values of the flux and the energy density, respectively, equal to half of these quantities in the corresponding complex wave, and

$$N_{oscillation} = N_{real} - \langle N_{real} \rangle , \quad \mathcal{E}_{oscillation} = \mathcal{E}_{real} - \langle \mathcal{E}_{real} \rangle . \tag{3.103}$$

Note that subscript x drops out after the integration of the corresponding quantity over the waveguide cross-section.

3.4.8 *Additivity of Energy Fluxes in Different Waves*

If several different waves can exist in a waveguide, an important question is whether the additivity is valid. The energy flux is a quadratic function of field variables, such as displacements and stresses, and the question concerns any two waves propagating simultaneously. Clearly, the additivity does not take place if the superposition of two identical waves propagating in the same direction, as two parts of a single wave, is considered. However, there exist some types of the additivity of *different* waves.

3.4.8.1 General Case

Consider the energy flux contributed by a sum of two complex waves of type (3.50). Following to (3.91) and (3.94), let

$$\begin{aligned} N &= N_1 + N_2 + N_{12}\,, \\ N_{12} &= -\int_S \Re\left(\boldsymbol{\sigma}_1 \frac{\partial \overline{\boldsymbol{u}_2}}{\partial t} + \overline{\boldsymbol{\sigma}_2} \frac{\partial \boldsymbol{u}_1}{\partial t}\right) \mathrm{d}S \\ &= \Im\left\{\mathcal{N} \exp\left[\mathrm{i}\left(\omega_1 - \overline{\omega_2}\right) t - \mathrm{i}\left(k_1 - \overline{k_2}\right) x\right]\right\}\,, \\ \mathcal{N} &= \int_S \left(\omega_1 \boldsymbol{A}_1 \overline{\boldsymbol{\sigma}_2^0} - \overline{\omega_2 \boldsymbol{A}_2} \boldsymbol{\sigma}_1^0\right) \mathrm{d}S\,. \end{aligned} \tag{3.104}$$

where N_1 and N_2 are the energy fluxes corresponding to the first and second waves separately and the subscript denotes the number of the wave considered. The question is when is the wave interaction energy flux, N_{12}, equal to zero.

In the case of real frequencies and wavenumbers, such that frequencies and/or wavenumbers of the two waves are different, the wave interaction energy flux averaged over the time-interval, $2\pi/|\omega_1 - \omega_2|$, or the space-interval, $2\pi/|k_1 - k_2|$, is zero. So, additivity is valid for the averaged energy fluxes of such different waves. However, if the corresponding values are close to each other the time-interval (or the space-interval) is large. Below some conditions are considered where stronger conclusions are valid.

3.4.8.2 Conjugate Frequencies

Consider the case $\omega_1 = \overline{\omega_2}$, $k_1 \neq \overline{k_2}$. Since each wave can exist separately, the energy and energy flux, $\mathcal{E}_1$ and N_1, satisfy (3.98) as well as $\mathcal{E}_2$ and N_2. Hence the interaction energy and energy flux must satisfy this relation. Here the interaction energy per unit length of the waveguide, which has the same structure as the energy flux (3.104), is independent of time. It follows immediately from (3.98) that

$$N_{12} = \mathcal{N} = 0\,. \tag{3.105}$$

As an example consider equation (3.80) for an elastic beam, axially compressed by force T and supported by an elastic foundation of stiffness $\varkappa$. From dispersion relation (3.81) it follows that for a given frequency in the range

$$\sqrt{\frac{\varkappa - T^2/(4D)}{\varrho S}} < \omega \leq \sqrt{\frac{\varkappa}{\varrho S}}, \tag{3.106}$$

there are two different positive wavenumbers

$$k_{1,2} = \sqrt{\frac{T}{2D}\left[1 \pm \sqrt{1 - 4D(\varkappa - \varrho S\omega^2)/T^2}\right]}, \tag{3.107}$$

In this problem, the 'generalized' displacement components are the cross-section rotation, $\partial u/\partial x$, and the transversal displacement, u. The corresponding traction components are the bending moment, $-D\partial^2/\partial x^2$, and the total force acting in the direction of the displacement. The latter is a sum of the transversal force, $-D\partial^3 w/\partial x^3$, and the transversal projection of the compressive force, $P\partial w/\partial x$. Thus

$$\begin{aligned} \boldsymbol{u} &= \left(\frac{\partial u}{\partial x}, u\right) = (-\mathrm{i}k\,,\ 1)u\,, \\ \boldsymbol{\sigma} &= \left(-D\frac{\partial^2 u}{\partial x^2}, D\frac{\partial^3 u}{\partial x^3} + T\frac{\partial u}{\partial x}\right) = \left(Dk^2\,,\ \mathrm{i}Dk^3 - \mathrm{i}Tk\right)u\,, \end{aligned} \tag{3.108}$$

The integrant in the expression for $\mathcal{N}$ is thus proportional to

$$\mathrm{i}(k_1 + k_2)[D(k_1^2 + k_2^2) - T], \tag{3.109}$$

that, with refer to (3.107), is equal to zero; hence $N_{12} = 0$ as it should.

Comment 3.2 As the frequency approaches the lower limit shown in (3.106), where the waves unite, the wave interaction energy flux remain to be zero. It the limit, where these two waves are identical, it is zero since the group speed is zero and $N_1 = N_2 = N_{1,2} = 0$. It is a *resonant frequency*; the resonant waves and the energy behavior at resonant regimes are discussed in Chap. 16.

3.4.8.3 Conjugate Frequencies and Conjugate Wavenumbers

Consider the case where $\omega_1 = \overline{\omega_2}$ and $k_1 = \overline{k_2}$, but $\mathrm{d}\omega_1/\,\mathrm{d}k\ (k = k_1) \neq \overline{\mathrm{d}\omega_2/\,\mathrm{d}k\ (k = k_2)}$. In this case, the additivity equality (3.105) is valid as well. Indeed, consider a variation of the wavenumbers resulting in a nonzero difference $\delta k_1 - \overline{\delta k_2}$, while the equality $\omega_1 = \overline{\omega_2}$ is held. This is possible since $\mathrm{d}\omega_1/\,\mathrm{d}k \neq \overline{\mathrm{d}\omega_2/\,\mathrm{d}k}$. This particular variation leads to the previous case and hence the additivity is valid in the present case as well.

For two sinusoidal waves, that is for real frequencies and wavenumbers, this case corresponds to the crossing-point of two dispersion curves (Fig. 3.2). The above variation perturbs the level of the horizontal line, $\omega = \omega_1 = \omega_2$, as shown in Fig. 3.2. It is clear that since the additivity is valid for a perturbed case, it is still valid in the considered case as the limit.

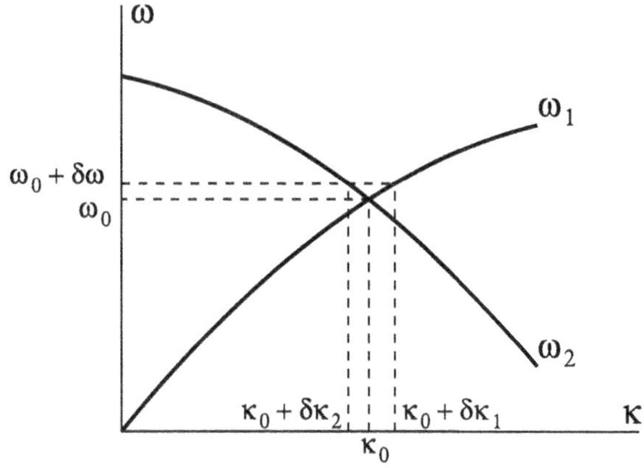

Fig. 3.2 Dispersion curves cross each other

3.4.8.4 Conjugate Wavenumbers

Considerations similar to those in the case of conjugate frequencies show that if $k_1 = \overline{k_2}$ and $\omega_1 \neq \overline{\omega_2}$, then the energy conservation law (3.98) leads to the conclusion that the wave interaction energy is zero in this case as well.

Thus, the *orthogonality relation* (3.105) is valid in the following cases:

$$\begin{gathered} \omega_1 = \overline{\omega_2}\,, \quad k_1 \neq \overline{k_2}\,, \\ \omega_1 \neq \overline{\omega_2}\,, \quad k_1 = \overline{k_2}\,, \\ \omega_1 = \overline{\omega_2}\,, \quad k_1 = \overline{k_2}\,, \quad \mathrm{d}\omega_1/\,\mathrm{d}k \neq \overline{\mathrm{d}\omega_2/\,\mathrm{d}k}\,. \end{gathered} \tag{3.110}$$

Recall that in the general case of real frequencies and wavenumbers, such that the frequencies are different as well as the wavenumbers, the wave interaction energy flux averaged by the time-interval, $2\pi/|\omega_1 - \omega_2|$, or the space-interval, $2\pi/|k_1 - k_2|$, is also zero. If $\omega_1 \to \omega_2$ (or $k_1 \to k_2$), the corresponding interval tends to infinity, however, as follows from the above-shown orthogonality conditions, in this case the interaction energy vanishes.

Conclusion 3.1 Thus, from the long-term estimation point of view, the additivity of energy fluxes in *different* exponential complex waves is valid.If the waves are different by both the frequency and wavenumber, the period (or an approximate *effective period*) of the wave interaction oscillations can be large, especially if the waves differ not too much. In special cases, where frequencies *or* wavenumbers coincide, or frequencies *and* wavenumbers coincide but the group speeds are different, a generalized type orthogonality of the wave modes take place. In this latter case, the wave interaction energy flux is zero independently of other wave parameters. In particular, this conclusion is true for the case of sinusoidal waves, $\Im\omega = \Im k = 0$.

3.4.9 Floquet Wave

Floquet or Floquet-Bloch wave, Floquet (1883), Bloch (1928), is a modified sinusoidal wave propagating in a waveguide of a periodic structure. Consider a general one-dimensional periodic structure and choose reference coordinates $x = am$, $m = 0, \pm 1, \ldots$, where a is the period. Consider two fundamental linear problems for a cell, say, $0 < x < a$, under boundary conditions

$$u_I(0,t) = U_l e^{i\omega t}\,, \quad u_I(a,t) = 0 \quad \text{(problem I)} \tag{3.111}$$

and

$$u_{II}(0,t) = 0\,, \quad u_{II}(a,t) = U_r e^{i\omega t} \quad \text{(problem II)}\,, \tag{3.112}$$

where $u_{I,II}(x,t)$ are the (generalized) displacements for the above problems, respectively

$$u_I(x,t) = U_1(x)U_l e^{i\omega t}\,, \quad u_{II}(x,t) = U_2(x)U_r e^{i\omega t}\,. \tag{3.113}$$

Let $q_{I,II}(0,t)$ and $q_{I,II}(a,t)$ are the frequency-dependent tractions acting on the left, $x = 0$, and on the right, $x = a$, ends of the waveguide section, respectivelyto the above problems

$$\begin{aligned} q_I(0,t) &= Q_{ll}U_l e^{i\omega t}\,, \quad q_I(a,t) = Q_{lr}U_l e^{i\omega t}\,, \\ q_{II}(0,t) &= Q_{rl}U_r e^{i\omega t}\,, \quad q_{II}(a,t) = Q_{rr}(a)U_r e^{i\omega t}\,. \end{aligned} \tag{3.114}$$

Note that *Betti's reciprocity theorem* yields

$$Q_{rl} = Q_{lr}\,. \tag{3.115}$$

Similar relation are valid for any section, $am < x < a(m+1)$, $m = 0, \pm 1, \ldots$.

The condition of the internal force continuity at $x = am$ is

$$Q_{ll}U_m + Q_{rl}U_{m+1} = -[Q_{rr}U_m + Q_{lr}U_{m-1}] \tag{3.116}$$

or

$$Q_{lr}[u(a(m+1),t) + u(a(m-1),t)] + (Q_{ll} + Q_{rr})u(am,t) = 0\,, \tag{3.117}$$

where $u(x,t)$ is the displacement amplitude. Note that in a stable structure

$$Q_{ll} > 0\,, \quad Q_{rr} > 0\,. \tag{3.118}$$

Equation (3.117) is satisfied by a wave

$$u(x,t) = \mathrm{e}^{\mathrm{i}(\omega t - kx)}, \quad x = am, \tag{3.119}$$

with the dispersion relation

$$2Q_{rl} \cos ka + (Q_{ll} + Q_{rr}) = 0. \tag{3.120}$$

Referring to (3.113) we came to the Floquet wave

$$u(x,t) = f(\xi)\mathrm{e}^{\mathrm{i}(\omega t - kx)}, \quad f(\xi) = \left[U_1(\xi) + U_2(\xi)\mathrm{e}^{-\mathrm{i}ka}\right]\mathrm{e}^{\mathrm{i}k\xi}, \quad \xi/a = x/a - [x/a], \tag{3.121}$$

where $[x/a]$ is the integer part of x/a, with the dispersion relation (3.120).

Indeed, the solution inside the cell is valid by the definition, whereas the continuity conditions with respect to the displacements and forces at the interfaces, $x = am$, are satisfied in accordance with the expressions in (3.121) and (3.120), respectively. Thus the propagating waves exist in those ranges of ω where

$$-1 \le \frac{Q_{ll} + Q_{rr}}{2Q_{rl}} \le 1. \tag{3.122}$$

Comment 3.3 In a homogeneous waveguide, the complex exponential wave possesses the translation symmetry, such that an arbitrary shift in the coordinate, $x \longrightarrow x + l$ results in the corresponding change of the wave phase, $-kl$, for any l. In contrast, in a a-periodic waveguide, the Floquet wave possesses such symmetry only for a discrete set of the shift: $l = an, n = 0, \pm 1, \ldots$. So the latter wave can be considered as the complex exponential wave only for this discrete set of the waveguide cross-sections, which defined by the period, a, and the 'origin' as an arbitrary chosen coordinate of a single cross-section.

Further, since for the above-discussed set

$$\mathrm{e}^{\mathrm{i}[\omega t - (k + 2\pi/a)x]} = \mathrm{e}^{\mathrm{i}[\omega t - kx]}, \tag{3.123}$$

the dispersion relations for the Floquet waves are $(2\pi/a)$-periodic, and along with functions $U_{1,2}(x)$ (3.113) a finite segment, say, $-\pi/a, \pi/a$, is sufficient for the description of the wave.

All this makes evident the similarity between the Floquet waves and waves in periodic discrete systems.

3.4.10 Some Simple Examples

For a regular mass-spring chain

$$U_I(\xi) = 1 - \frac{\xi}{a}, \quad U_{II}(\xi) = \frac{\xi}{a}, \quad Q_{rr} = Q_{ll} = \mu - m_0\omega^2/2, \quad Q_{rl} = -\mu \tag{3.124}$$

and the dispersion relation is

$$\omega^2 = 2\omega_0^2(1 - \cos ka)\,, \quad \omega_0^2 = \frac{\mu}{m_0}\,. \tag{3.125}$$

In accordance with (3.121), the wave can be represented as

$$u = Cf(\xi)\mathrm{e}^{\mathrm{i}(\omega t - kx)}\,, \quad f(\xi) = \left[1 + \frac{\xi}{a}\left(\mathrm{e}^{-\mathrm{i}ka} - 1\right)\right]\mathrm{e}^{\mathrm{i}k\xi}\,, \quad \xi/a = x/a - [x/a]\,. \tag{3.126}$$

For an alternating masses chain, considered by Martinsson and Movchan (2003),

$$\begin{aligned} Q_{rr} = Q_{ll} &= \frac{\mu^2 - \mu(m_1 + m_2)\omega^2 + m_1 m_2 \omega^4/2}{2\mu - m_2\omega^2}\,, \\ Q_{rl} &= -\frac{\mu^2}{2\mu - m_2\omega^2}\,. \end{aligned} \tag{3.127}$$

With refer to (3.120) the dispersion relation is

$$m_1 m_2 \omega^4 - 2\mu(m_1 + m_2)\omega^2 + 2\mu^2(1 - \cos ka) = 0\,. \tag{3.128}$$

It follows that there are two wave modes, acoustic and optical

$$\omega^2 = \frac{\mu}{m_1 m_2}\left(m_1 + m_2 \mp \sqrt{m_1^2 + m_2^2 + 2m_1 m_2 \cos ka}\right)\,. \tag{3.129}$$

In the case of an asymmetrical cell, $\mu_1 \neq \mu_2$, the following relations are valid

$$\begin{aligned} U_I(0) &= \frac{2(\mu_1 + \mu_2 - m_2\omega^2)Q_{ll}}{(2\mu_1 - m_1\omega^2)(\mu_1 + \mu_2 - m_2\omega^2) - \mu_1^2}\,, \\ U_{II}(a) &= \frac{2(\mu_1 + \mu_2 - m_2\omega^2)Q_{rr}}{(2\mu_2 - m_1\omega^2)(\mu_1 + \mu_2 - m_2\omega^2) - \mu_2^2}\,, \\ m_1 m_2 \omega^4 - (\mu_1 &+ \mu_2)(m_1 + m_2)\omega^2 + 2\mu_1\mu_2(1 - \cos ka) = 0\,. \end{aligned} \tag{3.130}$$

For an elastic bar with concentrated masses, m_0, attached periodically at $x = am$, $m = 0, \pm 1, \ldots$, it can be found that

$$Q_{rr} = Q_{ll} = ES\frac{\omega}{c}\cot\frac{a\omega}{c} - \frac{m_0\omega^2}{2}\,, \quad Q_{rl} = -ES\frac{\omega}{c}\frac{1}{\sin a\omega/c}\,, \tag{3.131}$$

where ES is the bar stiffness and c is the wave speed in the uniform bar without the concentrated masses. Hence the dispersion relation is

$$\frac{m_0\omega^2}{2} - ES\frac{\omega}{c}\frac{\cos a\omega/c - \cos ka}{\sin a\omega/c} = 0\,. \tag{3.132}$$

It is interesting to retrace the dispersion picture evolution with the change of the mass value. Waves in periodic systems are considered in more detail in Sect. 18.7.7.

Lastly, consider an elastic beam, Eq. (3.66) with no tensile force, $T = 0$, simply supported at $x = 0, \pm a, \ldots$. For the considered wave the homogeneous equation becomes

$$\frac{\mathrm{d}^4 W(X)}{\mathrm{d}X^4} - \lambda^4 W(X) = 0 \tag{3.133}$$

with

$$X = \frac{x}{a}\,, \quad W(X) = w(X)\mathrm{e}^{-\mathrm{i}\omega t}\,, \quad \lambda = \left(\frac{a^4\omega^2\varrho S}{EI}\right)^{1/4}\,. \tag{3.134}$$

The dispersion relation follows as

$$\begin{aligned}\cos K &= -\frac{Q_{ll} + Q_{rr}}{2Q_{lr}} = \frac{\cosh\lambda\sin\lambda - \sinh\lambda\cos\lambda}{\sinh\lambda - \sinh\lambda} \\ &\sim \sin\lambda - \cos\lambda \quad (\lambda \to \infty)\,, \quad K = ak\,.\end{aligned} \tag{3.135}$$

Due to presence of fast growing hyperbolic functions in the 'exact' relation, it is, in fact, very close to the above-presented asymptotic even for the lower, minimal-frequency branch of the dispersion relation. Using the asymptotic relation we can express it in an explicit form with respect to the frequency, namely

$$\begin{aligned}\omega(K) &= \pm\lambda^2\sqrt{\frac{EI}{a^4\varrho S}}\,, \\ \lambda &= 2\pi n - \frac{\pi}{4} \pm \arccos\left(-\frac{\cos K}{\sqrt{2}}\right)\,, \quad 0 \le \arccos(..) < \pi\,.\end{aligned} \tag{3.136}$$

First four dispersion curves are plotted in Fig. 3.136.

3.4.11 Kinetic and Potential Energies in Periodic Waves

Consider a periodic steady-state wave, where the displacement depends on $\eta = x - vt$ and possibly on cross-section coordinates. In this case,

$$\dot{\boldsymbol{u}} = \frac{\partial \boldsymbol{u}}{\partial t} = -v\boldsymbol{u}' = -v\frac{\partial \boldsymbol{u}}{\partial x} \tag{3.137}$$

and the kinetic energy density is

$$\mathcal{K}_0(\boldsymbol{u}, \dot{\boldsymbol{u}}) = \mathcal{P}_k(\boldsymbol{u}, \boldsymbol{u}') \,. \tag{3.138}$$

In this section, we consider waves where the kinetic and potential energies, or the energies averaged over the period, are equal to each other. The below statements are based on Hamilton's principle of stationary action.

3.4.11.1 One-Dimensional Linear Wave Equation

First consider non-dispersive waves which can be, in general, described by displacement $u(\eta)$ with $\eta = x - ct$, where x is the propagation direction, c is a constant wave speed, and an arbitrary dependence on η is assumed. This concerns one-dimensional waves in a string (3.45) or a rod, and plane waves in a homogeneous elastic body, Chap. 7.

The key point is that $u = u(\eta)$. It follows that $\dot{u}(\eta) = -cu'(\eta)$ and

$$\mathcal{K} = \frac{1}{2}\varrho c^2 (u')^2 \,, \quad \Pi = \frac{1}{2} T (u')^2 \,. \tag{3.139}$$

From this Hamilton's principle yields

$$\frac{\mathrm{d}(\mathcal{K} - \Pi)}{\mathrm{d}u'} = 0 \tag{3.140}$$

and referring to (3.139) we obtain

$$c^2 = T/\varrho \quad \text{and} \quad \mathcal{K} = \Pi \,. \tag{3.141}$$

Thus, in the wave obeying the wave equation and propagating in one direction, the kinetic and potential energies are equal to each other. Note that this statement does not hold if the superposition of waves propagating in opposite directions is considered.

3.4.11.2 Arbitrary Shaped String Under Uniform Tensile Force

The above conclusion remains valid for the wave in an arbitrary shaped string under a constant tensile force, $F = \varrho D^2$, Sect. 10.2. Recall that, in this nonlinear problem, the position vector $\boldsymbol{R} = \boldsymbol{R}_0(S - Dt) + v_0 t\boldsymbol{k}$, where S is the coordinate placed along the stretched string and $\boldsymbol{k}$ is a constant unit vector. To obtain a solitary transversal wave we now choose $v_0 = D, \boldsymbol{k} = \boldsymbol{k}_x$ and take $\boldsymbol{R}' = \boldsymbol{k}_x$ outside the moving support which is assumed to be finite. In addition, using the current notation, denote $D = c_T$, $F = T$. Thus we have a wave propagating with speed c_T to the right along the unmoving string placed on the x-axis outside of the wave support.

The string is under a constant tensile force, and hence the strain energy measured from the 'initial' state, where the straight string is stretched, is equal to zero. However, the curved string possesses potential energy which is equal to the work required for the creation of the curvilinear part of the string which is under tension. Thus, the kinetic and potential energies per unite length of the string are

$$\begin{gathered}\mathcal{K} = \frac{\varrho}{2}(-c_T \boldsymbol{R}_0' + c_T \boldsymbol{k}_x)^2 = \varrho c_T^2 (1 - \boldsymbol{R}_0' \boldsymbol{k}_x) = \varrho c_T^2 \left(1 - \frac{\mathrm{d}x}{\mathrm{d}S}\right), \\ \Pi = T\frac{\mathrm{d}S - \mathrm{d}x}{\mathrm{d}S} = T\left(1 - \frac{\mathrm{d}x}{\mathrm{d}S}\right).\end{gathered} \tag{3.142}$$

Note that

$$\frac{\mathrm{d}x}{\mathrm{d}S} = \pm\left(1 + (w'(x))^2\right)^{-1/2}, \tag{3.143}$$

where w is the transversal coordinate of the string. Since the functions in (3.142) may differ only by constant coefficients, it now follows from the Hamilton's principle that $c_T^2 = T/\varrho$ and $\mathcal{K} = \Pi$.

3.4.12 Sinusoidal Waves in a Homogeneous Waveguide

Now consider linear sinusoidal waves. In the partial case of small amplitude sea waves, the equality between the potential and kinetic energies averaged over the period was mentioned by Lighthill (1978). We show below that the equality is valid for a general case of a linear sinusoidal wave propagating in a perfect waveguide.

Consider a complex wave

$$\boldsymbol{u}(y, z)\mathrm{e}^{\mathrm{i}(\omega t - kx)} \tag{3.144}$$

propagating along an x-uniform waveguide with the phase velocity $c = \omega/k$. Recall that the energy density calculated based on this complex wave is the doubled energy of the corresponding sinusoidal wave averaged over the period as in (3.95). The waveguide can be of a finite or infinite cross-section area. The latter case corresponds to surface or interface waves which amplitude decreases fast enough as $y^2 + z^2 \to \infty$. In the case of a finite cross-section, zero boundary conditions are assumed, the conditions which restrict any energy flux to the wave. Components of vector $\boldsymbol{u}(y, z)$ may be not only projections of it on the coordinate axes but also reflect different degrees of freedom. For example, in the case of a beam with a layer of uniformly distributed oscillators, there are two (or possibly four) components: the beam transversal displacements (and possibly rotations) and the oscillators distributed mass transversal displacements (and possibly longitudinal displacements).

The key point is that for such waves the kinetic and potential energies averaged over the wave length (or, what is the same, over the time-period) are quadratic functions of the displacements with coefficients depending on ω and k, respectively. Hamilton's principle leads to a homogeneous system of equations the main determinant of which represents the dispersion relation for the wave. The system looks as

$$\frac{\partial \mathcal{K}}{\partial u_n} - \frac{\partial \Pi}{\partial u_n} = 0\,, \quad n = 1, 2, \ldots, N\,, \tag{3.145}$$

where N is the number of the degrees of freedom. By Euler's theorem on homogeneous functions it directly follows from this that

$$\mathcal{K} = \frac{1}{2}\sum_n u_n \frac{\partial \mathcal{K}}{\partial u_n}\,, \quad \Pi = \frac{1}{2}\sum_n u_n \frac{\partial \Pi}{\partial u_n} \quad \Longrightarrow \quad \mathcal{K} = \Pi\,. \tag{3.146}$$

Note that the conclusion is valid for wave (3.144) propagating in one direction.

3.4.13 *Floquet Linear Waves in a Periodic Waveguide*

Consider Floquet wave in a one-dimensional waveguide of a periodic structure. Let $x = am, m = 0, \pm 1, \ldots$, where a is the period, be the set of reference points. As discussed in Sect. 3.4.9 there exists a complex wave with respect to these points

$$u_m(t) = u\mathrm{e}^{\mathrm{i}(\omega t - kx)}\,, \quad x = am\,. \tag{3.147}$$

Dynamics of each part of the given waveguide between the neighboring reference points is uniquely defined by the amplitude, u, frequency, ω, and the wavenumber k. Due to the linearity the kinetic and potential energies, $\mathcal{K}_a$ and Π_a, of the wave in the period, $an \le x < a(n+1)$, are still quadratic functions of u as in the previous case of a uniform waveguide, however, now each of the coefficients may depend on both the frequency and the wavenumber. The following considerations are similar to those used for the uniform waveguide: Eqs. (3.145) and (3.146) with $N = 1$ and with respect to $\mathcal{K}_a$ and Π_a instead of $\mathcal{K}$ and π, respectively. Thus, this case differs by the averaging: now the kinetic and potential energies averaged over both the waveguide period, a, and the wave period, $2\pi/\omega$, are equal to each other. (In fact, since sinusoidal waves are considered, while the energies are quadratic functions, the averaging over a half-period of the wave can be assumed.)

Part II
Wave Orthogonality and Localization

Chapter 4
Generalized Orthogonality of Waves

There exist a similar to (3.105) orthogonality relation based on *Betti's reciprocity theorem* (1872) and the translation and reflection symmetries.

4.1 Betti's Theorem and Orthogonality of Oscillations

4.1.1 Betti's Theorem

The well-known Betti's theorem is based on the linearity and the existence of a potential, that is a functional as

$$W = \int_\Omega \int_0^{\boldsymbol{u}} \boldsymbol{P} \boldsymbol{u} \mathrm{d}\boldsymbol{u} \mathrm{d}\Omega \,, \tag{4.1}$$

that depends only on the final vector field $\boldsymbol{u}$ in domain Ω but is independent of the integration path. Displacements $\boldsymbol{u}$ are induced by quasi-statically applied external forces $\boldsymbol{P}$, distributed or/and concentrated, which are assumed to be self-equilibrated. Consider two different states, $\boldsymbol{u}_1$ and $\boldsymbol{u}_2$, and the corresponding forces, $\boldsymbol{P}_1$ and $\boldsymbol{P}_2$. The final state, $\boldsymbol{P}_1 + \boldsymbol{P}_2$, can be achieved, for instance, by the application first $\boldsymbol{P}_1$ and then $\boldsymbol{P}_2$, or vice versa. In these two ways of the loading

$$\begin{aligned} W &= \int_\Omega \left[\int_0^{\boldsymbol{u}_1} \boldsymbol{P} \mathrm{d}\boldsymbol{u} + \left(\int_0^{\boldsymbol{u}_2} \boldsymbol{P} \mathrm{d}\boldsymbol{u} + \boldsymbol{P}_1 \boldsymbol{u}_2 \right) \right] \mathrm{d}\Omega \\ &= \int_\Omega \left[\int_0^{\boldsymbol{u}_2} \boldsymbol{P} \mathrm{d}\boldsymbol{u} + \left(\int_0^{\boldsymbol{u}_1} \boldsymbol{P} \mathrm{d}\boldsymbol{u} + \boldsymbol{P}_2 \boldsymbol{u}_1 \right) \right] \mathrm{d}\Omega \,, \end{aligned} \tag{4.2}$$

where the order of the terms corresponds to the loading order (the expressions in parentheses correspond to the second step), and the latter equality is a consequence of the potentiality of the considered system. The integrated terms reflect the fact that

L. I. Slepyan, *Mechanical Waves: Free, Forced, and Transition*,
https://doi.org/10.1007/978-3-032-15864-2_4

the forces, applied in the first step of the loading, work on the displacements of the second step. Note that all these values may be complex. Betti's theorem immediately follows from (4.2). It states that *The work of the forces of the first state on the displacements of the second state is equal to the work of the forces of the second state on the displacements of the first state*

$$\int_{\Omega} \boldsymbol{P}_1 \boldsymbol{u}_2 \mathrm{d}\Omega = \int_{\Omega} \boldsymbol{P}_2 \boldsymbol{u}_1 \mathrm{d}\Omega \,. \tag{4.3}$$

Comment 4.1 Vector functions $\boldsymbol{u}$ and $\boldsymbol{P}$ are called here the displacements and forces, respectively; these functions, however, can have different nature. The only conditions must be satisfied: the existence of the potential formed by these linearly dependent variables. Also there is a certain freedom in the definition of the potential for an elastic body. For example, in the case of monochromatic waves or oscillations, the inertia forces, $-\varrho\omega^2\boldsymbol{u}$, can be referred to as the internal forces, as the manifestation of a distributed negative Winkler elastic foundation (in this case this manifests itself in the force-displacement dependence), or this can be treated as external forces.

4.1.2 *Orthogonality of Free Oscillations*

A straightforward consequence of Betti's theorem is that on the orthogonality of free oscillation forms. Consider two modes, $\boldsymbol{u}_{1,2}$, with different frequencies, $\omega_1^2 \neq \omega_2^2$, keeping the latter treatment. Betti's theorem says

$$-\int_{\Omega} \boldsymbol{u}_1 \boldsymbol{u}_2 \varrho \omega_2^2 \mathrm{d}\Omega = -\int_{\Omega} \boldsymbol{u}_2 \boldsymbol{u}_1 \varrho \omega_1^2 \mathrm{d}\Omega \,; \tag{4.4}$$

hence the orthogonality condition is valid as

$$\int_{\Omega} \varrho \boldsymbol{u}_1 \boldsymbol{u}_2 \mathrm{d}\Omega = 0 \,. \tag{4.5}$$

4.1.3 *Orthogonality of Wave Modes*

A similar orthogonality condition is obtained in another way, using Betti's theorem and the translation-rotation symmetries. Consider a uniform waveguide as a cylinder or a circular torus of arbitrary cross-section, which material possesses the potential in the above-considered sense. The longitudinal coordinate, x, is the cylinder axis or the polar angle in the case of the torus.

It is assumed that there exist the translation and reflection symmetries, and identical complex waves can propagate in the waveguide in both directions.

We now consider the inertia forces as the internal forces. Three waves are introduced

$$\{\boldsymbol{u}_1 = (u'_x, \boldsymbol{u}'_t)e^{k_1 x}\,,\ \ (\sigma'_x, \boldsymbol{\sigma}'_t)e^{k_1 x}\},\ \ \{\boldsymbol{u}_2 = (u''_x, \boldsymbol{u}''_t)e^{k_2 x}\,,\ \ (\sigma''_x, \boldsymbol{\sigma}''_t)e^{k_2 x}\}\,,$$
$$\{\boldsymbol{u}_3 = (-u''_x, \boldsymbol{u}''_t)e^{-k_2 x}\,,\ \ (\sigma''_x, -\boldsymbol{\sigma}''_t)e^{-k_2 x}\}\ \ (k_1^2 \neq k_2^2)\,, \tag{4.6}$$

where the second and the third waves are identical, but propagate in opposite directions. Vectors $\boldsymbol{u}_t$, $\boldsymbol{\sigma}_t$ are the transversal components; the reflection symmetry is taken into account (see (3.56)).

Select a portion of the waveguide, $a < x < b$. External forces act only on the end cross-section, and Betti's theorem written with respect to the first-second and the first-third states yields

$$M_+ \int_\Omega (u'_x\sigma''_x + \boldsymbol{u}'_t\boldsymbol{\sigma}''_t - u''_x\sigma'_x - \boldsymbol{u}''_t\boldsymbol{\sigma}'_t)\mathrm{d}\Omega = 0\,,$$
$$M_- \int_\Omega (u'_x\sigma''_x - \boldsymbol{u}'_t\boldsymbol{\sigma}''_t + u''_x\sigma'_x - \boldsymbol{u}''_t\boldsymbol{\sigma}'_t)\mathrm{d}\Omega = 0\,, \tag{4.7}$$
$$M_\pm = \exp[(k_1 \pm k_2)b] - \exp[(k_1 \pm k_2)a]\,.$$

In these relations, the coordinates, a, b, are arbitrary numbers, and $M_\pm \neq 0$ since $k_1^2 \neq k_2^2$. The orthogonality relation (Saraikin and Slepyan, 1979) follows as

$$\int_\Omega (u'_x\sigma''_x - \boldsymbol{u}''_t\boldsymbol{\sigma}'_t)\mathrm{d}\Omega = 0\,. \tag{4.8}$$

Earlier this relation was derived for a uniform elastic layer with free surfaces by means of the analysis of the plane problem of elasticity (Zilbergleit and Nuller, 1977).

It can be seen that if the potential is defined as a real functional for the complex exponential waves, that is, if

$$W = \int_\Omega \Re\left[\int_0^u \overline{\boldsymbol{P}}\boldsymbol{u}\mathrm{d}\boldsymbol{u}\right]\mathrm{d}\Omega\,, \tag{4.9}$$

while the wave interaction energy flux (3.104) is also considered for the backward waves, $q = q_1$ and $q = -q_2$, then for $\omega_1 = \overline{\omega_2}$ the result (4.8) following from Betti's theorem coincides with that in (3.105).

4.1.4 Representation of a Complex Wave in Orthogonal Axes

Consider a complex wave

$$u = \exp(\mathrm{i}(\omega t - \mathbf{kr}))\,, \tag{4.10}$$

where **r** is an arbitrary oriented axis, and **k** is the relater wavevector. In the orthogonal axes, k_x, k_y, k_z, the wave looks as

$$u = \exp(\mathrm{i}(\omega t - k_x x - k_y y - k_z z))\,. \tag{4.11}$$

The wave orientation in the coordinates is defined by the **k** projections. In particular, in the 2D case ($k_z = 0$),

$$x = l\cos\theta\,,\ k_x = k\cos\theta\,,\quad y = l\sin\theta\,,\ k_y = k\sin\theta\,, \tag{4.12}$$

where θ is the angle between the wave propagation direction and the x-axis. The wave orientation follows as

$$\theta = \arctan\frac{k_y}{k_x}\,,\quad \sin\theta = \frac{vk_y}{\omega} = \frac{k_y}{k}\,,\quad \cos\theta = \frac{vk_x}{\omega} = \frac{k_x}{k}\,, \\ v = \frac{\omega}{k}\,,\quad k = \sqrt{k_x^2 + k_y^2}\,, \tag{4.13}$$

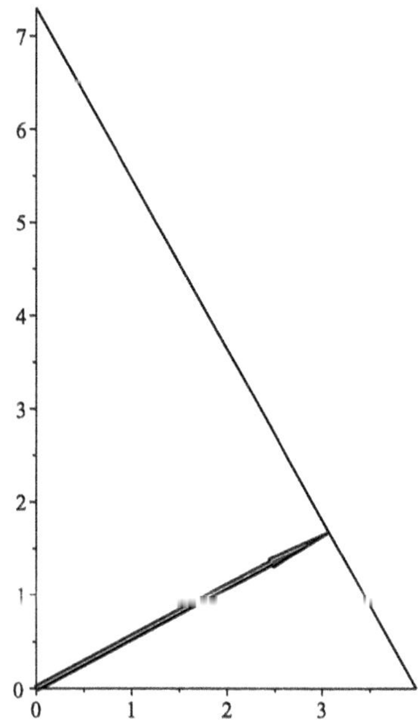

Fig. 4.1 Wave in 2D space. The arrow shows the wave propagation direction; its length corresponds to the wave speed, v. The thin line is a wave crest; the distances from the coordinate origin to its cross-sections with the coordinate axes define the wave phase speeds along the axes, v_x, v_y, Eq. (4.14). Here $v = 3.5$, $v_x = 3.99$, $v_y = 7,30$ ($\theta = 0.5$)

The wave speeds along the axes follow from Eq. (4.11) ($k_z = 0$)

$$v_x = \frac{\omega}{k_x} = \frac{v}{\cos\theta} \quad (y = 0)\,, \quad v_y = \frac{\omega}{k_y} = \frac{v}{\sin\theta} \quad (x = 0)\,. \tag{4.14}$$

So, these speeds are not projections of v on the axes, but exceed it. Graphically this fact is illustrated in Fig. 4.1.

Chapter 5
Wave Localization

We consider infinite media with lower-speed, nonzero-wide waveguides and wave localization in such structures. General considerations and studying of particular problems for a membrane, sound channel in fluid, square-cell lattice, and a bending plate are presented. Localized waves are found to consist of two parts. One propagates in the waveguide. It has a sinusoidal transverse mode (that increases the speed), and another exponentially localizes with the distance from the waveguide (that decreases the speed—to be the same). The localized exponents as functions of waveguide-to-outside owned speed ratio (we use the density ratio) are determined. Multiple localizations under other fixed conditions are found. This new phenomenon, valid for the nonzero-wide waveguide only, manifests itself in all the problems considered, and the corresponding localized-exponent functions are presented in the figures. All results are obtained using an elementary technique.

The first analytical description of a localized wave was that the water gravity wave exponentially decreased with the depth of Airy (1841), and Stokes (1847). The following well-known wave is the Rayleigh wave, also exponentially decreasing with the distance from the surface, Lord Rayleigh (1885).

The in-plane-localized wave has long been observed and discussed, beginning with water surface waves, Garipov et al. (1965), and Babich et al. (1979). The manifestation of this phenomenon in seismic waves was considered in Adams et al. (2007), Fu et al. (2013), Larose et al. (2004). The papers by Munk (1974) and Barros and Gendron (2019) are devoted to localized sound waves. The localized wave was also observed in fracture, Mishuris et al. (2009).

The steady-state waves considered localize at a waveguide of a lower phase speed. The universality of that fact was noted in Mishuris et al. (2020), where a general approach is discussed along with the analytical study of specific problems for continuous and discrete models.

Two kinds of localized waves can be distinguished. In one of them, the waveguides, boundaries, or lower-wave-speed singular lines lead to decreased wave speeds in all the body areas outside the singularities. In turn, the wave-speed decrease occurs

L. I. Slepyan, *Mechanical Waves: Free, Forced, and Transition*,
https://doi.org/10.1007/978-3-032-15864-2_5

together with and because of the exponential localization, where the localization exponent is an imaginary part of the wavenumber.

Another more complicated wave is formed in the below-considered structure with a nonzero-width strip instead of a line. In this case, the wave also exponentially localizes outside the strip but has a sinusoidal dependence within the strip. Such a dependence leads to multiple wave modes.

We consider plane-localized steady-state complex waves propagating in a plane with a waveguide strip. The x, r wave with a cylindrical waveguide is also presented. The main defining equality is

$$\omega^2(k) = \omega^2 = \text{const} > 0\,, \tag{5.1}$$

where $\omega(k)$ is the dispersion relation, and k is the wavenumber. Since the frequency is fixed, what we need is only a cross-section of the dispersion relation figure. The above equality serves as a condition the wavenumber components must satisfy independently of whether they all are real or some of them are complex.

5.1 General Considerations

5.1.1 Preliminaries

For a complex wave in a perfect isotropic media, the dispersion relation and one of the possible inverse equalities are

$$\omega^2 = \omega^2(k^2) \quad k^2(\omega^2) > 0\,. \tag{5.2}$$

We assume $\omega^2 > 0$ and consider the inverse relation. It follows that the wave phase velocity (squared)

$$v^2 = \frac{\omega^2}{k^2(\omega^2)}\,. \tag{5.3}$$

We also assume the frequency given

$$\omega^2 = \text{const} > 0 \implies k^2 = \text{const} > 0\,, \quad v^2 = \text{const} > 0\,. \tag{5.4}$$

Now consider the above discussed wave expressed as propagating in 2D space

$$u = \exp(\mathrm{i}(\omega t - k_x x - k_y y)\,, \tag{5.5}$$

where x, y are the orthogonal coordinates, and k_x, k_y are the wavenumbers. Due to the uniformity

$$k^2 = k_x^2 + k_y^2 \,. \tag{5.6}$$

One of the wavenumbers (wavenumber components) can be imaginary, say $k_y^2 < 0$. The equality (5.6) only is essential. Indeed, the dispersion relation is independent of k_x, k_y separately but only on k^2.

If both wavenumbers are real the wave coordinates and the wavenumbers are

$$x = vt\cos\theta \,,\ k_x = k\cos\theta \,, \quad y = vt\sin\theta \,,\ k_y = k\sin\theta \,, \tag{5.7}$$

where θ is the angle between the wave propagation direction and the x-axis. The wave orientation follows as

$$\theta = \arctan\frac{k_y}{k_x} \,, \quad \sin\theta = \frac{vk_y}{\omega} = \frac{k_y}{k} \,, \quad \cos\theta = \frac{vk_x}{\omega} = \frac{k_x}{k} \,, \tag{5.8}$$

and phase speeds along the axes, as it follows from Eq. (5.5), are

$$v_x = \frac{\omega}{k_x} = \frac{v}{\cos\theta} \ (y = 0) \,, \quad v_y = \frac{\omega}{k_y} = \frac{v}{\sin\theta} \ (x = 0) \,. \tag{5.9}$$

So, the wavenumber is a vector, but not the phase speeds. They are not projections of v on the axes but exceed them, (5.9), Fig. 5.1.

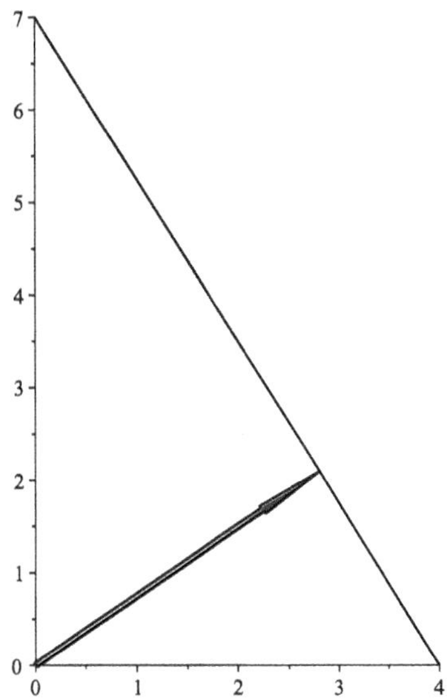

Fig. 5.1 Wave in 2D space. The arrow shows the wave propagation direction; its length corresponds to the wave speed, v. The thin line is a wave crest; the distances from the coordinate origin to its intersections with the coordinate axes define the wave phase speeds along the axes, v_x, v_y, Eq. (5.9). Here $v = 3.5$, $v_x = 3.99$, $v_y = 7, 30$ $(\theta = 0.5)$

Returning to Eq. (5.6) we have

$$k_x^2 = k^2 - k_y^2\,, \quad v_x = \frac{\omega}{\sqrt{k^2 - k_y^2}}\,, \quad k_y^2 < k^2\,, \tag{5.10}$$

and

$$k_x < k\,,\ v_x > v\ (k_y^2 > 0)\,, \quad k_x > k\,,\ v_x < v\ (k_y^2 < 0)\,, \tag{5.11}$$

where the latter, ($k_y^2 < 0$), corresponds to the wave localization, while the former ($k_y^2 > 0$)—to the wave propagating at an angle to the x-axis, Eqs. (5.8) and (5.9).

The single fundamental equality (5.6) does not define both wavenumber components. So, there is some freedom to vary them while building a combined wave propagating in a composite structure.

5.1.2 *The Wave in the Composite Structure*

Consider a structure consisting of a *waveguide*, the strip $-\infty < x < \infty, -a < y < a$, and the outside part of the unbounded plane, $a < |y|$, both uniform and isotropic, Fig. 5.2.

Let the owned phase speed, v_1, in the waveguide is below that, v_2, in the outside area

$$0 < v_1 = \omega/k_1 < v_2 = \omega/k_2\,, \tag{5.12}$$

which can be called *the localization condition*. The task is to build a combined given-frequency wave propagating along the x-axis in the waveguide and outside of it with the same phase speed, v and the wavenumber, $k_x = k$, such that

$$v_{1x} \le v \le v_{2x}\,, \quad k_{2x} \le k \le k_{1x}\,. \tag{5.13}$$

This can be achieved by introducing proper-value wavenumber components k_{1y} and k_{2y}. For an y-symmetric wave we take two waves propagating in the waveguide

$$u_1 = \frac{1}{2} A_1 e^{i(\omega t - kx} \left(e^{-ik_{1y}y)} + e^{ik_{1y}y)}\right) = A_1 e^{i(\omega t - kx)} \cos(k_{1y}y), \quad (|y| \le a)\,. \tag{5.14}$$

while the part of the combined wave propagating outside the waveguide

$$u_2 = A_2 e^{i(\omega t - kx - k_{2y}(|y| - a))} \quad (|y| \ge a)\,. \tag{5.15}$$

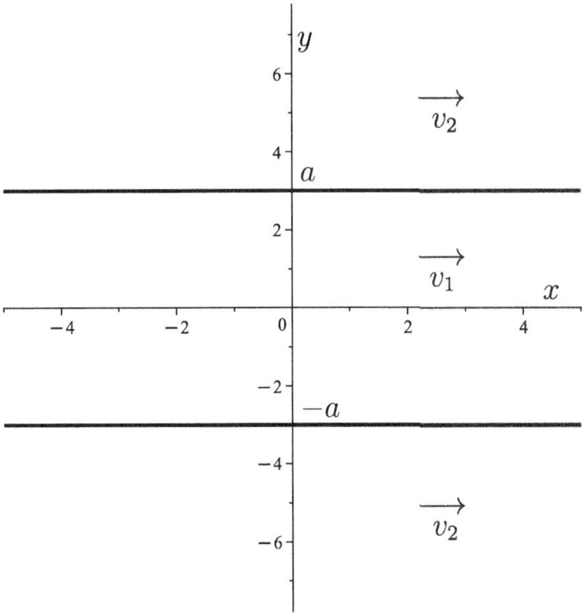

Fig. 5.2 Lower-wave-speed strip (the waveguide), $-\infty < x < \infty$, $-a < y < a$, in a 2D infinite body. The owned phase speeds, v_1, v_2, and the wavenumbers $k_1 = \omega/v_1$, $k_2 = \omega/v_2$, correspond to 1D waves in the strip and outside areas considered separately as unbounded structures

As noted, the k_y-components are introduced to create a combined wave propagating faster than the wave in the waveguide and/or slower than outside it. In other words, we have to decrease the wavenumber in the waveguide and increase it outside. The latter can only be achieved by $k_{2y}^2 < 0$ (while $k_{1y}^2 > 0$). So, we represent u_2 as

$$u_2 = A_2 \mathrm{e}^{\mathrm{i}(\omega t - kx) - \alpha(|y| - a)}, \quad k_{2y} = -\mathrm{i}\alpha\,, \quad \alpha > 0 \quad (|y| \geq a)\,. \tag{5.16}$$

There are four unknowns there, k, k_{1y}, α and one of the wave amplitudes, A_2 or A_1 (the other can be taken arbitrary). Generally we have two relations

$$k^2 + k_{1y}^2 = k_1^2 \quad \left(k_1^2 = \frac{\omega^2}{v_1^2}\right), \quad k^2 - \alpha^2 = k_2^2 \quad \left(k_2^2 = \frac{\omega^2}{v_2^2}\right). \tag{5.17}$$

In addition, we can use the continuity relations at $|y| = a$. In particular, if the displacement is continuous, we have the third relation

$$A_1 \cos(k_{1y} a) = A_2\,. \tag{5.18}$$

The *localization exponent* α is determined below for some structures based on the continuation of internal forces at the separation lines, $|y| = a$.

5.1.3 On the Nonuniqueness in the Wave Localization

As may be seen below, a number of localized waves can exist in a considered structure with a lower-speed waveguide. Each of the dependencies, of the localization exponent starts at points corresponding to a non-localized wave, $\alpha = 0$, propagating in such a structure in spite of its non-uniformity. These points can be determined in a general case.

Consider the wave (5.14), in the waveguide, $|y| \leq a$. Based on (5.17) for the non-localized wave we have

$$\alpha = 0 \Longrightarrow k^2 = k_2^2 \Longrightarrow k_{1y}^2 = k_1^2 - k_2^2, \quad (k_{1,2} = \omega^2/v_{1,2}^2, \quad v_1 < v_2), \tag{5.19}$$

and the combine wave is

$$\begin{gathered} u = u_1 = A1 \cos\left(y\sqrt{k_1^2 - k_2^2}\right) e^{i(\omega t - k_2 x)} \quad (|y| \leq a), \\ u = u_2 = A_2 e^{i(\omega t - k_2 x)} \quad (|y| \geq a), \quad A_2 = A_1 \cos\left(a\sqrt{k_1^2 - k_2^2}\right). \end{gathered} \tag{5.20}$$

Now, assuming a discontinuity of $\partial u/\partial y$ is not the case, and since in the 1D wave outside the waveguide $\partial u_2/\partial y = 0$ we have

$$\frac{\partial u_1}{\partial y} = 0 \Longrightarrow \sin\left(a\sqrt{k_1^2 - k_2^2}\right) = 0 \quad (|y| = a). \tag{5.21}$$

If there are no other conditions, it follows that a non-localized wave can propagate in the considered nonuniform structure with the waveguides

$$k_1^2 = k_2^2 + \frac{n^2\pi^2}{a^2}, \quad n = 0, 1, 2, \ldots, \tag{5.22}$$

The above relation can also be represented in terms of the phase speed ($k = \omega/v$)

$$\frac{1}{v_1^2} = \frac{1}{v_2^2} + \frac{n^2\pi^2}{\omega^2 a^2}. \tag{5.23}$$

The case $n = 0$ is trivial. It corresponds to the uniform structure, $v_1 = v_2, k_{y1} = 0$. Otherwise, $k_{y1} > 0$ ($n > 0$), and in contrast to the outside area, the wave in the waveguide propagating with the same phase speed is nonuniform. In Fig. 5.4, the starting points coordinates match those defined in (5.21).

5.2 Particular Problems

5.2.1 Localized Wave in a Membrane

Consider a membrane equation under tensile forces $T_x = T_y = T = \text{const}$

$$\varrho\frac{\partial^2 u(x,y,t)}{\partial t^2} - T\nabla^2 u(x,y,t) = 0\,, \quad \nabla^2 = \frac{\partial^2}{\partial x^2} + \frac{\partial^2}{\partial y^2}\,. \tag{5.24}$$

In the wave, (5.14), (5.15),

$$\begin{aligned} \omega^2 = v^2(k^2 + k_{y1}^2)\,, \;\; k_{y1}^2 > 0 \;\; (|y| \le a)\,, \\ k_{y2}^2 = -\alpha^2\,, \;\; \alpha^2 > 0 \;\; (|y| \ge a)\,. \end{aligned} \tag{5.25}$$

The fourth equation (in addition to Eqs. (5.17), (5.18)) is

$$\frac{\partial u_1}{\partial y} = \frac{\partial u_2}{\partial y} \quad (|y| = a)\,, \tag{5.26}$$

that is (also see Eq. (5.18))

$$A_1 k_{y1}\sin(k_{y1}a) = A_2\alpha \quad \Longrightarrow \quad k_{y1}\sin(k_{y1}a) = \alpha\cos(k_{y1}a)\,, \tag{5.27}$$

It follows that

$$v^2(k_x^2 + k_y^2) = \omega^2\,, \quad v^2 = \frac{T}{\varrho}\,, \tag{5.28}$$

where ϱ_1 and ϱ_2 are the densities in the waveguide and outside of it, respectively.

In the above notations, the functions (symmetric by y and equal to each other at $|y| = a$) are

$$\begin{aligned} u = u_1 = A\frac{\cos(k_{y1}y)}{\cos(k_{y1}a)}\mathrm{e}^{\mathrm{i}(\omega t - k_x x)} \quad (|y| \le a)\,, \\ u = u_2 = A\mathrm{e}^{\mathrm{i}(\omega t - k_x x) - \alpha(|y| - a)} \quad (a \le |y|)\,. \end{aligned} \tag{5.29}$$

Since $T_1 = T_2$, the y-derivatives at $y = a$ are also equal to each other

$$\frac{\mathrm{d}u_2}{\mathrm{d}y} = -\alpha A\mathrm{e}^{\mathrm{i}(\omega t - k_x x)} = \frac{\mathrm{d}u_1}{\mathrm{d}y} = -k_{y1}\tan(k_{y1}a)A\mathrm{e}^{\mathrm{i}(\omega t - k_x x)}\,. \tag{5.30}$$

Two equalities follow from Eqs. (5.17) and (5.30) (remind that $k_{y2}^2 = -\alpha^2$)

$$\alpha^2 + k_{y1}^2 = \Lambda\,, \quad \Lambda = k_1^2 - k_2^2 = \frac{\omega^2}{v_1^2} - \frac{\omega^2}{v_2^2} > 0\,, \quad \alpha = k_{y1}\tan(k_{y1}a)\,. \tag{5.31}$$

We have

$$\frac{k_{y1}^2}{\cos^2(k_{y1}a)} = \Lambda\,, \quad 0 < k_{y1}^2 < \Lambda\,, \quad \alpha^2 = \Lambda\sin^2(k_{y1}a) > 0\,. \tag{5.32}$$

5.2.2 *From the Strip to a Singular Line*

Consider a simple example. Let the mass in the narrowing waveguide be fixed, while the growing density, ϱ, still be uniform, $\varrho = (a_0/a)\varrho_1$, where $2a_0$ and ϱ_1 are the 'initial' waveguide width and density. Then

$$\begin{gathered} v_1 = \sqrt{\frac{aT}{a_0\varrho_1}}\,, \quad \Lambda = \frac{\omega^2}{T}\left(\frac{a_0}{a}\varrho_1 - \varrho_2\right) \sim \frac{\omega^2}{T}\frac{a_0}{a}\varrho_1\,, \quad k_{y1} \sim \sqrt{\Lambda}\,, \\ \varrho \sim a_0\varrho_1\delta(y)\,, \quad \alpha \to \frac{\omega^2}{T}a_0\varrho_1 \quad (a \to 0)\,. \end{gathered} \tag{5.33}$$

The localization exponent in its dependence on the waveguide width, $\alpha(a)$, $0 < a/a_0 < 1$, is shown in Fig. 5.3.

The starting points of the localized number in its dependencies on the waveguide density, $\alpha(\varrho_1)$, are defined by the relation (5.23). In terms of the density, since $v^2 = T/\varrho$,

$$\varrho_1 = \varrho_2 + \frac{n^2\pi^2 T}{\omega^2 a^2}\,. \tag{5.34}$$

5.2.3 *Sound Channel in Fluid*

Let there exist a lower-speed (say, increase-salinity) underwater cylindrical channel, $r < a$. From the wave equation

$$\frac{\partial^2\varphi}{\partial r^2} + \frac{1}{r}\frac{\partial\varphi}{\partial r} + \frac{\partial^2\varphi}{\partial x^2} - \frac{1}{v^2}\frac{\partial^2\varphi}{\partial t^2} = 0\,, \tag{5.35}$$

and the wave potential

$$\varphi = R(r)\mathrm{e}^{\mathrm{i}(\omega t - kx)}\,, \tag{5.36}$$

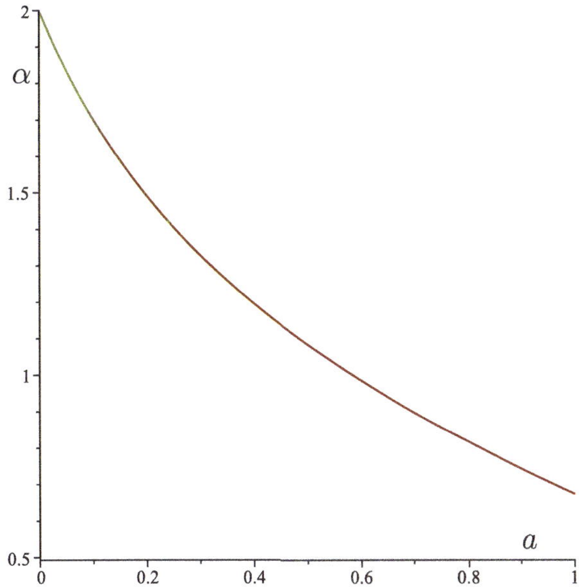

Fig. 5.3 The localized wave in a membrane. The localization exponent, α, as a function of the waveguide width, a, under the fixed mass in the waveguide: $\varrho = \varrho_1 a_0/a$. The structure parameter are $\varrho_1 = 2$, $\varrho_2 = 1$, $\omega = 1$, $T = 1$, $a_0 = 1$

we have

$$\frac{\mathrm{d}^2 R(r)}{\mathrm{d}r^2} + \frac{1}{r}\frac{\mathrm{d}R(r)}{\mathrm{d}r} + \left(\frac{\omega^2}{v^2} - k^2\right) R(r) = 0\,. \tag{5.37}$$

It follows that regular in $0 \le r \le a$

$$R(r) = A_1\,\mathrm{BesselJ}(0, k_1\, r)\,, \quad k_1^2 = \frac{\omega^2}{v_1^2} - k^2 > 0\,, \quad A_1 = \mathrm{const}\,, \tag{5.38}$$

and outside of the waveguide, $a \le r < \infty$, exponentially decreasing for $r \to \infty$,

$$R(r) = A_2\,\mathrm{BesselK}(0, k_2 r)\,, \quad k_2^2 = k^2 - \frac{\omega^2}{v_2^2} > 0\,, \quad A_2 = \mathrm{const}\,, \tag{5.39}$$

where $v_1 < v_2$, $v_{1,2} = \sqrt{B_{1,2}/\varrho_{1,2}}$; B and ϱ are the bulk modulus and density, respectively.

There are two unknowns in the above relations, one of the wave amplitudes, say A_2 (while A_1 can be taken arbitrarily), and the wavenumber, k. To this end, we use the continuity of the potential and the r-velocity, $\partial\varphi/\partial r$, at $r = a$. We have

$$\begin{aligned} A_1\mathrm{BesselJ}(0, k_1\, a) &= A_2\mathrm{BesselK}(0, k_2\, a)\,, \\ A_1 k_1\mathrm{BesselJ}(1, k_1\, a) &= A_2 k_2\mathrm{BesselK}(1, k_2\, a)\,. \end{aligned} \tag{5.40}$$

Thus, the wavenumber, k, is defined by

$$\begin{aligned}\Phi &= k_2\text{BesselJ}(0, k_1\, a)\text{BesselK}(1, k_2\, a)\\ &k_1\text{BesselJ}(1, k_1\, a)\text{BesselK}(0, k_2\, a) = 0\,.\end{aligned} \tag{5.41}$$

Also see Eqs. (5.38), (5.39), where the wave frequency, ω, is a given constant.

5.2.4 About the Branches' Starting Points

Similar to that for the membrane, Fig. 5.4, here are several branches of the localization exponent, $k_2(\varrho_1)$ (in a range of the waveguide density), Fig. 5.5. The branches start points, $k_2(\varrho) = 0$. The corresponding densities follow directly from Eq. (5.41). Indeed, the first term of the Φ-expression is finite at $k_2 = 0 \quad (k_1^2 = \omega^2/v_1^2)$

$$k_2\text{BesselJ}(0, k_1\, a)\text{BesselK}(1, k_2\, a) \to \frac{1}{a}\text{BesselJ}(1, k_1\, a) \quad (k_2 \to 0)\,. \tag{5.42}$$

The function BesselK$(1, k_2\, a)$ has a logarithmic singularity

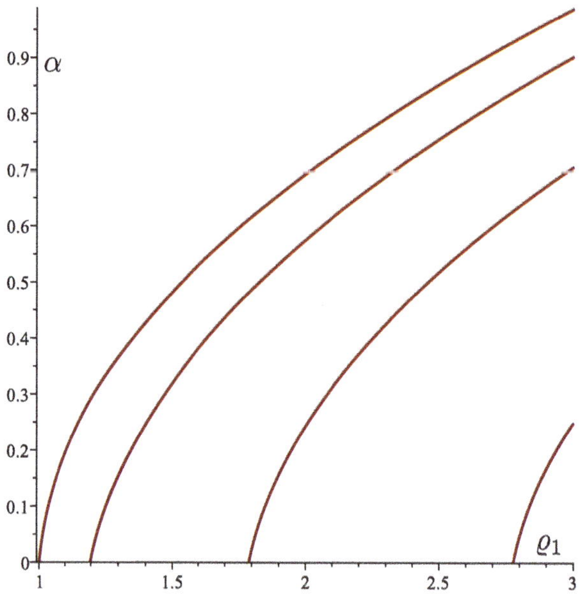

Fig. 5.4 The localized wave in a membrane 2. The localization exponent, α, as a function of the waveguide density, ϱ_1 ($a = \text{const}$). Here $\omega^2 = 1/2$, $\varrho_2 = 1$, $a = 10$, the membrane stiffness $T = 1$. A follows from (5.34), the starting points ($\alpha = 0$) are $\varrho = 1, 1.197, 1.790, 2.777$, which is in line with the graph

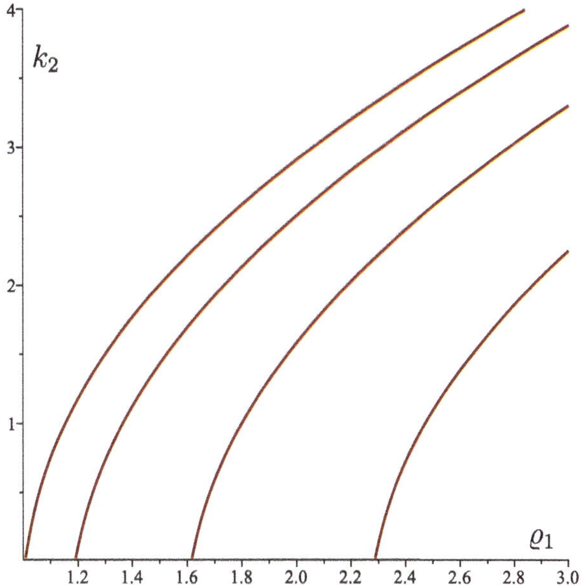

Fig. 5.5 The sound channel in fluid, $-\infty < x < \infty, -a < y < a$, in an infinite area. The localization exponent, k_2, is shown as a function of the waveguide density, ϱ_1. Here we take the fluid bulk modulus $B = 1$, the density $\varrho_2 = 1\ (r > a)\ a = \omega = 3$

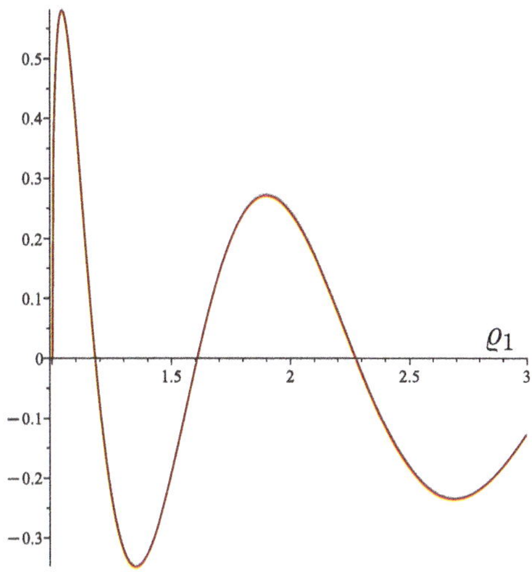

Fig. 5.6 BesselJ$(1, k_1\, a)$ as a function of the fluid density, ϱ_1, in the channel, $k_1 = \omega\sqrt{\varrho_1/B}$. The zero points correspond to the starting points of the localization exponent branches, Fig. 5.5

$$\text{BesselK}(0, k_2\, a) \quad \sim \quad \ln(k_2 a) \quad (k_2 \to 0)\,. \tag{5.43}$$

It follows that $\Phi = 0$ at each point of the density, ϱ_1, where BesselJ$(1, k_1\, a) = 0$. This statement is illustrated by comparing Fig. 5.4 with Fig. 5.5 (Fig. 5.6).

5.2.5 Localized Wave in a Square-Call Lattice

Consider a complex wave in a squire-cell lattice

$$u_{j,n} = A\mathrm{e}^{\mathrm{i}(\omega t - k_x x - k_y y)}, \quad x = a_0 j, \quad y = a_0 n, \quad (j, n) = 0, \pm 1, \ldots. \tag{5.44}$$

Here a_0 is the cell size; the frequency and wavenumber k_x are positive, and, as the above for continuous media, k_y is real for the lower-wave-speed strip, $|n| \leq N$, while

$$k_y = -\mathrm{i}\alpha, \alpha > 0 \quad (n > N); \quad u_{j,-n} = u_{j,n}. \tag{5.45}$$

The lattice dynamic equation and the particle masses are

$$\begin{aligned} &m(n)\frac{\mathrm{d}^2 u_{j,n}}{\mathrm{d}t^2} = \varkappa(u_{j+1,n} + u_{j-1,n} + u_{j,n+1} + u_{j,n-1} - 4u_{j,n}), \\ &m(n) = m_1 \ (|n| \leq N), \quad m(n) = m_2 \ (|n| > N), \quad m_1 > m_2. \end{aligned} \tag{5.46}$$

The dispersion relations for the strip (also see Fig. 5.7) and the outside area, respectively, follow as

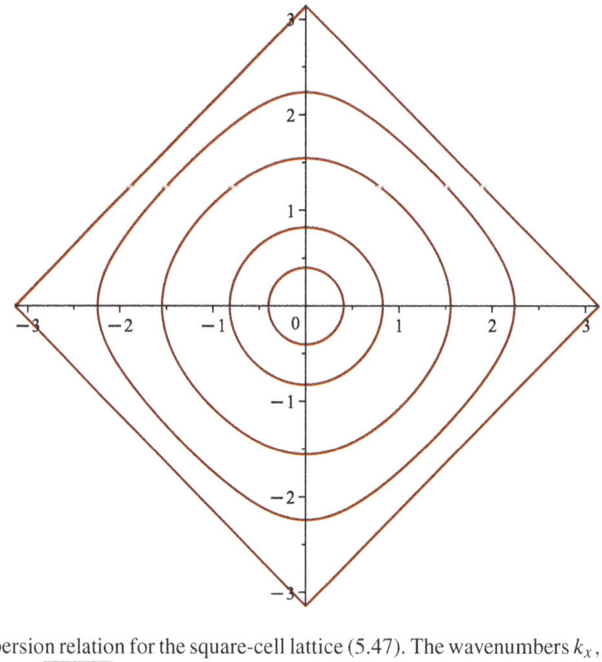

Fig. 5.7 Dispersion relation for the square-cell lattice (5.47). The wavenumbers k_x, k_y contours are shown at $\Omega = \omega\sqrt{m/(4\varkappa)} = 0.2, 0.4, 0.7, 0.9, 1.0$. As the frequencies decrease (the wavelength increases), the discreteness role disappears, and the contours approach circles. At the maximal frequency, $\Omega = 1$, it is a squire with the corners at $\pm\pi$ on both axes

$$\Omega_1^2 = \frac{m_1}{4\varkappa}\omega^2 = \left(\sin^2\frac{a_0 k_{x1}}{2} + \sin^2\frac{a_0 k_y}{2}\right), \tag{5.47}$$

and

$$\Omega_2^2 = \frac{m_2}{4\varkappa}\omega^2 = \left(\sin^2\frac{a_0 k_{x2}}{2} - \sinh^2\frac{a_0\alpha}{2}\right), \quad k_y = -\mathrm{i}\alpha\,, \quad \alpha > 0\,. \tag{5.48}$$

For the combined wave, $-\infty < n < \infty$, and $k_{x1} = k_{x2} = k$. In accordance with the above dispersion relations,

$$k_y = \frac{2}{a_0}\arcsin\sqrt{\Omega_1^2 - \sin^2\frac{a_0 k}{2}}\,, \quad \alpha = \frac{2}{a_0}\operatorname{arcsinh}\sqrt{\sin^2\frac{a_0 k}{2} - \Omega_2^2}\,,$$
$$k = \frac{2}{a_0}\arcsin\sqrt{\sinh^2\frac{a_0\alpha}{2} + \Omega_2^2} \quad (a_0 = 1)\,. \tag{5.49}$$

Note that

$$\Omega_2 < \sin\frac{a_0 k}{2} < \Omega_1\,. \tag{5.50}$$

Thus, we have the wave consisting of two parts, which can be represented as

$$u = u_1 = A_1\frac{\cos(k_y y)}{\cos(k_y a_0 N)}e^{\mathrm{i}(\omega t - kx)}\,, \quad (|y| \le a_0\,N)\,,$$
$$u = u_2 = A_2 e^{\mathrm{i}(\omega t - kx) - \alpha(|y| - a_0(N+1))} \quad (a_0\,(N+1) \le |y|)\,. \tag{5.51}$$

With the account of (5.49), there are two unknowns here, A_1/A_2 ratio (one of these amplitudes we can take arbitrarily) and the wavenumber k. Those values are defined by two additional equations combining the separate parts of the wave. Namely, the dynamic equations for the lattice masses on the interface lines, $n = N$ and $n = N + 1$. (In connection to the lattice interface, also see Brun et al., 2010.)

$$m_1\frac{\mathrm{d}^2 u_{j,N}}{\mathrm{d}t^2} = \varkappa(u_{j-1,N} + u_{j+1,N} + u_{j,N-1} + u_{j,N+1} - 4u_{j,N})\,.$$
$$m_2\frac{\mathrm{d}^2 u_{j,N+1}}{\mathrm{d}t^2} = \varkappa(u_{j-1,N+1} + u_{j+1,N+1} + u_{j,N} + u_{j,N+2} - 4u_{j,N+1})\,. \tag{5.52}$$

That is

$$\sin^2\frac{a_0\,k}{2}u_{m,N} + \frac{1}{4}(2u_{m,N} - u_{m,N-1} - u_{m,N+1}) = \Omega_1^2 u_{j,N}\,,$$
$$\sin^2\frac{a_0\,k}{2}u_{j,N+1} + \frac{1}{4}(2u_{j,N+1} - u_{j,N} - u_{j,N+2}) = \Omega_2^2 u_{j,N+1}\,. \tag{5.53}$$

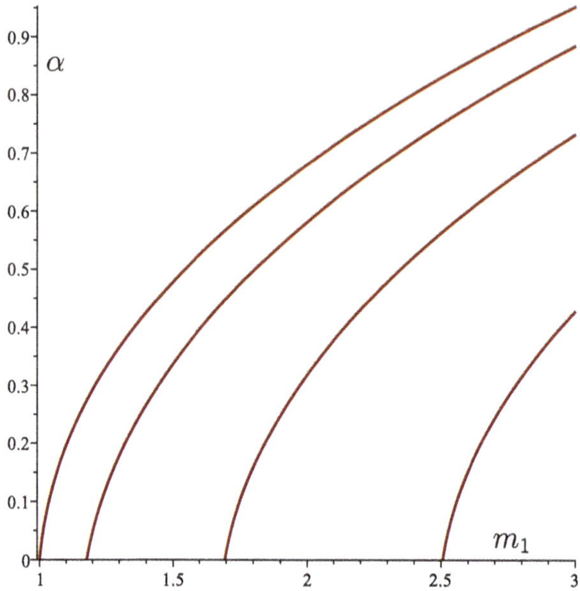

Fig. 5.8 Localized wave in a square-cell lattice. The localization exponent, α, as a function of the lattice mass m_1 $(m_2 = 1)$

With account of Eqs. (5.51) and (5.49), it follows

$$\Phi_1 A_1 - A_2 = \Phi_2 A_2 - A_1 = 0\,,$$
$$\Phi_1 = 4\sin^2\frac{a_0\,k}{2} - 4\Omega_1^2 + 2 - \frac{\cos(a_0\,k_y(N-1))}{\cos(a_0\,k_y N)}\,, \tag{5.54}$$
$$\Phi_2 = 4\sin^2\frac{a_0\,k}{2} - 4\Omega_2^2 + 2 - \mathrm{e}^{-a_0\,\alpha} = \mathrm{e}^{a_0\,\alpha}\,.$$

Finally, we obtain the wavenumber-defining equation (also see Fig. 5.8)

$$\Phi_1\Phi_2 = 1\,, \quad \text{while} \quad A_1/A_2 = \Phi_2\,. \tag{5.55}$$

5.2.6 *Localized Wave in a Bending Plate*

The waves in the bending plate problem defined by a parabolic equation differ, in some aspects, from those in the previous examples based on wave equations.

The dispersion relation as it follows from the bending plate dynamic equation

$$\varrho h \frac{\partial^2 u(x,y,t)}{\partial t^2} + D\,\nabla^2\nabla^2 u(x,y,t) = 0\,, \quad D = \frac{Eh^3}{12(1-\nu^2)}\,, \tag{5.56}$$

where h is the plate thickness, end ν is the Poisson's ratio. We have

$$\lambda\omega^2 = (k_x^2 + k_y^2)^2\,, \quad \lambda = \frac{\varrho h}{D}\,. \tag{5.57}$$

As for the membrane, we assume the plate consists of the lower wave-speed waveguide, $-a < y < a$, and the outside area, Fig. 5.2, of the densities

$$\varrho = \varrho_1 \quad (-a < y < a)\,, \quad \varrho = \varrho_2 \quad (|y| > a)\,, \quad \lambda_{1,2} = \frac{\varrho_{1,2} h}{D} \quad \varrho_1 > \varrho_2\,. \tag{5.58}$$

The wave consists of two parts

$$\begin{aligned} u_1 &= \left(A_{11}\cos(k_{y1}y) + A_{12}\cosh(\alpha_1 y)\right) \mathrm{e}^{\mathrm{i}(\omega t - kx)} \quad (|y| \le a)\,. \\ u_2 &= \left(A_{21}\mathrm{e}^{-\alpha_2(|y|-a)} + A_{22}\mathrm{e}^{-\alpha_3(|y|-a)}\right) \mathrm{e}^{\mathrm{i}(\omega t - kx)} \quad (|y| \ge a)\,, \end{aligned} \tag{5.59}$$

with

$$\lambda_2\omega < k^2 < \lambda_1\omega\,. \tag{5.60}$$

The wavenumber and localization exponents are

$$\begin{aligned} k_{y1} &= \sqrt{\lambda_1\,\omega - k^2}\,, \quad \alpha_1 = \sqrt{\lambda_1\,\omega + k^2}\,, \\ \alpha_2 &= \sqrt{k^2 - \lambda_2\,\omega}\,, \quad \alpha_3 = \sqrt{k^2 + \lambda_2\,\omega}\,. \end{aligned} \tag{5.61}$$

The continuity equations are

$$u_1 = u_2\,, \quad \frac{\partial u_1}{\partial y} = \frac{\partial u_2}{\partial y}\,, \quad M_{y1} = M_{y2}\,, \quad Q_{y1} = Q_{y2} \quad (|y| = a))\,, \tag{5.62}$$

where, in the bending plate 2D formulation, the bending moment and shear force are

$$M_y = -D\left(\frac{\partial^2 u}{\partial y^2} + \nu\frac{\partial^2 u}{\partial x^2}\right)\,, \quad Q_y = -D\frac{\partial}{\partial y}\left(\frac{\partial^2 u}{\partial x^2} + \frac{\partial^2 u}{\partial y^2}\right)\,. \tag{5.63}$$

In the following, we base on the determinant, $\mathcal{A}$, of the equation system (5.62) (Its expression is too large to show it here.) The equation $\mathcal{A} = /$ allowed us to determine the localization exponent, α_2 as a function of λ_1. As in the other particular problems considered here, there is a set of the $\alpha_2(\varrho_1)$ functions, Fig. 5.9.

Fig. 5.9 The localized wave in a bending plate. The localization exponent, α, as a function of the waveguide density, ϱ_1. Here we take $\omega = 1$, $\varrho_2 = 1$, $a = 3$, the beam ratio $h/D = 1$. Also see Fig. 5.10, where some values of ϱ_1 are shown explicitly as the zeros of the determinant $\mathcal{A}$. The first starting point is at $\varrho_1 = \varrho_{11} \approx 4$. There are no waves in $1 < \varrho_1 < \varrho_{11}$

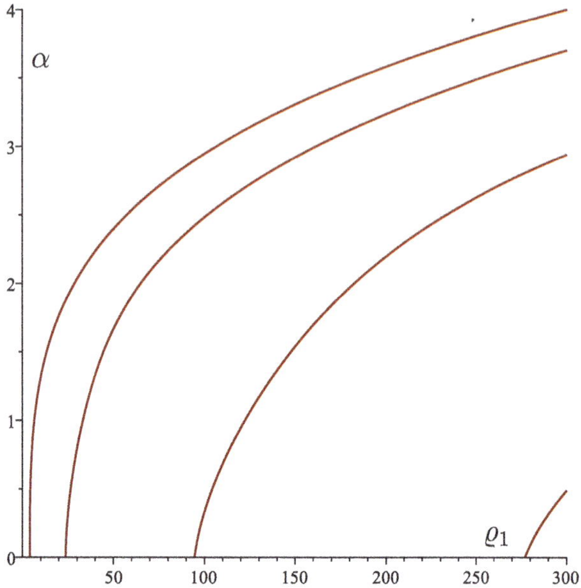

Fig. 5.10 The localized wave in a bending plate. The determinant $\mathcal{A}$ as a function of the waveguide density, ϱ_1, at $\alpha_2 = 0$. Here $\omega = 1$, $\varrho_2 = 1$, $a = 3$, and $h/D = 1$. The zero points of the determinant correspond to the waves

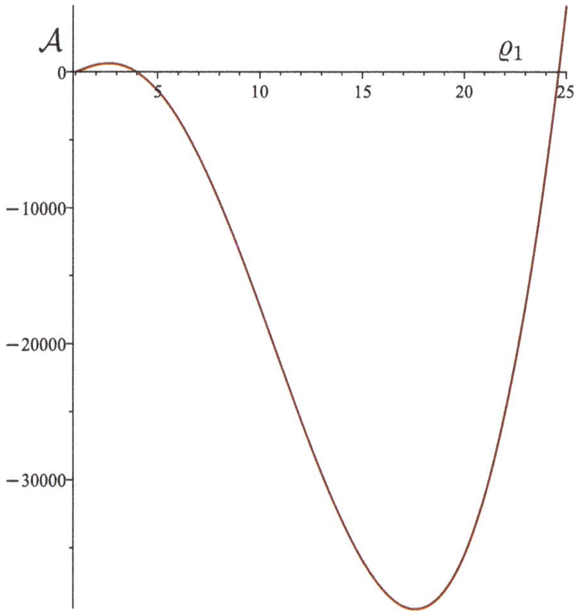

In this connection, note that in contrast to the wave-equation-related wavenumber k_{y1} and localization exponent α_2, the exponents α_1 and α_3 do not disappear as $\lambda_1 \to \lambda_2$ (that is when the considered plate becomes uniform over the space), see (5.61). Any way, clearly, the 1D-wave-corresponding point $\varrho_1 = 1, \alpha_2 = 0$ (and $\alpha_1 = \alpha_3 = 0$) does exist. However, it is a separate point, not a starting one of a curve, as in the other cases. Moreover, there is a range, $1 < \varrho_1 <\approx 4$, adjacent to this point, where no wave can exist.

5.3 Concluding Remarks

We discussed wave localization in a composite of an infinite medium with a waveguide of a lower owned phase speed. General considerations and the analysis of some particular problems are presented. We show the following.

A decrease of the owned phase speed outside the waveguide (by introducing wave transverse dependencies) results in wave localization and vice versa.

There are some peculiarities in the wave localization at a nonzero-wide waveguide, which does not appear in a model of a singular-line one. The main of them is the existence of multiple waves of different localization exponents under the composite structure and the wave frequency fixed. This phenomenon (caused by transverse sinusoidality in the waveguide) demonstrates itself in the figures for all the problems studied.

Concerning the owned phase speeds in such a composite, the lower-speed-waveguide condition of localization is necessary but not sufficient.

Firstly, there is a discrete set of the required speed ratios where the wave in the composite is not localized. Next, there can be a range of the waveguide-to-outside density ratio, where no wave can propagate.

The latter phenomenon appears in a bending plate model, which obeys a parabolic dynamic equation. There is no such stopband in the other problems considered, the membrane, sound channel in fluid, and the square-cell lattice, all obeying wave equations.

Part III
Dynamic Elasticity

Chapter 6
Elastic Isotropic Space and Perfect Compressible Fluid

6.1 Elastic Constants and Dynamic Equations

A linear elastic isotropic material is characterized by the density ϱ and a couple of elastic constants: the Lamé elastic constants λ and μ, or Young's modulus E and shear modulus G ($G \equiv \mu$), or the the bulk modulus B and Poisson's ratio ν. The velocities of the longitudinal and shear waves, c_1 and c_2, respectively, can also be considered as material constants. Below some useful relationships between the constants are presented (Table 6.1).

Note that Young's modulus, E, corresponds to the material stiffness under uniaxial stress state, whereas the bulk modulus, B, defines that corresponding to variation of the material volume.

In the linear theory, strains and stresses are expressed as

$$\varepsilon_{ij} = \frac{1}{2}\left(\frac{\partial u_i}{\partial x_j} + \frac{\partial u_j}{\partial x_i}\right), \quad \sigma_{ij} = 2\mu\varepsilon_{ij} + \lambda\varepsilon\delta_{ij}, \quad i, j = 1, 2, 3$$
$$\varepsilon = \varepsilon_{ii} \equiv \varepsilon_{11} + \varepsilon_{22} + \varepsilon_{33}, \quad \delta_{ij} = 1\ (i = j), \quad \delta_{ij} = 0\ (i \neq j)\,. \tag{6.1}$$

Here u_i are the displacement components, ε is the *dilatation* or *cubic strain*, x_i are the Cartesian rectangular coordinates and δ_{ij} is the *Kronecker delta*. The above presented stress-strain relation is the *Hooke's law*. In particular

$$\sigma_{11} = E\varepsilon_{11} \ \ (\varepsilon_{22} = \varepsilon_{33} = 0)\,, \quad \sigma \equiv \frac{1}{3}\sigma_{ii} = B\varepsilon\,. \tag{6.2}$$

The dynamic equilibrium equations are

$$\frac{\partial\sigma_{ij}(\boldsymbol{x}, t)}{\partial x_i} - \varrho\frac{\partial^2 u_j(\boldsymbol{x}, t)}{\partial t^2} = -q_j(\boldsymbol{x}, t)\,, \tag{6.3}$$

L. I. Slepyan, *Mechanical Waves: Free, Forced, and Transition*,
https://doi.org/10.1007/978-3-032-15864-2_6

Table 6.1 Relations between the elastic constants

	λ, μ	B, μ	E, μ	E, ν	μ, ν	c_1^2, c_2^2
λ	λ	$B - \frac{2}{3}\mu$	$\frac{\mu(E-2\mu)}{3\mu-E}$	$\frac{\nu E}{(1+\nu)(1-2\nu)}$	$\frac{2\nu\mu}{1-2\nu}$	$\varrho(c_1^2 - 2c_2^2)$
μ	μ	μ	μ	$\frac{E}{2(1+\nu)}$	μ	ϱc_2^2
B	$\lambda + \frac{2}{3}\mu$	B	$\frac{\mu E}{3(3\mu-E)}$	$\frac{E}{3(1-2\nu)}$	$\frac{2\mu(1+\nu)}{3(1-2\nu)}$	$\varrho\left(c_1^2 - \frac{4}{3}c_2^2\right)$
E	$\frac{\mu(3\lambda+2\mu)}{\lambda+\mu}$	$\frac{9\mu B}{3B+\mu}$	E	E	$2\mu(1+\nu)$	$\frac{\varrho c_2^2(3c_1^2-4c_2^2)}{c_1^2-c_2^2}$
ν	$\frac{\lambda}{2(\lambda+\mu)}$	$\frac{3B-2\mu}{6B+2\mu}$	$\frac{E}{2\mu} - 1$	ν	ν	$\frac{c_1^2-2c_2^2}{2(c_1^2-c_2^2)}$
c_1^2	$\frac{\lambda+2\mu}{\varrho}$	$\frac{B+4\mu/3}{\varrho}$	$\frac{\mu(4\mu-E)}{\varrho(3\mu-E)}$	$\frac{(1-\nu)E}{\varrho(1+\nu)(1-2\nu)}$	$\frac{2\mu(1-\nu)}{\varrho(1-2\nu)}$	c_1^2
c_2^2	$\frac{\mu}{\varrho}$	$\frac{\mu}{\varrho}$	$\frac{\mu}{\varrho}$	$\frac{E}{2\varrho(1+\nu)}$	$\frac{\mu}{\varrho}$	c_2^2

where $\boldsymbol{q}$ is the *body force* and the $\boldsymbol{x}$ is the position vector. Using Hooke's law the latter system can be rewritten in terms of the displacements as the Navier–Lamé vector equation

$$\mu\Delta\boldsymbol{u}(\boldsymbol{x}, t) + (\lambda + \mu)\text{grad div}\,\boldsymbol{u}(\boldsymbol{x}, t) - \varrho\frac{\partial^2\boldsymbol{u}(\boldsymbol{x}, t)}{\partial t^2} = -\boldsymbol{q}(\boldsymbol{x}, t)\,. \tag{6.4}$$

6.1.1 *The Sobolev Wave Equations*

Any vector can be presented by the sum of gradient of a scalar function and rotor of a vector function:

$$\boldsymbol{u} = \nabla\varphi + \nabla \times \psi\,, \qquad \boldsymbol{q} = \nabla f + \nabla \times \boldsymbol{p}\,, \tag{6.5}$$

where φ and ψ (as well as f and $\boldsymbol{p}$) are the *scalar and vector functions*, respectively. The operator ∇ being applied to a scalar (vector) function acts as gradient (divergence) of the latter, and $\nabla\times$ results in rotor of a vector function, see (17.1). Substituting the decomposition (6.5) in (6.4) and taking into account that $div\,rot = 0$ we obtain

$$\nabla((\lambda + 2\mu)\Delta\varphi - \varrho\frac{\partial^2\varphi}{\partial t^2} + f) + \nabla \times (\mu\Delta\psi - \varrho\frac{\partial^2\psi}{\partial t^2} + \boldsymbol{p}) = 0\,. \tag{6.6}$$

This equality is satisfied if the potentials satisfy the following scalar and vector wave equations (Sobolev, 1937),

$$(\lambda + 2\mu)\Delta\varphi - \varrho\frac{\partial^2\varphi}{\partial t^2} = -f - \Phi\,, \qquad \Delta\Phi = 0\,,$$
$$\mu\Delta\psi - \varrho\frac{\partial^2\psi}{\partial t^2} = -\boldsymbol{p} - \boldsymbol{P}\,, \qquad \nabla\times\boldsymbol{P} = -\nabla\Phi\,. \tag{6.7}$$

Here Φ and $\boldsymbol{P}$ are (scalar and vector) functions satisfying the above conditions and arbitrary in other respects. As follows from (17.4), the scalar potential defines a rotor-less portion of the displacements, while the vector potential defines a field with zero dilatation.

Four scalar functions are introduced instead of three projections of u. This also pertains to the body force function $\boldsymbol{p}$ and $\boldsymbol{P}$. So, an additional condition on the vector functions can be imposed, namely, without loss of generality in the description of the displacement field, it can be assumed that

$$\nabla\psi = \nabla\boldsymbol{p} = \nabla\boldsymbol{P} = 0\,. \tag{6.8}$$

Note that these equalities are satisfied automatically in the case of the plane problem where each of the vector potentials possesses only a normal-to-the-plane component.

The formulation in terms of wave equations (6.7) is quite general. This can be shown based on the original equation (6.4). To this end we consider the homogeneous equations. Since div rot = rot grad =0, it follows that

$$\Delta\left[(\lambda + 2\mu)\Delta\varphi - \varrho\frac{\partial^2\varphi}{\partial t^2}\right] = 0\,, \tag{6.9}$$

and we can write

$$(\lambda + 2\mu)\Delta\varphi - \varrho\frac{\partial^2\varphi}{\partial t^2} = -\varrho f\,, \quad \Delta f = 0\,, \quad \mathrm{rot}\left(\mu\Delta\psi - \varrho\frac{\partial^2\psi}{\partial t^2}\right) = \varrho\,\mathrm{grad}\, f\,. \tag{6.10}$$

Put

$$\varphi = \varphi_1 + \int\!\!\int f\,\mathrm{d}t\mathrm{d}t\,, \quad \mathrm{rot}\,\psi = \mathrm{rot}\,\psi_1 - \mathrm{grad}\int\!\!\int f\,\mathrm{d}t\mathrm{d}t\,. \tag{6.11}$$

Now the uniform scalar wave equation in (6.10) with respect to φ_1 is satisfied, and

$$\mathrm{rot}\left(\mu\Delta\psi_1 - \varrho\frac{\partial^2\psi_1}{\partial t^2}\right) = 0 \implies \mu\Delta\psi_1 - \varrho\frac{\partial^2\psi_1}{\partial t^2} = -\varrho\,\mathrm{grad}\, g\,. \tag{6.12}$$

where, in accordance with (6.8), $\nabla \psi_1 = 0$; hence $\Delta g = 0$. Lastly, we put

$$\psi_1 = \psi_2 + \text{grad} \int\int g \mathrm{d}t \mathrm{d}t \,. \tag{6.13}$$

Now the uniform resulting vector equation in (6.12) with respect to ψ_2 is satisfied. Thus the statement is proved.

The Sobolev equations (17.4) developed on account of a microstructure imbedded in the elastic medium are presented in Sect. 6.2.

6.1.2 The Stress Components

In terms of the potentials, the stress components are

$$\begin{aligned}
\sigma_{11} &= \lambda\Delta\varphi + 2\mu\left(\frac{\partial^2\varphi}{\partial x_1^2} + \frac{\partial^2\psi_3}{\partial x_1\partial x_2} - \frac{\partial^2\psi_2}{\partial x_3\partial x_1}\right),\\
\sigma_{22} &= \lambda\Delta\varphi + 2\mu\left(\frac{\partial^2\varphi}{\partial x_2^2} + \frac{\partial^2\psi_1}{\partial x_2\partial x_3} - \frac{\partial^2\psi_3}{\partial x_1\partial x_2}\right),\\
\sigma_{33} &= \lambda\Delta\varphi + 2\mu\left(\frac{\partial^2\varphi}{\partial x_3^2} + \frac{\partial^2\psi_2}{\partial x_3\partial x_1} - \frac{\partial^2\psi_1}{\partial x_2\partial x_3}\right),\\
\sigma_{12} &= \mu\left(2\frac{\partial^2\varphi}{\partial x_1\partial x_2} + \frac{\partial^2\psi_3}{\partial x_2^2} - \frac{\partial^2\psi_3}{\partial x_1^2} + \frac{\partial^2\psi_1}{\partial x_3\partial x_1} - \frac{\partial^2\psi_2}{\partial x_2\partial x_3}\right),\\
\sigma_{23} &= \mu\left(2\frac{\partial^2\varphi}{\partial x_2\partial x_3} + \frac{\partial^2\psi_1}{\partial x_3^2} - \frac{\partial^2\psi_1}{\partial x_2^2} + \frac{\partial^2\psi_2}{\partial x_1\partial x_2} - \frac{\partial^2\psi_3}{\partial x_3\partial x_1}\right),\\
\sigma_{31} &= \mu\left(2\frac{\partial^2\varphi}{\partial x_3\partial x_1} + \frac{\partial^2\psi_2}{\partial x_1^2} - \frac{\partial^2\psi_2}{\partial x_3^2} + \frac{\partial^2\psi_3}{\partial x_2\partial x_3} - \frac{\partial^2\psi_1}{\partial x_1\partial x_2}\right).
\end{aligned} \tag{6.14}$$

Note that the *mean stress* is completely defined by the scalar potential since div rot=0

$$\sigma = \frac{1}{3}(\sigma_{11} + \sigma_{22} + \sigma_{33}) = B\varepsilon = B\nabla\boldsymbol{u} = B\Delta\varphi = \frac{B}{c_1^2}\left(\frac{\partial^2\varphi}{\partial t^2} - f\right), \tag{6.15}$$

where ε and B are the dilatation and the bulk modulus, respectively, $\varepsilon = \varepsilon_{11} = \varepsilon_{22} + \varepsilon_{33}$, $B = \lambda + \frac{2}{3}\mu$.

6.1.3 Dynamic Equations in the Cylindrical Coordinates

In terms of the corresponding components of the displacement, the strains are

$$\begin{aligned}
\varepsilon_{rr} = \frac{\partial u_r}{\partial r}, \quad \varepsilon_{\theta\theta} &= \frac{1}{r}\left(\frac{\partial u_\theta}{\partial \theta} + u_r\right), \quad \varepsilon_{zz} = \frac{\partial u_z}{\partial z}, \\
\varepsilon_{r\theta} = \varepsilon_{\theta r} &= \frac{1}{2}\left[r\frac{\partial}{\partial r}\left(\frac{u_\theta}{r}\right) + \frac{1}{r}\frac{\partial u_r}{\partial \theta}\right], \\
\varepsilon_{\theta z} = \varepsilon_{z\theta} &= \frac{1}{2}\left[\frac{1}{r}\frac{\partial u_z}{\partial \theta} + \frac{\partial u_\theta}{\partial z}\right], \\
\varepsilon_{zr} = \varepsilon_{rz} &= \frac{1}{2}\left[\frac{\partial u_z}{\partial r} + \frac{\partial u_r}{\partial z}\right].
\end{aligned} \tag{6.16}$$

The dynamic equations in terms of stresses and displacements are

$$\begin{aligned}
\frac{\partial \sigma_{rr}}{\partial r} + \frac{1}{r}\frac{\partial \sigma_{\theta r}}{\partial \theta} + \frac{\partial \sigma_{zr}}{\partial z} + \frac{1}{r}(\sigma_{rr} - \sigma_{\theta\theta}) - \varrho\frac{\partial^2 u_r}{\partial t^2} &= -\varrho q_r , \\
\frac{\partial \sigma_{r\theta}}{\partial r} + \frac{1}{r}\frac{\partial \sigma_{\theta\theta}}{\partial \theta} + \frac{\partial \sigma_{z\theta}}{\partial z} + \frac{2}{r}\sigma_{r\theta} - \varrho\frac{\partial^2 u_\theta}{\partial t^2} &= -\varrho q_\theta , \\
\frac{\partial \sigma_{rz}}{\partial r} + \frac{1}{r}\frac{\partial \sigma_{\theta z}}{\partial \theta} + \frac{\partial \sigma_{zz}}{\partial z} + \frac{1}{r}\sigma_{rz} - \varrho\frac{\partial^2 u_z}{\partial t^2} &= -\varrho q_z .
\end{aligned} \tag{6.17}$$

The stresses in terms of the potentials

$$\begin{aligned}
\sigma_{rr} &= \lambda\Delta\varphi + 2\mu\left[\frac{\partial^2\varphi}{\partial r^2} + \frac{\partial}{\partial r}\left(\frac{1}{r}\frac{\partial \psi_z}{\partial \theta} - \frac{\partial \psi_\theta}{\partial z}\right)\right], \\
\sigma_{\theta\theta} &= \lambda\Delta\varphi + 2\mu\frac{1}{r}\left(\frac{1}{r}\frac{\partial^2\varphi}{\partial \theta^2} + \frac{\partial \varphi}{\partial r} + \frac{\partial^2 \psi_r}{\partial \theta \partial z} - \frac{\partial \psi_\theta}{\partial z} + \frac{1}{r}\frac{\partial \psi_z}{\partial \theta} - \frac{\partial^2 \psi_z}{\partial r \partial \theta}\right), \\
\sigma_{zz} &= \lambda\Delta\varphi + 2\mu\left[\frac{\partial^2\varphi}{\partial z^2} + \frac{1}{r}\left(\frac{\partial}{\partial r}\left(r\frac{\partial \psi_\theta}{\partial z}\right) - \frac{\partial^2 \psi_r}{\partial \theta \partial z}\right)\right], \\
\sigma_{r\theta} &= \mu\left[2\frac{\partial}{\partial r}\left(\frac{1}{r}\frac{\partial \varphi}{\partial \theta}\right) + r\frac{\partial}{\partial r}\left(\frac{1}{r}\left(\frac{\partial \psi_r}{\partial z} - \frac{\partial \psi_z}{\partial r}\right)\right) + \frac{1}{r^2}\frac{\partial^2 \psi_z}{\partial \theta^2} - \frac{1}{r}\frac{\partial^2 \psi_\theta}{\partial \theta \partial z}\right], \\
\sigma_{\theta z} &= \mu\left[\frac{2}{r}\frac{\partial^2 \varphi}{\partial \theta \partial z} + \frac{1}{r^2}\left(\frac{\partial}{\partial r}\left(r\frac{\partial \psi_\theta}{\partial \theta}\right) - \frac{\partial^2 \psi_r}{\partial \theta^2}\right) + \frac{\partial^2 \psi_r}{\partial z^2} - \frac{\partial^2 \psi_z}{\partial z \partial r}\right], \\
\sigma_{zr} &= \mu\left[2\frac{\partial^2 \varphi}{\partial z \partial r} - \frac{\partial}{\partial r}\left(\frac{1}{r}\frac{\partial \psi_r}{\partial \theta}\right) + \frac{\partial}{\partial r}\left(\frac{\psi_\theta}{r}\right) + \frac{\partial^2 \psi_\theta}{\partial r^2} - \frac{\partial^2 \psi_\theta}{\partial z^2} + \frac{1}{r}\frac{\partial^2 \psi_z}{\partial \theta \partial z}\right].
\end{aligned} \tag{6.18}$$

6.1.4 Dynamic Equations in the Spherical Coordinates

In spherical coordinates, the strains are expressed as

$$
\begin{aligned}
\varepsilon_{rr} &= \frac{\partial u_r}{\partial r}, \quad \varepsilon_{\alpha\alpha} = \frac{1}{r}\left(\frac{\partial u_\alpha}{\partial \alpha} + u_r\right), \\
\varepsilon_{\beta\beta} &= \frac{1}{r}\left(\frac{1}{\sin\alpha}\frac{\partial u_\beta}{\partial \beta} + u_\alpha \cot\alpha + u_r\right), \\
\varepsilon_{r\alpha} &= \frac{1}{2}\left[\frac{1}{r}\left(\frac{\partial u_r}{\partial \alpha} - u_\alpha\right) + \frac{\partial u_\alpha}{\partial r}\right], \\
\varepsilon_{\alpha\beta} &= \frac{1}{2r}\left(\frac{1}{\sin\alpha}\frac{\partial u_\alpha}{\partial \beta} + \frac{\partial u_\beta}{\partial \alpha} - u_\beta \cot\alpha\right), \\
\varepsilon_{\beta r} &= \frac{1}{2}\left[\frac{1}{r}\left(\frac{1}{\sin\alpha}\frac{\partial u_r}{\partial \beta} - u_\beta\right) + \frac{\partial u_\beta}{\partial r}\right].
\end{aligned}
\tag{6.19}
$$

The dynamic equations in terms of stresses and displacements are

$$
\begin{aligned}
&\frac{\partial \sigma_{rr}}{\partial r} + \frac{1}{r}\frac{\partial \sigma_{r\alpha}}{\partial \alpha} + \frac{1}{r\sin\alpha}\frac{\partial \sigma_{\beta r}}{\partial \beta} \\
&+\frac{1}{r}(2\sigma_{rr} - \sigma_{\beta\beta} - \sigma_{\alpha\alpha} + \sigma_{r\alpha}\cot\alpha) - \varrho\frac{\partial^2 u_r}{\partial t^2} = -\varrho q_r, \\
&\frac{\partial \sigma_{r\alpha}}{\partial r} + \frac{1}{r}\frac{\partial \sigma_{\alpha\alpha}}{\partial \alpha} + \frac{1}{r\sin\alpha}\frac{\partial \sigma_{\alpha\beta}}{\partial \beta} \\
&+\frac{1}{r}[(\sigma_{\alpha\alpha} - \sigma_{\beta\beta})\cot\alpha + 3\sigma_{r\alpha}] - \varrho\frac{\partial^2 u_\alpha}{\partial t^2} = -\varrho q_\alpha, \\
&\frac{\partial \sigma_{\beta r}}{\partial r} + \frac{1}{r}\frac{\partial \sigma_{\alpha\beta}}{\partial \alpha} + \frac{1}{r\sin\alpha}\frac{\partial \sigma_{\beta\beta}}{\partial \beta} \\
&+\frac{1}{r}(3\sigma_{\beta r} + 2\sigma_{\alpha\beta}\cot\alpha) - \varrho\frac{\partial^2 u_r}{\partial t^2} = -\varrho q_\beta.
\end{aligned}
\tag{6.20}
$$

The stresses in terms of the potentials

$$\sigma_{rr} = \lambda\Delta\varphi + 2\mu\left[\frac{\partial^2\varphi}{\partial r^2} + \frac{1}{\sin\alpha}\left(\frac{\partial^2}{\partial r\partial\alpha}\left(\frac{\psi_\beta \sin\alpha}{r}\right) - \frac{\partial}{\partial r}\left(\frac{1}{r}\frac{\partial\psi_\alpha}{\partial\beta}\right)\right)\right],$$

$$\sigma_{\alpha\alpha} = \lambda\Delta\varphi + \frac{2\mu}{r}\left[\frac{1}{r}\frac{\partial^2\varphi}{\partial\alpha^2} + \frac{\partial\varphi}{\partial r} + \frac{1}{r}\frac{\partial}{\partial\alpha}\left(\frac{1}{\sin\alpha}\frac{\partial\psi_r}{\partial\beta} - \frac{\partial(r\psi_\beta)}{\partial r}\right)\right.$$
$$\left. + \frac{1}{r\sin\alpha}\left(\frac{\partial}{\partial\alpha}(\psi_\beta\sin\alpha) - \frac{\partial\psi_\alpha}{\partial\beta}\right)\right],$$

$$\sigma_{\beta\beta} = \lambda\Delta\varphi + \frac{2\mu}{r^2}\left[\frac{1}{\sin^2\alpha}\frac{\partial^2\varphi}{\partial\beta^2} + \frac{\partial\varphi}{\partial\alpha}\cot\alpha + r\frac{\partial\varphi}{\partial r} + \frac{1}{\sin\alpha}\left(\frac{\partial}{\partial\alpha}(\psi_\beta\sin\alpha) - \frac{\partial\psi_\alpha}{\partial\beta}\right)\right.$$
$$\left. + \frac{\cos\alpha}{\sin^2\alpha}\frac{\partial\psi_r}{\partial\beta} - \frac{\partial(r\psi_\beta)}{\partial r}\cot\alpha + \frac{1}{\sin\alpha}\left(\frac{\partial}{\partial r}\left(r\frac{\partial\psi_\alpha}{\partial\beta}\right) - \frac{\partial^2\psi_r}{\partial\alpha\partial\beta}\right)\right],$$

$$\sigma_{r\alpha} = \frac{\mu}{r}\left[2\frac{\partial^2\varphi}{\partial r\partial\alpha} - \frac{2}{r}\frac{\partial\varphi}{\partial\alpha} + \frac{1}{r}\frac{\partial}{\partial\alpha}\left(\frac{1}{\sin\alpha}\left(\frac{\partial}{\partial\alpha}(\psi_\beta\sin\alpha) - \frac{\partial\psi_\alpha}{\partial\beta}\right)\right)\right. \tag{6.21}$$
$$\left. - \frac{2}{r}\left(\frac{1}{\sin\alpha}\frac{\partial\psi_r}{\partial\beta} - \psi_\beta\right) + \frac{1}{\sin\alpha}\frac{\partial^2\psi_r}{\partial\beta\partial r} - r\frac{\partial^2\psi_\beta}{\partial r^2}\right],$$

$$\sigma_{\alpha\beta} = \frac{\mu}{r^2}\left[\frac{2}{\sin\alpha}\left(\frac{\partial^2\varphi}{\partial\alpha\partial\beta} - \frac{\partial\varphi}{\partial\beta}\cot\alpha\right) + \frac{1}{\sin\alpha}\left(\frac{1}{\sin\alpha}\frac{\partial^2\psi_r}{\partial\beta^2} - \frac{\partial}{\partial r}\left(r\frac{\partial\psi_\beta}{\partial\beta}\right)\right)\right.$$
$$\left. + \frac{\partial}{\partial r}\left(r\frac{\partial\psi_\alpha}{\partial\alpha}\right) - \frac{\partial^2\psi_r}{\partial\alpha^2} - \left(\frac{\partial(r\psi_\alpha)}{\partial r} - \frac{\partial\psi_r}{\partial\alpha}\right)\cot\alpha\right],$$

$$\sigma_{\beta r} = \frac{\mu}{r}\left[\frac{2}{\sin\alpha}\left(\frac{\partial^2\varphi}{\partial\beta\partial r} - \frac{1}{r}\frac{\partial\varphi}{\partial\beta}\right) + \frac{1}{r\sin^2\alpha}\left(\frac{\partial}{\partial\alpha}\left(\frac{\partial\psi_\beta}{\partial\beta}\sin\alpha\right) - \frac{\partial^2\psi_\alpha}{\partial\beta^2}\right)\right.$$
$$\left. + \frac{2}{r}\left(\frac{\partial\psi_r}{\partial\alpha} - \psi_\alpha\right) + r\frac{\partial^2\psi_\alpha}{\partial r^2} - \frac{\partial^2\psi_r}{\partial r\partial\alpha}\right].$$

6.1.5 The Plane and Antiplane Problems

In particular, in the *plane strain problem* defined in (x, y)-plane, it is assumed that

$$u_z = \sigma_{zx} = \sigma_{zy} = \psi_x = \psi_y = 0 \tag{6.22}$$

and the other variable are functions of x, y, t. The wave equations (6.7) become

$$\Delta\varphi - \frac{1}{c_1^2}\frac{\partial^2\varphi}{\partial t^2} = -\frac{f}{c_1^2} \quad \left(\Delta = \frac{\partial^2}{\partial x^2} + \frac{\partial^2}{\partial y^2}\right),$$
$$\Delta\psi - \frac{1}{c_2^2}\frac{\partial^2\psi}{\partial t^2} = -\frac{p}{c_2^2} \quad (\psi = \psi_z,\ p = p_z). \tag{6.23}$$

Next, in the so called *antiplane problem* defined in a plane with respect to the single displacement component normal to the plane,

$$\begin{aligned} u_x &= u_y = \sigma_{xx} = \sigma_{yy} = \sigma_{zz} = \sigma_{xy} = 0\,, \\ u_z &= u_z(x, y)\,, \quad \sigma_{zx} = \mu \frac{\partial u_z}{\partial x}\,, \quad \sigma_{zy} = \mu \frac{\partial u_z}{\partial y}\,, \end{aligned} \tag{6.24}$$

and the wave equation can be directly written with respect to the displacement

$$\Delta u - \frac{1}{c_2^2}\frac{\partial^2 u}{\partial t^2} = -\frac{q}{c_2^2} \quad (u = u_z\,,\ q = q_z)\,. \tag{6.25}$$

6.1.6 Steady-State Formulation and the Static Limit

In the *steady-state formulation*, it is assumed that for an observer moving uniformly with a certain speed the state looks steady. Let the movement is parallel to the x-axis and v be the speed. The field variables depend on $\eta = x - vt$ instead of x and t separately. In the moving coordinate system, η, y, z, the homogeneous wave equations become

$$\begin{aligned} \alpha_1^2 \frac{\partial^2 \varphi}{\partial \eta^2} + \frac{\partial^2 \varphi}{\partial y^2} + \frac{\partial^2 \varphi}{\partial z^2} &= 0\,, \\ \alpha_2^2 \frac{\partial^2 \psi}{\partial \eta^2} + \frac{\partial^2 \psi}{\partial y^2} + \frac{\partial^2 \psi}{\partial z^2} &= 0\,, \quad \alpha_{1,2}^2 = 1 - \frac{v^2}{c_{1,2}^2}\,. \end{aligned} \tag{6.26}$$

For an *intersonic velocity*, $c_2 < v < c_1$, or a *supersonic velocity*, $c_1 < v$, the second equation or both equations in (6.26), respectively, remain to be wave equations, where the first term plays the role of the inertia term. In this case, arbitrary *plane waves* satisfy the equation, namely

$$\begin{aligned} \varphi &= \varphi(\eta + a\beta_1 y + b\beta_1 z)\,, \quad \beta_1^2 = -\alpha_1^2 > 0\,, \quad a^2 + b^2 = 1\,, \\ \psi &= \psi(\eta + a\beta_2 y + b\beta_2 z)\,, \quad \beta_2^2 = -\alpha_2^2 > 0\,, \quad a^2 + b^2 = 1\,. \end{aligned} \tag{6.27}$$

These waves can also be represented as

$$\begin{aligned} &\varphi = \varphi(c_1 t - \nu_x x - \nu_y y - \nu_z z)\,, \quad \psi = \psi(c_2 t - \nu_x x - \nu_y y - \nu_z z)\,, \\ &\nu_x^2 = \frac{c_{1,2}^2}{v^2}\,, \quad \nu_y^2 = a^2\beta_{1,2}^2 \frac{c_{1,2}^2}{v^2}\,, \quad \nu_z^2 = b^2\beta_{1,2}^2 \frac{c_{1,2}^2}{v^2}\,, \quad \nu_x^2 + \nu_y^2 + \nu_z^2 = 1\,, \end{aligned} \tag{6.28}$$

where real numbers ν_x, ν_y, ν_z are the directional cosines of the plane wave vector, and in terms of relations (6.27).

In the opposite cases where $\alpha_1^2 \geq 0$ or, in addition, $\alpha_2^2 \geq 0$, the steady-state versions of the wave equations become 'anisotropic' Laplace equations. The latter are satisfied by the *complex plane waves* (Sobolev, 1937)

$$
\begin{aligned}
\varphi &= \varphi(\eta + \mathrm{i}a\alpha_1 y + \mathrm{i}b\alpha_1 z)\,, \quad \alpha_1^2 > 0\,, \quad a^2 + b^2 = 1\,, \\
\psi &= \psi(\eta + \mathrm{i}a\alpha_2 y + \mathrm{i}b\alpha_2 z)\,, \quad \alpha_2^2 > 0\,, \quad a^2 + b^2 = 1\,.
\end{aligned}
\tag{6.29}
$$

which real or/and imaginary parts correspond to real dynamic states.

6.1.6.1 The Static Limit

The question is of interest whether there is a static limit of the wave equations (6.7) or their steady-state version (6.26), that is, whether one can omit the inertia term in (6.7) or put $v = 0$ in (6.26). The answer is no. For example, the limiting scalar equations defined zero dilatation, $\Delta\varphi = 0$, that is not true in general. Nevertheless, it is seems obvious that the dynamic state must approach the static one as the inertia forces vanish.

The contradiction is resolved by the fact that the potentials do not have a limit, while it exists for the field variables. The latter, however, cannot be obtained from the discussed equations in a general view but from a specific solution in terms of the field variables. We show below how it works by an example of a plane steady-state problem.

Consider such a problem defined in x, y-plane. From the complex plane waves (6.29) with $b = 0$ or directly from equations (6.26) for $\varphi = \varphi(x, y)$, $\psi = \psi_z(\psi_x = \psi_y = 0)$ with $\alpha_{1,2} > 0$ we can choose a solution as

$$
\varphi = A\mathrm{e}^{-|k|\alpha_1 y} \cos kx\,, \quad \psi = B\mathrm{e}^{-|k|\alpha_2 y} \sin kx\,,
\tag{6.30}
$$

where A, B are arbitrary constants. It follows that for $y \geq 0$

$$
\begin{aligned}
u_x &= -\left(Ak\mathrm{e}^{-|k|\alpha_1 y} + B|k|\alpha_2 \mathrm{e}^{-|k|\alpha_2 y}\right) \sin kx\,, \\
u_y &= -\left(A|k|\alpha_1 \mathrm{e}^{-|k|\alpha_1 y} + Bk\mathrm{e}^{-|k|\alpha_2 y}\right) \cos kx\,.
\end{aligned}
\tag{6.31}
$$

We take the boundary conditions at $y = 0$ as

$$
u_x = 0\,, \quad u_y = \cos kx\,,
\tag{6.32}
$$

and find the constants and the displacements:

$$
\begin{aligned}
A &= \frac{\alpha_2}{|k|(1 - \alpha_1\alpha_2)}\,, \quad B = -\frac{1}{k(1 - \alpha_1\alpha_2)}\,, \\
u_x &= \left(\mathrm{e}^{-|k|\alpha_2 y} - \mathrm{e}^{-|k|\alpha_1 y}\right) \frac{\alpha_2 \sin |k|x}{1 - \alpha_1\alpha_2}\,, \\
u_y &= \left(\mathrm{e}^{-|k|\alpha_2 y} - \alpha_1\alpha_2 \mathrm{e}^{-|k|\alpha_1 y}\right) \frac{\cos kx}{1 - \alpha_1\alpha_2}\,.
\end{aligned}
\tag{6.33}
$$

Now it can be seen that the constants and hence the potentials tend to infinity as $v \to 0$ ($\alpha_{1,2} \to 1$). At the same time the displacements tend to finite limits:

$$\begin{aligned} \lim_{v\to 0} u_x &= \frac{c_1^2 - c_2^2}{c_1^2 + c_2^2} k y \mathrm{e}^{-|k|y} \sin kx \,, \\ \lim_{v\to 0} u_y &= \left(1 + \frac{c_1^2 - c_2^2}{c_1^2 + c_2^2} |k| y\right) \mathrm{e}^{-|k|y} \cos kx \,. \end{aligned} \tag{6.34}$$

Note that such structure of the solution is characteristic for the case when two eigenvalues coincide (here the eigenvalues for the potentials coincide since the equations for φ and ψ become the same when $v = 0$).

This result is also valid as the static limit of the corresponding transient solution. The laplace transform on time applied to the wave equations (6.7) yields equations (6.26) with the Laplace parameter, s, instead the speed, v. The static limit, if exists, corresponds to $s \to 0$, and the limit is the same as above. (The Laplace transform and the double, LF, transform are considered in detail in Chapt. 18.)

6.2 Three-Dimensional Elastic Medium with Embedded Microstructure

6.2.1 *Derivation of the Dynamic Equations*

Mishuris et al. (2019). The Navier–Lamé dynamic equation for a homogeneous, isotropic, linearly-elastic medium is

$$\mu \Delta \boldsymbol{u} + (\lambda + \mu) \nabla (\nabla \boldsymbol{u}) - \varrho \frac{\partial^2 \boldsymbol{u}}{\partial t^2} = -\boldsymbol{p} \,, \tag{6.35}$$

where $\boldsymbol{u}$ and $\boldsymbol{p}$ are the displacements and a body force, respectively. We represent the latter as a sum of an external body force $\boldsymbol{p}_1$ acting directly on the medium, and a force from the embedded microstructure due to its displacements. In turn, the latter is defined by the displacement $\boldsymbol{u}$ and the external force $\boldsymbol{p}_2$ acting on the microstructure

$$\boldsymbol{p} = \boldsymbol{p}_1 - \mathcal{R} ** \boldsymbol{u} + \mathcal{M} ** \boldsymbol{p}_2 \,, \tag{6.36}$$

where $\mathcal{R}$ and $\mathcal{M}$ are the convolution operators. Assuming the latter independent of space and time explicitly

$$\begin{aligned} \mathcal{R}(\boldsymbol{x}, t) ** \boldsymbol{u}(\boldsymbol{x}, t) &= \int_{-\infty}^{\infty} \int_{-\infty}^{\infty} \boldsymbol{u}(\boldsymbol{\xi}, \tau) \mathcal{R}(\boldsymbol{x} - \boldsymbol{\xi}, t - \tau) \mathrm{d}\boldsymbol{\xi} \mathrm{d}\tau \,, \\ \mathcal{M}(\boldsymbol{x}, t) ** \boldsymbol{p}_2(\boldsymbol{x}, t) &= \int_{-\infty}^{\infty} \int_{-\infty}^{\infty} \boldsymbol{p}_2(\boldsymbol{\xi}, \tau) \mathcal{M}(\boldsymbol{x} - \boldsymbol{\xi}, t - \tau) \mathrm{d}\boldsymbol{\xi} \mathrm{d}\tau \,. \end{aligned} \tag{6.37}$$

Note that $-\mathcal{R}$ and $\mathcal{M}$ are the microstructure related inverse and regular dynamic Green's functions, respectively. Some specified expressions of these operators are derived in the following sections.

Thus, in these terms, the Navier–Lamé developed equation is

$$\mu\Delta \boldsymbol{u} + (\lambda+\mu)\nabla(\nabla \boldsymbol{u}) - \left(\varrho\frac{\partial^2}{\partial t^2} + \mathcal{R}**\right)\boldsymbol{u} = -\boldsymbol{p}_1 - \mathcal{M}**\boldsymbol{p}_2\,, \quad (6.38)$$

where $\mathcal{R}$ can be considered as added non-local inertia, it plays a crucial role in how microstructure affects wave propagation.

With refer to (6.37), Sobolev's wave equations with account taken of the embedded structure follows from this single vector equation in the same way as in the case of a homogeneous elastic medium with no microstructure. We represent vectors $\boldsymbol{u}$ and $\boldsymbol{p}_{1,2}$ (as any vector fields can be expressed) through the scalar and vector potentials

$$\begin{aligned} \boldsymbol{u} &= \nabla\varphi + \nabla\times\boldsymbol{\psi}\,, \\ \boldsymbol{p}_{1,2} &= \nabla\Phi_{1,2} + \nabla\times\boldsymbol{\Psi}_{1,2}\,. \end{aligned} \quad (6.39)$$

Substitution of these expressions into (6.38) leads to

$$\begin{aligned} &\nabla\left[(\lambda+2\mu)\Delta\varphi - \left(\varrho\frac{\partial^2}{\partial t^2} + \mathcal{R}**\right)\varphi + \Phi_1 + \mathcal{M}**\Phi_2\right] \\ &+\nabla\times\left[\mu\Delta\boldsymbol{\psi} - \left(\varrho\frac{\partial^2}{\partial t^2} + \mathcal{R}**\right)\boldsymbol{\psi} + \boldsymbol{\Psi}_1 + \mathcal{M}**\boldsymbol{\Psi}_2\right] = 0\,. \end{aligned} \quad (6.40)$$

In this connections, recall that rotor of gradient and divergence of rotor are zero: $\nabla\times(\nabla\varphi) = \nabla(\nabla\times\boldsymbol{\psi}) = 0$. Also note that the potentials φ and $\boldsymbol{\psi}$ are defined with the accuracy to a constant and a gradient of a function, respectively, since these additions do not affect the displacements. From this note and the orthogonality relations it follows that each of two terms in (6.40) are at zero.

Thus, the wave Sobolev's equations with account of the microstructure are

$$\begin{aligned} (\lambda+2\mu)\Delta\varphi - \left(\varrho\frac{\partial^2}{\partial t^2} + \mathcal{R}**\right)\varphi &= -\Phi_1 - \mathcal{M}**\Phi_2\,, \\ \mu\Delta\boldsymbol{\psi} - \left(\varrho\frac{\partial^2}{\partial t^2} + \mathcal{R}**\right)\boldsymbol{\psi} &= -\boldsymbol{\Psi}_1 - \mathcal{M}**\boldsymbol{\Psi}_2\,. \end{aligned} \quad (6.41)$$

Under zero initial conditions, the LF transform (the Laplace transform in time t and the Fourier transform in spatial variable $\boldsymbol{x}$ leads to

$$\begin{aligned} (\lambda+2\mu)\boldsymbol{k}^2\varphi^{LF} - \left(\varrho s^2 + \mathcal{R}^{LF}\right)\varphi^{LF} &= -\Phi_1^{LF} - \mathcal{M}^{LF}\Phi_2^{LF}\,, \\ \mu\boldsymbol{k}^2\boldsymbol{\psi}^{LF} - \left(\varrho s^2 + \mathcal{R}^{LF}\right)\boldsymbol{\psi}^{LF} &= -\boldsymbol{\Psi}_1^{LF} - \mathcal{M}^{LF}\boldsymbol{\Psi}_2^{LF}\,. \end{aligned} \quad (6.42)$$

In particular, if the tensor-operator $\mathcal{R}$ is isotropic and acts locally in space

$$\mathcal{R} = \mathcal{R}_0 \prod_{i=1}^{3} \delta(x_i)\,, \tag{6.43}$$

where $\mathcal{R}_0$ is a time-operator, then the convolution on the space coordinates is an identical transformation, and only the time-convolution remains.

Note that the superposition principle is valid, and a sum of different structures and related operators each can be introduced. So the equations (6.41) can be expressed in a more general view as

$$\begin{aligned} (\lambda + 2\mu)\Delta\varphi - \left(\varrho\frac{\partial^2}{\partial t^2} + \sum_i \mathcal{R}_i ** \right)\varphi &= -\Phi_1 - \sum_i \mathcal{M}_i ** \Phi_{2i}\,, \\ \mu\Delta\boldsymbol{\psi} - \left(\varrho\frac{\partial^2}{\partial t^2} + \sum_i \mathcal{R}_i ** \right)\boldsymbol{\psi} &= -\boldsymbol{\Psi}_1 - \sum_i \mathcal{M}_i ** \boldsymbol{\Psi}_2\,, \end{aligned} \tag{6.44}$$

where the sums can also be replaced by integrals if these sets are distributed continuously (see Slepyan, 1967).

Homogenisation approach is not used here, and instead the embedded system of resonators is taken into account via the forcing term $\boldsymbol{p}$ in the equations of motion. Prior to the Green's function formulation, we discuss an example of a centrosymmetric wave, characterised by an exponential decay.

6.3 Non-viscous Compressible Fluid

In contrast to solids, fluid dynamics is usually described in terms of Eulerian variables. Consider *perfect fluid*, i.e. fluid which does not resist to shear. In such fluid, shear stresses are zero and the *Navier–Stokes* equation (17.50) is reduced to

$$\frac{\partial \boldsymbol{v}}{\partial t} + (\boldsymbol{v}\nabla)\boldsymbol{v} = -\frac{1}{\varrho}\nabla p + \boldsymbol{q}\,. \tag{6.45}$$

In addition to this equation, the mass conservation law (17.55)

$$\frac{\mathrm{d}\varrho}{\mathrm{d}t} + \varrho\nabla\boldsymbol{v} = 0 \tag{6.46}$$

and constitution relations must be taken into account. Below potential flow is considered where $\nabla \times v = 0$. In this case, the particle velocities and body forces can be expressed in terms of scalar potentials, namely

$$v = \nabla\varphi, \quad q = \nabla f, \tag{6.47}$$

and for the potential flow relation (17.5) yields

$$(\boldsymbol{v}\nabla)\boldsymbol{v} = \frac{1}{2}\nabla v^2 . \tag{6.48}$$

Further, it is assumed that pressure is a function of density only, $p = p(\varrho)$. Note that the *local wave speed* squired is

$$c^2 = \frac{\mathrm{d}p}{\mathrm{d}\varrho} = \frac{B}{\varrho}, \tag{6.49}$$

where B is the tangent bulk modulus. In these terms, the *pressure function* is introduced as

$$P = \int_0^p \frac{\mathrm{d}p}{\varrho} \tag{6.50}$$

with

$$\nabla P = \frac{1}{\varrho}\nabla p, \quad \frac{1}{\varrho}\frac{\mathrm{d}\varrho}{\mathrm{d}t} = \frac{1}{c^2}\frac{\mathrm{d}P}{\mathrm{d}t} . \tag{6.51}$$

Equation (6.45) becomes

$$\nabla\left(\frac{\partial\varphi}{\partial t} + \frac{v^2}{2}\right) = -\nabla(P - f) . \tag{6.52}$$

From this and with account of Eq. (6.46) it follows that

$$P = f - \frac{\partial\varphi}{\partial t} - \frac{v^2}{2} + P_0, \quad P_0 = \text{const.} \tag{6.53}$$

Equation (6.46) with account of (6.51) becomes

$$\nabla v = -\frac{1}{c^2}\left(\frac{\partial P}{\partial t} + \boldsymbol{v}\nabla P\right) . \tag{6.54}$$

Referring to the relations for $\boldsymbol{v}$ (17.3) and P (6.53) this equation can be rewritten in the form

$$\Delta\varphi - \frac{1}{c^2}\frac{\partial^2\varphi}{\partial t^2} = \frac{1}{c^2}\left(\frac{\partial}{\partial t}v^2 + \frac{1}{2}\boldsymbol{v}\nabla v^2\right) - \frac{1}{c^2}\left(\frac{\partial f}{\partial t} + \boldsymbol{v}\boldsymbol{q}\right) . \tag{6.55}$$

6.3.1 A Semi-linear Formulation

If pressure and particle velocities are small enough, $p \ll \varrho c^2$, $v^2 \ll c^2$, the nonlinear terms in this equation can be neglected and the wave velocity, c, and the density, ϱ, can be considered as constants. This leads to the linear wave equation

$$\Delta\varphi - \frac{1}{c^2}\frac{\partial^2\varphi}{\partial t^2} = -\frac{1}{c^2}\frac{\partial f}{\partial t}, \quad c = \sqrt{\frac{B}{\varrho}}, \tag{6.56}$$

whereas the expression for pressure remains nonlinear

$$p = \varrho\left(f - \frac{\partial\varphi}{\partial t} - \frac{v^2}{2}\right) + p_0 . \tag{6.57}$$

Note that the constant, p_0, can be included in the potential, φ; however, it is convenient to keep it separately.

Usual, in the linearized theory of the compressible fluid dynamics, the nonlinear term in this expression is also neglected; however, it must be preserved if, in a space-time domain considered, both the wave propagation and the flow of the fluid are important. To show this consider a spherical wave problem for $r > r_0$, where under zero initial conditions a radial particle velocity, $v = v_0(t)$, is given at $r = r_0, t > 0$. It is assumed that there is no body forces.

The linear wave equation (6.56) yields

$$\varphi = \frac{f(r - ct)}{r}, \tag{6.58}$$

where f is an arbitrary function, and the boundary condition reads

$$v = \frac{\partial\varphi}{\partial r} = \frac{f'(r_0 - ct)}{r_0} - \frac{f(r_0 - ct)}{r_0^2} = v_0(t) . \tag{6.59}$$

The solution is

$$f = -r_0^2\int_0^{\psi} v_0(r_0\xi/c)\mathrm{e}^{\xi}\mathrm{d}\xi\,\mathrm{e}^{-\psi}H(\psi), \quad \psi = (ct - r + r_0)/r_0 . \tag{6.60}$$

In particular, for $v_0(t) = v_0H(t)$, $v_0 = \text{const}$,

$$\frac{\partial\varphi}{\partial t} = -cv_0\mathrm{e}^{-ct/r_0} \quad (r = r_0, t > 0) . \tag{6.61}$$

It can be seen that that nonlinear term $\varrho v_0^2/2$ in the expression for pressure (6.57) is negligible in the beginning when the wave propagation is essential. In the following, the flux of the fluid becomes dominant, and the role of the first term, $\varrho\partial\varphi/\partial t$, vanishes. Thus, both terms are important. As for the nonlinear terms in the wave equation (6.55), they are still negligibly small (if $v/c \ll 1$) in both regimes. In particular

$$\frac{v}{c^2}\nabla v^2 \sim -\frac{4v_0^3}{c^2 r_0} \ll \Delta\varphi \sim \frac{2v_0}{r_0} \quad (r = r_0\,, t \to \infty)\,. \tag{6.62}$$

Thus, this *semi-linear formulation* is not useless.

Lastly, if compressibility of the fluid can be neglected the the wave equation is reduced to the Laplace equation

$$\Delta\varphi = 0\,. \tag{6.63}$$

6.3.2 Surface Conditions for Potential Movement of Fluid Under Gravity

We here denote Eulerian variables by $x,\ y,\ z$. For the theory of *gravity waves* in fluids conditions at the disturbed surface (or interface) are essential. In the case of gravitaty body forces, the corresponding potential

$$\varrho f = -\varrho g z\,, \tag{6.64}$$

where g is the gravity acceleration and $-z$ corresponds to the direction of gravity (it is assumed here that this direction and g are constants). Hence the pressure (6.57) is

$$p(x, y, z, t) = -\varrho\left(gz + \frac{\partial\varphi}{\partial t} + \frac{v^2}{2}\right) + p_0 \tag{6.65}$$

with $z = h(x, y, t)$ at the boundary.

Besides, there exists a kinematic condition. To proceed we note that the surface can be considered as formed by fluid particles, which do not leave the surface (the normal speed of the surface coincide with that of the particle). It follows that the vertical component of the particle velocity at the surface, $z = h$, is

$$v_z = \frac{\partial\varphi}{\partial z} = \frac{\partial h}{\partial t} + \frac{\partial h}{\partial x}v_x + \frac{\partial h}{\partial y}v_y\,. \tag{6.66}$$

In a linear theory, this relation is reduced to $\partial h/\partial t = \partial\varphi/\partial z$, and for a surface the linearized relation (6.65) becomes

$$\frac{\partial^2\varphi}{\partial t^2} + g\frac{\partial\varphi}{\partial z} = -\frac{1}{\varrho}\frac{\partial p}{\partial t}\,. \tag{6.67}$$

Thus, for a free surface of the deep liquid

$$\frac{\partial^2 \varphi}{\partial t^2} + g\frac{\partial \varphi}{\partial z} = 0 \quad (z = 0)\,, \quad \Delta\varphi = 0 \quad (z < 0)\,, \quad \Delta = \frac{\partial^2}{\partial x^2} + \frac{\partial^2}{\partial z^2}\,. \tag{6.68}$$

The complex wave with a real wavenumber follows

$$\varphi = \mathrm{e}^{\mathrm{i}(\omega t - kx)} Z(z)\,, \quad Z(z) = A\mathrm{e}^{|k|z}\,, \tag{6.69}$$

with the dispersion relation

$$\omega = \pm\sqrt{g|k|}\,. \tag{6.70}$$

For a relatively short waves, where surface tension is not negligible, the corresponding surface pressure is

$$p = -T_0(\kappa_1 + \kappa_2), \tag{6.71}$$

where $\kappa_{1,2}$ are main curvatures of the surface. In the linear theory, this leads to the dynamic equation for *capillary waves* (compare with (6.67))

$$\frac{\partial^2 \varphi}{\partial t^2} + g\frac{\partial}{\partial z}\left(1 - \frac{T_0}{\varrho g}\Delta\right)\varphi = 0,\,. \tag{6.72}$$

Note that for water at 25 °C the surface tension $T_0 \approx 0.0720\,\mathrm{N/m}$. It follows from the latter equation that for sinusoidal waves the surface tension is negligible if the wave length $\lambda \gg 1\,\mathrm{cm}$.

Chapter 7
Waves in Homogeneous Isotropic Space and Half-Space

7.1 Longitudinal and Shear Plane Waves

7.1.1 The Waves

In a linear homogeneous elastic space, *longitudinal and shear plane waves* can exist. The former are defined by the scalar potential, that is, by the first wave equation in (6.7), whereas the latter is defined by the vector potential satisfying the second equation. The corresponding *D'Alembert solution* for the longitudinal wave potential, φ, and particle displacements, $\boldsymbol{u}$, is

$$\varphi = \varphi(c_1 t - \boldsymbol{kx}), \quad \boldsymbol{u} = \nabla\varphi = \varphi'(c_1 t - \boldsymbol{kx})\boldsymbol{k}\nabla\boldsymbol{x} = \varphi'(c_1 t - \boldsymbol{kx})\boldsymbol{k},$$
$$|\boldsymbol{k}|^2 = 1, \quad c_1 = \sqrt{\frac{\lambda + 2\mu}{\varrho}}, \tag{7.1}$$

where φ is an arbitrary function of its argument and $\boldsymbol{k}$ is the unit vector of the direction of the wave propagation, the unit *wave vector*, φ' means the derivative of φ with respect to its argument as usual. This is a true plane wave: the dynamic state is independent of the coordinates in the plane normal to that direction, $\boldsymbol{kx} = \text{const}$. It propagates with velocity $c_1\boldsymbol{k}$, and this is the wavefront velocity if the front exists, $\varphi' = 0$ for $c_1 t - \boldsymbol{kx} +$ const < 0. The wave is called longitudinal since the particle displacements are parallel to the wave vector. In such a wave, material elements which are parallel and normal to the wave vector do not rotate, and the longitudinal wave is also called *irrotational wave*. Other names are *dilatational wave* and *pressure wave* (P-wave).

The kinetic and potential energy densities are

$$\mathcal{K} = \frac{\varrho}{2}\left(\frac{\partial u}{\partial t}\right)^2 = \pi = \frac{\lambda + 2\mu}{2}\left(\frac{\partial u}{\partial x_n}\right)^2 = \frac{1}{2}\varrho c_1^2(\varphi'')^2, \tag{7.2}$$

L. I. Slepyan, *Mechanical Waves: Free, Forced, and Transition*,
https://doi.org/10.1007/978-3-032-15864-2_7

where x_n is the coordinate parallel to the wave vector.

The shear waves are defined by the vector potential ψ which components are

$$\psi_i = \psi_i(c_2 t - \mathbf{k}\mathbf{x}), \quad c_2 = \sqrt{\frac{\mu}{\varrho}}, \tag{7.3}$$

where the constant, C, is taken the same for each component with the aim to have a single wave. Without loss of generality we can take x_1-axis as the external normal to the wave front. Now

$$\psi_i = \psi_i(c_2 t - x_1) \tag{7.4}$$

and displacements are

$$\begin{aligned} u_1 &= \frac{\partial \psi_3}{\partial x_2} - \frac{\partial \psi_2}{\partial x_3} = 0, \\ u_2 &= \frac{\partial \psi_1}{\partial x_3} - \frac{\partial \psi_3}{\partial x_1} = \psi_3'(c_2 t - x_1), \\ u_3 &= \frac{\partial \psi_2}{\partial x_1} - \frac{\partial \psi_1}{\partial x_2} = -\psi_2'(c_2 t - x_1). \end{aligned} \tag{7.5}$$

In this wave, particle displacements are normal to wave vector, and the shear wave, S-wave, is also called the *transversal wave*. The kinetic and potential energy densities are

$$\mathcal{K} = \frac{\varrho}{2}\left(\frac{\partial u}{\partial t}\right)^2 = \Pi = \frac{\mu}{2}\left(\frac{\partial u}{\partial x_1}\right)^2 = \frac{1}{2}\varrho c_2^2\left[(\psi_2'')^2 + (\psi_3'')^2\right] \quad (u = |\mathbf{u}|). \tag{7.6}$$

7.1.2 Reflection and Refraction at a Plane Interface

Consider the steady-state problem for a plane wave propagating at an angle α (to the x-axis), Fig. 1.2. The speeds of the wavefront–interface intersection point for P and S incident waves are

$$v = v_P = \frac{c_1}{\cos\alpha}, \quad v = v_S = \frac{c_2}{\cos\alpha}. \tag{7.7}$$

Thus

$$c_1 \le v_P < \infty, \quad c_2 \le v_S < \infty, \tag{7.8}$$

Three types of waves can be considered: the longitudinal P-wave, the shear wave SH, where the displacements are directed parallel to z-axis, and the shear wave SV

with displacements normal to the wave vector and z-axis. In a general case, there are four interface conditions: continuity of two components of the displacements and two components of the stress vector acting on the interface. The conditions must be satisfied by four waves, two reflected and two refracted waves. The speeds of these waves along the interface must be the same as that of the incident wave (7.7). This condition together with the continuity conditions define the reflected and refracted waves type, intensity and orientation.

For P and PV incident waves we have the plane strain steady-state problem (see (6.22)), and the wave equations can be written as

$$\Delta\varphi - \frac{1}{c_1^2}\frac{\partial^2\varphi}{\partial x^2} = 0, \quad \Delta\psi_z - \frac{1}{c_2^2}\frac{\partial^2\psi_z}{\partial x^2} = 0 \quad \left(\Delta = \frac{\partial^2}{\partial x^2} + \frac{\partial^2}{\partial y^2}\right) \tag{7.9}$$

with

$$\begin{gathered} u_x(x, y) = \frac{\partial\varphi}{\partial x} + \frac{\partial\psi}{\partial y}, \quad u_y(x, y) = \frac{\partial\varphi}{\partial y} - \frac{\partial\psi}{\partial x}, \quad u_z(x, y) = 0 \quad (\psi = \psi_z), \\ \sigma_{yx} = \mu\left(2\frac{\partial^2\varphi}{\partial x\partial y} + \frac{\partial^2\psi_z}{\partial y^2} - \frac{\partial^2\psi_z}{\partial x^2}\right), \quad \sigma_{yy} = \lambda\Delta\varphi + 2\mu\left(\frac{\partial^2\varphi}{\partial y^2} - \frac{\partial^2\psi_z}{\partial x\partial y}\right). \end{gathered} \tag{7.10}$$

These stresses can also be represented in terms of the normal and shear stresses in the incident wave

$$\begin{aligned} \sigma_{xy} &= \frac{1}{2}(\sigma_{11} - \sigma_{22})\sin 2\alpha + \sigma_{12}\cos 2\alpha, \\ \sigma_{yy} &= \frac{1}{2}(\sigma_{11} + \sigma_{22}) + \frac{1}{2}(\sigma_{11} - \sigma_{22})\cos 2\alpha + \sigma_{12}\sin 2\alpha, \end{aligned} \tag{7.11}$$

where the right-hand side stresses correspond to $x_{1,2}$, oriented in the x, y-plane as the incident wave vector and the normal to it, respectively.

In the case of SH incident wave, we have the antiplane problem (6.24), (6.25).

Homogeneous solutions with respect to reflected and refracted waves follow from the steady-state versions of the wave equations (6.26). In the moving coordinate system, $\eta = x - vt$, y, where the speed v is defined in (7.7), the two-dimensional equations for the plane and antiplane problems are

$$\left(1 - \frac{v^2}{c_1^2}\right)\frac{\partial^2\varphi}{\partial\eta^2} + \frac{\partial^2\varphi}{\partial y^2} = 0, \quad \left(1 - \frac{v^2}{c_2^2}\right)\frac{\partial^2\psi}{\partial\eta^2} + \frac{\partial^2\psi}{\partial y^2} = 0, \tag{7.12}$$

and

$$\left(1 - \frac{v^2}{c_2^2}\right)\frac{\partial^2 u}{\partial\eta^2} + \frac{\partial^2 u}{\partial y^2} = 0, \tag{7.13}$$

respectively. Consider possible scenarios.

7.1.3 P Incident Wave

$$\varphi = \varphi_i(c_1 t - x \sin\alpha + y \cos\alpha). \tag{7.14}$$

In the case of a normal incidence, $\alpha = 0$, both reflected and refracted waves are also longitudinal. Otherwise, for an oblique incidence, $v > c_1 > c_2$, Eq. (7.12), are still wave equations and they admit the plane wave configurations. Namely,

$$\varphi_{1,2} = f(\eta \pm \beta_1 y), \quad \psi_{1,2} = g(\eta \pm \beta_2 y), \quad u = w(\eta \pm \beta_2 y), \quad \beta_{1,2} = \sqrt{v^2/c_{1,2}^2 - 1}. \tag{7.15}$$

The reflected waves are directed from the interface, where they can differ from the incident wave only by a constant multiplier. Accordingly, the reflected waves are

$$\varphi = C_1\varphi_i(c_1 t - x \sin\alpha - y \cos\alpha), \quad \psi = C_2\varphi_i\left(c_1 t - x \sin\alpha - y\sqrt{c_1^2/c_2^2 - \sin^2\alpha}\right), \tag{7.16}$$

where, in general, the constants, $C_{1,2}$, together with those for the refracted waves are defined by the interface conditions.

The refracted waves can be constructed in the same way if the inequalities, $v > c_{1+} > c_{2+}$, are satisfied on the other side of the interface, $y < 0$, where the wave speeds $c_{1+,2+}$ can differ from the speeds in the incident-wave half-space, $y > 0$. Otherwise, if $c_{1+} > v > c_{2+}$ there is no real plane refracted P-wave, and if $c_{1+} > c_{2+} > v$ this is true for both P and S refracted waves. In such a case, the equations in the other side of the interface become 'anisotropic' Laplace equation. For example, the steady-state version of the wave equation for φ becomes (7.13)

$$\gamma_1^2\frac{\partial^2\varphi}{\partial\eta^2} + \frac{\partial^2\varphi}{\partial y^2} = 0, \quad \gamma_{1,2}^2 = 1 - \frac{v^2}{c_{1+,2+}^2} > 0. \tag{7.17}$$

This equation is satisfied by the complex 'plane' wave as an arbitrary function of $\eta \pm \mathrm{i}\gamma_1 y$. Similar solutions correspond to function ψ if $v < c_{2+}$. Thus, the refracted waves can be found as linear forms of real or/and imaginary parts of the complex waves

$$\begin{aligned} \varphi &: \ \varphi_i(vt - x \pm \mathrm{i}\gamma_1 y) \quad (v = c_1/\sin\alpha < c_{1+}), \\ \psi &: \ \psi_i(vt - x \pm \mathrm{i}\gamma_2 y) \quad (v = c_1/\sin\alpha < c_{2+}). \end{aligned} \tag{7.18}$$

In an important partial case of a sinusoidal incident wave, it follows that for $\omega > 0$

$$\varphi = C_3 \exp[\mathrm{i}\omega(t - x/v) + \omega\gamma_1 y/v], \quad \psi = C_4 \exp[\mathrm{i}\omega(t - x/v) + \omega\gamma_2 y/v], \tag{7.19}$$

respectively. Here solutions vanishing at infinity, $y \to -\infty$, are accepted. Recall that the real or imaginary part of these complex waves are the real solutions.

The latter case corresponds to the *total internal reflection*, when there is no energy flux through the interface. This occurs if

$$c_1/c_{2+} < 1 \quad \text{and} \quad \alpha > \alpha_*, \quad \alpha_* = \arcsin(c_1/c_{2+}). \tag{7.20}$$

Note that this conclusion is valid only for the steady-state solution corresponding to the uniform half-space on the other side of the interface. In the case of a finite-thickness interface as a layer of such high-wave-speed material between the lower-wave-speed half-spaces, it can conduct a part of the energy flux of the incident wave.

7.1.4 *SV* Incident Wave and a Special Case $v = \sqrt{2}\, c_2$

The reflected and refracted waves can be constructed in the same way as above. The main difference is that if

$$\alpha > \arcsin(c_2/c_1) \tag{7.21}$$

then a complex P wave reflects from the interface.

There is, however, a special case where only the shear wave reflects. For the intersonic speed, $c_2 < v < c_1$, in the plane above the interface, the potentials are

$$\varphi = \varphi(vt - x \pm \mathrm{i}y\sqrt{1 - v^2/c_1^2}), \quad \psi = \psi(vt - x \pm y\sqrt{v^2/c_2^2 - 1}), \tag{7.22}$$

and the corresponding stress components (7.10) are

$$\begin{aligned} \sigma_{xy} &= \mu\left[\mp 2\mathrm{i}\sqrt{1 - \frac{v^2}{c_1^2}}\,\varphi'' + \left(\frac{v^2}{c_2^2} - 2\right)\psi''\right], \\ \sigma_{yy} &= \mu\left[\left(\frac{v^2}{c_2^2} - 2\right)\varphi'' \pm 2\sqrt{\frac{v^2}{c_2^2} - 1}\,\psi''\right]. \end{aligned} \tag{7.23}$$

It is remarkable that for the speed $v = \sqrt{2}\, c_2$ the potentials separate, namely

$$\sigma_{xy} = \mp 2\mathrm{i}\mu \sqrt{\frac{\lambda}{\lambda + 2\mu}}\, \varphi'', \quad \sigma_{yy} = \pm 2\mu\, \psi''. \tag{7.24}$$

Let SV wave

$$\psi_{incident} = \psi(c_2 t - x \sin\alpha + y \cos\alpha) \tag{7.25}$$

be incident on a free plane boundary at the angle $\alpha = \pi/4$ ($v = \sqrt{2}\, c_2$). Then from (7.24) it follows that

$$\psi_{reflected} = \psi(c_2 t - x \sin\alpha - y \cos\alpha), \quad \varphi_{reflected} = 0. \tag{7.26}$$

The incident SV wave potential in (7.25) with $\alpha = \pi/4$ as a quadratic function of its argument, $\psi(..) = (..)^2$, corresponds to a step wave with respect to the shear stresses. In this case, in the area covered by both incident and reflected waves, there are no stresses but only a uniform particle velocity parallel to the y-axis.

7.1.5 *SH Waves*

7.1.6 *Complex Waves in Crack Dynamics*

In the plane and antiplane problems for straight-line cracks, there exists three fracture modes

$$\begin{aligned}
&\text{Mode I}: \ u_x(\eta, -y) = u_x(\eta, y), \quad u_y(\eta, -y) = -u_y(\eta, y), \quad u_z = 0,\\
&\text{Mode II}: \ u_x(\eta, -y) = -u_x(\eta, y), \quad u_y(\eta, -y) = u_y(\eta, y), \quad u_z = 0,\\
&\text{Mode III}: \ u_x = u_y = 0, \quad u_z(\eta, -y) = -u_z(\eta, y).
\end{aligned} \tag{7.27}$$

In the homogeneous problems of the uniform subsonic, $0 < v < c_R$, crack propagation in an unbounded homogeneous isotropic body the field variables are defined by the complex waves

$$\begin{gathered}
\varphi = \Re\left[A(\eta + \mathrm{i}\gamma_1 y)^{3/2}\right], \quad \psi = \Im\left[B(\eta + \mathrm{i}\gamma_2 y)^{3/2}\right],\\
\eta = x - vt, \quad \gamma_2 = \sqrt{1 - v^2/c_2^2},
\end{gathered} \tag{7.28}$$

with

$$B = -\frac{2\gamma_1}{\gamma_2^2+1}A, \ \ \Im A = 0 \ \ (\text{mode I})\,; \quad A = -\frac{2\gamma_2}{\gamma_2^2+1}B, \ \ \Re B = 0 \ \ (\text{mode II}), \tag{7.29}$$

and for mode III

$$u_z = C\Im\,(\eta + \mathrm{i}\gamma_2 y)^{1/2}, \quad \Im C = 0. \tag{7.30}$$

In these relations, the multi-valued functions are assumed positive at $y = 0, \eta > 0$. This formulation results in the corresponding symmetry, in zero traction on the crack faces, $\eta < 0, y = \pm 0$, and in a positive energy flux in the propagating crack tip, $\eta = y = 0$, for positive crack speed v.

The intersonic and supersonic crack speeds are impossible, except mode II for the above-considered special case, $v = \sqrt{2}\,c_2$, where the positive energy flux in the crack tip is only maintained. Instead of the solution considered in Sect. 7.1.4, which is represented by the z-component of the vector potential, now relations (7.24) lead to the solution represented by a single scalar potential

$$\varphi = A\Im(\eta + \mathrm{i}\gamma y)^{3/2}, \quad \psi = 0, \tag{7.31}$$

which satisfy the mode II conditions

$$\sigma_{yy} = 0 \ \ (y = 0, -\infty < \eta < \infty), \quad \sigma_{xy} = 0 \ \ (y = \pm 0, \eta < 0). \tag{7.32}$$

Note that for different intersonic speeds the exists SV wave which carry all the energy, delivered by the complex wave, out of the crack.

7.2 Rayleigh Wave

A combination of the complex plane waves also results in a homogeneous solution for a free-boundary elastic half-space. Now there is no need in separating the real and imaginary parts of the potentials, and we consider the complex potentials for a subsonic speed, $0 < v < c_2$

$$\varphi = \varphi(\eta \pm \mathrm{i}\gamma_1 y), \quad \psi = \psi(\eta \pm \mathrm{i}\gamma_2 y). \tag{7.33}$$

The boundary conditions

$$\sigma_{xy} = \sigma_{yy} = 0 \ \ (y = 0) \tag{7.34}$$

with expressions for the stresses (7.10) result in the equations

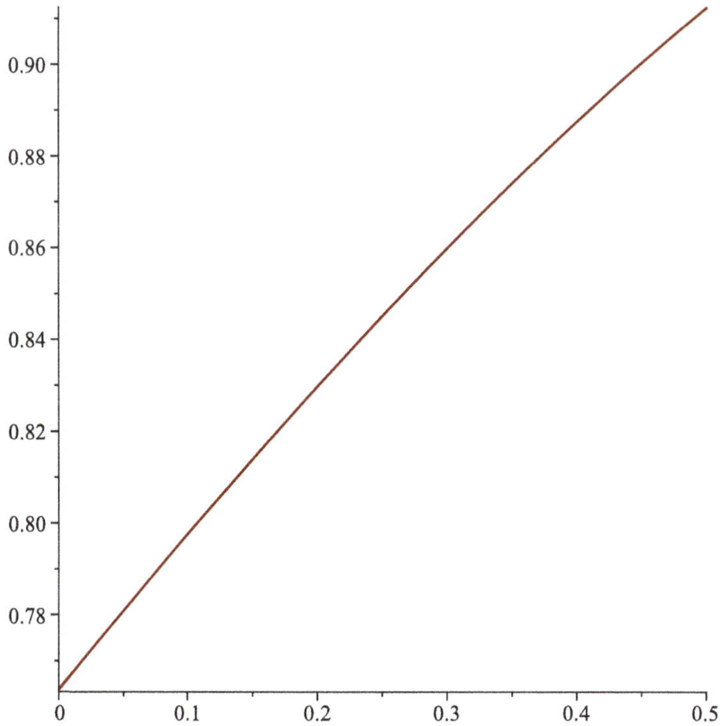

Fig. 7.1 The normalized Rayleigh wave speed, c_R/c_2, as a function of Poisson's ratio

$$\begin{aligned} &\pm 2\mathrm{i}\gamma_1\varphi'' - (\gamma_2^2 + 1)\psi'' = 0, \\ &-(\gamma_2^2 + 1)\varphi'' \mp 2\mathrm{i}\gamma_2\psi'' = 0. \end{aligned} \tag{7.35}$$

This system has a nontrivial solution if and only if

$$(\gamma_2^2 + 1)^2 - 4\gamma_1\gamma_2 = 0 \quad \left(\gamma_{1,2} = \sqrt{1 - \frac{v^2}{c_{1,2}^2}}\right). \tag{7.36}$$

The latter equation is satisfied by the Rayleigh wave, $v = c_R$. The ratio, c_R/c_2, as a function of Poisson's ratio, ν, is plotted in Fig. 7.1.

The expression

$$\frac{c_R}{c_2} \approx 0.8740 + 0.2004\nu - 0.07567\nu^2 \tag{7.37}$$

is an accurate approximation. In a partial case where $\nu = 1/4$, $c_R/c_2 = \sqrt{2 - 2/\sqrt{3}}$ ≈ 0.9194.

Usually the Rayleigh wave is assumed as a sinusoidal wave. In this case, for the upper half-plane

$$\varphi = \frac{\mathcal{A}}{k\sqrt{\alpha_1^2 - \alpha_2^2}} \exp[-k\,\alpha_1 y + \mathrm{i}(\omega t - kx)],$$
$$\psi_z = -\frac{\mathrm{i}(1+\alpha_2^2)\mathcal{A}}{2k\alpha_2\sqrt{\alpha_1^2 - \alpha_2^2}} \exp[-k\,\alpha_2 y + \mathrm{i}(\omega t - kx)], \quad \psi_x = \psi_y = 0, \tag{7.38}$$

where $\mathcal{A}$ is the wave amplitude at $y = 0$,

$$\alpha_1 = \sqrt{1 - (c_R/c_1)^2}, \quad \alpha_2 = \sqrt{1 - (c_R/c_2)^2}, \quad \omega = c_R k. \tag{7.39}$$

The displacement vector, $\boldsymbol{u}(u_x, u_y)$, is defined by its components as

$$u_x = \frac{\mathrm{i}\mathcal{A}}{2\sqrt{\alpha_1^2 - \alpha_2^2}} \left[-2\mathrm{e}^{-k\,\alpha_1 y} + (1+\alpha_2^2)\mathrm{e}^{-k\,\alpha_2 y}\right]\mathrm{e}^{\mathrm{i}(\omega t - kx)},$$
$$u_y = \frac{\mathcal{A}}{2\alpha_2\sqrt{\alpha_1^2 - \alpha_2^2}} \left[-2\alpha_1\alpha_2\mathrm{e}^{-k\,\alpha_1 y} + (1+\alpha_2^2)\mathrm{e}^{-k\,\alpha_2 y}\right]\mathrm{e}^{\mathrm{i}(\omega t - kx)}. \tag{7.40}$$

Using identity (7.36) with $v = c_R$ the amplitudes of these components at $y = 0$ can be found as

$$A = A_x = |u_x| = \sqrt{\frac{\alpha_2}{\alpha_1 + \alpha_2}}\,\mathcal{A}, \quad A = A_y = |u_y| = \sqrt{\frac{\alpha_1}{\alpha_1 + \alpha_2}}\,\mathcal{A}. \tag{7.41}$$

The stresses are

$$\sigma_{xx} = \frac{\mu k\mathcal{A}}{\sqrt{\alpha_1^2 - \alpha_2^2}}$$
$$\left[-(1 - \alpha_2^2 + 2\alpha_1^2)\mathrm{e}^{-k\,\alpha_1 y} + (1+\alpha_2^2)\mathrm{e}^{-k\,\alpha_2 y}\right]\mathrm{e}^{\mathrm{i}(\omega t - kx)},$$
$$\sigma_{yy} = \frac{\mu k\mathcal{A}(1+\alpha_2^2)}{\sqrt{\alpha_1^2 - \alpha_2^2}}\left[\mathrm{e}^{-k\,\alpha_1 y} - \mathrm{e}^{-k\,\alpha_2 y}\right]\mathrm{e}^{\mathrm{i}(\omega t - kx)}, \tag{7.42}$$
$$\sigma_{xy} = \frac{2\mathrm{i}\mu k\mathcal{A}\alpha_1}{\sqrt{\alpha_1^2 - \alpha_2^2}}\left[\mathrm{e}^{-k\,\alpha_1 y} - \mathrm{e}^{-k\,\alpha_2 y}\right]\mathrm{e}^{\mathrm{i}(\omega t - kx)}.$$

There is no wave dispersion in this case too, and the energy flux velocity in the wave, which is equal to the group velocity, coincides with the Rayleigh wave speed, c_R.

7.2.1 Energy Flux in the Rayleigh Wave

The energy flux associated with a complex wave localized at the upper half-plane boundary is

$$N = \int_0^\infty \Im(\omega \boldsymbol{u} \overline{\boldsymbol{\sigma}}_x) \mathrm{d}y, \tag{7.43}$$

where σ_x is the corresponding traction vector (the stress acting from the right on the line $x = \text{const}$). Recall that the complex wave is comprised by two real waves, and the energy flux (7.43) corresponds to that in two waves of the same amplitude and frequency. In the considered fracture problem, there also exist two waves localized at the upper and the lower crack faces. Thus the total energy flux in the two real incident waves is just equal to that in the complex wave (7.43). Using identity (7.36) with $v = c_R$ the energy flux can be found in a rather compact form

$$N = A^2 \mu\omega \frac{1-\alpha_2^2}{\alpha_1+\alpha_2}\left(1+\frac{\alpha_1-\alpha_2}{2\alpha_1\alpha_2^2}\right), \quad \alpha_1 = \sqrt{1-c_R^2/c_1^2}, \quad \alpha_2 = \sqrt{1-c_R^2/c_2^2}. \tag{7.44}$$

7.3 Steady-State Plane Dynamic Problem

We based on Eq. (6.26) assuming $v \neq 0, -c_R < v < c_R$. The equations are satisfied by complex potentials, $\varphi(z_1), \psi(z_2) = \psi_z(z_2)$, as arbitrary analytic functions of their arguments

$$z_1 = \eta + \mathrm{i}\alpha_1 y, \quad z_2 = \eta + \mathrm{i}\alpha_2 y, \quad \alpha_{1,2} = \sqrt{1 - v^2/c_{1,2}^2}, \quad \eta = x - vt. \tag{7.45}$$

We assume the real potentials to be $\Re\varphi$ and $\Im\psi$. Referring to (6.14) we obtain that in these terms

$$\begin{aligned} u_x &= \Re\varphi'(z_1) + \alpha_2 \Re\psi'(z_2), \quad u_y = -\alpha_1 \Im\varphi'(z_1) - \Im\psi'(z_2), \\ \sigma_{xx} &= [(\lambda+2\mu) - \lambda\alpha^2]\Re\varphi''(z_1) + 2\mu\alpha_2\Re\psi''(z_2), \\ \sigma_{yy} &= [-(\lambda+2\mu)\alpha_2^2 + \lambda)]\Re\varphi''(z_1) - 2\mu\alpha_2\Re\psi''(z_2), \\ \sigma_{xy} &= -\mu[2\alpha_1\Im\varphi''(z_1) + (1+\alpha_2^2)\Im\psi''(z_2)]. \end{aligned} \tag{7.46}$$

Now consider elastic half-plane, $-\infty < x < \infty, 0 < y$, under the traction

$$\sigma_{yx}(\eta) = \sigma(\eta), \quad \sigma_{yy}(\eta) = \tau(\eta) \quad (y = 0) \tag{7.47}$$

uniformly moving with a nonzero subcritical speed, $-c_R < v < c_R$. The equations follows for $y = 0$ as

$$\left(1 + \alpha_2^2\right) \Re\varphi''(\eta) + 2\alpha_2 \Im\psi''(\eta) = -\frac{\sigma}{\mu}, \qquad (7.48)$$
$$-2\alpha_1.$$

7.4 Spherical Waves

7.4.1 The Scalar Potential

Referring to Eqs. (17.10) and (6.7), the homogeneous wave equation for the scalar potential in spherical coordinates has the form

$$\Delta\varphi = \frac{1}{r^2}\frac{\partial}{\partial r}\left(r^2\frac{\partial\varphi}{\partial r}\right) + \frac{1}{r^2 \sin\alpha}\frac{\partial}{\partial\alpha}\left(\frac{\partial\varphi}{\partial\alpha}\sin\alpha\right) + \frac{1}{r^2 \sin^2\alpha}\frac{\partial^2\varphi}{\partial\beta^2} - \frac{1}{c_1^2}\frac{\partial^2\varphi}{\partial t^2} = 0. \qquad (7.49)$$

For the central symmetry waves it can be rearranged to the form

$$\frac{\partial^2(r\varphi)}{\partial r^2} - \frac{1}{c_1^2}\frac{\partial^2(r\varphi)}{\partial t^2} = 0. \qquad (7.50)$$

This yields

$$\varphi = \frac{f_1(c_1 t - r)}{r} + \frac{f_2(c_1 t + r)}{r}, \qquad (7.51)$$

where $f_{1,2}$ are arbitrary functions. Thus, two spherical waves can exist, one propagating to the center and another propagating from the center.

In the propagation, only the amplitude of the potential is changing (as $1/r$) but not the shape. It follows that the field variables have the same support as the potential. The inverse statement, however, in general is not valid. The mean stress (6.15) is an exception. It can be represented as

$$\sigma = \frac{B}{c_1^2}\frac{\partial^2\varphi}{\partial t^2} = B\left[\frac{f_1''(c_1 t - r)}{r} + \frac{f_2''(c_1 t + r)}{r}\right], \qquad (7.52)$$

hence it propagates in the same way as the potential. At the same time, the radial stresses

$$\sigma_{rr} = \lambda \Delta \varphi + 2\mu \frac{\partial^2 \varphi}{\partial r^2} \tag{7.53}$$

may have a more complicated behavior.

7.4.2 The Central Spherical Expansion

Consider the central spherical expansion as the source, which displaces the material of a given time-dependent volume $V(t)H(t)$. It induces an outgoing spherical wave defined by the potential and the central condition as

$$\varphi(r,t) = \frac{f(c_1 t - r)}{r}, \quad \lim_{r \to 0} 4\pi r^2 u_r(r,t) = V(t)H(t). \tag{7.54}$$

It follows that

$$u_r(r,t) = \frac{\partial \varphi(r,t)}{\partial r} = \frac{1}{4\pi r}\left(\frac{V(t - r/c_1)}{r} + \frac{V'(t - r/c_1)}{c_1}\right) H(c_1 t - r). \tag{7.55}$$

Consider the case of a uniform expansion rate, $V(t) = At$, $A = \text{const}$. We have

$$u_r(r,t) = \frac{At}{4\pi r^2} H(c_1 t - r), \quad \dot{u}_r(r,t) = \frac{A}{4\pi r^2} H(c_1 t - r) + \frac{A}{4\pi r}\delta(c_1 t - r). \tag{7.56}$$

First we note that this wave resembles a step wave in a bar. Behind the wavefront each particle moves uniformly, and the state satisfies the static equilibrium conditions. Only the displacement jump at the wavefront is the dynamic phenomenon.

Next, it can be seen that there is an overlap in the particle positions. Indeed, the actual coordinate of a particle is

$$R(r,t) = r + u_r(r,t). \tag{7.57}$$

7.4.3 Elastic Sphere

Consider an elastic sphere, $0 < r < r_0$, under a suddenly applied center-symmetric normal traction. In addition to this boundary condition on the surface, $r = r_0$, we have to impose a condition at the center, where no source is assumed to exist. Thus we have

$$\sigma_{rr}(r_0, t) = \sigma_0 H(t), \quad \lim_{r\to 0} r^2 u_r(r, t) = \lim_{r\to 0} r^2 \frac{\partial \varphi(r, t)}{\partial r} = 0. \tag{7.58}$$

Using the Laplace transform on t we obtain from relations (7.51), (7.53) and (7.58) the following solution

$$\sigma_{rr}^L(r, s) = \frac{Q(r, s) r_0 \sigma_0}{Q(r_0, s) r s},$$
$$Q(r, s) = \left[\left(s^2 + \frac{4c_2^2}{r^2}\right)\left(1 - \mathrm{e}^{-2sr/c_1}\right) - \frac{4c_2^2 s}{c_1 r}\left(1 + \mathrm{e}^{-2sr/c_1}\right)\right] \mathrm{e}^{-s(r_0 - r)/c_1}. \tag{7.59}$$

As discussed in Sect. 18.6.2 there are two ways of the inverse transform, namely, the solution in terms of the original functions can be represented as a series expansion in the free oscillations or as the exponent-series expansion. The former is a series by residues at the poles, $s = \pm \mathrm{i}\omega_n$, $n = 0, 1, 2, \ldots$. In the considered case

$$\sigma(r, t) = \frac{2r_0\sigma_0}{r} \sum_{n=1}^{\infty} \Re \frac{Q(r, \mathrm{i}\omega_n)}{\mathrm{i}\omega_n \partial Q(r_0, \mathrm{i}\omega_n)/\partial s} \mathrm{e}^{\mathrm{i}\omega_n t}, \tag{7.60}$$

where the non-dimensional frequencies, $\Omega_n = r_0\omega_n/c_1$, are the roots of the equation

$$(\Omega^2 - 1)\sin\Omega + \Omega\cos\Omega = 0, \quad (\Omega \neq 0). \tag{7.61}$$

The frequencies are

$$\Omega_1 \approx 2.743707270, \quad \Omega_2 \approx 6.116764264, \quad \Omega_3 \approx 9.316615629,$$
$$\Omega_n \sim \frac{(2n-1)\pi}{2} \quad (n \to \infty). \tag{7.62}$$

The latter way of the inverse transform results in explicit expressions for the waves propagating to the center and reflecting from the center and the surface. The first two waves are defined by the terms containing $\exp(-s(r_0 \mp r)/c_1)$ as the multiplier, that is

$$\sigma_{rr}^L(r, s) = \sigma_i^L(r, s) = \frac{r_0\sigma_0 Q_0(r, s)}{rsQ_0(r_0 s)} \quad \text{the incident wave},$$
$$\sigma_{rr}^L(r, s) = \sigma_r^L(r, s) = \frac{r_0\sigma_0(Q(r, s) - Q_0(r_0 s))}{rsQ_0(r_0 s)} \quad \text{the first reflected wave}, \tag{7.63}$$
$$Q_0(r, s) = \left(s^2 - \frac{4c_2^2 s}{c_1 r} + \frac{4c_2^2}{r^2}\right) \mathrm{e}^{-s(r_0 - r)/c_1}.$$

Now

$$\sigma_i(r,t) = \frac{r_0\sigma_0}{3r^3}$$
$$(3r_0^2(H(c_1t - r_0 + r) - (3(r_0^2 - r^2)\cos\alpha - \sqrt{3}(r_0 - r)^2 \sin\alpha)\exp(R_1)) \quad (7.64)$$
$$0 < t < \frac{2r_0}{c_1}, \quad \alpha = \frac{\sqrt{3}(c_1t - r_0 + r)}{2r_0}, \quad R_1 = \frac{c_1t - r_0 + r}{2r_0}.$$

and

$$\sigma_r(r,t) = -\frac{r_0\sigma_0}{3r^3}$$
$$(3r_0^2 H(c_1t - r_0 - r) + (3(r_0^2 - r^2)\cos\beta - \sqrt{3}(r_0 + r)^2 \sin\beta)\exp(R_2)(H(\gamma) - 1)) \quad (7.65)$$
$$\frac{r_0}{c_1} < t < f2r_0c_1, \quad \beta = \frac{\sqrt{3}(c_1t - r_0 - r)}{2r_0}, \quad R_2 = \frac{c_1t - r_0 + r}{2r_0}, \quad \gamma = r_0 + r - c_1t.$$

It follows that the normal-stress wave differs from that of the mean stress (7.52). Recall that, in the propagation the latter, only the amplitude is changing (as $1/r$) but not the shape. This is not the case for the normal-stress wave. However, in spite of the presence of r^3 in the denominator, the wave amplitude at the wavefront varies as $1/r$, whereas the central point, $r = 0$ is regular. Indeed, it follows from the above-presented solution that

$$\sigma_i(r,t) \to \frac{\sigma_0 r_0}{r} H(c_1t - r_0 + r) \quad \left(0 < t < \frac{r_0 - r}{c_1} < \frac{r_0}{c_1}\right),$$
$$\sigma_r(r_f + 0, t) + \sigma_i(r_f + 0, t) - \sigma_r(r_f - 0, t) - \sigma_i(r_f - 0, t) = \frac{\sigma_0 r_0}{r},$$
$$r_f = c_1t - r_0 \quad \left(\frac{r_0}{c_1} < t < \frac{2r_0}{c_1}\right), \quad (7.66)$$
$$\sigma_{rr}(0,t) = \frac{4\sigma_0}{9}(3\cos\frac{\sqrt{3}(c_1t - r_0)}{2r_0} - \sin\frac{\sqrt{3}(c_1t - r_0)}{2r_0})e^{(c_1t - r_0)/(2r_0)}$$
$$(r_0/c_1 < t < 3r_0/c_1).$$

In addition, the incident wave corresponding to an impulse load is

$$\sigma_{zz}^0(r,t) = \frac{I_0r_0c_1}{r}(\delta(c_1t - r_0 + r) - \frac{r_0^2}{r^2}\left(3\frac{r}{r_0}\left(1 - \frac{r}{r_0}\right)\cos\frac{\sqrt{3}(c_1t - r_0 + r)}{2r_0}\right)$$
$$-\sqrt{3}\left(2 - \frac{r}{r_0}\left(1 + \frac{r}{r_0}\right)\right)\sin\frac{\sqrt{3}(c_1t - r_0 + r)}{2r_0} \times e^{(c_1t - r_0 + r)/(2r_0)} H(c_1t - r_0 + r)), \quad (7.67)$$

where I_0 is the impulse. Note that the wave for the load $\sigma_{zz} = \sigma_0(t)$ is defined by a convolution of the latter function with the above-shown solution.

Chapter 8
D'Alembert's Paradox in Elasticity

8.1 The Paradox

For potential flow of unbounded incompressible perfect fluid d'Alembert proved that *The solid body moving with constant velocity meets zero drag force*. We discuss general conditions under which the *paradox* takes place, and examine some moving-contact problems of elasticity. The steady-state solution for an elastic half-plane under a moving indenter is derived based on the corresponding transient problem and on a condition concerning energy fluxes. The resulting stresses and displacements are found explicitly starting from their expressions in terms of a single analytical function. This solution incorporates all speed ranges, including the super-Rayleigh subsonic and intersonic speed regimes, which received no final description to date. Next, the wedging of an elastic plane is considered for a finite wedge moving at a distance of the crack tip. Finally, we solve the problem for a wedge moving along the interface of two elastic half-planes compressed together. It is found that in addition to the sub-Rayleigh speed regime, where the *paradox* takes place for a smooth frictionless indenter of arbitrary shape, there exists a sharp decrease in the resistance in a vicinity of the longitudinal wave speed with zero limit at this speed. For other speed regimes we determine the driving forces caused by the main underling factors: stress field singular points on the contact area (the super-Rayleigh subsonic speed regime), the wave radiation (intersonic and supersonic regimes) and the fracture resistance (the wedging problem).

For potential flow of unbounded incompressible perfect fluid d'Alembert proved that *The solid body moving with constant velocity meets zero drag force*. (see, e.g., Darrigol (2002)). This follows from the Laplace equation as a highly simplified mathematical model of the fluid dynamic. At the same time, this *paradox* has more wide underlying physics. In particular, it can be met in elasticity, which is also an appropriate model. By comparing and considering the viscosity and instability of potential flow of fluid, the elasticity looks even more adequate.

L. I. Slepyan, *Mechanical Waves: Free, Forced, and Transition*,
https://doi.org/10.1007/978-3-032-15864-2_8

For example, we can consider a smooth indenter uniformly moving without friction along an elastic half-plane boundary or along an infinite layer rested on a rigid foundation. Of course, contrary to fluids, a macroscopic rigid body can move *through* an intact elastic medium only if the movement is accompanied by fracture, that creates resistance to the movement. In this case, the paradox-related question remains as whether the contact interaction itself gives a contribution to the resistance in addition to the fracture resistance alone.

So, under certain conditions, the d'Alembert's statement remains valid in a more general case of the steady-state regime. Namely, it can be true with respect to any medium or structure, where such regime can exist, if no energy can be dissipated or released in the medium in which the body moves, in the contact zones and other boundaries if exist and at infinity. In other words, this means that (a) the medium and the boundaries are perfect, (b) no singular point exists at which the energy can be dissipated or released, including such point at infinity, that is, there exist no configurational and remote forces, and (c) there are no waves, which could carry the energy away from the body or to the body. Since no energy is radiated or absorbed while in motion, no driving force is required to support the steady-state regime. In the present paper, we assume the strains to be sufficiently small so that strain-associated thermal effects can be neglected. Along with the 'paradoxical' regimes, we study in detail the others, where driving forces arise caused by the singular points, wave radiation and fracture resistance.

It can happen that a couple of singular points exists, one of them, the *active singular point* radiates energy, while the other, the *passive singular point* absorbs the radiated energy. In this energy-balance case, the regime might be considered as a free motion, which does not require any driving force. This, however, could be true if the energy-source point is physically justified as a real 'micro-level' source. Otherwise, the energy source point must be eliminated. This consideration has a crucial importance for the super-Rayleigh subsonic speed regime. An indenter uniformly moving along the elastic half-plane boundary with subsonic speeds (except the Rayleigh wave speed) excites no wave. Thus, in this speed range, only the existence of friction or/and energy-flux-significant singular points can prohibit *d'Alembert's paradox*. Such singular points really appear at the super-Rayleigh subsonic speeds. This also concerns the movement of a body along the interface between two elastic half-planes compressed together. (Slepyan and Brun, 2012.)

The steady-state dynamic contact problem was repeatedly considered. In particular, the problem for a parabolic indenter was considered by Craggs and Roberts (1967), Georgiadis and Barber (1993), Brock (2002) and in Galin and Gladwell (2008). While the problem for the sub-Rayleigh speed range received the complete solution, the super-Rayleigh sub- and intersonic regimes are still under question. These regimes were discussed in Georgiadis and Barber (1993) and Brock (2002), where no way was found how to make the solutions to be unique and/or physically accepted. Note that dynamic problems for straight-propagating cracks are closely related to the dynamic contact problems. The super-critical regimes for cracks are considered in many works, mainly for mode II shear fracture, see, e.g., Freund (1990), Huang and Liu (1998), Broberg (1999), Needleman and Rosakis (1999), Gao et al.

(2001), Geubelle and Kubair (2001), Samudrala et al. (2002), Slepyan (2002). Mode I opening crack, intersonic speed problem was considered by Radi and Loret (2008) for a porous, liquid-saturated material, where, however, 'intersonic' does not mean the shear - longitudinal range in the uniform solid material considered below.

The above-mentioned non-uniqueness suggests that there is a missed point in the problem formulation. In this connection, in addition to the well-known Signorini contact conditions, we add the condition concerning the energy-flux significant singular points in the stress - particle velocity field: no *active* singular point can exist, as a source of energy, unless such a point source is physically justified. Thus, we base our solutions on the following three conditions

$$\begin{aligned} &(a)\ \textit{positive (tensile) normal contact stresses are not allowed,}\\ (b) &= \textit{penetration of the elastic material into the indenter is not allowed,}\\ (c) &= \textit{energy-source (active) singular points are not allowed.} \end{aligned} \tag{8.1}$$

With respect to the latter condition we note that an angular point of the body surface can give rise of an energy absorbing singular point. In the super-Rayleigh speed regimes, singular points arise in the case of a smooth body shape as well. As an example of an energy-transfer singular point a moving semi-infinite crack can be considered. Under sub-Rayleigh speeds the crack tip appears to be a passive singular point. In the hypothetic case of super-Rayleigh subsonic speeds, the crack tip is an energy source. Usually, there is no such source in the elastic material, and if the crack tip is free of external actions, this regime is impossible. However, if a hypothetic tool exists, which can act directly on the crack tip just giving the required energy, the super-Rayleigh crack speed seems to be possible. In any case, the semi-infinite crack growth requires external energy.

In the case of the Yoffe model of the propagating constant-length crack, Yoffe (1951), there are two singularities, the front and the rear end points of the crack. For sub-Rayleigh speeds the former is the energy absorber, while the latter is the energy source. The energy radiated by the rear point goes to the front point and thereby supports the crack movement. A body embedded inside the crack can thus move freely together with the crack. Is this the 'paradox' manifestation? It would be so if there was a sort of "channel" through which the energy could be transferred from one point to the other. Otherwise, a real source of energy must exist at the rear point to support the process, that is, a nonzero driving force must exist. In any case, if no such energy source is assumed, the formal solution for the moving finite-length crack cannot be accepted. Nevertheless, a configuration like a propagating constant-length crack can exist as a finite separation zone between two elastic half-planes, or two layers, which are compressed together. Such 'nonsingular' moving zone was considered with respect to a different problem by Comninou and Dundurs (1977).

As we show in the present paper, the above-mentioned three conditions (15.1) are sufficient for the half-plane contact problem, and the solutions, obtained here for a smooth frictionless contact area moving at any speed, are uniquely defined. In the super-Rayleigh subsonic speed range, the leading point of the contact area appears a passive singular point. Thus, for the motion in this range, a speed-dependent driving

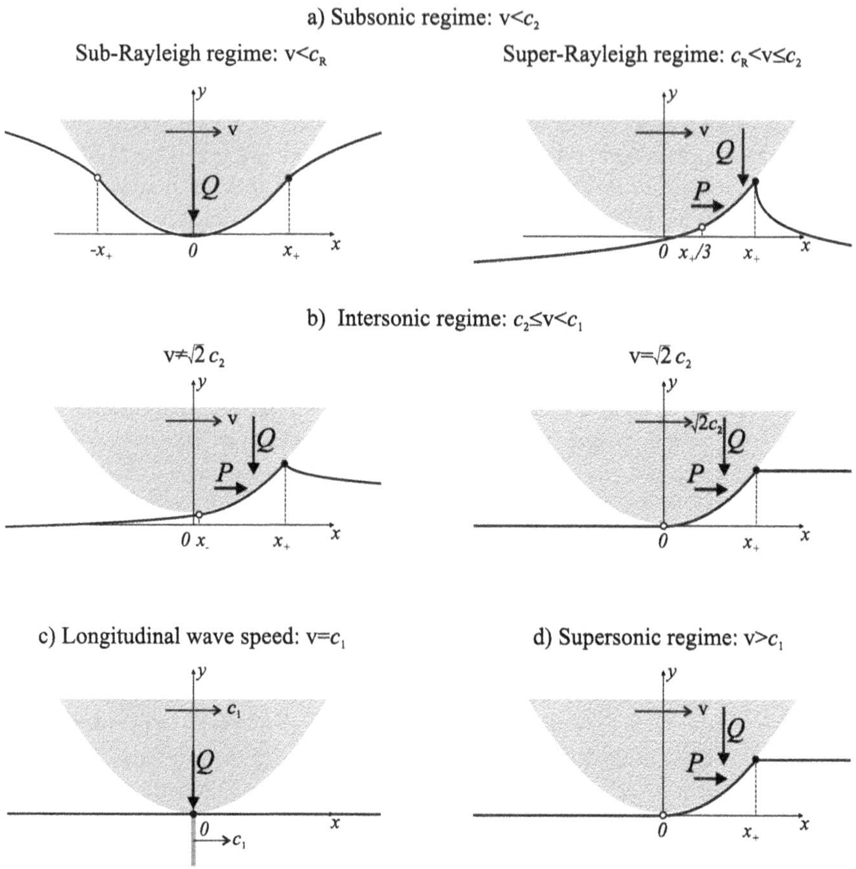

Fig. 8.1 The contact configurations in different speed regimes (Sect. 8.3). The bold curve is the deformed elastic half-plane boundary. The leading, x_+, and the rear, x_-, end points of the contact region are marked by black and white circles, respectively. In this figure, notation x is used instead of η as accepted below beginning from Sect. 8.2.3. The localized longitudinal wave in equilibrium with the indenter is sketched in (**c**)

force is required. In the intersonic and supersonic speed regimes, the wave radiation creates the resistance.

The contact configurations for the sub- and super-Rayleigh subsonic, intersonic and supersonic speed regimes can be compared in Fig. 8.1, where we anticipate the results of Sect. 8.3. It is of interest that not only for subsonic speeds but also for the longitudinal wave speed there is no resistance to the movement in the steady-state regime.

Besides the half-plane contact problem, the related problem of the steady-state wedging of an elastic plane by a smooth rigid body is investigated below. This problem for sub- and super-Rayleigh speeds was considered by Barenblatt and Cherepanov (1960) and Barenblatt and Goldstein (1972), respectively. In the former work, a semi-infinite, 'moving-with-friction' rigid wedge placed at a distance from the crack tip was considered, while in the latter a finite-length, 'moving-without-friction' wedge in contact with the crack tip was analyzed, and condition (a) in (15.1) was used. In both cases, only one simply connected stress-free region was present in the problem formulation. Here, we consider this problem for a finite-length wedge placed at a distance from the crack tip. In this case, there are two stress-free regions, and the problem appears more complicated but analytically solvable. The solution obtained presents relations between all the input and output quantities: the fracture resistance, the wedge shape and width and the driving force (or the speed), on one hand, and the wedge/crack speed (or the driving force), the wedge position, the contact area and the contact stress distribution, on the other.

We also consider the movement of a finite rigid body along the interface of two elastic half-planes compressed together, where no energy significant singularities exist at sub-Rayleigh speeds, and where we meet d'Alembert's paradox manifestation. The papers by Comninou and Dundurs (1977; 1978) prompted the authors to address this problem.

The steady-state solution of these problems is preceded by an analysis of the transient problem for the normal load moving along the elastic half-plane boundary. We construct the solution corresponding to the load distributed as a pre-delta function that results in a more transparent solution, the limit of which represents the corresponding Green's function. Then we derive the steady-state limit and the asymptotic representations. The transient solution allows us, in particular, to extract explicitly the general expressions of the normal traction and the derivative of the displacement trough a single analytical function. These expressions valid for all ranges of the speed are then used to address the above-mentioned mixed problems. The use of the transient problem with zero initial conditions also allows us to exclude automatically the energy flux from infinity, which is important for intersonic and supersonic regimes.

The problems are considered in the framework of linear isotropic elasticity. This implies, in particular, that the strains and rotations are assumed to be small enough, and the contact normal stresses can be considered as a true traction, σ_{yy}, acting on the half-plane boundary. Mathematically, the considered mixed problems are formulated and solved (a) for a single region where the real part of an analytical function is given, whereas the imaginary part of this function is given outside of this region (the dynamic contact problem for the half-plane), (b) for two such regions (the wedging problem for the elastic plane) and (c) for three regions (the two half-planes compressed together). Note that the analytical technique used in this paper is applicable for any number of the 'real-part' regions and for a periodic array of the real/imaginary (stress/displacement) regions. So the considered problems can also be solved, in the same way, for the contact zone consisting of several simply connected regions or of a periodic array of such regions.

The results evidence that in the wedging problem the driving force is required to force the crack to grow as it should be. However, in the sub-Rayleigh speed regime, the driving force is just equal to the crack resistance; the contact interaction gives no additional contribution to the driving force. D'Alembert's paradox is associated with the sub-Rayleigh speed regimes for the other two problems.

In the case of the Rayleigh wave speed there is no steady-state solution of the half-plane problem. In this resonant regime, the displacement grows in proportion to time if a constant vertical force is given, or the contact stresses rapidly vanish in time if a constant vertical displacement of the indenter is given. For the super-Rayleigh subsonic speeds, $c_R < v \leq c_2$, there is energy release though a singular point at the front end of the contact zone. In this case, a driving force is required to support the motion. It can be equally treated as a Newtonian force or as a configurational force of the same power (as in the case where the indenter is a self-propelled body):

$$P = -\int_{x_-}^{x_+} \sigma_{yy}(x)u_y'(x)\,\mathrm{d}x = \frac{1}{v}\int_{x_-}^{x_+} \sigma_{yy}(x)\dot{u}_y(x)\,\mathrm{d}x, \tag{8.2}$$

respectively, where $[x_-, x_+]$ is the contact segment and the equality follows from the fact that, in the steady-state regime, $\partial u_y/\partial t = -v\partial u_y/\partial x$. Note that the first equality is valid for the transient regime too. Indeed, let the half-plane boundary be defined as $y(x,t)$. The partial t-derivative, $\partial y(x,t)/\partial t$, corresponds to a given point on the half-plane boundary, whereas the 'total' derivative, $\mathrm{d}y(x(t),t)/\mathrm{d}t$, corresponds to the vertical speed of the rigid body moving with horizontal speed $v(t) = \mathrm{d}x(t)/\mathrm{d}t$. The latter derivative, the work rate due to the vertical movement of the body, N_Q, and the total rate of the work, N, are

$$\begin{gathered}\frac{\mathrm{d}y(x(t),t)}{\mathrm{d}t} = \frac{\partial y(x,t)}{\partial t} + v(t)\frac{\partial y(x,t)}{\partial x},\\ N_Q = -\int_{x_-}^{x_+} \sigma_{yy}(x,t)\frac{\mathrm{d}y(x(t),t)}{\mathrm{d}t}\,\mathrm{d}x, \quad N = -\int_{x_-}^{x_+} \sigma_{yy}(x,t)\frac{\partial y(x,t)}{\partial t}\,\mathrm{d}x,\end{gathered} \tag{8.3}$$

and the driving force is

$$P = -\frac{N - N_Q}{v(t)} = -\int_{x_-}^{x_+} \sigma_{yy}(x,t)\frac{\partial y(x,t)}{\partial x}\,\mathrm{d}x. \tag{8.4}$$

(The Green's function for the latter derivative is presented below in (8.19).)

In the intersonic regime, $c_2 < v < c_1$, the singular point becomes weaker, it can neither absorb nor release energy; however, a shear wave arises which carries energy from the indenter. At a special speed, $v = c_2\sqrt{2}$, the field is represented by this wave only. It is the speed of the point of intersection of the x-axis with the wave front tilted at an angle of 45°C. In this orientation, the shear wave is supported by the normal pressure, $\sigma_{yy}(x)$, only, and no other field can arise. This is why the half-plane outside of the wave, $x - y < x_-, x - y > x_+$, remains at rest (with a shift of the displacement). Note that by the same reason no wave is excited in the case of a tangential action, which could be referred to mode II symmetry. In the supersonic case, $v > c_1$, both the shear wave and the longitudinal wave are radiated.

Note that the body—elastic half-plane interaction at these speeds, $v = \sqrt{2}c_2$ and $v > c_1$, resembles the aquaplaning, and this is not a coincidence: these are similar wave phenomena.

Lastly, at the longitudinal wave speed limit the resistance appears equal to zero with a sharp decrease in the vicinity of this speed. In this limit, the only localized longitudinal wave exists in the half-plain, in equilibrium with the indenter. This is a special example of the d'Alembert's paradox manifestation.

8.2 Transient Problem for a Moving Load

First we consider the transient problem corresponding to zero initial conditions. The solution will allow us to predict possible steady-state regimes, to extract the steady-state limits, if exist, and to see the transient process otherwise.

8.2.1 General Solution

Consider an elastic half-plane, $-\infty < x < \infty$, $y < 0$, initially resting and subjected at $t = 0$ to a hereafter invariable normal load uniformly moving along the x-axis. The load can be represented as

$$\sigma_{yy}(x,t) = \sigma(\eta)H(t), \quad \eta = x - vt, \quad v = \text{const}, \tag{8.5}$$

where H is the Heaviside step function. In the following, we use the Fourier transform on x ($(..)^F$) and on η ($(..)^{F_\eta}$), and the Laplace transform on time, t ($(..)^L$). Note that in these terms

$$\begin{gathered}\sigma_{yy}^{LF}(k,s) = \int_{-\infty}^{\infty}\int_0^{\infty} \sigma_{yy}(x,t)\mathrm{e}^{ikx-st}\,\mathrm{d}t\,\mathrm{d}x = \frac{\sigma^{F_\eta}(k)}{s-\mathrm{i}kv},\\ \sigma^{F_\eta}(k) = \int_{-\infty}^{\infty}\sigma(\eta)\mathrm{e}^{\mathrm{i}k\eta}\,\mathrm{d}\eta.\end{gathered} \tag{8.6}$$

To avoid singularities and thus to make the corresponding Green's function more transparent we choose the load to be distributed as a regular pre-delta function, namely

$$\sigma(\eta) = \frac{\varepsilon Q_y}{\pi(\eta^2+\varepsilon^2)} \to Q_y\delta(\eta)\ (\varepsilon \to 0),\quad \sigma^{F_\eta}(k) = Q_y\mathrm{e}^{-\varepsilon|k|}\quad (\Im k = 0, \varepsilon > 0), \tag{8.7}$$

where the constant $Q_y = -Q\ (Q \geq 0)$ is the total force. Note that in the plane problem the dimension of Q is N/m.

The Laplace and Fourier, LF-transform of the prelimiting Green's function (multiplied by Q_y) for the normal particle velocity at $y = 0$ is (see, e.g., Slepyan (2002), p. 300)

$$(\dot{u}_y)^{LF}(k,s) = G_\varepsilon^{LF}(k,s) = \frac{s^3 n_1}{c_2^2 \mu R_n} \sigma_{yy}^{LF}(k,s) = \frac{Q_y s^3 n_1 \mathrm{e}^{-\varepsilon|k|}}{c_2^2 (s - \mathrm{i}kv)\mu R_n}, \tag{8.8}$$

where

$$R_n = (k^2 + n_2^2)^2 - 4k^2 n_1 n_2, \quad n_{1,2} = \sqrt{k^2 + s^2/c_{1,2}^2}, \tag{8.9}$$

and the longitudinal and shear wave speeds are

$$c_1 = \sqrt{(\lambda + 2\mu)/\varrho}, \quad c_2 = \sqrt{\mu/\varrho}, \tag{8.10}$$

respectively, with λ, μ the Lamé elastic constants and ϱ the mass density. In addition to these waves, an important role is played by the Rayleigh surface wave. Its speed, $c_R < c_2$, is defined by $R(c_R) = 0$, in which the Rayleigh function

$$\begin{aligned} R(v) &= (1 + \alpha_2^2(v))^2 - 4\alpha_1(v)\alpha_2(v), \ \alpha_{1,2} = \sqrt{1 - v^2/c_{1,2}^2} \ \ (0 < v < c_2), \\ &= (1 - \beta_2^2(v))^2 + 4\mathrm{i}\alpha_1(v)\beta_2(v), \ \beta_2 = \sqrt{v^2/c_2^2 - 1} = \mathrm{i}\alpha_2 \ \ (c_2 < v < c_1), \\ &= (1 - \beta_2^2(v))^2 + 4\beta_1(v)\beta_2(v), \ \beta_1 = \sqrt{v^2/c_1^2 - 1} = \mathrm{i}\alpha_1 \ \ (v > c_1). \end{aligned} \tag{8.11}$$

The expression in (8.8) satisfies the conditions which allow the original expression for function $G_\varepsilon(x,t)$ to be obtained using the Cagniard-de Hoop method, Cagniard (1939) and De Hoop (1958/1959). A modified version of this method, which does not require any transformation of the integration path, was presented in Slepyan (1972, pp. 80–85, 2002, pp. 61–63), where the result was derived explicitly for a general case. Here, for the convenience of the reader, we do not use the final result but the method itself in its application to the considered double transform.

First, we introduce parameter q by equality $k = qs$, assuming $s > 0$. We now have $\mathrm{d}k = s\,\mathrm{d}q$ and

$$\begin{aligned} G_\varepsilon^{LF}(k,s) &= \frac{Q_y}{s} f(q)\mathrm{e}^{-\varepsilon s|q|}, \quad f(q) = \frac{m_1}{c_2^2 \mu (1 - \mathrm{i}qv) R_m(q)}, \\ R_m(q) &= (2q^2 + 1/c_2^2)^2 - 4q^2 m_1 m_2, \quad m_{1,2} = \sqrt{q^2 + 1/c_{1,2}^2}. \end{aligned} \tag{8.12}$$

Next, we represent L-transform of the Green's function as the generalized inverse Fourier transform of $G_\varepsilon^{LF}(k,s)$

$$\begin{aligned} G_\varepsilon^L(z,s) &= \frac{Q_y}{2\pi}[\lim_{z \to -\mathrm{i}x} G_{\varepsilon+}^L(z,s) + \lim_{z \to \mathrm{i}x} G_{\varepsilon-}^L(z,s)], \\ G_{\varepsilon+}^L(z,s) &= \int_{-\infty}^0 f(q)\mathrm{e}^{(\varepsilon+z)sq}\,\mathrm{d}q = \int_0^\infty f(-q)\mathrm{e}^{-(\varepsilon+z)sq}\,\mathrm{d}q \quad (\Re z > 0), \\ G_{\varepsilon-}^L(z,s) &= \int_0^\infty f(q)\mathrm{e}^{-(\varepsilon+z)sq}\,\mathrm{d}q \quad (\Re z > 0). \end{aligned} \tag{8.13}$$

In the third step, we substitute $q(\varepsilon + z) = t$, assuming for the moment that z lies on the real positive half-axis, $z > 0$. This yields

$$G^L_{\varepsilon+}(z, s) = \frac{1}{\varepsilon + z}\int_0^\infty f(-t/(\varepsilon + z))\mathrm{e}^{-st}\,\mathrm{d}t,$$
$$G^L_{\varepsilon-}(z, s) = \frac{1}{\varepsilon + z}\int_0^\infty f(t/(\varepsilon + z))\mathrm{e}^{-st}\,\mathrm{d}t. \tag{8.14}$$

Further, since each of the integrals represent the Laplace transform, it follows that

$$G_{\varepsilon\pm}(z, t) = \frac{m_1}{c_2^2\mu(\varepsilon + z)(1 \pm \mathrm{i}qv)R_m(q)}. \tag{8.15}$$

Now, in accordance with the generalized Fourier transform, we make analytical continuation from $z > 0$ to the imaginary axis, $z \to \mp\mathrm{i}x$ for $G_{\varepsilon\pm}(z, t)$, respectively. We obtain

$$G_\varepsilon(z, t) = \frac{Q_y}{2\pi}[\lim_{z\to-\mathrm{i}x} G_{\varepsilon+}(z, t) + \lim_{z\to\mathrm{i}x} G_{\varepsilon-}(z, t)] \tag{8.16}$$

with

$$q = \pm\frac{\mathrm{i}t}{x \pm \mathrm{i}\varepsilon}, \quad m_{1,2} = \pm\frac{\mathrm{i}t}{x \pm \mathrm{i}\varepsilon}\alpha_{1,2}, \quad (1 \pm \mathrm{i}qv) = 1 - \frac{vt}{x \pm \mathrm{i}\varepsilon},$$
$$R_m = \frac{t^4}{(x \pm \mathrm{i}\varepsilon)^4}R((x \pm \mathrm{i}\varepsilon)/t), \tag{8.17}$$

where symbol $\pm$ corresponds to $G_{\varepsilon\pm}(z, t)$, respectively.

Finally, referring to (8.12) and (8.15) we obtain

$$G_\varepsilon(x, t) = \dot{u}_y(x, t) = \frac{Q_y}{\pi}\Re[\lim_{z\to-\mathrm{i}x} G_{\varepsilon+}(z, t)]$$
$$= -\frac{Q_y}{\pi\mu c_2^2 t^3}\Re\left[\frac{(x + \mathrm{i}\varepsilon)^3\sqrt{1 - (x + \mathrm{i}\varepsilon)^2/(c_1 t)^2}}{(x - vt + \mathrm{i}\varepsilon)R[(x + \mathrm{i}\varepsilon)/t]}\right]. \tag{8.18}$$

This fundamental relation between the stress and time-derivative of the displacement is valid for any speed, v. Below it is extended by superposition for an arbitrary stress distribution and used in the formulation of the contact problems. Note that the above used technique is valid for the determination of the field for $y < 0$ as well as for $y = 0$. We recall that the Green's function itself is the limit of $G_\varepsilon(x, t)$ ($\varepsilon \to +0$).

The Green's functions corresponding the displacement and its x-derivative follow from (8.18) as

$$
\begin{aligned}
G_u(x,t) &= -\frac{Q_y}{\pi\mu c_2^2}\Re\int_{x/t}^{\infty}\frac{\xi^2\sqrt{1-\xi^2/c_1^2}}{(\xi-v)R(\xi)}\,\mathrm{d}\xi \quad (\xi=\lim_{\varepsilon\to+0}\xi+\mathrm{i}\varepsilon),\\
\frac{\partial G_u(x,t)}{\partial x} &= \frac{Q_y}{\pi\mu c_2^2}\Re\frac{(x/t)^2\sqrt{1-(x/c_1t)^2}}{(x-vt)R(x/t)} \quad (x=\lim_{\varepsilon\to+0}x+\mathrm{i}\varepsilon).
\end{aligned}
\tag{8.19}
$$

8.2.2 *The Steady-State Limits and the Asymptotic Representations*

8.2.2.1 The Limits

The steady-state limit, if exists, follows from (8.18) with $\varepsilon > 0$ as

$$
\begin{aligned}
&\lim G_\varepsilon(vt+\eta,t) = G_\varepsilon(\eta) = \dot{u}_y(\eta) = -vu'_y(\eta)\\
&= -\frac{Q_y v^3}{\pi\mu c_2^2}\Re\left[\frac{\sqrt{1-v^2/c_1^2-\mathrm{i}0}}{(\eta+\mathrm{i}\varepsilon)R(v-\mathrm{i}0)}\right] \quad (t\to\infty,\ \eta/t\to 0).
\end{aligned}
\tag{8.20}
$$

For the subsonic speed regime, including the shear wave speed and except the Rayleigh wave speed, c_R, it follows that

$$
G_\varepsilon(\eta) = -\frac{Q_y v^3\alpha_1(v)}{\pi\mu c_2^2 R(v)}\frac{\eta}{\eta^2+\varepsilon^2} \quad (0<v<c_R,\ c_R<v\le c_2). \tag{8.21}
$$

In this speed regime, there is no limit for the displacement.

For the intersonic speed region, $c_2 \le v \le c_1$, we have

$$
G_\varepsilon(\eta) = -\frac{Q_y v^3\alpha_1(v)}{\pi\mu c_2^2}\frac{(1+\alpha_2^2(v))^2\eta-4\alpha_1(v)\beta_2\varepsilon}{(1+\alpha_2^2(v))^4+16\alpha_1^2(v)\beta_2^2(v)}\frac{1}{\eta^2+\varepsilon^2}. \tag{8.22}
$$

Finally, for the supersonic regime, $v \ge c_1$,

$$
G_\varepsilon(\eta) = \frac{Q_y v^3}{\pi\mu c_2^3}\frac{\beta_1(v)}{(1+\alpha_2^2(v))^2+4\beta_1(v)\beta_2(v)}\frac{\varepsilon}{\eta^2+\varepsilon^2}. \tag{8.23}
$$

8.2.2.2 The Asymptotic Representations

There are two special rays in this problem: $x = c_R t$ and $x = c_1 t$, where there is no steady-state limit for $\varepsilon \ge 0$ and for $\varepsilon = 0$, respectively. For these cases we consider asymptotic representations, $t \to \infty$, for the particle velocities and for the displacements. Also we consider the asymptotic difference between the transient and the corresponding steady-state limit for some cases where the latter exists. From the solution (8.18) it follows that

$$
\begin{aligned}
G_\varepsilon &\sim -\frac{Q_y c^3 \alpha_1}{\pi \mu c_2^2 (c-v) R(c) t} \quad (x = ct \neq vt, 0 < c < c_2, c \neq c_R),\\
G_\varepsilon &\sim -\frac{Q_y}{\pi \mu c_2^2} \left(U_\eta + U_t\right) \quad (x = vt + \eta, c_2 > v \neq c_R),\\
U_\eta &= \frac{v^3 \alpha_1}{R(v)} \frac{\eta}{\eta^2 + \varepsilon^2}, \quad U_t = \frac{v^2 \alpha_1}{R(v) t} \left(3 - \frac{v^2}{c_1^2 \alpha_1^2} - \frac{v R'(v)}{R(v)}\right)
\end{aligned}
\tag{8.24}
$$

with

$$
R'(v) = \frac{\mathrm{d}R(v)}{\mathrm{d}v} = \frac{4v}{c_2^2} \left(\frac{c_2^2 \alpha_2}{c_1^2 \alpha_1} + \frac{\alpha_1}{\alpha_2} - 1 - \alpha_2^2\right). \tag{8.25}
$$

The latter asymptotic representation consists of two very different terms. In the moving coordinate system, the first term, $U_\eta = U_\eta(v, \eta)$, is independent of time; it corresponds to the steady-state regime. This term is orthogonal to the load and the corresponding rate of the work and hence the driving force is zero. Indeed, for sub-Rayleigh speeds, the load is symmetric but the particle velocities are antisymmetric. Note that this statement concerning the driving force is still valid for the contact stress distribution corresponding to a smooth indenter of an arbitrary shape (see Sect. 8.3.2).

The second term, $U_t = U_t(v, t)$, is asymptotically independent of the coordinate but depends on time. It corresponds to the subsidence of the elastic half-plane under the load. Contrary to the former the latter has the same signum in both sub-Rayleigh and super-Rayleigh subsonic regimes; namely, $U_t < 0\,(0 < v < c_R, c_R < v < c_2)$. Besides, it is symmetric, and the load does work on this vertical motion. There exists the steady-state limit with respect to the particle velocities since U_t vanishes in time; however, the decrease rate does not result in the limit for the displacement which grows unboundedly. Nevertheless, the limit exists in term of relative displacement between different points in a finite region. Note that this is associated with the 2D infinite domain; the displacement has a finite limit in the 3D case.

Thus, the steady-state formulation is acceptable for the determination of the particle velocities and the disturbed shape of the half-plane boundary. The asymptotic representation of the vertical displacement at the center of the load support, $x = vt$, which is valid for nonzero subsonic speeds, except the Rayleigh wave speed, follows from (8.18) as

$$
\begin{aligned}
u(vt, t) &= \int_0^{x/v} G_\varepsilon(x, t)\, \mathrm{d}t \sim -\frac{Q_y v^2}{\pi \mu c_2^2} \left(u_\infty + \Re \frac{\sqrt{1 - v^2/c_1^2 - \mathrm{i}0}}{R(v + \mathrm{i}0)} \ln \frac{vt}{\varepsilon}\right),\\
&\qquad (x = vt,\ vt/\varepsilon \to \infty),\\
u_\infty &= \Re \int_0^1 \frac{f(\tau) - f(1)}{1 - \tau}\, \mathrm{d}\tau, \quad f(\tau) = \frac{\sqrt{1 - v^2/(c_1^2 \tau^2) - \mathrm{i}0}}{\tau^3 R(v/\tau + \mathrm{i}0)}.
\end{aligned}
\tag{8.26}
$$

In this asymptotic representation, we take into account the constant term along with the logarithmic one because the latter increases very slowly.

It can be seen in (8.26) that the logarithmic term does not exist for $v = \sqrt{2}\,c_2$ and for the supersonic regime (in these cases, the coefficient of the logarithm is pure imaginary). Thus, at these speeds the displacement have the steady-state limit as well as its derivative.

For the Rayleigh wave speed

$$
\begin{aligned}
G_\varepsilon &\sim -\frac{Q_y c_R^3 \sqrt{1 - c_R^2/c_1^2}}{\pi \mu c_2^2 (c_R - v) R'(c_R)} \frac{\eta}{\eta^2 + \varepsilon^2} \quad (x = c_R t + \eta,\ v \neq c_R,\ 0 < v < c_2),\\
G_\varepsilon &\sim -\frac{Q_y c_R^3 \sqrt{1 - c_R^2/c_1^2}}{\pi \mu c_2^2 R'(c_R)} \frac{(\eta^2 - \varepsilon^2) t}{(\eta^2 + \varepsilon^2)^2} \quad (x = c_R t + \eta,\ v = c_R),
\end{aligned}
\tag{8.27}
$$

the steady-state regime at the Rayleigh wave speed, $v = c_R$, cannot be reached.

For $v = c_1$

$$
G_\varepsilon \sim \frac{Q_y c_1^3 c_2^2}{\pi \mu (c_1^2 - 2c_2^2)^2 \sqrt{c_1 t}} \frac{\sqrt{\sqrt{\eta^2 + \varepsilon^2} - \eta}}{\sqrt{\eta^2 + \varepsilon^2}}. \tag{8.28}
$$

It follows that the particle velocities and the displacements in a vicinity of $x = c_1 t$ tend to zero. It is remarkable that this statement is entirely valid for a distributed load ($\varepsilon > 0$), but not for the concentrated one ($\varepsilon = 0$). Indeed, at the limit, $\varepsilon = 0$, the square-root type singular point arises as const$\times H(-\eta)/\sqrt{-t\eta}$, which, however, vanishes by its intensity as $t \to \infty$. At this speed, the true Green's function for the displacements ($\varepsilon = +0$) can be found from (8.18) as

$$
u_y(x) = \frac{Q_y c_1^2 H(c_1^2 t^2 - x^2)}{\pi \mu c_2^2} \int_{x/(c_1 t)}^{1} \sqrt{\frac{1 + \xi}{1 - \xi}}\, \Re\left[\frac{1}{R(c_1 \xi)}\right] \xi^2 \,\mathrm{d}\xi. \tag{8.29}
$$

The corresponding plot, $U(X) = u_y(X)\pi\mu/Q$ with $X = x/(c_1 t)$, is presented in Fig. 8.2. As shown below, Sect. 8.3, Fig. 8.6, the steady-state regime for a smooth indenter, moving at this speed, corresponds to zero driving force.

8.2.3 *The Steady-State Limit for a General Load*

In the following, considering the steady-state regime, we express the solution in the moving coordinates. With this in mind we below keep notation x instead of η as x belongs to the moving coordinate system, the origin of which is attached to the lower point of the boundary of the moving body as indicated in Figs. 8.1, 8.7 and 8.11. Recall that in these terms, $\partial/\partial t = -v\partial/\partial x$.

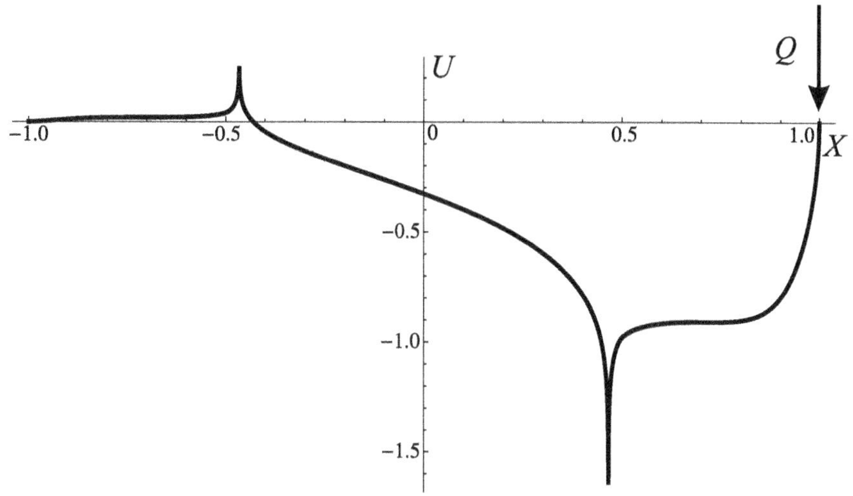

Fig. 8.2 The normalized displacement caused by a concentrated normal force moving at the longitudinal wave speed ($v = c_1$), $U(X) = u_y(X)\pi\mu/Q$ with $X = x/(c_1 t)$. The transient problem

Consider the relation between the load (8.7) and the steady-state limit of $u_y'(x)$ (8.20). It can be seen that both quantities are expressed in terms of one analytical function as follows

$$\sigma_{yy}(x) = -Q_y \Im \frac{1}{\pi\zeta}, \quad u_y'(x) = \frac{Q_y}{\mu} \Re \frac{1}{\pi\Omega(v)\zeta},$$
$$\Omega(v) = \frac{c_2^2 R(v + \mathrm{i}0)}{v^2\sqrt{1 - v^2/c_1^2 - \mathrm{i}0}}, \quad \zeta = x + \mathrm{i}\varepsilon. \tag{8.30}$$

In the limit, $\varepsilon \to +0$, the stress corresponds to the Dirac delta function (multiplied by Q_y); hence the expression for $u_y'(x)$ becomes the corresponding Green's function. So we write

$$u_y'(x) = \lim_{\varepsilon\to+0} \Re \frac{1}{\pi\Omega(v)\zeta} \quad \text{corresponds to} \quad \sigma_{yy}(x) = -\mu \lim_{\varepsilon\to+0} \Im\,[1/(\pi\zeta)] = \mu\delta(x). \tag{8.31}$$

Let an arbitrarily given stress distribution be $\sigma_0(x)$. We make a convolution on x of $\sigma_0(x)$ with each part of the relation in (8.31), or, which is the same, the convolution with the pre-delta. We find that if

$$\varphi(\zeta) = -\frac{1}{\mu\pi}\int_{-\infty}^{\infty} \frac{\sigma_0(\xi)}{\xi - \zeta}\,\mathrm{d}\xi \quad (\varepsilon > 0), \tag{8.32}$$

then

$$\sigma_{yy}(x) = -\mu \lim_{\varepsilon \to +0} \Im \varphi(\zeta), \quad u'_y(x) = \lim_{\varepsilon \to +0} \Re \frac{\varphi(\zeta)}{\Omega(v)}. \tag{8.33}$$

We define

$$\Omega(v) = \frac{c_2^2 R(v)}{v^2 \alpha_1(v)} = \omega_1(v) + \mathrm{i}\omega_2(v), \quad \frac{1}{\Omega(v)} = \psi_1(v) + \mathrm{i}\psi_2(v), \tag{8.34}$$

where $\omega_{1,2}(v)$, $\psi_{1,2}(v)$ are real.

In these terms

$$u'_y(x) = -\frac{\psi_1(v)}{\mu\pi} V.p. \int_{-\infty}^{\infty} \frac{\sigma_0(\xi)}{\xi - x} \mathrm{d}\xi + \frac{\psi_2(v)}{\mu} \sigma_{yy}(x). \tag{8.35}$$

Replacing in (8.33) $\varphi(\zeta) \to \varphi(\zeta)\Omega(v)$ we obtain the inverse relation

$$\begin{gathered} \sigma_{yy}(x) = -\mu \lim_{\varepsilon \to +0} \Im[\Omega(v)\varphi(\zeta)] = \mu \left[\frac{\omega_1(v)}{\pi} V.p. \int_{-\infty}^{\infty} \frac{u'_y(\xi)}{\xi - x} \mathrm{d}\xi - \omega_2 u'_y(x) \right], \\ u'_y(x) = \lim_{\varepsilon \to +0} \Re \varphi(\zeta). \end{gathered} \tag{8.36}$$

with

$$\varphi(\zeta) = \frac{1}{\pi \mathrm{i}} \int_{-\infty}^{\infty} \frac{u'_y(\xi)}{\xi - \zeta} \mathrm{d}\xi \quad (\varepsilon > 0). \tag{8.37}$$

These relations, where $\sigma_{yy}(x)$ and $u'_y(x)$ are expressed through one analytical function, represent the base for the solution of the mixed problems considered below. Note that a homogeneous solution can be added, which must satisfy conditions at infinity and other conditions of the problem, as done, for example, in (8.42).

8.3 The Half-Plane Steady-State Contact Problem

We consider the steady-state regime of a frictionless-contact of a smooth rigid body moving along the half-plane boundary. The plane strain dynamic problem is examined with the conditions at the boundary, $y = 0$

$$\begin{gathered} u'(x) = y'(x), \quad x \in (x_-, x_+); \quad \sigma_{yy} = 0, \quad x \notin (x_-, x_+); \\ u'(x) = O(1/|x|) \quad |x| \to \infty. \end{gathered} \tag{8.38}$$

Recall that the moving coordinate system is used.

The solution is obtained for any speed and for an arbitrary shape of the indenter but such that it corresponds to a single simply connected contact domain. For the parabolic indenter the distributions of the contact stresses and the displacements are presented explicitly. Then each speed range is examined in more detail. Although the solutions to this problem for the sub-Rayleigh and supersonic speed regimes and for the special intersonic speed $v = \sqrt{2}\, c_2$ are known, for completeness we present the corresponding simplest results among the others.

8.3.1 General Speed Solution

Referring to relations (8.36)-(8.37) and (8.11) we introduce a new analytical function of the complex variable $\zeta = x + \mathrm{i}\varepsilon$, $\varepsilon > 0$

$$\Phi(\zeta) = \Psi(\zeta)\varphi(\zeta),\quad \Psi(\zeta) = (\zeta - x_-)^{1/2+\nu}(x_+ - \zeta)^{1/2-\nu},$$
$$0 \le \nu = \frac{1}{\pi}\arctan\frac{\omega_2(v)}{\omega_1(v)} \le \frac{1}{2}, \tag{8.39}$$

with[1]

$$\begin{aligned}\Psi(x) &= (x - x_-)^{1/2+\nu}(x_+ - x)^{1/2-\nu} \quad (x_- < x < x_+),\\ &= -\mathrm{i}(x - x_-)^{1/2+\nu}(x - x_+)^{1/2-\nu}\mathrm{e}^{\mathrm{i}\pi\nu} \quad (x > x_+)\\ &= \mathrm{i}(x_- - x)^{1/2+\nu}(x_+ - x)^{1/2-\nu}\mathrm{e}^{\mathrm{i}\pi\nu} \quad (x < x_-).\end{aligned} \tag{8.40}$$

The real part of this function is known on the whole x-axis

$$\Re\Phi(x) = \Psi(x)y'(x) \quad (x_- < x < x_+), \quad \Re\Phi(x) = 0 \quad (x < x_-, x > x_+), \tag{8.41}$$

where $y(x)$ is the coordinate of the indenter lower surface. We define this function by the Cauchy type integral with an additional imaginary constant

$$\Phi(\zeta) = \frac{1}{\pi\mathrm{i}}\int_{x_-}^{x_+}\frac{\Psi(\xi)y'(\xi)}{\xi - \zeta}\,\mathrm{d}\xi + \mathrm{i}C. \tag{8.42}$$

It follows from this and (8.36) that the contact stresses[2] are

$$\sigma_{yy}(x) = \mu\left[\frac{\omega_1(v)}{\Psi(x)}\left(\frac{1}{\pi}V.p.\int_{x_-}^{x_+}\frac{\Psi(\xi)y'(\xi)}{\xi - x} - C\right) - \omega_2(v)y'(x)\right]. \tag{8.43}$$

Quantities $\omega_{1,2}$ and ν as functions of the speed, v, are plotted in Fig. 8.3. Outside the contact region

---

[1] Note that ν is not Poisson's ratio!.

[2] In the following part of this section, for brevity, we will not indicate the contact region, $x \in (x_-, x_+)$, for the contact stresses.

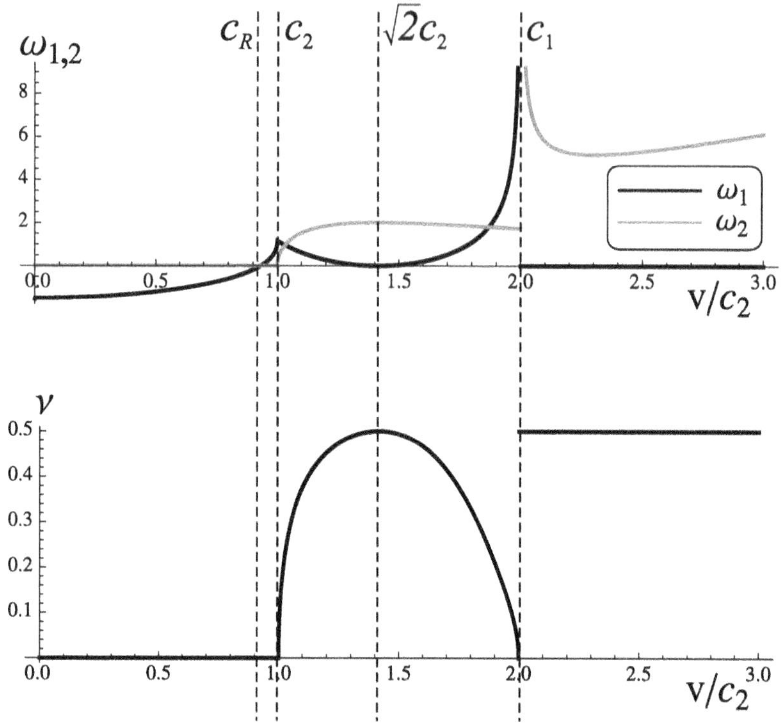

Fig. 8.3 The real and imaginary parts of the function $\Omega(v)$ (8.34) and the parameter ν (8.39), computed for Poisson's ratio equal to $1/3$ ($c_1 = 2c_2$)

$$
\begin{aligned}
u_y'(x) &= -\frac{\cos\pi\nu}{(x-x_+)^{1/2-\nu}(x-x_-)^{1/2+\nu}}\left[\frac{1}{\pi}\int_{x_-}^{x_+}\frac{\Psi(\xi)y'(\xi)}{x-\xi}\,\mathrm{d}\xi + C\right] \quad (x > x_+),\\
&= \frac{\cos\pi\nu}{(x_+-x)^{1/2-\nu}(x_--x)^{1/2+\nu}}\left[C-\frac{1}{\pi}\int_{x_-}^{x_+}\frac{\Psi(\xi)y'(\xi)}{\xi-x}\,\mathrm{d}\xi\right] \quad (x < x_-).
\end{aligned}
\tag{8.44}
$$

Then we have to eliminate the singularity at $x = x_-$. For $\nu > 0$ it is too strong and it cannot be accepted because it corresponds to an infinite energy. For $\nu = 0$ it leads to violation of condition (b) in (15.1) (if $v < c_R$) or condition (c) in (15.1) (for $c_R < v \le c_2$). Note that this requirement is also in agreement with the finite-stress condition introduced in Barenblatt and Cherepanov (1960) for the leading separation point (in the case of a smooth semi-infinite wedge—for sub-Rayleigh speeds) and in Barenblatt and Goldshtein (1972) for the rear end point (in the case of a smooth wedge adjacent to the crack tip—for super-Rayleigh subsonic speeds). However, this condition cannot be applied to the leading separation point for an indenter moving with the super-Rayleigh speed at a distance from the crack tip.

So we define constant C as

$$C = \frac{1}{\pi}\int_{x_-}^{x_+}\left(\frac{x_+ - \xi}{\xi - x_-}\right)^{1/2-\nu} y'(\xi)\,\mathrm{d}\xi. \tag{8.45}$$

It is now found that

$$\begin{aligned}
\frac{\sigma_{yy}(x)}{\mu} &= \left(\frac{x - x_-}{x_+ - x}\right)^{1/2-\nu}\frac{\omega_1(v)}{\pi}V.p.\int_{x_-}^{x_+}\left(\frac{x_+ - \xi}{\xi - x_-}\right)^{1/2-\nu}\frac{y'(\xi)}{\xi - x}\,\mathrm{d}\xi - \omega_2(v)y'(x),\\
u_y'(x) &= -\left(\frac{x - x_-}{x - x_+}\right)^{1/2-\nu}\frac{\cos\pi\nu}{\pi}\int_{x_-}^{x_+}\left(\frac{x_+ - \xi}{\xi - x_-}\right)^{1/2-\nu}\frac{y'(\xi)}{x - \xi}\,\mathrm{d}\xi \quad (x > x_+),\\
u_y'(x) &= \left(\frac{x_- - x}{x_+ - x}\right)^{1/2-\nu}\frac{\cos\pi\nu}{\pi}\int_{x_-}^{x_+}\left(\frac{x_+ - \xi}{\xi - x_-}\right)^{1/2-\nu}\frac{y'(\xi)}{\xi - x}\,\mathrm{d}\xi \quad (x < x_-).
\end{aligned} \tag{8.46}$$

The asymptotic representation for the contact zone ends are

$$\begin{aligned}
u_y'(x) &\sim -\left(\frac{x_+ - x_-}{x - x_+}\right)^{1/2-\nu}\frac{\cos\pi\nu}{\pi}\int_{x_-}^{x_+}\frac{y'(\xi)\,\mathrm{d}\xi}{(\xi - x_-)^{1/2-\nu}(x_+ - \xi)^{1/2+\nu}}\\
\frac{\sigma_{yy}}{\mu} &\sim -\left(\frac{x_+ - x_-}{x_+ - x}\right)^{1/2-\nu}\frac{\omega_1(v)}{\pi}\int_{x_-}^{x_+}\frac{y'(\xi)\,\mathrm{d}\xi}{(\xi - x_-)^{1/2-\nu}(x_+ - \xi)^{1/2+\nu}} + \omega_2(v)y'(x_+).
\end{aligned} \tag{8.47}$$

for $x \to x_+ \pm 0$, respectively, and

$$u_y'(x) \to y'(x_-) \quad (x \to x_- - 0), \quad \sigma_{yy} \to 0 \quad (x \to x_- + 0). \tag{8.48}$$

These results are valid for any ν, $0 \le \nu \le 1/2$, that is for any speed regime, $0 < v$.

In the case of a symmetric indenter, $y(-x) = y(x)$ and a symmetric contact zone, $x_- = -x_+$, (that corresponds to the sub-Rayleigh speed regime, where $\nu = 0$, $\omega_1(v) = \Omega(v) < 0$, $\omega_2(v) = 0$, see Fig. 8.3) the results can be simplified. Symmetrization leads to

$$\begin{aligned}
\sigma_{yy}(x) &= \mu\Omega(v)\sqrt{x_+^2 - x^2}\,\frac{2}{\pi}V.p.\int_0^{x_+}\frac{\xi y'(\xi)}{\sqrt{x_+^2 - \xi^2}\,(\xi^2 - x^2)}\,\mathrm{d}\xi,\\
u_y'(x) &= \sqrt{x^2 - x_+^2}\,\frac{2}{\pi}\int_0^{x_+}\frac{\xi y'(\xi)}{\sqrt{x_+^2 - \xi^2}\,(x^2 - \xi^2)}\,\mathrm{d}\xi\ \mathrm{sign}\,x \quad (|x| > x_+).
\end{aligned} \tag{8.49}$$

For a symmetric parabolic indenter, $u_y'(x) = x/r$, results in (8.49) specialize to

$$\begin{aligned}
\sigma_{yy}(x) &= \frac{\mu\Omega(v)}{r}\sqrt{x_+^2 - x^2},\\
u_y'(x) &= \frac{1}{r}\left[x - \mathrm{sign}\,x\sqrt{x^2 - x_+^2}\right] \quad (|x| \ge x_+).
\end{aligned} \tag{8.50}$$

The displacement profile has the form

$$u(x) = \frac{1}{2r}\left[x^2 - |x|\sqrt{x^2 - x_+^2} + x_+^2 \ln\left(|x/x_+| + \sqrt{(x/x_+)^2 - 1}\right)\right] \quad (|x| \geq x_+),$$
$$u(x) = \frac{x^2}{2r} \quad (|x| \leq x_+). \tag{8.51}$$

The indenter—half-plain configurations are shown in Fig. 8.1a left.

For **greater speeds**, $v > c_R$, when $\omega_1 \geq 0$, for a parabolic indenter, $u_y'(x) = x/r$, it follows from (8.46) that

$$\frac{\sigma_{yy}(x)}{\mu} = -\frac{\omega_1(v)}{\cos\pi\nu}\left(\frac{x - x_-}{x_+ - x}\right)^{1/2-\nu}\frac{x - x_- + \gamma}{r},$$
$$u_y'(x) = \frac{x}{r} + \left(\frac{x_- - x}{x_+ - x}\right)^{1/2-\nu}\frac{x_- - x - \gamma}{r} \quad (x \leq x_-),$$
$$u_y'(x) = \frac{x}{r} - \left(\frac{x - x_-}{x - x_+}\right)^{1/2-\nu}\frac{x - x_- + \gamma}{r} \quad (x > x_+),$$
$$\gamma = (1/2 - \nu)x_+ - (3/2 - \nu)x_- \quad (x_- \leq x < x_+). \tag{8.52}$$

Note that, in the derivation of these expressions, we have used the identity

$$\int_0^1 \left(\frac{1-\xi}{\xi}\right)^{1/2-\nu}\frac{\mathrm{d}\xi}{\xi - \mathrm{i}\varepsilon} = \frac{\pi}{\cos\pi\nu}\left[\left(1 + \frac{\mathrm{i}}{\varepsilon}\right)^{1/2-\nu} - 1\right] \quad (\Re\varepsilon > 0, 0 \leq \nu < 1/2) \tag{8.53}$$

with the corresponding analytical continuation.

We find from the expression for $\sigma_{yy}(x)$ in (8.52) that the condition (a) in (15.1) is satisfied only if $\gamma \geq 0$. On the other hand, as can be seen in the expression for $u_y'(x)$ $(x \leq x_-)$ the condition (b) in (15.1) is satisfied only if $\gamma \leq 0$. Thus, to satisfy both conditions the only possibility remains that $\gamma = 0$. Now the expressions in (8.52) [also see (8.46)] can be rewritten in the form

$$\sigma_{yy}(x) = -\frac{\mu}{r}\frac{\omega_1(v)}{\cos\pi\nu}\frac{(x - x_-)^{3/2-\nu}}{(x_+ - x)^{1/2-\nu}} \quad (\omega_1 \neq 0),$$
$$\sigma_{yy}(x) = -\mu\,\omega_2(v)\frac{x}{r} \quad (\omega_1 = 0,\ \omega_2 \neq 0),$$
$$u_y'(x) = \frac{x}{r} + \frac{(x_- - x)^{3/2-\nu}}{r(x_+ - x)^{1/2-\nu}} \quad (x \leq x_-),$$
$$u_y'(x) = \frac{x}{r} - \frac{(x - x_-)^{3/2-\nu}}{r(x - x_+)^{1/2-\nu}} \quad (x > x_+),$$
$$x_- = \frac{1 - 2\nu}{3 - 2\nu}x_+. \tag{8.54}$$

Note that the latter relation also results in elimination of constant terms in the asymptotic representation of $u_y'(x)$ for $x \to \mp\infty$, respectively. For $\nu < 1/2$ they become of the order of $1/|x|$, for $\nu = 1/2$ the derivatives outside the contact region

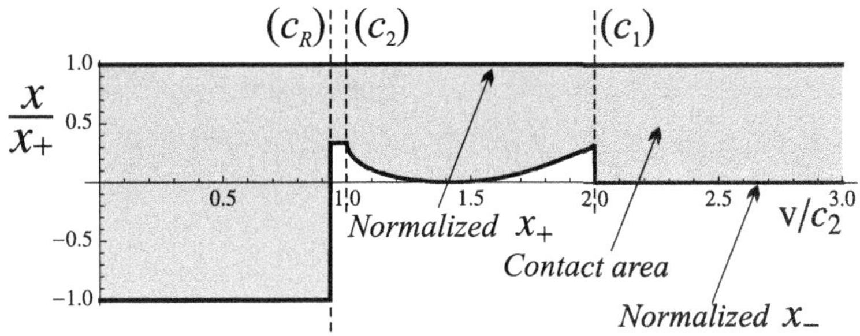

Fig. 8.4 The normalized contact zone, $x_-/x_+ < x/x_+ < 1$, as a function of the speed, v. Results correspond to Poisson's ratio equal to 1/3 ($c_1 = 2c_2$). Note that the value of x_+ depends on both v and Q_y (8.55), (8.60), (8.74) and (8.81). In the antiresonance case $v = c_1$, $x_+ = x_- = 0$

are zero ($x_- = 0$). The plots of x_-/x_+ and of these stress distributions are presented in Figs. 8.4 and 8.5, respectively.

8.3.2 *Subsonic Regime*

In subsonic regime, function $\Omega(v)$ is real, $\Omega(v) < 0$ in the sub-Rayleigh speed regime, $0 < v < c_R$, and $\Omega(v) > 0$ in the super-Rayleigh subsonic speed regime, $c_R < v < c_2$.

Sub-Rayleigh regime. In the sub-Rayleigh speed regime, the stress and displacement derivative distributions are presented in (8.50). The remaining parameter, that is, the contact zone half-length follows from the equilibrium relation as

$$Q_y = 2\int_0^{x_+} \sigma_{yy}(x)\,\mathrm{d}x = \frac{\pi\mu x_+^2\Omega(v)}{2r} \quad \Longrightarrow \quad x_+ = \sqrt{\frac{2rQ_y}{\pi\mu\Omega(v)}}. \tag{8.55}$$

(Note that $Q_y < 0$ and $R(v)$ is also negative in this speed regime.) There is no resistance to the movement, D'Alembert's paradox is in force.

8.3.2.1 Zero Driving Force in the Sub-Rayleigh Speed Regime. A General Statement

In the case of the sub-Rayleigh speed regime, $0 < v < c_R$, there is no resistance to the movement of a smooth indenter of any shape. Although this statement follows directly from energy considerations, here we present a formal proof. We first note that, in this speed range, in accordance with condition (b) in (15.1), the remaining singularity at $x = x_+$ must be eliminated. Referring to (8.46) with $x \to x_+$ we thus have to put

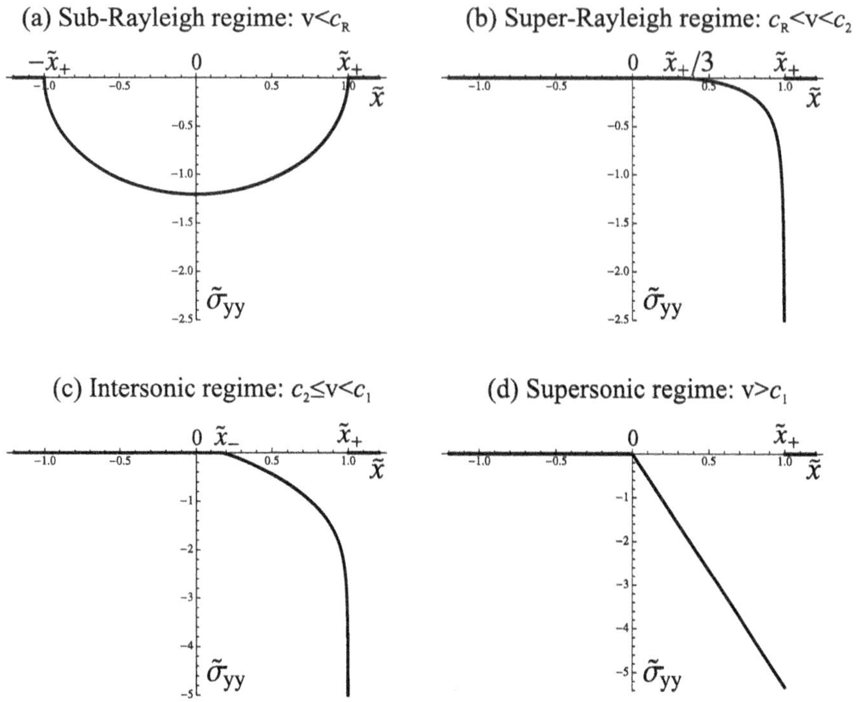

Fig. 8.5 The contact stress distributions as a function of the normalized coordinate $\tilde{x} = x/x_+$. The plots are based on (8.50) for sub-Rayleigh speed regime, (8.54) for super-Rayleigh subsonic and intersonic speed regimes and (8.81) for the supersonic speed regime. The special case $v = \sqrt{2}c_2$ (8.77) is the same as in the supersonic regime. At $v = c_1$ the contact zone is localized, $\sigma_{yy} = Q_y\,\delta(x)$ (8.83). In the super-Rayleigh subsonic (**b**) and in the intersonic (**c**) speed regimes, the leading end points of the contact zone, x_+, are singular, $\sigma_{yy} \to -\infty$ as $x \to x_+ - 0$. In (**b**) it is a passive square-root type point, whereas in (**c**) it is a weak point, which can neither absorb nor release energy

$$\int_{x_-}^{x_+} \frac{y'(\xi)\,\mathrm{d}\xi}{\sqrt{(x_+ - \xi)(\xi - x_-)}} = 0\,. \tag{8.56}$$

It also follows from (8.46) that the driving force is

$$\begin{aligned} P &= -\int_{x_-}^{x_+} \sigma_{yy}(x)y'(x)\,\mathrm{d}x \\ &= -\frac{\mu\Omega(v)}{\pi}\int_{x_-}^{x_+}\sqrt{\frac{x - x_-}{x_+ - x}}\;P.v.\!\int_{x_-}^{x_+}\sqrt{\frac{x_+ - \xi}{\xi - x_-}}\,\frac{y'(\xi)y'(x)}{\xi - x}\,\mathrm{d}\xi\,\mathrm{d}x\,. \end{aligned} \tag{8.57}$$

Consider a sum of the integrant corresponding to points which are symmetric relatively to the singular line, $\xi = x$, that is, points (x, ξ) and (ξ, x). It is

$$\frac{y'(\xi)y'(x)}{\xi - x}\left[\sqrt{\frac{(x-x_-)(x_+-\xi)}{(x_+-x)(\xi-x_-)}} - \sqrt{\frac{(\xi-x_-)(x_+-x)}{(x_+-\xi)(x-x_-)}}\right] = -\frac{y'(\xi)y'(x)(x_+-x_-)}{\sqrt{(x_+-\xi)(\xi-x_-)(x-x_-)(x_+-x)}}. \tag{8.58}$$

The latter expression is symmetric relatively this line, and we can preserve the integration over the initial square domain, $x_- < (x, \xi) < x_+$. Now we find from (8.58) and (8.56) that $P = 0$, as it should be.

Super-Rayleigh regime. In the super-Rayleigh subsonic regime, $c_R < v < c_2$, where $\Omega(v) > 0$, the contact zone is defined in (8.54) with $\nu = 0$

$$x_- = \frac{x_+}{3}. \tag{8.59}$$

The stress and displacement's derivative distributions are also presented in (8.54). This solution corresponds to the vertical force

$$Q_y = \int_{x_+/3}^{x_+} \sigma_{yy}\, \mathrm{d}x = -\frac{\pi\Omega(v)\mu x_+^2}{6r}, \tag{8.60}$$

applied at position

$$x = x_Q = \int_{x_+/3}^{x_+} x\,\sigma_{yy}\, \mathrm{d}x \Big/ \int_{x_+/3}^{x_+} \sigma_{yy}\, \mathrm{d}x = \frac{8}{9}x_+. \tag{8.61}$$

In this case, the front end point of the contact zone, x_+, is the singular point of the square root type, and there exists the energy flux to this point. The corresponding stress intensity factor is

$$K_I \equiv \lim_{x\to x_+-0} \sigma_{yy}\sqrt{2\pi(x_+-x)} = -\left(\frac{2x_+}{3}\right)^{3/2}\frac{\sqrt{2\pi}\,\mu\Omega(v)}{r} \tag{8.62}$$

and the energy release rate is

$$G = \frac{K_I^2}{4\mu\Omega(v)} = \frac{4\pi\mu\Omega(v)x_+^3}{27r^2} = \frac{8}{9}Q\sqrt{\frac{6Q}{\pi\Omega(v)\mu r}}. \tag{8.63}$$

(The relation between the energy release rate and the stress intensity factors for crack dynamics was introduced in Kostrov et al. (1969).) This relation (see Fig. 8.6) just corresponds to the horizontal driving force

$$P = -\int_{x_+/3}^{x_+} \sigma_{yy}(x)u_y'(x)\, \mathrm{d}x = G, \tag{8.64}$$

which is not surprising since there is no wave radiation at these speeds, and the energy produced by the contact stresses completely goes to the singular point. Recall that this driving force can be equally treated as a Newtonian horizontal force or as a configurational force. The non-dimensional $P - Q$ relation looks as

$$\mathcal{P} = \frac{P}{Q}\sqrt{\frac{\mu r}{Q}} = \frac{8}{9}\sqrt{\frac{6}{\pi\Omega(v)}}. \tag{8.65}$$

The equality, $P = G$, confirms that no wave is radiated in this speed regime. In this connection, note that the radiation mentioned in Barenblatt and Goldstein (1960) can be caused by the active singular point existing at the crack tip in the super-Rayleigh speed regime.

The displacement profile has the form

$$\begin{gathered} u(x) = \frac{x_+^2}{2r}\left\{\tilde{x}^2 + (\tilde{x} + 2/3)\sqrt{(1-\tilde{x})(1/3-\tilde{x})}\right. \\ \left. + \frac{2}{3}\ln\left[\sqrt{3/2}\left(\sqrt{1-\tilde{x}} + \sqrt{1/3-\tilde{x}}\right)\right]\right\} \quad (x \le x_+/3), \\ u(x) = \frac{x^2}{2r} \quad (x_+/3 \le x \le x_+), \\ u(x) = \frac{x_+^2}{2r}\left\{\tilde{x}^2 - (\tilde{x} + 2/3)\sqrt{(\tilde{x}-1)(\tilde{x}-1/3)}\right. \\ \left. - \frac{2}{3}\ln\left[\sqrt{3/2}\left(\sqrt{\tilde{x}-1} + \sqrt{\tilde{x}-1/3}\right)\right]\right\} \quad (x \ge x_+), \end{gathered} \tag{8.66}$$

where $\tilde{x} = x/x_+$. The indenter—half-plain configurations are shown in Fig. 8.1a right.

What happens at the front singular points, through which the energy releases under compression? What happens to this energy, where is it spent? The classical elasticity cannot answer these questions, whereas the 'micro'-structure mechanisms of the energy dissipation can be different. One of them is plasticity. Indeed, in a vicinity of the singular point, the normal stresses are (see, e.g., Slepyan (2002), pp. 303–304)

$$\sigma_{xx}(x) \sim -\sigma_{yy}\left[1 + \frac{2(\alpha_1^2(v) - \alpha_2^2(v))(1+\alpha_2^2(v))}{R(v)}\right], \quad \sigma_{yy}(x) \sim \frac{K_I}{\sqrt{2\pi(x_+ - x)}}. \tag{8.67}$$

In this case, $\sigma_{yy}(x) < 0$ ($K_I < 0$), $\sigma_{xx}(x) > 0$ and $\sigma_{xx}(x) > |\sigma_{yy}(x)|$. Thus this point is singular with respect to shear stresses. This suggests that plastic deformations possibly with melting can occur in a surface layer. For a brittle material the high level tensile stresses can result in cracking and fragmentation of the surface layer. Note that latter inequality was mentioned in Barenblatt and Cherepanov (1960) for the case of a moving wedge approaching the Rayleigh wave speed from below.

8.3.2.2 The Crack Healing Limit

Expressions (8.54) valid for the parabolic indenter can be reduced to match the super-Rayleigh subsonic speed regime of the crack closure. For this regime, where $\nu = 0,\ \omega_1(v) = \Omega(v) > 0$, we put

$$r \to \infty, \quad x_+ \to \infty, \quad \frac{((2/3)x_+)^{3/2}}{r} \to \sqrt{a}\ (a > 0), \\ u_y(x_+) = 0, \quad x \to x_+ + x. \tag{8.68}$$

For any finite x we find

$$u_y(x) = 0 \quad (x \le 0), \quad u_y(x) = -\sqrt{2ax} \quad (x \ge 0), \\ \sigma_{yy}(x) = -\mu\Omega(v)\sqrt{a/(-x)}H(-x) \quad (c_R < v \le c_2), \tag{8.69}$$

where the displacement corresponds to the lower half-plane. The equilibrium is achieved by adding the symmetrically disturbed upper half-plane. As a result we have the super-Rayleigh subsonic speed regime of the crack closure with the energy absorbing singular point, $x = 0$ (recall that x is the moving coordinate), the end point of the semi-infinite crack, $0 < x < \infty$. Now the energy is coming from the infinity; the energy release rate is

$$G = \pi\mu a\Omega(v). \tag{8.70}$$

If G is large enough the crack healing may be expected. The limit, a, depends on the speed. Assume the healing criterion to be $G = G_c = \text{const}$; it follows that the critical value of the above limit and the critical stress intensity factor are

$$a = a_c(v) = \frac{G_c}{\pi\mu\Omega(v)}, \quad K_{Ic}(v) = -\sqrt{2\mu G_c\Omega(v)}. \tag{8.71}$$

The latter relation is in agreement with the Kostrov-Nikitin-Flitman formula (Kostrov et al. (1969)). Thus the critical stress intensity factor increases by its modulus from zero to

$$K_{max} = K_{Ic}(c_2) = -\sqrt{\frac{2\mu G_c}{\sqrt{1 - c_2^2/c_1^2}}}. \tag{8.72}$$

as the speed growth from c_R to c_2.

Note that this crack related solution follows, of course, directly from the corresponding problem formulation. We here have shown only the connection with the contact problem.

8.3.3 Intersonic Regime

The general relations for this regime and those for a parabolic indenter are presented in Sect. 8.3.1. In the transition from the subsonic speed range to the intersonic one, the solution varies continuously. Parameter ν grows from zero to 1/2 as the speed increases from c_2 to $\sqrt{2}\,c_2$; then it decreases to zero as $v \to c_1$. Thus $0 < \nu \leq 1/2$ for $c_2 < v < c_1$. Under this inequalities the most strong singular point at $x = x_+$ is weak in the sense that it cannot conduct energy. However, in this speed range, the shear wave arises excited by the moving body. Thus, in the intersonic range, the driving force is required to support the movement. It is

$$P = \frac{4\pi\mu(1-\nu)(1-2\nu)\omega_1(v)x_+^3}{3(3-2\nu)^2\cos^2(\pi\nu)\,r^2}, \tag{8.73}$$

whereas the vertical force is

$$Q_y = -\frac{\pi\mu(1-2\nu)\omega_1(v)x_+^2}{2(3-2\nu)\cos^2(\pi\nu)\,r}. \tag{8.74}$$

The indenter—half-plane configuration is shown in Fig. 8.1b. The non-dimensional ratio

$$\mathcal{P} = \frac{P}{Q}\sqrt{\frac{\mu r}{Q}} = \frac{16(1-\nu)\cos\pi\nu}{3\sqrt{2\pi}(1-2\nu)(3-2\nu)\omega_1(v)} \tag{8.75}$$

as a function of v is plotted in Fig. 8.6 for Poisson's ratio equal 1/3 ($c_1 = 2c_2$). For the super-Rayleigh subsonic range the plot is based on (8.65).

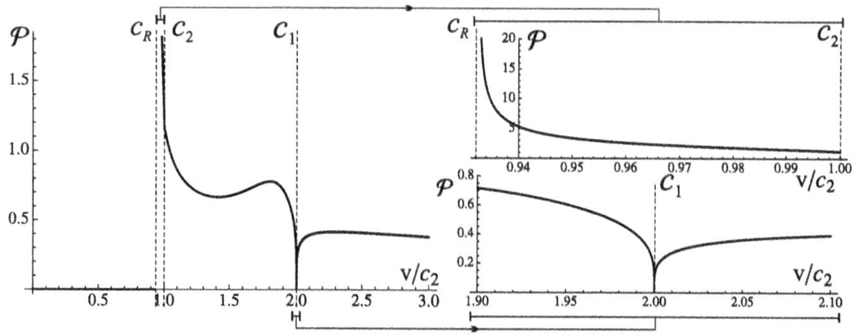

Fig. 8.6 The driving force—vertical force relation: $\mathcal{P} = P/[Q\sqrt{\mu r/Q}]$ as a function of the speed, v. It can be seen that the resistance to the movement is zero at sub-Rayleigh speeds. It drops down sharply in a vicinity of the longitudinal wave speed, c_1 ($v \to c_1 \pm 0$)

8.3.4 *Special Speed Value, $v = \sqrt{2}c_2$, and the Supersonic Regime*

At this **intersonic speed** $\nu = 1/2$, $\omega_1 = 0$, $\omega_2 = 2$, $x_- = 0$, and as it follows directly from (8.43)

$$\sigma_{yy}(x) = -2\mu u_y'(x), \quad Q_y = -2\mu y(x_+). \tag{8.76}$$

For the parabolic indenter the contact stresses, the vertical force and the driving force are

$$\sigma_{yy}(x) = -\frac{2\mu x}{r}, \quad Q_y = -\frac{\mu x_+^2}{r}, \quad P = \frac{2\mu x_+^3}{3r^2}, \tag{8.77}$$

respectively, which is also in agreement with the limit ($\nu \to 1/2$) following from (8.73), (8.74). The half-plane boundary outside of the contact zone is horizontal, whereas there is a shift in the displacement

$$u_y'(x) = 0 \quad (x < 0, x > x_+), \quad u_y(x_+) - u_y(0) = y(x_+). \tag{8.78}$$

The solution of the same type corresponds to the **supersonic regime**, which differs by the presence of both the shear and the longitudinal waves. In the transition to this regime, parameter ν has a jump discontinuity from zero at $v = c_1 - 0$ to 1/2 at $v > c_1$. For the latter regime

$$\omega_1(v) = 0, \quad \omega_2(v) = \frac{c_2^2 R(v)}{v^2\sqrt{v^2/c_1^2 - 1}}, \tag{8.79}$$

where the corresponding expression for the Rayleigh function, $R(v)$ is shown in (8.11). Thus the stresses and the vertical force are

$$\sigma_{yy}(x) = -\mu\omega_2(v)u_y'(x), \quad Q_y = -\frac{\mu\omega_2(v)y(x_+)}{r}. \tag{8.80}$$

For the parabolic indenter the contact stresses, the vertical force and the driving force are

$$\begin{gathered}\sigma_{yy}(x) = -\frac{\mu\omega_2(v)x}{r}, \quad Q_y = -\frac{\mu\omega_2(v)x_+^2}{2r}, \quad P = \frac{\mu\omega_2(v)x_+^3}{3r^2}, \\ \mathcal{P} = \frac{P}{Q}\sqrt{\frac{\mu r}{Q}} = \frac{2}{3}\sqrt{\frac{2}{\omega_2(v)}}.\end{gathered} \tag{8.81}$$

8.3.5 *The Limiting, Antiresonance, Resistance-Free, Steady-State Regime, $v \to c_1$*

It follows from (8.73) and (8.81) that, for nonzero Poisson's ratio,

$$\mathcal{P} \sim \frac{16}{3\sqrt{6\pi\omega_1(v)}} \to 0 \quad (\omega_1 \to \infty \text{ as } v \to c_1 - 0),$$
$$\mathcal{P} \to 0 \quad (\omega_2 \to \infty \text{ as } v \to c_1 + 0). \tag{8.82}$$

In this case $x_+ = x_- = 0$, and the indenter being under the external vertical force, Q_y, is in equilibrium with a concentrated longitudinal wave propagating in the half-plane. Thus, this is a special case of d'Alembert's paradox manifestation

$$x_+ = x_- = 0, \quad \sigma_{yy} = Q_y\delta(x), \quad P = 0 \quad (v = c_1). \tag{8.83}$$

Recall that this result corresponds to the limit, $v \to c_1$, of the established steady-state regimes for $v < c_1$ and $v > c_1$.

8.4 Wedging

The plane problem for the motion of a rigid symmetric wedge through the elastic material is examined in this section. To avoid geometrical nonlinearities we do not consider the case when the wedge is to close to the crack tip. We derive the exact solution for the sub-Rayleigh regime. A simplified solution is also given based on the assumption that the contact zone is much shorter than its distance to the crack tip.

First we note that there is no crack closure in the considered problem. This follows from relation (5.73) in Slepyan (2002), which remains qualitatively valid for the sub-Rayleigh speed regime. It can be seen in this relation that, for compressive crack face traction, the crack opening behind the contact zone ($x < x_-$) remains positive and it tends to zero as $x \to -\infty$. Thus the traction behind the contact region is zero. Denote the contact region end points and the crack tip coordinates as $x_- < x_+ < x_c$, respectively (see Fig. 8.7). The conditions on the lower half-plane boundary (and similar conditions on the upper half-plane boundary) are

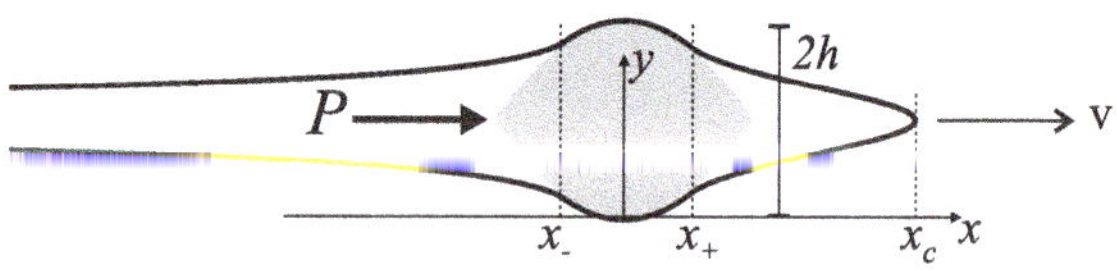

Fig. 8.7 Crack face contour in the wedging, $v < c_R$

$$
\begin{aligned}
&u_y(x) = y(x) \quad (x_- < x < x_+), \quad u_y(x) = h \quad (x \geq x_c),\\
&u_y'(x) = \mathrm{O}\left(|x|^{-3/2}\right) \quad (x \to -\infty), \quad \sigma_{yy} = 0 \quad (x < x_- \text{ and } x_+ < x < x_c),
\end{aligned} \tag{8.84}
$$

where h is the wedge half-width (the origin of the coordinates coincides with the lower point of the wedge as above). We use the same representation as above, namely for $v < c_R$,

$$
u_y'(x) = \Re\varphi(x), \quad \sigma_{yy}(x) = -\mu\Omega(v)\Im\varphi(x), \tag{8.85}
$$

where $\varphi(\zeta)$ is an analytical function of the complex variable $\zeta = x + \mathrm{i}\varepsilon$, $\varepsilon > 0$. Next, another analytical function is introduced as

$$
\Phi(\zeta) = \Psi(\zeta)\varphi(\zeta), \quad \Psi(\zeta) = \sqrt{(\zeta - x_-)(\zeta - x_+)(\zeta - x_c)}, \tag{8.86}
$$

where

$$
\Psi(x) = \begin{cases} \sqrt{(x - x_-)(x - x_+)(x - x_c)} \ (x > x_c) \\ \mathrm{i}\sqrt{(x - x_-)(x - x_+)(x_c - x)} \ (x_+ < x < x_c) \\ -\sqrt{(x - x_-)(x_+ - x)(x_c - x)} \ (x_- < x < x_+) \\ -\mathrm{i}\sqrt{(x_- - x)(x_+ - x)(x_c - x)} \ (x < x_-) \end{cases} \tag{8.87}
$$

Since outside the contact region $\Re\Phi(x) = 0$ we can express this function in the same way as above

$$
\Phi(\zeta) = \frac{\mathrm{i}}{\pi}\int_{x_-}^{x_+} \frac{y'(\xi)\sqrt{(\xi - x_-)(x_+ - \xi)(x_c - \xi)}}{\xi - \zeta}\,\mathrm{d}\xi + \mathrm{i}C. \tag{8.88}
$$

Now consider the stress intensity at the end points. It follows from (8.85) and (8.88) that

$$
\begin{aligned}
&\lim_{x \to x_- + 0} \sigma_{yy}(x)\sqrt{x - x_-}\\
&= \frac{\mu\Omega(v)}{\sqrt{(x_+ - x_-)(x_c - x_-)}}\left(\frac{1}{\pi}\int_{x_-}^{x_+}\sqrt{\frac{(x_+ - \xi)(x_c - \xi)}{\xi - x_-}}\,y'(\xi)\,\mathrm{d}\xi + C\right),\\
&\lim_{x \to x_+ - 0} \sigma_{yy}(x)\sqrt{x_+ - x}\\
&= -\frac{\mu\Omega(v)}{\sqrt{(x_+ - x_-)(x_c - x_+)}}\left(\frac{1}{\pi}\int_{x_-}^{x_+}\sqrt{\frac{(\xi - x_-)(x_c - \xi)}{x_+ - \xi}}\,y'(\xi)\,\mathrm{d}\xi - C\right),\\
&\lim_{x \to x_c + 0} \sigma_{yy}(x)\sqrt{x - x_c}\\
&= \frac{\mu\Omega(v)}{\sqrt{(x_c - x_-)(x_c - x_+)}}\left(\frac{1}{\pi}\int_{x_-}^{x_+}\sqrt{\frac{(x_+ - \xi)(\xi - x_-)}{x_c - \xi}}\,y'(\xi)\,\mathrm{d}\xi - C\right).
\end{aligned} \tag{8.89}
$$

For the sub-Rayleigh speed regime the stress intensities at the contact region ends, $x = x_\pm$, must be zero to avoid interpenetration. Let the stress intensity factor at the crack tip be given as K_I. These three conditions lead to the following relations

$$
\begin{aligned}
C &= -\frac{1}{\pi}\int_{x_-}^{x_+}\sqrt{\frac{(x_+-\xi)(x_c-\xi)}{\xi-x_-}}\,y'(\xi)\,\mathrm{d}\xi, \\
&\int_{x_-}^{x_+}\sqrt{\frac{x_c-\xi}{(\xi-x_-)(x_+-\xi)}}\,y'(\xi)\,\mathrm{d}\xi = 0, \\
\mu\Omega(v)&\sqrt{\frac{x_c-x_-}{x_c-x_+}}\frac{1}{\pi}\int_{x_-}^{x_+}\sqrt{\frac{x_+-\xi}{(\xi-x_-)(x_c-\xi)}}\,y'(\xi)\,\mathrm{d}\xi = \frac{K_I}{\sqrt{2\pi}},
\end{aligned}
\tag{8.90}
$$

An additional condition concerning the wedge maximal width $2h$ must be taken into account; recalling that the coordinate of the maximal width is taken as the origin of the x-axis the following equation is obtained

$$
y(x_+) + \int_{x_+}^{x_c} u_y'(x)\,\mathrm{d}x = h, \tag{8.91}
$$

where, in this region, $x_+ < x < x_c$

$$
u_y'(x) = \sqrt{\frac{x-x_-}{(x-x_+)(x_c-x)}}\frac{1}{\pi}\int_{x_-}^{x_+}\sqrt{\frac{(x_+-\xi)(x_c-\xi)}{\xi-x_-}}\frac{y'(\xi)}{\xi-x}\,\mathrm{d}\xi. \tag{8.92}
$$

Finally, the nonzero normal stress component is

$$
\sigma_{yy}(x) = \mu\Omega(v)\sqrt{\frac{x-x_-}{(x_+-x)(x_c-x)}}\frac{1}{\pi}V.p.\int_{x_-}^{x_+}\sqrt{\frac{(x_+-\xi)(x_c-\xi)}{\xi-x_-}}\frac{y'(\xi)}{\xi-x}\,\mathrm{d}\xi, \tag{8.93}
$$

in the contact region ($x_- < x < x_+$) and

$$
\sigma_{yy}(x) = -\mu\Omega(v)\sqrt{\frac{x-x_-}{(x-x_+)(x-x_c)}}\frac{1}{\pi}\int_{x_-}^{x_+}\sqrt{\frac{(x_+-\xi)(x_c-\xi)}{\xi-x_-}}\frac{y'(\xi)}{\xi-x}\,\mathrm{d}\xi \tag{8.94}
$$

ahead of the crack tip.

Thus, three unknowns remain: x_-, x_+ and x_c; they can be obtained form (8.91) and the last two relations in (8.90). In particular, it follows from the second relation in (8.90) that $x_- < 0$ and $x_+ > 0$.

8.4.1 The Parabolic Wedge

For the parabolic wedge, $y(x) = x^2/(2r)$, the second condition in (8.90) and Eq. (8.91) lead to the following relations

$$(x_c - x_+)\,K\left(\frac{x_+ - x_-}{x_c - x_-}\right) - (x_c - 2x_+ - 2x_-)\,E\left(\frac{x_+ - x_-}{x_c - x_-}\right) = 0, \quad (8.95)$$

$$\frac{\mu\,\Omega(v)(x_c - x_-)}{3\pi r}\left[K\left(\frac{x_+ - x_-}{x_c - x_-}\right)(2x_c + x_-) - E\left(\frac{x_+ - x_-}{x_c - x_-}\right)(2x_c + 2x_- - x_+)\right] = \frac{K_I}{\sqrt{2\pi}}, \quad (8.96)$$

where $K(m) = \int_0^{\pi/2}(1 - m\,\sin^2\theta)^{-1/2}\,d\theta$ and $E(m) = \int_0^{\pi/2}(1 - m\,\sin^2\theta)^{1/2}\,d\theta$ are the complete elliptic integrals of the first and second kind, respectively.

To proceed we make some simplification assuming the contact zone to be much shorter than its distance to the crack tip, $x_+ - x_- \ll x_c - x_*$, $x_* = (x_+ + x_-)/2$, namely, for the integrals over the contact region we replace $\sqrt{x_c - \xi}$, where $\xi \in (x_-, x_+)$, by its two-term asymptote as

$$\sqrt{x_c - \xi} \sim \sqrt{x_c - x_*} + \frac{x_* - \xi}{2\sqrt{x_c - x_*}}, \quad \frac{1}{\sqrt{x_c - \xi}} \sim \frac{1}{\sqrt{x_c - \xi}} - \frac{x_* - \xi}{2(x_c - x_*)^{3/2}} \quad (8.97)$$

The second condition in (8.90) of zero stress intensity at x_+ leads now to an explicit expression for x_c in terms of $x_\pm$

$$x_c = \frac{5x_+^2 + 6x_+x_- + 5x_-^2}{8(x_+ + x_-)}. \quad (8.98)$$

Note that $x_c \to \infty$ when $x_- \to -x_+ + 0$ as it should be (in the limit, we have a half-plane under the moving indenter, that is we return the previous problem), and $x_- \to -3/5x_+$ when $x_c \to x_+ + 0$.

The relative position of the crack tip $\tilde{x}_c = x_c/x_+$ as a function of the ratio $\tilde{x}_- = x_-/x_+$ is shown in Fig. 8.8, where the approximate result (8.98) is compared with the exact one following from (8.95). It is evident that the approximation can be acceptable for $\tilde{x}_c > 2$.

Next, the last condition in (8.90) concerning the stress intensity at x_c yields

$$\frac{\mu\Omega(v)}{2r}\,\frac{7x_+^2 + 18x_+x_- + 7x_-^2}{\sqrt{(5x_+ + 3x_-)(-3x_+ - 5x_-)}}\sqrt{x_+ + x_-} = \frac{K_I}{\sqrt{\pi}}. \quad (8.99)$$

We also represent this relation in a normalized form as

$$\Omega(v)\hat{x}_+^{3/2}\,g_w(\tilde{x}_-) = M_w \quad (8.100)$$

with

$$M_w = 2\left(\frac{r}{h^3}\right)^{\frac{1}{4}}\frac{K_I}{\sqrt{\pi}\,\mu}, \quad g_w(\tilde{x}_-) = \frac{7 + 18\tilde{x}_- + 7\tilde{x}_-^2}{\sqrt{(5 + 3\tilde{x}_-)(-3 - 5\tilde{x}_-)}}\sqrt{1 + \tilde{x}_-} \quad (8.101)$$

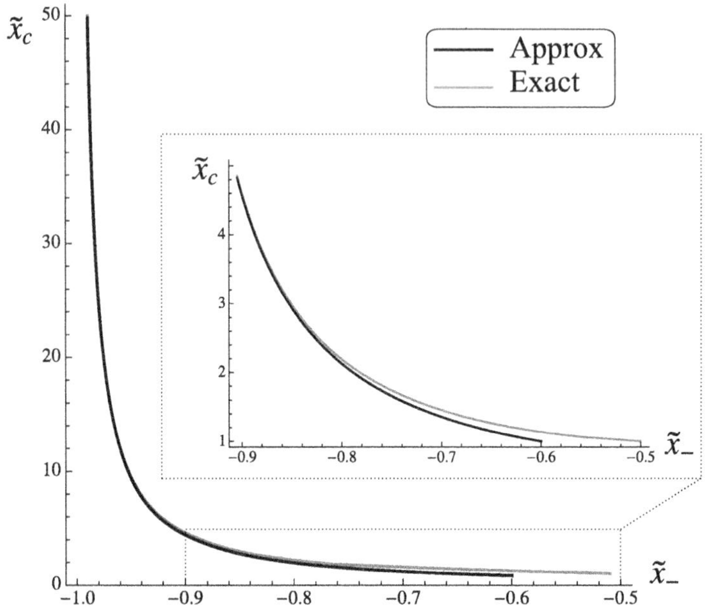

Fig. 8.8 Distribution of the normalized position of the crack tip, $\tilde{x}_c = x_c/x_+$, as a function of the ratio $\tilde{x}_- = x_-/x_+$ for the parabolic indenter. Approximate results (in black) are compared with the exact one (in gray)

where M_w is supposed to be given and $\hat{x}_+ = x_+/\sqrt{r\,h}$. Note that $g_w(\tilde{x}_-)$ is a monotonically decreasing function spanning the negative real axis in the admissible range $-1 < \tilde{x}_- < -3/5$, so that Eq. (8.100) admits a unique solution in term of $\tilde{x}_-$ for each values of the positive quantities $\hat{x}_+$ and M_w.

Finally, the condition (8.91) on the opening displacement imposed by the wedge can be expressed in the normalized form

$$\hat{x}_+^2\, h_w(\tilde{x}_-) = 1 \tag{8.102}$$

with

$$\begin{gathered} h_w(\tilde{x}_-) = -\frac{1}{120}\frac{\sqrt{1-\tilde{x}_-}}{(1+\tilde{x}_-)^2}\Big\{2\sqrt{1+\tilde{x}_-}\Big[2(1-\tilde{x}_-)(17+22\tilde{x}_-+17\tilde{x}_-^2)\,E(\varkappa)+ \\ (5+3\tilde{x}_-)(1+26\tilde{x}_-+13\tilde{x}_-^2)\,K(\varkappa)\Big]+(3+67\tilde{x}_-+63\tilde{x}_-^2+25\tilde{x}_-^3)\sqrt{-5-3\tilde{x}_-}\Big\}, \\ \varkappa = (3+5\tilde{x}_-)/(8(1+\tilde{x}_-)). \end{gathered} \tag{8.103}$$

where, similarly to the previous condition, $h_w(\tilde{x}_-)$ is now a positive monotonically decreasing real function of $\tilde{x}_-$ spanning the positive real axis in the admissible range $-1 < \tilde{x}_- < -3/5$. Thus Eq. (8.102) admits a unique solution in term of $\tilde{x}_-$ for each value of the positive quantity $\hat{x}_+$.

Further, using equations (8.100) and (8.102) we can eliminate $\hat{x}_+$ and find $\tilde{x}_-$ numerically from the obtained relation:

$$f_w(\tilde{x}_-) = \frac{h_w(\tilde{x}_-)}{[-g_w(\tilde{x}_-)]^{4/3}} = \left[\frac{-\Omega(v)}{M_w}\right]^{4/3}. \tag{8.104}$$

After this x_+ is found from Eq. (8.102) as

$$\hat{x}_+ = h_w(\tilde{x}_-)^{-1/2}. \tag{8.105}$$

Then, the stresses for $x_- < x < x_+$ and for $x > x_c$ are

$$\begin{aligned} \sigma_{yy}(x) &= \frac{\mu\,\Omega(v)}{r}\frac{\sqrt{2(x_+ + x_-)}}{x_+ - x_-}\sqrt{\frac{(x_+ - x)(x - x_-)}{x_c - x}}\,f_1(x), \\ \sigma_{yy}(x) &= \frac{\mu\,\Omega(v)}{r}\frac{\sqrt{2(x_+ + x_-)}}{(x_+ - x_-)\sqrt{x_c - x}}\left[\sqrt{(x - x_+)(x - x_-)}\,f_1(x) - x\,f_2(x)\right], \end{aligned} \tag{8.106}$$

respectively, where

$$f_1(x) = x - \frac{(x_+ - x_-)^2}{4(x_+ + x_-)} \quad \text{and} \quad f_2(x) = x - \frac{3x_+^2 + 2x_+x_- + 3x_-^2}{4(x_+ + x_-)}. \tag{8.107}$$

The total vertical force in each separated wedge contact area and its position are

$$\begin{aligned} Q_y &= \int_{x_-}^{x_+} \sigma_{yy}(x)\,dx = -\frac{\pi}{4}\frac{\mu\,\Omega(v)}{r}(x_+^2 + 4x_+x_- + x_-^2), \\ x_Q &= \frac{(x_+ + x_-)(x_+ + 3x_-)(3x_+ + x_-)}{4(x_+^2 + 4x_+x_- + x_-^2)}, \end{aligned} \tag{8.108}$$

respectively. In addition, the total driving force P is given by

$$P = -2\int_{x_-}^{x_+} \sigma_{yy}(x)\,u_y'(x)\,dx = -\frac{\mu\,\Omega(v)}{r^2}\frac{\pi\,(x_+ + x_-)}{8}(x_+ + 3x_-)(3x_+ + x_-), \tag{8.109}$$

and it can be compared with the energy release rate $G = K_I^2/(2\mu\Omega(v))$, that assumes the approximate form

$$G = -\frac{\mu\,\Omega(v)}{r^2}\frac{\pi\,(x_+ + x_-)}{8}\frac{(7x_+ + 18x_+x_- + 7x_-^2)^2}{(5x_+ + 3x_-)(3x_+ + 5x_-)}, \tag{8.110}$$

and the exact one

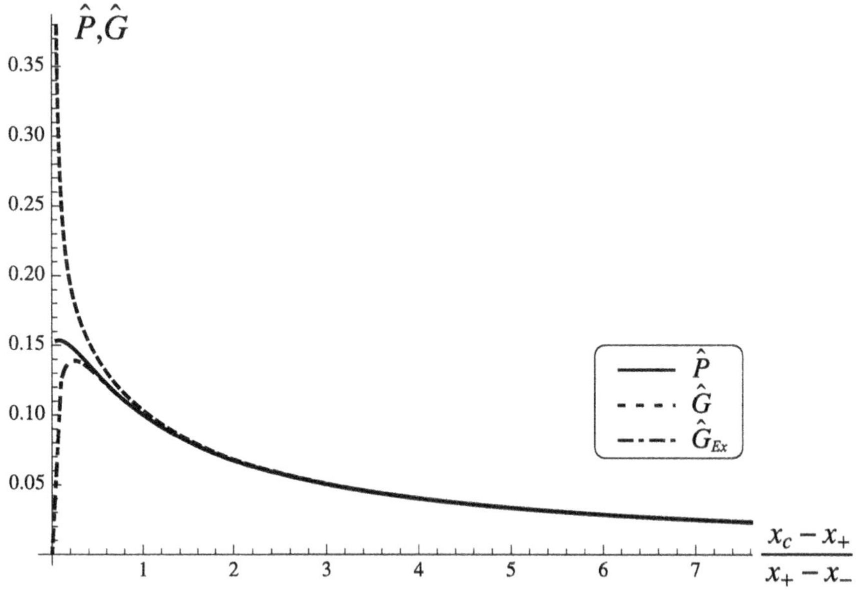

Fig. 8.9 The driving force as a function of the normalized crack-tip distance, $(x_c - x_+)/(x_+ - x_-)$. The exact solution $\hat{G}_{Ex} = G_{Ex}\, r^2/(\mu\Omega(v))$ (8.111), the approximate solutions $\hat{P} = P\, r^2/(\mu\Omega(v))$ (8.109) and $\hat{G} = G\, r^2/(\mu\Omega(v))$ (8.110). The plot corresponds to Poisson's ratio equal to 1/3 ($c_1 = 2c_2$)

$$G_{Ex} = \frac{\mu\,\Omega(v)}{r^2}\frac{(x_c - x_-)^2}{9\pi}\left[K\left(\frac{x_+ - x_-}{x_c - x_-}\right)(2x_c + x_-) - E\left(\frac{x_+ - x_-}{x_c - x_-}\right)(2x_c + 2x_- - x_+)\right]^2. \tag{8.111}$$

Results (8.109)–(8.111) are compared in Fig. 8.9, where they are given as functions of the relative crack-tip distance $(x_c - x_+)/(x_+ - x_-)$. The comparative analysis shows the precision of the approximation and the fact that, except when $x_c \to x_+$ the driving force decreases at increasing distance of the crack tip, as expected on physical ground.

The nonzero displacement derivative is

$$\begin{aligned} u'_y(x) &= \frac{1}{r}\frac{\sqrt{2(x_+ + x_-)}}{x_+ - x_-}\frac{1}{\sqrt{x_c - x}}[-\sqrt{(x_+ - x)(x_- - x)}\,f_1(x) - x\, f_2(x)] \quad (x < x_-),\\ &= \frac{x}{r} \quad (x_- < x < x_+), \\ &= \frac{1}{r}\frac{\sqrt{2(x_+ + x_-)}}{x_+ - x_-}\frac{1}{\sqrt{x_c - x}}[\sqrt{(x - x_+)(x - x_-)}\,f_1(x) - x\, f_2(x)] \quad (x_+ < x < x_c). \end{aligned} \tag{8.112}$$

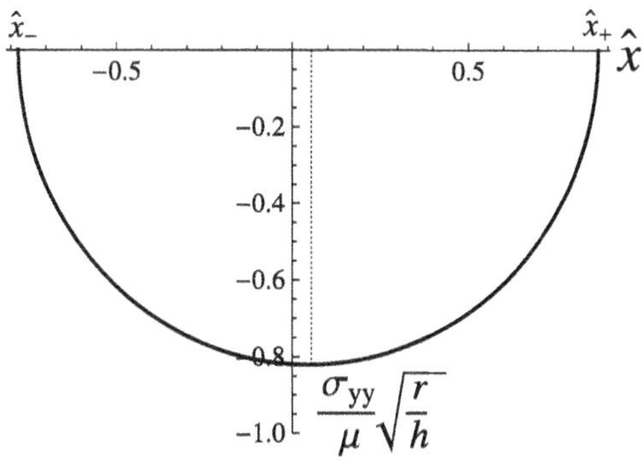

Fig. 8.10 Distribution of the normalized stress, $\sigma_{yy}/\mu\sqrt{r/h}$, as a function of $\hat{x} = x/\sqrt{r\,h}$, where $\hat{x}_+ = x_+/\sqrt{r\,h}$ and $\hat{x}_- = x_-/\sqrt{r\,h}$. Results are given for Poisson's ratio equal to $1/3$ and $(-\Omega(v)/M_w)^{4/3} = 2.5$ (8.104), which corresponds to $\tilde{x}_c = x_c/x_+ = 4.269$

Finally, the displacement $u_y(x)$ takes the form:

$$
\begin{aligned}
u_y(x) &= \frac{x_-^2}{2r} + \frac{1}{15}\frac{1}{\sqrt{2(x_+ - x_-)}}\frac{1}{x_+ - x_-}[g_+(x) - g_+(x_-)] \quad (x < x_-), \\
&= \frac{x^2}{2r} \quad (x_- < x < x_+), \\
&= \frac{x_+^2}{2r} + \frac{1}{15}\frac{1}{\sqrt{2(x_+ - x_-)}}\frac{1}{x_+ - x_-}[g_-(x) - g_-(x_+)] \quad (x_+ < x < x_c),
\end{aligned}
\tag{8.113}
$$

where

$$
\begin{aligned}
g_\pm(x) &= \sqrt{x_c - x}\left[12x^2(x_+ + x_-) - (x + 2x_c)(5x_+^2 - 2x_+x_- + 5x_-^2)\right. \\
&\quad \left.\pm\sqrt{(x_- - x)(x_+ - x)}\left(12x(x_+ + x_-) + x_+^2 + 14x_+x_- + x_-^2\right)\right] \\
&\quad -\sqrt{(x_c - x)(x_+ - x_-)}\left(13x_+^2 + 26x_+x_- + x_-^2\right) F\left[\arcsin(\chi(x)), \chi^{-2}(x_+)\right] \\
&\quad +\sqrt{(x_c - x)}\frac{(x_+ - x_-)^2}{4(x_+ + x_-)}\left(17x_+^2 + 22x_+x_- + 17x_-^2\right) E\left[\arcsin(\chi(x)), \chi^{-2}(x_+)\right],
\end{aligned}
\tag{8.114}
$$

$$
\chi(x) = \sqrt{(x_c - x)/(x_c - x_-)}
$$

with $F(\phi, m) = \int_0^\phi (1 - m\sin^2\theta)^{-1/2}\,\mathrm{d}\theta$ and $E(\phi, m) = \int_0^\phi (1 - m\sin^2\theta)^{1/2}\,\mathrm{d}\theta$ the incomplete elliptic integral of the first and second kind, respectively.

In Figs. 8.10 and 8.7 results are shown for $(-\Omega(v)/M_w)^{4/3} = 2.5$. Equations (8.104) and (8.105) yield

$$
\tilde{x}_- = -0.894, \quad \hat{x}_+ = 0.872, \quad (\hat{x}_- = x_-/\sqrt{r\,h} = -0.779, \ \hat{x}_c = x_c/\sqrt{r\,h} = 3.723), \tag{8.115}
$$

such that

$$
\tilde{x}_c = x_c/x_+ = 4.269, \quad (x_c - x_+)/(x_+ - x_-) = 1.726 \tag{8.116}
$$

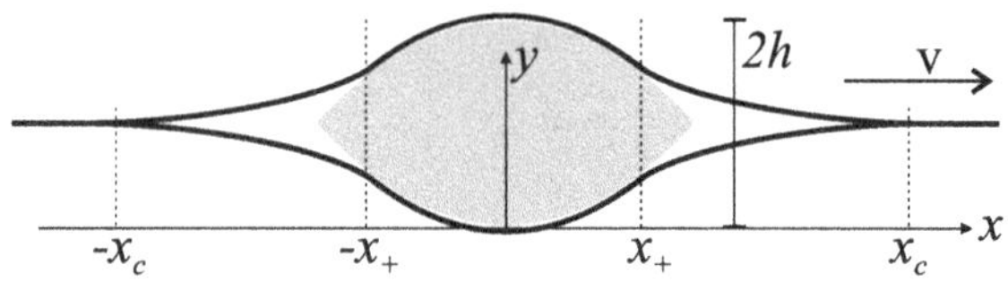

Fig. 8.11 The wedge moving between two prestress elastic half-planes, $v < c_R$

in Fig. 8.8 and in Fig. 8.9, respectively.

Distribution of the traction $\sigma_{yy}(x)$ is given in Fig. 8.10, whereas the deformed contour is represented in Fig. 8.7. We note that conditions (a) and (b) in (15.1) are fully respected. In particular, a compressive stress is present in the contact zone and there is no interpenetration between the wedge and the elastic material. The absence of stress singularity at the ends of the contact zone is associated to the continuity of u_y'. The presence of the crack tip and the constraint $u = 0$ in the crack continuation line induce a stress distribution which is nonsymmetric with respect to $x = 0$ and the consequent existence of the driving force $P = G$. It should be stressed that the driving force is generated by the crack resistance but not by the contact interaction, phenomenon demonstrated in Fig. 8.9.

8.5 Wedge Moving Between Two Elastic Half-Planes

The plane problem for the motion of a rigid wedge through the interface of two elastic half-planes compressed at infinity by a uniform stress $\sigma_{yy} = -\sigma_0 < 0$ is studied. As for the previous wedge problem, we consider the steady-state movement of a finite-length rigid indenter in the sub-Rayleigh regime.

The wedge is assumed symmetric relatively to both the longitudinal axis, $y = h$, and the normal y-axis, $y(-x, 0) = y(x, 0)$, $y(\pm x, 2h) = 2h - y(x, 0)$. We also assume that the deformed half-plane boundaries possess the same symmetry in the considered sub-Rayleigh speed regime. Denote the wedge contact region and the half-plane contact region ends as shown in Fig. 8.11. The wedge—half-plane boundaries configuration is assumed to consist of five regions, three regions where the displacement is given and two regions where the stresses are given. Namely, with refer to the conditions (a) and (b) in (15.1), the boundary conditions for the lower half-plane are

$$
\begin{aligned}
u_y(x) = y(x) \quad (|x| \le x_+), \quad u_y(x) = 0 \quad (|x| \ge x_c), \\
\sigma_{yy} = 0 \quad (x_+ \le |x| \le x_c), \quad \sigma_{yy} \to -\sigma_0 \quad (|x| \to \infty),
\end{aligned}
\tag{8.117}
$$

and the symmetric conditions are valid for the upper half plane.

Representation (8.85) is used for the displacement derivative and the normal traction, and the following analytical function is introduced

$$\Phi(\zeta) = \Psi(\zeta)\varphi(\zeta), \quad \Psi(\zeta) = \sqrt{(\zeta^2 - x_+^2)(\zeta^2 - x_c^2)}, \quad \zeta = x + \mathrm{i}\varepsilon \ \ (\varepsilon > 0), \tag{8.118}$$

where

$$\Psi(x) = \begin{cases} \sqrt{(x^2 - x_+^2)(x^2 - x_c^2)} & (|x| > x_c) \\ \mathrm{i}\,\mathrm{sign}\,x\,\sqrt{(x^2 - x_+^2)(x_c^2 - x^2)} & (x_+ < |x| < x_c) \\ -\sqrt{(x_+^2 - x^2)(x_c^2 - x^2)} & (|x| < x_+) \end{cases} \tag{8.119}$$

Outside the contact region $\Re\Phi(x) = 0$, and function $\Phi(\zeta)$ is now defined as

$$\Phi(\zeta) = \frac{\mathrm{i}}{\pi}\int_{-x_+}^{x_+} \frac{y'(\xi)\sqrt{(x_+^2 - \xi^2)(x_c^2 - \xi^2)}}{\xi - \zeta}\,\mathrm{d}\xi + \mathrm{i}(C_1 + C_2\,\zeta^2). \tag{8.120}$$

Note that the latter term, $C_2\,\zeta^2$, reflects the prestress, whereas the linear term is not present due to the symmetry of the problem.

Now consider the stress at remote points and the stress intensity at contact end points. It follows from (8.85) and (8.120) that

$$\begin{aligned}
&\sigma_{yy}(x) \sim -\mu\,\Omega(v)C_2 \quad (x \to \infty),\\
&\sigma_{yy}(x) \sim -\frac{\mu\Omega(v)}{\sqrt{2x_+\,(x_c^2 - x_+^2)(x_+ - x)}}\\
&\times\left[\frac{1}{\pi}\int_{-x_+}^{x_+} y'(\xi)\sqrt{\frac{(x_+ + \xi)(x_c^2 - \xi^2)}{x_+ - \xi}}\,\mathrm{d}\xi - (C_1 + C_2\,x_+^2)\right] \quad (x \to x_+ - 0),\\
&\sigma_{yy}(x) \sim \frac{\mu\Omega(v)}{\sqrt{2x_c\,(x_c^2 - x_+^2)(x - x_c)}}\\
&\times\left[\frac{1}{\pi}\int_{-x_+}^{x_+} y'(\xi)\sqrt{\frac{(x_c + \xi)(x_+^2 - \xi^2)}{x_c - \xi}}\,\mathrm{d}\xi - (C_1 + C_2\,x_c^2)\right] \quad (x \to x_c + 0),
\end{aligned} \tag{8.121}$$

The conditions in (8.117) lead to the following equalities

$$\begin{aligned}
C_2 &= \frac{\sigma_0}{\mu\,\Omega(v)},\\
C_1 &= \frac{2}{\pi}\int_0^{x_+} y'(\xi)\sqrt{\frac{x_c^2 - \xi^2}{x_+^2 - \xi^2}}\,\xi\,\mathrm{d}\xi - \frac{\sigma_0}{\mu\,\Omega(v)}x_+^2,\\
&\frac{2}{\pi}\int_0^{x_+} y'(\xi)\sqrt{\frac{x_+^2 - \xi^2}{x_c^2 - \xi^2}}\,\xi\,\mathrm{d}\xi + \frac{\sigma_0}{\mu\,\Omega(v)}(x_c^2 - x_+^2) = 0.
\end{aligned} \tag{8.122}$$

As for the previous wedge case, the condition concerning the wedge maximal width must be taken into account and Eq. (8.91) must be satisfied.

From (8.122) two different expressions for the contact stresses suitable for $|x| \leq x_+$ and $|x| \geq x_c$ are

$$\sigma_{yy}(x) = -\sqrt{\frac{x_+^2 - x^2}{x_c^2 - x^2}}\left[\sigma_0 - \frac{2\mu\Omega(v)}{\pi} V.p. \int_0^{x_+} \sqrt{\frac{x_c^2 - \xi^2}{x_+^2 - \xi^2}} \frac{y'(\xi)\xi \, d\xi}{\xi^2 - x^2}\right] \tag{8.123}$$

and

$$\sigma_{yy}(x) = -\sqrt{\frac{x^2 - x_c^2}{x^2 - x_+^2}}\left[\sigma_0 + \frac{2\mu\Omega(v)}{\pi} \int_0^{x_+} \sqrt{\frac{x_+^2 - \xi^2}{x_c^2 - \xi^2}} \frac{y'(\xi)\xi \, d\xi}{x^2 - \xi^2}\right], \tag{8.124}$$

respectively. Each of these relations can be analytically continued from one region to the others, and one expression appears equal to the other due to the last condition in (8.122). From the latter representation in (8.124) it follows that the analytical function $\varphi(\zeta)$ can be represented as

$$\varphi(\zeta) = \mathrm{i}\sqrt{\frac{\zeta^2 - x_c^2}{\zeta^2 - x_+^2}}\left[\frac{\sigma_0}{\mu\Omega(v)} + \frac{2}{\pi} \int_0^{x_+} \sqrt{\frac{x_+^2 - \xi^2}{x_c^2 - \xi^2}} \frac{y'(\xi)\xi \, d\xi}{\zeta^2 - \xi^2}\right], \tag{8.125}$$

and this yields

$$u_y'(x) = -\sqrt{\frac{x_c^2 - x^2}{x^2 - x_+^2}}\left[\frac{\sigma_0}{\mu\Omega(v)} + \frac{2}{\pi} \int_0^{x_+} \sqrt{\frac{x_+^2 - \xi^2}{x_c^2 - \xi^2}} \frac{y'(\xi)\xi \, d\xi}{x^2 - \xi^2}\right] \mathrm{sign}x \quad (x_+ \leq |x| \leq x_c) \tag{8.126}$$

The unknown values x_+ and x_c are defined by Eq. (8.91) and the last condition in (8.122).

8.5.1 The Parabolic Wedge

For this shape we consider the solution independently. When the shape of the indenter at the contact zone is parabolic the condition in (8.121) of zero stress intensity at x_+ gives

$$C_1 = x_+^2 \left\{\frac{2x_+}{3\pi r} \tilde{x}_c \left[(\tilde{x}_c^2 - 1) K(\tilde{x}_c^{-2}) - (\tilde{x}_c^2 - 2) E(\tilde{x}_c^{-2})\right] + \frac{\sigma_0}{\mu\Omega(v)}\right\}, \tag{8.127}$$

where $\tilde{x}_c = x_c/x_+$. The condition (8.121) of zero stress intensity at x_c takes the form

$$x_+^2(\tilde{x}_c^2-1)\left\{\frac{2x_+}{\pi\, r}\tilde{x}_c\left[K\left(\tilde{x}_c^{-2}\right)-E\left(\tilde{x}_c^{-2}\right)\right]+\frac{\sigma_0}{\mu\Omega(v)}\right\}=0, \tag{8.128}$$

where the solution $\tilde{x}_c = 1$ is associated with tensile stress in the region $|x| < x_+$ and it cannot be accepted. Then, Eq. (8.128) can be given the equivalent normalized form

$$\Omega(v)\,\hat{x}_+\, g_p(\tilde{x}_c) = M_p, \tag{8.129}$$

where

$$\hat{x}_+ = \frac{x_+}{\sqrt{r\,h}}, \quad g_p(\tilde{x}_c) = \frac{2\tilde{x}_c}{\pi}\left[E\left(\tilde{x}_c^{-2}\right)-K\left(\tilde{x}_c^{-2}\right)\right], \quad M_p = \frac{\sigma_0}{\mu}\sqrt{\frac{r}{h}} \tag{8.130}$$

and $g_p(\tilde{x}_c)$ is a monotonic increasing real function of $\tilde{x}_c$ spanning the negative real axis in the admissible range $1 < \tilde{x}_c < \infty$, so that Eq. (8.129) admits a unique solution in term of $\tilde{x}_c$ for each value of the positive quantities $\hat{x}_+$ and M_p.

Lastly, condition (8.91) is represented in the normalized form

$$\hat{x}_+^2\, h_p(\tilde{x}_c) = 1, \tag{8.131}$$

where

$$h_p(\tilde{x}_c) = 1+\frac{2}{\pi}\left\{-\frac{1}{\tilde{x}_c}\int_{-1}^{1}\sqrt{\frac{\tilde{x}_c^2-\xi^2}{1-\xi^2}}\,\Pi\left(-\frac{\tilde{x}_c^2-1}{\tilde{x}_c^2}\frac{\xi^2}{1-\xi^2}, \frac{\tilde{x}_c^2-1}{\tilde{x}_c^2}\right)\mathrm{d}\xi + \right.$$
$$\left. 2\,K\left(\frac{\tilde{x}_c^2-1}{\tilde{x}_c^2}\right)K\left(\tilde{x}_c^{-2}\right)-2\tilde{x}_c^2 E\left(\frac{\tilde{x}_c^2-1}{\tilde{x}_c^2}\right)\left[K\left(\tilde{x}_c^{-2}\right)-E\left(\tilde{x}_c^{-2}\right)\right]\right\}, \tag{8.132}$$

and $\Pi(n,m) = \int_0^{\pi/2}(1-n\sin^2\theta)^{-1}(1-m\sin^2\theta)^{-1/2}\,\mathrm{d}\theta$ is the complete elliptic integral of the third kind. Note that the integral in Eq. (8.132) has been computed numerically and that $h_p(\tilde{x}_c)$ is a monotonic increasing function of $\tilde{x}_c$ spanning the positive real axis in the admissible range $1 < \tilde{x}_c < \infty$, so that Eq. (8.131) admits a unique solution in term of $\tilde{x}_c$ for each value of the quantity $\hat{x}_+$.

Similarly to Sect. 8.4.1, a solution of Eqs. (8.129) and (8.131) can be found numerically. We define

$$f_p(\tilde{x}_c) = \frac{h_p(\tilde{x}_c)}{g_p(\tilde{x}_c)^2} = \left(\frac{\Omega(v)}{M_p}\right)^2 \tag{8.133}$$

in term of $\tilde{x}_c$, where M_p is a given quantity, and compute $\hat{x}_+$ as

$$\hat{x}_+ = h_p(\tilde{x}_c)^{-1/2}. \tag{8.134}$$

Then, the nonzero stresses for $|x| < x_+$ and $|x| > x_c$ are

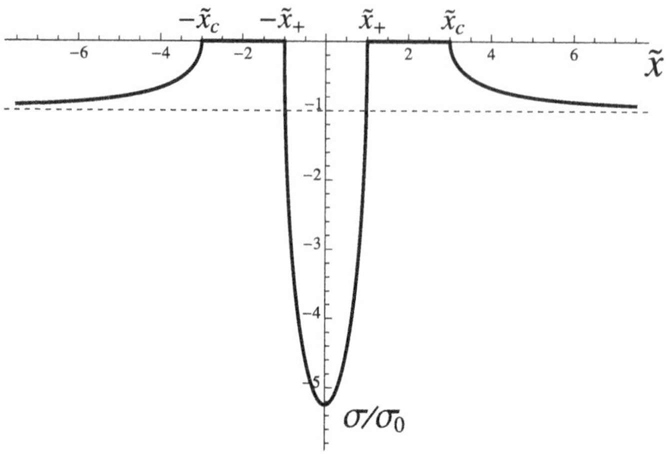

Fig. 8.12 Distribution of the normalized contact stresses σ_{yy}/σ_0, as a function of $\tilde{x} = x/x_+$ for $M_p = 0.122$, $\tilde{x}_c = 3.000$

$$\frac{\sigma_{yy}(\tilde{x})}{\sigma_0} = \sqrt{\frac{1-\tilde{x}^2}{\tilde{x}_c^2-\tilde{x}^2}}\left\{1+\frac{K\left(\tilde{x}_c^{-2}\right)\tilde{x}^2 - E\left(\tilde{x}_c^{-2}\right)\tilde{x}_c^2 + \Re\Pi\left(\tilde{x}^{-2},\tilde{x}_c^{-2}\right)\left(\tilde{x}_c^2-\tilde{x}^2\right)}{\tilde{x}_c^2\left[K\left(\tilde{x}_c^{-2}\right)-E\left(\tilde{x}_c^{-2}\right)\right]}\right\}$$
$$= -\sqrt{\frac{\tilde{x}^2-1}{\tilde{x}^2-\tilde{x}_c^2}}\left\{1-\frac{K\left(\tilde{x}_c^{-2}\right)\tilde{x}^2 - E\left(\tilde{x}_c^{-2}\right)\tilde{x}_c^2 - \Pi\left(\tilde{x}^{-2},\tilde{x}_c^{-2}\right)\left(\tilde{x}^2-\tilde{x}_c^2\right)}{\tilde{x}_c^2\left[K\left(\tilde{x}_c^{-2}\right)-E\left(\tilde{x}_c^{-2}\right)\right]}\right\}, \tag{8.135}$$

respectively, where $\tilde{x} = x/x_+$. The stress distribution is reported in Fig. 8.12 for $M_p = 0.122$, $\tilde{x}_c = 3.000$, $\hat{x}_+ = 0.703$.

Finally, the displacements are

$$\frac{u_y(\tilde{x})}{x_+^2/2r} = \tilde{x}^2 \quad (|x| < x_+), \qquad \frac{u_y(\tilde{x})}{x_+^2/2r} = \tilde{x}_c \quad (|x| > x_c),$$
$$\frac{u_y(\tilde{x})}{x_+^2/2r} = 1 + \frac{2}{\pi}\left\{2E\left(\tilde{x}_c^{-2}\right)F\left[\arcsin\left(\frac{\tilde{x}_c\sqrt{\tilde{x}^2-1}}{\tilde{x}\sqrt{\tilde{x}_c^2-1}}\right), \frac{\tilde{x}_c^2-1}{\tilde{x}_c^2}\right] - \frac{1}{\tilde{x}_c}L(\tilde{x},\tilde{x}_c)\right.$$
$$\left. -2\mathrm{i}\tilde{x}_c\left[K\left(\tilde{x}_c^{-2}\right)-E\left(\tilde{x}_c^{-2}\right)\right]\left[E\left(\arcsin\left(\tilde{x}_c^{-2}\right), \tilde{x}_c^2\right) - E\left(\arcsin\left(\frac{\tilde{x}}{\tilde{x}_c}\right), \tilde{x}_c^2\right)\right]\right\}$$
$$(x_+ < |x| < x_c), \tag{8.136}$$

where

$$L(\tilde{x}, \tilde{x}_c) = \int_{-1}^{1} \sqrt{\frac{\tilde{x}_c^2 - \xi^2}{1 - \xi^2}} \Pi \left[-\frac{(\tilde{x}_c^2 - 1)\xi^2}{\tilde{x}_c^2(1 - \xi^2)}, \arcsin\left(\frac{\tilde{x}_c \sqrt{\tilde{x}^2 - 1}}{\tilde{x}\sqrt{\tilde{x}_c^2 - 1}} \right), \frac{\tilde{x}_c^2 - 1}{\tilde{x}_c^2} \right] d\xi \tag{8.137}$$

with $\Pi(n, \phi, m) = \int_0^\phi (1 - n \sin^2 \theta)^{-1}(1 - m \sin^2 \theta)^{-1/2} \, d\theta$ the incomplete elliptic integral of the third kind.

Note that there is no driving force, and the wedge moves free inside the structure with any constant sub-Rayleigh speed. Thus, for this speed regime d'Alembert's paradox is fully valid in this example.

8.6 Concluding Remarks

So, we have discussed the general conditions under which a body moving in the contact with elastic medium meets zero driving force (see Introduction), and presented some examples of the resistance-free motion. In addition to the resistance-free sub-Rayleigh speed regimes in the half-plane problem and for the two half-planes compressed, we have found a sharp decrease in the resistance in a vicinity of the longitudinal wave speed, with zero limit at this speed.

We also have examined in detail all the other speed regimes, where a driving force is required to support the motion. In particular, we have derived a general solution of the moving-contact plane problem valid for all speed ranges, $0 < v < \infty$, including super-Rayleigh sub- and intersonic speeds (Sect. 8.3.1). In the case of a single simply connected contact zone, this solution is presented for a general smooth shape of the indenter. The general expressions are then deduced for the case of symmetry, and then for a parabolic (or a constant curvature) indenter. Also the dynamic wedging problem is considered for a finite length wedge moving at a sub-Rayleigh speed at a distance of the crack tip, and such problem for two half-planes compressed together.

In all these cases, we have determined the driving forces caused by the main underling factors: stress field singular points in the contact region (the super-Rayleigh subsonic regime), wave radiation (the intersonic and supersonic regimes) and crack resistance (the wedging problem). Note that in the latter sub-Rayleigh speed problem, the driving force is equal to the crack resistance, while the contact interaction gives no contribution to this force.

The contact problems are considered based on a general solution as a steady-state limit of the corresponding transient problem. In this solution, the stresses and displacements are expressed through a single analytical function for all the speed ranges. The uniqueness is achieved by this limit and by an energy condition. The latter states that active (energy-source) singular points, which can appear in a solution, must be eliminated, and only passive ones (energy absorbing) can be accepted.

Based on the conditions allowing the 'paradox' to exist, it is clear that the considered examples do not exhaust the class of structures where such resistance-free motion can exist. In particular, this concerns the various 1D models and more complicated 2D and 3D structures. Note that, in the latter case, the Rayleigh waves can be excited by a finite indenter moving with a super-Rayleigh speed along the boundary of an elastic half-space.

The considered problems are studied in the framework of linear elasticity. So the strains and rotations are assumed to be small (except those in a vicinity of a singular point). However, the conditions of the 'paradox' are valid independently of what the model of elastic medium is used, the exact or linearized ones. The only condition for the medium is that the strain energy must be potential (in the nonlinear elasticity terminology, this assumes the hyperelastic material). In this connection we recall that the energy fluxes can be determined using the Eshelby-Cherepanov-Rice integral over a suitably built contour or surface, as applicable for both the exact nonlinear elasticity and the linearized one. Along with this, if the contact strains and rotations are not small enough, the exact formulation may lead to essential changes of the contact area and the contact stress distribution.

Part IV
Some Nonlinear Waves

Chapter 9
Longitudinal Waves

9.1 Shock Wave

9.1.1 Shock Wave and Energy Dissipation

Now consider a step wave in an elastic bar made of a nonlinear material with hardening in compression, a *hardening type* nonlinear material ($\sigma \leq 0, \varepsilon \leq 0$). This means that $\sigma = \sigma(\varepsilon)$, and the *secant modulus* $E_s = \sigma/\varepsilon$ is an increasing function of the negative strain. Now the strain is not assumed to be small. It, however, must satisfy the inequality $-1 < \varepsilon$, where the lower boundary $\varepsilon = -1$ corresponds to the limiting compression: the bar length being initially nonzero becomes zero (there is no such a geometrical limit under extension). It follows that for the material under compression the secant modulus, E_s, the particle speed, v, and the material density in the wave, ϱ, satisfy the conditions

$$E_s > -\sigma, \quad 0 < v < c, \quad \varrho = \frac{\varrho_0}{1+\varepsilon}, \tag{9.1}$$

where c is the wavefront speed. Note that often, in the shock wave theory, the material properties are referred to the $p - V$ diagram, where V is the specific volume. We, however, still use the $\sigma - \varepsilon$ diagram as in the entire text. In the considered one-dimensional case, the correspondence is

$$p = -\sigma, \quad \frac{V}{V_0} = \frac{\varrho_0}{\varrho} = 1 + \varepsilon. \tag{9.2}$$

Fundamental relations, the momentum and mass conservation

$$\sigma = -\frac{\mathrm{d}}{\mathrm{d}t}(mv) = -\varrho_0 c v, \quad \varepsilon = -\frac{v}{c} \tag{9.3}$$

L. I. Slepyan, *Mechanical Waves: Free, Forced, and Transition*,
https://doi.org/10.1007/978-3-032-15864-2_9

are independent of the material properties and hence remain valid. Now

$$\sigma = E_s(\varepsilon)\varepsilon \tag{9.4}$$

and hence the *wavefront speed* and the particle speed are

$$c = \sqrt{\frac{E_s}{\varrho_0}}, \quad v = -\frac{\sigma}{\sqrt{\varrho_0 E_s}} = -\frac{\sigma}{\varrho_0 c}. \tag{9.5}$$

We now can find the *energy flux density*, N, in this wave. The energy flux per unit time through a unit area of the *material* cross-section from the left to the right is

$$N = -\sigma v = \frac{\sigma^2}{\sqrt{\varrho_0 E_s}} = \frac{\sigma^2}{\varrho_0 c}. \tag{9.6}$$

Thus there is a constant energy flux in the step wave. It transports the work of the applied stress to the wavefront, $x = ct$. The work per unit length is

$$A = \frac{N}{c} = \frac{\sigma^2}{E_s}. \tag{9.7}$$

The energy density is

$$\begin{gathered} W = \mathcal{K} + \Pi, \quad \mathcal{K} = \frac{\varrho_0 v^2}{2} = \frac{\sigma^2}{2E_s}, \\ \Pi = \int_0^{\varepsilon} \sigma(\varepsilon')\,\mathrm{d}\varepsilon' = \int_0^{\varepsilon} E_s(\varepsilon')\varepsilon'\,\mathrm{d}\varepsilon' < \frac{1}{2}E_s(\varepsilon)\varepsilon^2 = \frac{\sigma^2}{2E_s}, \end{gathered} \tag{9.8}$$

where $\mathcal{K}$ and Π are the kinetic and strain energies, and the inequality is valid for the hardening type of the material. Thus, there is an *excess energy*

$$A - W = \frac{\sigma^2}{2E_s} - \int_0^{\varepsilon} E_s(\varepsilon')\varepsilon'\,\mathrm{d}\varepsilon' > 0. \tag{9.9}$$

Such a contradiction with the energy conservation law reflects the *limitations of the continuum model*, which does not allow for a complete description of the processes at the moving discontinuity. The contradiction is evident not only in this case, but also, for example, in fracture, where the cracks growth is accompanied by the *energy release*.

In the step wave, behind the wavefront the material is stressed uniformly and it moves uniformly. Thus, it is in equilibrium, and the excess energy can dissipated only at the wave front. But does the jump at the wavefront really exist? To elucidate the shock wave phenomenon consider what is happen if the load, σ is not completely applied at a moment.

9.1.2 Shock Front Formation

Suppose that a step wave exists with the stress $\sigma_0 < 0$, the particle velocity $v_0 > 0$ and the wavefront speed $c > 0$. Consider the superposition of this wave and an infinitesimal one induced by an additional load $\delta\sigma < 0$ applied to the left end of the bar. In such a superposition, the initial wave represents a uniformly moving compressed material. The additional, infinitesimal wave, propagating relatively this material, can be expressed in the same way the above-used for the unstressed and unmoving material, taking into account the differences in the related expressions for the unstressed and stressed materials. The density of the latter (see (18.7))

$$\varrho = \frac{\varrho_0 c}{c - v} = \frac{\varrho_0}{1+\varepsilon}, \quad \varepsilon = -\frac{v}{c}, \quad c = \sqrt{\frac{E_s}{\varrho_0}}. \tag{9.10}$$

Hooke's law, valid for the infinitesimal variation of the stress, is defined by (9.4), where for the hardening type material $E_t > E_s$. It follows that the corresponding wavefront speed relatively the unmoving material is

$$c_a = v + \left(1 - \frac{v}{c}\right)\sqrt{\frac{E_t}{\varrho_0}} > v + \left(1 - \frac{v}{c}\right)\sqrt{\frac{E_s}{\varrho_0}} = c. \tag{9.11}$$

The particle velocity increment is

$$\delta v = -\frac{\delta\sigma}{\sqrt{E_t \varrho}} = -\frac{\delta\sigma}{\sqrt{E_t \varrho_0}}. \tag{9.12}$$

Thus, the infinitesimal-wave velocity, the so called *local sound velocity*, is greater than that of the main wave. Sooner or later the former reaches the wave front. However, it cannot propagate through the front since the infinitesimal-wave velocity in the unstressed material ahead of the front is lower than the front velocity. We come to the conclusion that the additional wavefront cannot remain behind the main front and cannot occur ahead of it. This means that a united wave is formed with the front propagating slightly faster than the initial wavefront and slower than the additional wavefront, namely, the united wavefront speed is

$$c_u = \sqrt{\frac{\sigma + \delta\sigma}{\varrho_0}}. \tag{9.13}$$

These considerations give evidence that, in a hardening type material, the wavefront, the *shock front*, with a jump of the material state is formed even in the case when the load, σ, is increasing from zero continuously.

9.1.3 Shock Wave and the Heat Pressure

The second question concerns the energy excess (9.9). Where is it going to? This question cannot be answered in the framework of the classical model of elasticity. It reasonable to expect that, in the hardening-type material, the jump in the medium state at the wave front is accompanied by micro-oscillations. The micro level as the level of a structure of the material is not described in the formulation. However, its existence must be assumed in order to satisfy the energy conservation law. *The existence of a microlevel, which accumulate a part of energy, is necessarily associated with a shock wave in a hardening type nonlinear material.*

It is said that the excess energy is wasted on heating the material. In turn, the heating manifests itself by a *heat pressure*, which increases the effective stiffness of the material. Thus, in the jump at the wavefront, the dynamic stiffness appears greater, and the *dynamic stress-strain relation*, the *shock adiabat* or the *Hugoniot adiabat*, appears harder than the corresponding quasi-static, *isothermal relation*. Note that the heat pressure arises due to the thermal expansion, which is closely related to the hardening under compression.

The dynamic effects become very strong in the case of a porous material. Dynamic compression of such a material in a shock wave leads to a sharp increase in temperature and to an abnormally high heat stress. As a result, the dynamic stiffness of the porous material can be higher than that for the same material without pores (Zel'dovich and Raizer, 2002; also see Barenblatt and Sivashinskii, 1970).

In connection with the step wave added by the oscillations, note that the momentum and mass conservation laws (9.3) are written in a global form; they correspond to the averaged values and, in this sense, are still valid. The latter values, however, are not uniquely defined by those laws. The system is closed by the dynamic stress-strain diagram (Hugoniot adiabat), which differs from the static one by the heat pressure caused by the oscillations. The Hugoniot adiabat can be derived, in principle, based on the quasi-static relation and considering the *the wavefront structure*, which is defined by the material structure. Below the simplest model of such a kind is considered that allows to elucidate the phenomena and to obtain a complete description of the nonlinear step wave, namely, to find (a) the wave speed as a function of the pressure, (b) the excess energy, (c) the Hugoniot adiabat and (d) the role of the material porosity.

9.1.4 Wave in a Line of Beads

In order to show the mechanism of dissipation in a shock wave and its influence on the dynamic stress-strain relation, consider the semi-infinite line of periodically placed rigid beads. Let d, a and m be the diameter of a bead, distance between the neighboring beads and the boll mass, respectively; $l = d + a$ is the period. Initially the beads are at rest. At $t = 0$ force $P = \text{const}$ is applied to the first bead to move it toward the second one. At the moment $t = T$ the first bead has moved at the distance a, reaches the second one and collides with it with the speed v_b

$$T = \sqrt{\frac{2am}{P}}, \quad v_b = \frac{P}{m}T = \sqrt{\frac{2aP}{m}}. \tag{9.14}$$

The collision is assumed to be perfectly elastic, so the first bead stops, while the second one takes its velocity. Then the momentum is transferred from the second bead to the third one and so on. This gives us the wave speed

$$c = \frac{lv_b}{a} = l\sqrt{\frac{2P}{am}}. \tag{9.15}$$

Then the process repeats, and at $t = 2T$ the first bead collides the second one again, and a new one-bead-in-moving wave propagates along the line. It follows that the averaged particle velocity and the strain are

$$v = \frac{a}{T} = \sqrt{\frac{aP}{2m}} = \frac{v_b}{2}, \quad \varepsilon = -\frac{v}{c} = -\frac{a}{2l}. \tag{9.16}$$

The averaged, K, and total, K_t, energies are

$$K = \frac{mv^2}{2}\frac{ct}{l} = \sqrt{\frac{a}{m}}\left(\frac{P}{2}\right)^{3/2} t, \quad K_t = Pvt = \sqrt{\frac{a}{2m}}P^{3/2}t = 2K. \tag{9.17}$$

In this process, at any time (except the collision moments) some of the beads move with the speed v_b, while the other, except the first one, are unmoving. From (9.16) it follows that half of the beads is in motion, and the other half is immobile. (In these considerations, assuming $ct \gg l$, the facts are neglected that the number of the beads is integer and that the first bead moves differently.)

Thus, the results of the above analysis are as follows. In the considered model, the wave speed is proportional to the pressure in power 3/2. The excess energy is equal to the *macrolevel energy* as that corresponding to the averaged particle velocity. The Hugoniot adiabat defines a constant strain as the static one; however, the dynamic strain, $-a/(2l)$, is equal to half the static one and both are independent of the pressure. The latter ratio gives evidence of role of the porosity: if the distance $a \gg d$, where d is the bead diameter, then the dynamic strain $\varepsilon \sim -1/2$; otherwise, $\varepsilon \to 0$ $(a/d \to 0)$. Note that zero resistance to pore closing corresponds to strong shock waves, where this resistance is really negligible.

Note that in a real bead-system, there is additional, relatively small dissipation due to oscillations in each elastic bead under the collision. This phenomenon was neglected in the above formulation where the collisions were assumed to be 'perfectly elastic'.

9.1.5 Wavefront Degradation

In the case of the above-considered step wave, the energy dissipation in the wavefront does not influence the wave itself. The wave looks as if it propagates in the linearly elastic material, and the dissipation is reflected only by the difference between the energy flux and the energy of the wave. This, however, is true only for the step wave. Otherwise, if the wave intensity decreases with the distance from the wavefront as in the case of collision of a rigid mass with a long elastic bar, the dissipation leads to wave degradation. This is due to the fact that in a hardening type material the unloading wave propagates faster the wavefront.

Consider such a wave in a bar made of the material with the tangent modulus monotonically growing with the stress: $\mathrm{d}^2\sigma/\mathrm{d}\varepsilon^2 > 0$. In this case, both the shock front and the unloading wave speeds grow with the stress and the latter exceeds the former

$$\frac{\mathrm{d}c}{\mathrm{d}\sigma} > 0, \quad \frac{\mathrm{d}c_t}{\mathrm{d}\sigma} > 0, \quad c < c_t, \tag{9.18}$$

where the infinitesimal unloading wave speed is

$$c_t = \sqrt{\frac{E_t(\sigma)}{\varrho}}, \quad E_t(\sigma) = \frac{\mathrm{d}\sigma}{\mathrm{d}\varepsilon}. \tag{9.19}$$

Note that with respect to the unloading wave the material appear as a softening one. Let at a moment the stress wave be

$$\sigma = \sigma_0 f(x) H(-x), \quad f'(x) > 0, \quad f'(0) = \alpha. \tag{9.20}$$

Consider an infinitesimal part of it adjacent to the wavefront, $-\delta x < x < 0$. The corresponding unloading wave speed is defined in (9.19) with $E_t(\sigma_0)$. It follows that during the time of the interaction of this wave with the shock front, δt, the intensity of the latter is changed by $\delta\sigma$, where

$$\delta t = \frac{\delta x}{c_t - c} \quad (c_t\delta t + \delta x = c\delta t), \quad \delta\sigma = -\sigma_0\alpha\delta x\,; \tag{9.21}$$

hence

$$\frac{\mathrm{d}\sigma}{\mathrm{d}t} = -\alpha\sigma_0(c_t - c) \quad (c_t = c_t(\sigma),\ c = c(\sigma),\ \sigma = \sigma_0). \tag{9.22}$$

In addition to the decrease of the shock front intensity, the wave stretches due to the decrease of the local speed, c_t, with the distance from the shock front. Hence α decreases in time as well as c_t and c. So the wave energy decreases as should be, while its pulse remains invariable.

9.1.6 Wavefront Conditions

Consider a more general case, where the plane wavefront propagates through a stressed and moving material, and denote the wavefront speed *relatively the latter* by D. The particle velocities are assumed not to be necessarily normal to the front, but the wavefront speed is normal to it. The mass per unit cross-section area, passing through the front per unit time, is $\varrho_+ D$ (the subscripts $\pm$ correspond to the values ahead of the front and behind it, respectively), and the mass conservation law is

$$\varrho_-(D - v_{n-} + v_{n+}) = \varrho_+ D, \tag{9.23}$$

where v_n is the normal-to-the-front component of vector $\boldsymbol{v}$. The velocity of this mass is incremented by $\boldsymbol{v}_- - \boldsymbol{v}_+$, and the momentum conservation law reads as a vector equation

$$\boldsymbol{\sigma}_+ - \boldsymbol{\sigma}_- = \varrho_+ D(\boldsymbol{v}_- - \boldsymbol{v}_+), \tag{9.24}$$

where $\boldsymbol{\sigma}_\mp$ are the stress vectors acting on the wavefront plane from behind and in front of it, respectively.

The third is a thermodynamic inequality, which concerns a regular, non-active material. It expresses the fact that the wavefront, as a moving jump in the material state, cannot be a positive source of energy. During the wavefront passes the mass, $\varrho_+ D$, the work of the stresses, acting on the mass from outside, must be equal to or greater than the change in the sum of the strain and kinetic energies, that is

$$\boldsymbol{\sigma}_+ \boldsymbol{v}_+ - \boldsymbol{\sigma}_- \boldsymbol{v}_- \geq \left[\Pi(\varrho_-) - \Pi(\varrho_+) + \frac{v_-^2 - v_+^2}{2} \right] \varrho_+ D, \tag{9.25}$$

where $\Pi(\varrho)$ is the specific strain energy as a function of density.

9.2 Continuous Longitudinal Waves

Now consider a *softening type stress-strain relation* $\sigma_{xx} = \sigma(\varepsilon)$ with the tangent modulus $E_t = \mathrm{d}\sigma / \mathrm{d}\varepsilon < E_s = \sigma/\varepsilon$. In this case, the local sound velocity decreases with the increase of the stress, and a suddenly applied load does not produce any step wave. Really, assume for a moment that a step wave exists. But in this hypothetic case, a small increase of the applied force leads to an additional wave propagating slower. So, the shock wave is not formed. The wave front, which arises at the moment when the load is suddenly applied, then spreads out. This conclusion can also be proved by the energy considerations. Indeed, for a softening nonlinear material the strain energy required for the stress jump is greater than that for the linear Hooke's law and hence there is a lack of the energy instead of the excess energy in the above-considered case of the hardening material.

Parameters of the wave induced by a suddenly applied load, $\sigma = \sigma_0 = \sigma(\varepsilon_0)$, can be determined using a *nonlinear superposition*, namely, by means of sequential superposition of elementary waves. Each following elementary wave propagates slower than any previous one, and hence it propagates along a region with a constant stress and a constant particle velocity. Let ε and v be the strain and the particle velocity in such a region. Then the 'initial' state is characterized by the values of $\varrho = \varrho_0/(1+\varepsilon)$, v, and the tangent modulus E_t. Substituting ϱ and E_t instead ϱ_0 and E in (9.3) and (9.4) the infinitesimal-wave speed are found as

$$c_a = (1+\varepsilon)c_t + v, \quad c_t = \sqrt{\frac{E_t}{\varrho_0}} \tag{9.26}$$

and the increment of the particle velocity

$$\delta v = -\frac{\delta\sigma}{\sqrt{\varrho E_t'}} = -\frac{\delta\sigma}{\sqrt{\varrho_0 E_t}} = -c_t\delta\varepsilon. \tag{9.27}$$

With refer to the discussion preceding to relation (17.18), now three different speeds of the additional infinitesimal wave can be noted: the 'absolute' speed relatively the unmoving material ahead of the wave, c_a (9.26), the speed relatively the moving deformed material in the main wave, $c_m' = c_t(1+\varepsilon)$, and the speed relatively the initial material frame, $c_m = c_t$. So the mass involved in the additional wave is $\varrho_0 c_t t$. It follows from this, (9.26) and (9.12) that the quasifront widening in time is defined in the space and material frames as

$$X(t) = c_a(\varepsilon)t \quad \text{and} \quad x(t) = c_t(\varepsilon)t. \tag{9.28}$$

The particle speed as a function of the strain is

$$v(\varepsilon) = -\int_0^{\varepsilon} c_t(\varepsilon)\,\mathrm{d}\varepsilon. \tag{9.29}$$

This solution represents a centered family of rays, each of a constant strain, in the segment

$$0 \le \varepsilon \le \varepsilon_{\max} = \frac{\sigma_0}{E_s(\varepsilon_{\max})} \quad (\sigma_0 = \sigma(\varepsilon_{\max})). \tag{9.30}$$

Correspondingly

$$c_t(\varepsilon_{\max}) \le c_t \le c_t(0). \tag{9.31}$$

The wave parameters are variable in the region

$$c_t(\varepsilon_{\max})\, t < x(t) < c_t(0)\, t \tag{9.32}$$

and invariable in the rest region where

$$\sigma = \sigma_0, \quad v = -\int_0^{\varepsilon_{\max}} c_t(\varepsilon)\, \mathrm{d}\varepsilon \quad (0 < x(t) < c_t(\varepsilon_{\max})\, t). \tag{9.33}$$

The momentum conservation law being applied to an infinitesimal element of the bar represents a one-dimensional *wave equation*

$$\varrho_0 \frac{\partial v(x,t)}{\partial t} - \frac{\partial \sigma(x,t)}{\partial x} = 0. \tag{9.34}$$

For an elastic material, where $\sigma = \sigma(\varepsilon)$, this equation becomes

$$\varrho_0 \frac{\partial v(x,t)}{\partial t} - E_t \frac{\partial \varepsilon(x,t)}{\partial x} = 0. \tag{9.35}$$

Finally, with refer to (9.26) and (9.29)

$$\frac{\partial \varepsilon(x,t)}{\partial t} - c_t(\varepsilon) \frac{\partial \varepsilon(x,t)}{\partial x} = 0 \quad \left(c_t(\varepsilon) = \sqrt{\frac{E_t(\varepsilon)}{\varrho_0}} \right). \tag{9.36}$$

The curves (now they are straight lines (9.28)), where a considered function is uniform, are called *characteristic curves*. The *method of characteristics* is widely used in solving wave-related partial differential equations. In general, for a first-order PDE (as, for example, (9.36)) the method of characteristics discovers curves (called characteristic curves or just characteristics) along which the PDE becomes an ordinary differential equation (ODE). It can be seen that centered family of rays (9.28) just represents the characteristics of the *quasi-linear wave equation* (9.36).

As an example, consider the dynamic extension of a bar made of an incompressible material. In this case, the dynamic extension of the bar is accompanied by the corresponding cross-section narrowing, such that the material volume remains constant:

$$A = \frac{A_0}{1+\varepsilon}. \tag{9.37}$$

where A is the cross-section area and A_0 is its initial value. As before, we take the thickness strain as a quasi-static. Assume that the tension force is proportional to both the cross-section area and the strain. In terms of stresses, this means that

$$\sigma = \frac{E_0 \varepsilon}{1+\varepsilon}, \quad E_t = \frac{E_0}{(1+\varepsilon)^2}, \quad c_t = \frac{c_0}{1+\varepsilon} \quad \left(c_0 = \sqrt{\frac{E_0}{\varrho_0}}\right). \tag{9.38}$$

In follows that

$$\begin{aligned} v(\varepsilon) &= c_0 \ln(1+\varepsilon), \quad \varepsilon(x,t) = \left(\frac{c_0 t}{x} - 1\right) H(c_0 t - x) \quad \left(x \geq \frac{c_0 t}{1+\varepsilon_{\max}}\right), \\ v &= c_0 \ln(1+\varepsilon_{\max}), \quad \varepsilon = \varepsilon_{\max} \quad \left(0 < x \leq \frac{c_0 t}{1+\varepsilon_{\max}}\right). \end{aligned} \tag{9.39}$$

Note that a stress-strain relation changes its type when loading changes to unloading. Hardening in compression is replaced by softening in tension and vice versa. Consequently, an *unloading wave* differs by the type from the wave of loading.

9.3 Failure Wave in a Highly Porous Structure

9.3.1 Homogeneous Model

We here consider a homogeneous model of a large-scale structure as a vertically oriented highly porous uniform rod. For example, it can be a tallest building. Let the height, cross-section area and density be H, S and ϱ, respectively. The x-axis with the origin at the top of the intact structure is directed downward. In this rod, a crushing wave is assumed to arise when a rigid body of mass M, moving with speed v_0 hits the upper end of the rod. Beginning from this moment the mass is in contact with the rod.

Neglecting elastic waves we consider two parts of the rod separated by a moving crushing front, $x = l(t)$, $v(t) = \mathrm{d}l(t)/\mathrm{d}t \geq 0$. The forward part, $x > l(t)$, is assumed to be unmoving, while the rest part moves down as a rigid body. Because of the high porosity the density of the crushed-material part is much greater than the initial one, and the speed of the moving top end of the rod is close to the front speed. We neglect the difference assuming that the crushed-material part moves with the same speed $v(t)$. The mass of the latter part increases by gradually involving in the movement the material from the unmoving part. So it is a continuous model of a series of inelastic collisions.

Under this formulation, the wave is completely defined by the momentum conservation law

$$\frac{\mathrm{d}}{\mathrm{d}t}[(M + \varrho S l(t)) v(t)] = (M + \varrho S l(t)) g - Q, \tag{9.40}$$

where g is the gravity acceleration, and Q is the resistance to crushing; in this homogeneous model it is assumed to be a constant force. The dynamic equation can be written as

$$\frac{\mathrm{d}^2 L(t)}{\mathrm{d}t^2} + \frac{1}{L(t)}\left(\frac{\mathrm{d}L(t)}{\mathrm{d}t}\right)^2 = g - \frac{Q}{\varrho S L(t)}, \quad L(t) = l(t) + L_0,\ L_0 = M/(\varrho S). \tag{9.41}$$

Denoting $y = (v(t))^2$ and considering it as a function of L we come to a linear equation with respect to the speed squired, $y(L)$

$$\frac{1}{2}\frac{\mathrm{d}y}{\mathrm{d}L} + \frac{y}{L} = g - \frac{Q}{\varrho S L(t)}, \tag{9.42}$$

whose solution is

$$y = \frac{2}{3}gL - \frac{Q}{\varrho S} + \frac{C}{L^2}, \tag{9.43}$$

where C is an arbitrary constant. It follows that

$$\begin{aligned} \frac{\mathrm{d}L(t)}{\mathrm{d}t} &= \sqrt{2gL(t)/3 - Q/(\varrho S) + C/L^2(t)}, \\ \frac{L\,\mathrm{d}L}{\sqrt{2gL^3/3 - QL^2/(\varrho S) + C}} &= \mathrm{d}t. \end{aligned} \tag{9.44}$$

From the first equation in (9.44) we find

$$C = \left(v_0^2 - \frac{2}{3}gL_0 + \frac{Q}{\varrho S}\right)L_0^2 \quad [v_0 = v(0),\ L_0 = L(0) = M/(\varrho S)]. \tag{9.45}$$

A relation between L and t follows as

$$L_0 \int_1^{L/L_0} \left[\frac{2}{3}gL_0(\xi^3 - 1) + v_0^2 - \frac{Q}{\varrho S}(\xi^2 - 1)\right]^{-1/2} \xi\,\mathrm{d}\xi = t \quad (L/L_0 = 1 + l(t)/L_0). \tag{9.46}$$

In this relation, it was taken into account that $l(0) = L(0) - L_0 = 0$.

We now can express the rates of the work of gravity forces, of the resistance to crashing, of the kinetic energy and of the dissipation in the wavefront. These quantities are

$$\begin{aligned} A' = \frac{\mathrm{d}A}{\mathrm{d}L} = \varrho g S L(t), \quad R' = -Q, \quad \mathcal{K}' = \varrho g S L(t) - Q - \varrho S\frac{v^2(t)}{2}, \\ G = A' + B' - \mathcal{K}' = \varrho S\frac{v^2(t)}{2}, \end{aligned} \tag{9.47}$$

respectively. Recall that the dissipation arises due to the inelasticity of the continuous crushing as assumed in this model.

9.3.2 Fracture Criterion and Possible Scenarios

Contrary to the above-presented general description of the wave, a criterion of fracture cannot be set without invoking the 'microlevel' as the level of a specific structure. We here assume the structure to be a-periodic and the corresponding force-displacement diagram to be 'rigid-plastic'

$$q(u) = Q/\lambda \quad (0 < u < \lambda a), \quad q(u) = 0 \quad (\lambda a < u < a), \tag{9.48}$$

where the force, $q(u)$, corresponds to the displacement of the upper bound of the cell while the lower one is fixed. We assume that in the considered high-porous structure $\lambda \ll 1$. Under these conditions, the energy barrier, Qa, can be overcome if the kinetic energy of the moving part of the body and its weight, in aggregate, are sufficient, namely if

$$\varrho S(L + a)\left[\left(\frac{L}{L + a}\right)^2 \frac{v^2}{2} + g\lambda a\right] \geq Qa = G_c a. \tag{9.49}$$

Note that, in general, the fracture energy, G_c, is not a constant but depends on the mass and velocity at impact.

When this criterion is satisfied at the beginning, that is for $L = L_0$, the following scenarios are possible. If the kinetic energy just before the second collision is greater or equal to the initial one, that is if

$$\mathcal{K}(0) = \frac{1}{2}\varrho S L_0 v_0^2 \leq (L_0 + a)[\varrho g S(L_0 + a) - Q], \tag{9.50}$$

then it will increase in the following. Otherwise, it first decreases. As follows from the homogeneous model, Eqs. (9.42), (9.43), (9.45), the speed reaches minimum at the moment when

$$L = L_* = \left(\frac{3C}{g}\right)^{1/3}, \quad v^2 = v_{\min}^2 = gL_* - \frac{Q}{\varrho S}. \tag{9.51}$$

Then the speed increases. This, of course, is true if the criterion (9.49) is satisfied; otherwise, the wavefront stops.

The increasing wavefront speed approaches an asymptotic representation, which can be written as

$$v \sim \frac{gt}{3} + v_0, \quad L \sim \frac{gt^2}{6} + v_0 t + L_0. \tag{9.52}$$

following from (9.43) and (9.46) In the case of a specific connection between the mass, M, and the velocity at impact, v_0, this asymptotic representation exactly satisfies not only the initial conditions but also Eq. (9.43). The connection is

$$v_0^2 = \frac{2gM}{3\varrho S} - \frac{Q}{\varrho S}. \tag{9.53}$$

Note that in this case $C = 0$.

A detailed consideration of the Nine-Eleven progressive collapse mechanics can be found in Bažant et al. (2007); aslo see Bažant and Jirásek (2008) and Le et al. (2011).

9.3.3 Discrete Model

We now consider the structure as a a-periodic line of rigid masses, m, connected by equal bonds, which are destroyed as the crushing wave propagates. Inelastic collisions of the masses are still assumed. Below, as in the case of the above-considered homogeneous model, an explicit expression for the wave speed as a function of the wavelength is obtained.

If the criterion (9.49) is satisfied at the beginning, the wave starts to propagate. Denote the initial speed of nth mass, as its speed immediately after the collision with the preceding, $(n-1)$-mass and its speed at the moment just before the collision with the next mass, by v_{n-} and v_{n+}, respectively. The momentum conservation law yields

$$v_{n-} = \frac{M + (n-1)m}{M + nm} v_{(n-1)+}, \quad n = 1, 2, \ldots . \tag{9.54}$$

Then the moving masses meet the resistance and accelerate under gravity, and

$$\begin{aligned}(M+nm)^2 v_{n+}^2 &= [M+(n-1)m]^2 v_{(n-1)+}^2 + (M+nm)^2 2ga - (M+nm)2Qa \\ &= M^2 v_0^2 + [(M+nm)^2 + (M+(n-1)m)^2 + \cdots + (M+n)^2]2ga \\ &\quad -[(M+nm+M+(n-1)m+\cdots+M+m]2Qa.\end{aligned} \tag{9.55}$$

Thus

$$\begin{aligned}(M+nm)^2 v_{n-}^2 &= M^2 v_0^2 + (n-1)\left[M^2 + nMm + \frac{1}{6}n(2n-1)m^2\right]2ga \\ &\quad -(n-1)\left[M + \frac{n}{2}m\right]2Qa, \\ (M+nm)^2 v_{n+}^2 &= M^2 v_0^2 + n\left[M^2 + (n+1)Mm + \frac{1}{6}(n+1)(2n+1)m^2\right]2ga \\ &\quad -n\left[M + \frac{n+1}{2}m\right]2Qa.\end{aligned} \tag{9.56}$$

The energy dissipated under the inelastic collision is

$$G_n a = \frac{1}{2}[(M + (n-1)m)v_{(n-1)+}^2 - (M + nm)v_{n-}^2]$$
$$= \frac{M + nm}{M + (n-1)m} \frac{mv_{n-}^2}{2} = \frac{M + (n-1)m}{M + nm} \frac{mv_{(n-1)+}^2}{2}. \tag{9.57}$$

In the continuous-model terms, for the moment when $v = v_{n+}$ we have

$$M = L_0 \varrho S, \quad m = a\varrho S, \quad M + nm = L\varrho S, \quad L = L_0 + na. \tag{9.58}$$

In these terms

$$v_{n+}^2 = \left(\frac{L_0}{L}\right)^2 v_0^2 + \frac{2gl}{L^2}\left[L_0^2 + (l+a)L_0 + \frac{1}{6}(l+a)(2l+a)\right] - \frac{2Ql}{\varrho SL^2}\left[L_0 + \frac{l+a}{2}\right]$$
$$\sim \left(\frac{L_0}{L}\right)^2 v_0^2 + \frac{2gL}{3}\left(1 - \frac{L_0^3}{L^3}\right) - \frac{Q}{\varrho S}\left(1 - \frac{L_0^2}{L^2}\right) \quad (a \ll l = L - L_0). \tag{9.59}$$

The latter asymptotic representation, $a \ll l$, coincides with the corresponding result for the continuous structure (9.43), (9.45). Thus, the continuous model is valid when the length of the wave is relatively large as should be.

Chapter 10
Transient Nonlinear Waves in Strings

10.1 Dynamic Equations

A *string* is considered as a one-dimensional elastic system of negligible bending and torsion stiffness and compressive strength. Note that such model of the string is a long-wave approximation for a flexible rod (cable, pipe, etc.). The dynamic state of this *material curve* is described by the tensile force, F, and the position vector $\boldsymbol{R}$, as functions of the curvilinear *Lagrangian coordinate* s and time t

$$F = F(\lambda)H(\lambda - 1) = F(s, t), \quad \boldsymbol{R} = \boldsymbol{R}(s, t) \quad (\lambda = 1 + \varepsilon). \tag{10.1}$$

The dynamic equation for the string is similar by its sense to that for the bar (9.34) but now it is a vector equation

$$\varrho_0 \frac{\partial^2 \boldsymbol{R}(s, t)}{\partial t^2} - \frac{\partial \boldsymbol{F}(s, t)}{\partial s} = \boldsymbol{q}(s, t), \tag{10.2}$$

where ϱ_0 is the string mass per unit length in an initial unstrained state, $\boldsymbol{F}(s, t) = F(s, t)\boldsymbol{\tau}$ is the force directed as the coordinate s, $\boldsymbol{\tau}$ is the unit tangent vector, and the right-hand side, $\boldsymbol{q}(s, t)$, is a distributed external force, if exists. Recall that the Lagrangian formulation is used and hence $\partial^2 \boldsymbol{R}(s, t)/\partial t^2$ is the particle acceleration.

The material curve, $\boldsymbol{R}(s, t)$, is characterized by the stretch, λ, curvature, $\varkappa$, and torsion, χ. In terms of the Lagrangian coordinate, s, and the *Frenet triad* unit vectors, the tangent vector $\boldsymbol{\tau}$, the normal, $\boldsymbol{n}$ and the bi-normal, $\boldsymbol{b}$, we have

$$\boldsymbol{R}' = \frac{\partial \boldsymbol{R}}{\partial s} = \lambda \boldsymbol{\tau}, \ \boldsymbol{\tau}' = \frac{\partial \boldsymbol{\tau}}{\partial s} = \lambda \varkappa \boldsymbol{n}, \ \boldsymbol{n}' = \frac{\partial \boldsymbol{n}}{\partial s} = \lambda(-\varkappa \boldsymbol{\tau} + \chi \boldsymbol{b}), \ \boldsymbol{b}' = \frac{\partial \boldsymbol{b}}{\partial s} = -\lambda \chi \boldsymbol{n}. \tag{10.3}$$

L. I. Slepyan, *Mechanical Waves: Free, Forced, and Transition*,
https://doi.org/10.1007/978-3-032-15864-2_10

In these terms, Eq. (10.2) can be rewritten as

$$\varrho_0 \ddot{\boldsymbol{R}} - (F\boldsymbol{R}'/\lambda)' = \boldsymbol{q}, \quad F = F(\lambda)H(\lambda - 1), \quad \lambda = \left|\boldsymbol{R}'\right|, \tag{10.4}$$

where a dot above the symbol denotes the partial derivative with respect to time.

10.2 The String Under a Constant Tensile Force

An important partial case corresponds to a constant tensile force. The dynamic Eq. (10.4) can be represented as (see (10.3))

$$\varrho_0 \ddot{\boldsymbol{R}}(s,t) - \frac{F}{\lambda}\boldsymbol{R}''(s,t) = \boldsymbol{q}(s,t), \quad (F,\lambda) = \text{const}, \quad |\boldsymbol{R}'(s,t)| = \lambda \tag{10.5}$$

or, in terms of the uniform strain state of the string, $\varrho = \varrho_0/\lambda$ and $S = s\lambda$

$$\varrho \ddot{\boldsymbol{R}}(S,t) - F\boldsymbol{R}''(S,t) = \boldsymbol{q}(S,t), \quad F = \text{const}, \quad \boldsymbol{R}'(S,t) = \boldsymbol{\tau} \ (|\boldsymbol{\tau}| = 1), \tag{10.6}$$

where $\boldsymbol{\tau}$ is the unit tangent vector. The general homogeneous solution to the equation is

$$\boldsymbol{R} = \boldsymbol{R}_0(S - Dt) + v_0 t\boldsymbol{k} \quad \text{or} \quad \boldsymbol{R} = \boldsymbol{R}_0(S + Dt) + v_0 t\boldsymbol{k} \quad (v_0 = \text{const}, \ \boldsymbol{q} = 0), \tag{10.7}$$

where $\boldsymbol{k}$ is a constant unit vector and

$$D = \sqrt{\frac{F}{\varrho}}. \tag{10.8}$$

The equation in (10.6) looks as the linear one-dimensional wave equation. The variable S is, however, the curvilinear coordinate placed along the actual string line, rather than an independent straight-line coordinate as in the case of the linear wave equation. This imposes the nonlinear condition, $\boldsymbol{R}'(S,t) = \boldsymbol{\tau}$, which dramatically narrows the set of D'Alembert solutions. In particular, the superposition principle is not valid here (except a constant-velocity rigid-body movement which can be added to the nontrivial first term $\boldsymbol{R}_0$ (10.7)).

It follows from (10.7) that

$$\dot{\boldsymbol{R}}_0 = -D\boldsymbol{R}_0' = -D\boldsymbol{\tau} \quad \text{or} \quad \dot{\boldsymbol{R}}_0 = D\boldsymbol{R}_0' = D\boldsymbol{\tau}, \tag{10.9}$$

respectively.Each of these solutions corresponds to the string uniformly flowing with speed D along an arbitrary given rigid frame: $S = \mp Dt$, respectively.

Conclusion 10.1 A uniform curvilinear string of the actual linear density ϱ, stretched by a constant force F, is in dynamic equilibrium if it moves with velocity $\boldsymbol{v} = \pm(F/\varrho)^{1/2}\boldsymbol{\tau}$ so that its shape remains fixed.

It seems hard to accept this statement intuitively; however, it becomes clear if we pay attention to the fact that the tensile force (10.8) is necessary and sufficient to counterbalance the forces of inertia—independently of the generally nonuniform curvature. Note that in the uniform curvilinear motion the inertia forces are directed as the normal

$$\varrho\dot{\boldsymbol{v}} = \varrho D\dot{\boldsymbol{\tau}} = \varrho D^2\varkappa\boldsymbol{n}, \tag{10.10}$$

and the same but opposite normal action corresponds to the tensile force (10.8).

Now consider the first of the solutions in (10.7). Suppose that the frame is infinite, receding to the infinity along a straight line with a constant directional unit vector $\boldsymbol{k}$, $\boldsymbol{\tau} = \boldsymbol{k}$. In this case, $v_0\boldsymbol{k}$ is the wave velocity as the velocity of the curved part of the string. We denote this transversal wave speed c_T ($v_0 = c_T$). The particle velocity ahead of the wave is thus

$$\boldsymbol{v} = (-D + c_T)\boldsymbol{k}, \tag{10.11}$$

or

$$c_T = D + v, \tag{10.12}$$

and, in accordance with (10.8), the tensile force is

$$F = \varrho(c_T - v)^2. \tag{10.13}$$

It is remarkable that such a *nonlinear wave of arbitrary shape and amplitude* can exist in a uniformly stretched string.

Comment 10.1 These two relations define the *transversal wave speed*, c_T, as the 'absolute' uniform speed of the curvilinear part of the string, and the corresponding equilibrium tensile force, F, in a general case where a uniform particle speed, v, in the straight-line region of the string ahead of the wave is nonzero.

In particular case where the string is immobile ahead of the wave, the transversal wave speed is equal to $D = \sqrt{F/\varrho}$ (recall that here ϱ is the string mass per unit length, and F is the tensile force). So it can differ much from the longitudinal wave speed, $c = \sqrt{EA/\varrho}$, where E and A are the elastic modulus and the cross-section area, respectively. Usually $EA/F \gg 1$.

Note that the dynamic state remains the same for a semi-infinite string subjected to the action of constant-value external force (10.13) at the end. The direction of the force can vary in time. Clearly, there is the dynamic equilibrium since the effect of this force is equivalent to the action of a removed part of the infinite string.

10.2.1 Linear String Equation

Equation (10.5) becomes linear in the case of small deflections of a straight string, namely, where the derivatives dy/dx, dz/dx are so small that their squares are negligible compared with unity. In this case

$$\begin{gathered} \varrho\ddot{\boldsymbol{R}}(x,t) - F\boldsymbol{R}''(x,t) = \boldsymbol{q}(x,t), \\ (\varrho, F) = \text{const}, \quad x \sim S. \end{gathered} \tag{10.14}$$

This *one-dimensional linear wave equation* admits the superposition and the well-known homogeneous *D'Alembert solution*

$$\boldsymbol{R} = \boldsymbol{R}_+(x - Dt) + \boldsymbol{R}_-(x + Dt), \quad D = \sqrt{\frac{F}{\varrho}}, \tag{10.15}$$

where $\boldsymbol{R}_\pm(x)$ are arbitrary vector functions, which do not contradict to the above-discussed condition of linearity.

10.3 The Wavefront Conditions

Consider a wavefront as a moving point, where among the wave parameters, the tensile force, $\boldsymbol{F} = F\boldsymbol{\tau}$, the particle velocity, $\boldsymbol{v}$, and the unit tangent vector, $\boldsymbol{\theta}$, at least one is discontinuous. Recall that $F \geq 0$. Let the unit tangent vector ahead of the wavefront and behind it be $\boldsymbol{\tau}_+$ and $\boldsymbol{\tau}_-$, respectively. Respectively

$$\boldsymbol{F} = \boldsymbol{F}_\pm, \quad \boldsymbol{v} = \boldsymbol{v}_\pm, \quad \varepsilon = \varepsilon_\pm, \quad \varrho = \varrho_\pm. \tag{10.16}$$

Initial and actual densities are connected by the equalities

$$\varrho_-\lambda_- = \varrho_+\lambda_+ = \varrho_0. \tag{10.17}$$

Consider the string in a coordinate system moving uniformly together with the wavefront. In this system, at least in a close vicinity of the unmoving wavefront, the particle velocities are oriented along the string

$$\boldsymbol{v}_\pm = v_\pm\boldsymbol{\tau}_\pm. \tag{10.18}$$

Since the wavefront is neither a source nor sink of mass, the flux of the mass, N_ϱ, is continuous

$$N_\varrho = \varrho_+ v_+ = \varrho_- v_- = \varrho_0 v_f, \tag{10.19}$$

where v_f is defined by the latter equality. This is the mass conservation law. The momentum conservation law yields

$$N_\varrho(v_+\boldsymbol{\tau}_+ - v_-\boldsymbol{\tau}_-) = F_+\boldsymbol{\tau}_+ - F_-\boldsymbol{\tau}_-. \tag{10.20}$$

The last equation evidences that a kink, $\boldsymbol{\tau}_- \neq \boldsymbol{\tau}_+$, cannot propagate along an unstressed string, that is, under the condition $F_+ = 0$. Indeed, in this hypothetic case, $v_+ = -D < 0$, $\varrho_+ = \varrho_0 > 0$. Hence both N_ϱ and v_+ are nonzero, and the left-hand side of the momentum Eq. (10.20) contains a $\boldsymbol{\tau}_+$-component, while the right-hand side does not. Thus F_+ must be positive.

It follows from relations (10.19) and (10.20) that

$$F_+ = N_\varrho v_+ = N_\varrho v_f \lambda_+, \quad F_- = N_\varrho v_- = N_\varrho v_f \lambda_-. \tag{10.21}$$

In the case of a linear or a hardening type material, no more than one value of the stretch, λ, can satisfy such an equation, $F(\lambda) = \text{const}\cdot\lambda$ (see Fig. 1.5), and hence, in this case $F_+ = F_-$.

In the case of a softening type material, two different values of λ can satisfy the latter equation (see Fig. 1.5). In this case the energy flux to the front is

$$N = F_+v_+ - F_-v_- = N_\varrho(v_+^2 - v_-^2) = \frac{F_+^2 - F_-^2}{N_\varrho}. \tag{10.22}$$

However, there is no dissipation of energy in the wavefront propagating in such a material, Sect. 17.2.2.2, and the tensile force is nonnegative. Thus, the tensile force and hence the particle velocities are continuous at the kink and the tensile force – particle speed relation (10.8), considered in the moving, kink-associated coordinate system, is valid here.

Now consider a wavefront in a straight string,

$$\boldsymbol{k}_- = \boldsymbol{k}_+. \tag{10.23}$$

It follows from (10.21) that

$$\varrho_0 v_f^2(\varepsilon_+ - \varepsilon_-) = F_+ - F_-. \tag{10.24}$$

We have

$$F_\pm = E_{s\pm}\varepsilon_\pm, \quad v_+ = v_+^0 - c, \tag{10.25}$$

where v_+^0 is the particle velocity ahead of the front in an immobile coordinate system. It follows that the wave speed in a linear or a hardening type material is

$$c = v_+^0 + \lambda_+ \sqrt{\frac{E_{s+\varepsilon_+} - E_{s+\varepsilon_+}}{\varrho_0(\varepsilon_+ - \varepsilon_-)}}, \tag{10.26}$$

where the secant modules correspond to the Hugoniot adiabat (see Sect. 9.1.3). In particular, for a wavefront propagating with velocity c in an unstressed unmoving string, the relations (10.24) and (10.27) yield the formulas (18.7) for the 'absolute' speeds, the wave speed and the particle speed

$$c = \sqrt{\frac{E_s}{\varrho_0}}, \quad v = -\frac{F}{\sqrt{\varrho_0 E_s}} = -\frac{F}{\varrho_0 c}. \tag{10.27}$$

This is not surprising, since the wavefront condition (10.27) is derived based on the step-wave configuration which is assumed to exist.

Thus, there exist two different waves in a string, the *transversal wave* as a moving kink, or another curvilinearly shaped part of the string, and the longitudinal wave. Contrary to the former, the latter is characterized by a jump of the tensile force at the wavefront. The wavefront speeds (10.12) and (10.27) relatively the straight part of the string are

$$c_T = D = \sqrt{\frac{F}{\varrho}}, \quad c = \sqrt{\frac{E_s}{\varrho_0}}. \tag{10.28}$$

Note that, in this relation, c_T is not the 'absolute' speed if there exists a nonzero particle velocity ahead of the wave, that is, in the straight-line region of the string. In any case, the D-to-c ratio is

$$\frac{D}{c} = \sqrt{(1+\varepsilon)\varepsilon}. \tag{10.29}$$

The longitudinal wave speed, c, can differ much from the transversal wave speed, D. For a regular elastic material, which strength limit is much lower than the modulus, $D \ll c$. However, for an ultrastrong material the ratio can be not too small.

10.4 Inextensible String Model

If the tensile force does not vary too much, the string can be assumed to be inextensible. The above-considered string also behaves as inextensible but under a constant tension. In the inextensible string model, there is no such a restriction. For this case dynamic Eq. (10.4) splits into two equations

$$\varrho_0 \ddot{\boldsymbol{R}}(s,t) - (F(s,t)\boldsymbol{R}'(s,t))' = \boldsymbol{q}(s,t), \quad |\boldsymbol{R}'| = 1, \tag{10.30}$$

which admit a broader class of solutions with variable tensile forces. A simple example is a 'string pendulum', Subsect. 10.4.1, where the tensile force is constant except discrete moments, when the particle velocity changes abruptly, and when this model is no longer valid. Another example is a solitary wave in a helical string (Sect. 10.6), which is then also considered for a generally extensible string (Sect. 10.8). These examples, which are interesting in themselves, elucidate, to some extent, the strength and weakness of this model.

10.4.1 String Pendulum

Consider an inextensible string of a finite length, $0 < s < l$, with a fixed end point, $s = 0$. Initially it is at rest and placed horizontally. At $t = 0$ a vertically directed force, Q, is applied at $s = l$. What happens further? Under these conditions the string behaves as a nonlinear pendulum. During the oscillations the string is characterized by the constant tensile force $F = Q$, except the moments when the string is placed vertically. Almost all the time, except those moments, where the string directs vertically, the kink solution (10.21) (and of the same type solution for the opposite wave propagation) is valid here. Since the string ahead of the wave is at rest, the transversal wave speed relation (10.28) is valid here as

$$c_T = D = \sqrt{\frac{Q}{\varrho_0}}, \tag{10.31}$$

while $c = \infty$, and the particle velocities behind the kink is

$$\boldsymbol{v} = -D(\boldsymbol{\tau} + \boldsymbol{k}), \tag{10.32}$$

where both $\boldsymbol{\tau}$ and $\boldsymbol{k}$ are here constant. Relatively the unmoving rectangular coordinates, x, y ($x = 0$, $y = l$ at $s = 0$, see Fig. 10.4.1), the string particle velocities during the first half-cycle are

$$\begin{gathered} v_x = v_y = -D/\sqrt{2} \quad (Dt < s < l,\ 0 < t < l/D), \\ v_x = v_y = 0 \quad (0 < s < Dt,\ 0 < t < l/D), \\ v_x = -v_y = -D/\sqrt{2} \quad (2l - Dt < s < l,\ l/D < t < 2l/D), \\ v_x = v_y = 0 \quad (0 < s < 2l - Dt,\ l/D < t < 2l/D). \end{gathered} \tag{10.33}$$

Then it goes back in the same way. At the particular moments, $t = t_n$, the particle velocity reverses, and the tensile force becomes infinite, namely

$$F = 2\varrho D(l - s)\,\delta(t - t_n), \quad t_n = \frac{(2n-1)l}{D}, n = 1, 2, \ldots, \tag{10.34}$$

where $\delta(t)$ is Dirac delta-function and $\varrho = \varrho_0$.

Thus, to determine the 'true' maximum tensile force at such a collision moments, the string elasticity cannot be neglected. Consider the collision problem in a linear formulation. When the pendulum string approaches the vertical position it is under tension force F and has the uniformly distributed particle velocity $v = D$ directed downward. The condition $v = 0$ at $x = 0$ can be satisfied by adding a solution to the string dynamics equation

$$u(x,t)'' - \frac{1}{c^2}\ddot{u}(x,t) = 0 \tag{10.35}$$

with the conditions

$$u(0,t) = -Dt, \quad u'(l,t) = 0, \tag{10.36}$$

where u is the displacement and x are the coordinate both directed downward (the coordinate is denoted here by x instead of s since the latter is used for the Laplace transform parameter). The Laplace transform of u can be found as

$$u^L(x,s) = -A\left(\mathrm{e}^{-sx/c} + \mathrm{e}^{-s(2l-x)/c}\right), \quad A = -\frac{D}{s^2}\left(1+\mathrm{e}^{-2sl/c}\right)^{-1}. \tag{10.37}$$

In particular, for $x = 0$

$$u^L = -\frac{D}{s^2}, \quad \left(u^L\right)' = \frac{D}{cs}\tanh(-2sl/c). \tag{10.38}$$

It follows that the conditions (10.36) are satisfied and the corresponding tensile force is

$$F_u = \varrho c^2 u'(0,t) = \varrho c D\left[H(t) - 2H(t-2l/c) + 2H(t-4l/c) - \cdots\right]. \tag{10.39}$$

Thus the maximal tension force during the impart is

$$F_{max} = \varrho D^2 + F_u = \varrho D(D+c) = Q\left(1 + \frac{1}{\sqrt{\varepsilon}}\right) \quad (\varepsilon = Q/E), \tag{10.40}$$

where E is the elastic modulus. Note that the momentum, $-2\varrho l D$, imparted to the string during the period, $2l/c$, when only the first term of the series in (10.39) is in force, just equal to that required to revers its speed from D to $-D$. Thus, the collision shock can be very strong, and this may be a cause for rupture of a rope under a relatively weak flapping wave.

For an extensible string this result should be considered as a true solution if the string is initially under tension by the horizontally oriented force Q, which at $t = +0$ becomes vertical. Otherwise, the transversal wave is accompanied by a longitudinal wave. Besides, the solution evidences that after the above-mentioned period, $2l/c$, the tensile force oscillates around zero taking negative values as well as the positive ones. This, the solution is really true in the finite time-interval, $0 < t < l/D + 2l/c$, when the tensile force is positive. Further the string shows a more sophisticated behavior. During some time-intervals it continues dynamic deformations under a zero tension, the *flapping waves* show their nature.

10.5 Transversal-Longitudinal Wave Configuration

10.5.1 Preliminaries

A nonlinear problem for a combined transversal-longitudinal wave, is considered. First, we note that the transversal wave propagation along a straight part of the string can be added by a longitudinal wave. Let a linear or a hardening type elastic string, $0 < s < \infty$, be initially at rest, stressless, and placed horizontally as the direction unit vector $\boldsymbol{k}$. At $t = 0$ a constant external force, $\boldsymbol{Q} = Q\boldsymbol{\tau}$, is applied at its end, $s = 0$. If the force is directed as $\boldsymbol{\tau} = -\boldsymbol{k}$, the longitudinal wave starts to propagate with the speed and particle velocities as

$$c = \sqrt{\frac{E_s}{\varrho_0}}, \quad v = -\frac{Q}{\sqrt{\varrho_0 E_s}}. \tag{10.41}$$

Now let the external tension force be oriented at an angle to the initial string orientation, $\boldsymbol{\tau} \neq -\boldsymbol{k}$. It is remarkable that, in this case, the dynamic equations are satisfied by the same longitudinal wave together with the transversal wave, propagating with the speed c_T (10.12)

$$c_T = D + v = \sqrt{\frac{Q}{\varrho}} - \frac{Q}{\sqrt{\varrho_0 E_s}}, \tag{10.42}$$

while the string is under the uniform tensile force $F = Q$.

It follows from (10.42) and (10.29) that

$$c_T = \left(\sqrt{(1+\varepsilon)\varepsilon} - \varepsilon\right) c, \quad 0 < c_T < c. \tag{10.43}$$

Comment 10.2 This relation for the transversal wave speed, as well as (10.42), corresponds to a linear or a hardening type force-strain relation. The transversal wave propagates behind a straight-line longitudinal wave, while the tensile force is uniform over the whole wave. It can be seen that inequalities are valid as

$$0 < c_T < c. \tag{10.44}$$

The whole wave can also be considered as longitudinal with a curvilinear part, which propagates slower at a growing distance behind the longitudinal wavefront.

The latter point of view suggests that, in the case of a softening type force-strain relation, the longitudinal quasifront can be placed not only ahead of the transversal part of the wave. Consider an angled shape of the string, propagating with the speed c_T along a straight line. The vertex of the angle can also be contained inside the quasifront. The latter state is realized if the vertex speed is inside the region of the local sound speeds in the string

$$(c_t)_{min} < c_T < (c_t)_{max}. \tag{10.45}$$

Indeed, a moving angled shape of the string can exist if the nonzero tensile force and the particle speed are continuous, and this is the case, see Sect. 10.3. Besides, the dynamic equations impose no restriction for the quasifront to be in such angled configuration.

10.5.2 *The Transversally Impacted String*

Now consider an 'infinite' string, which is initially at rest, unstressed and placed along a horizontal line, x-axis. Beginning from $t = 0$ a point of it, $x = 0$, is forced to move in the vertical direction with the constant speed v_0. This is the Rakhmatulin problem (Rakhmatulin, 1947) related to the transversal impact of a barrage beadoon cable.

When the string is impacted, two types of waves propagate outwards in each direction from the point of impact as in the above-considered auxiliary problem, see (10.42) and (10.43). The difference is that, in this more realistic problem, the string directions in the central part of it do not coincide with the external force orientation. This is due to the interaction between the right and left parts of the string, which compensates the normal to the string components of the force. In the *longitudinal waves*, the particle velocities are directed along the string toward the point of impact. Since the particle speed uniform, it follows from the symmetry that in the *transversal waves* the particle velocities are directed downward. The task is to describe these waves and to find the resistance to the impact. First, examine the string of a linear or the hardening-type nonlinear stress-strain relation. Due to the symmetry the only right part of the string can considered.

The domain consists of three regions. In the first one, $x > ct$, the string is horizontal, stressless and unmoving. In the second region, $(D + v)t < x < ct$ $(D + v = c_T < c)$, the string remains to be horizontal. The particle velocities, $v = \text{const} < 0$, are directed toward the central point. The third region, $0 < x < c_T t$, is characterized by transversal particle velocity, the same as the velocity of the central point, v_0. The tensile force, $F = \text{const} > 0$, is uniformly distributed over double-wave support $|x| < ct$.

It follows from the mass conservation law that

$$\sqrt{v_0^2 + c_T^2} = c_T - v = c_T + c\varepsilon. \tag{10.46}$$

This and formula (10.43) yield the strain—vertical speed relation as

$$2\varepsilon\sqrt{(1+\varepsilon)\varepsilon} - \varepsilon^2 = \frac{v_0^2}{c^2}. \tag{10.47}$$

This equation uniquely defines the strain in the wave in terms of the ratio v_0/c. Now the external vertical force, Q, applied to the central point, $x = 0, t > 0$, can be determined, such that provides the considered wave configuration and corresponds to the vertical speed v_0. It is

$$Q = 2F\sin\alpha = 2E_s\varepsilon\frac{v_0}{\sqrt{c_T^2 + v_0^2}}, \quad \alpha = \arcsin\frac{v_0}{\sqrt{c_T^2 + v_0^2}}. \tag{10.48}$$

Asymptotic relations for low and high v-to-c ratios are as follows. In the case $v_0/c \to 0$

$$\begin{gathered}\varepsilon \sim 2^{-2/3}\left(\frac{v_0}{c}\right)^{4/3}, \quad F \sim 2^{-2/3}E_s\left(\frac{v_0}{c}\right)^{4/3}, \\ c_T \sim D \sim c\sqrt{\varepsilon} \sim 2^{-1/3}c\left(\frac{v_0}{c}\right)^{2/3}, \quad \alpha \sim \left(\frac{2v_0}{c}\right)^{1/3}, \quad Q \sim 2^{2/3}E_s\left(\frac{v_0}{c}\right)^{5/3}.\end{gathered} \tag{10.49}$$

In the opposite case, $v_0/c \to \infty$, the asymptotes are

$$\varepsilon \sim \frac{v_0}{c}, \quad F = E_s\frac{v_0}{c}, \quad D \sim v_0, \quad c_T \sim \frac{c}{2}, \quad \alpha \sim \frac{\pi}{2} - \frac{c}{2v_0}, \quad Q \sim 2E_s\frac{v_0}{c}. \tag{10.50}$$

10.5.3 Continuation. The Softening-Type Nonlinear String

In the case of such a force-strain diagram, the relation between the local sound speed, c_a (9.26) and the transversal wave speed, c_T (10.42), is important. It is

$$c_a - c_T = \frac{\lambda}{\sqrt{\varrho_0}}\left(\sqrt{\frac{\mathrm{d}F}{\mathrm{d}\lambda}} - \sqrt{\frac{F}{\lambda}}\right). \tag{10.51}$$

Since the tangent modulus is a decreasing function of λ the following statement is valid:

Conclusion 10.2 Let $\lambda = \lambda_0$ ($\varepsilon = 1 - \lambda_0 = \varepsilon_0$) be the stretch corresponding to equality $c_a = c_T$. Then

$$c_a > c_T \quad (\lambda < \lambda_0), \quad c_a < c_T \quad (\lambda > \lambda_0). \tag{10.52}$$

If the local sound speed, corresponding to a given tensile force, exceeds the transversal wave speed, that is, if $\lambda < \lambda_0$

$$\frac{\mathrm{d}F}{\mathrm{d}\lambda} > \frac{F}{\lambda} \quad (F > 0), \tag{10.53}$$

then, in the transversal wave speed relation (10.12), D is defined by (10.8) or (10.29); however the particle speed, v, is now equal to that for this type of the material (9.29). Thus

$$c_T = c\sqrt{(1+\varepsilon)\varepsilon} - \hat{c}\varepsilon, \tag{10.54}$$

where

$$c = \sqrt{\frac{E_s}{\varrho_0}}, \quad \hat{c} = \frac{1}{\varepsilon}\int_0^{\varepsilon} c_t(\varepsilon)\,\mathrm{d}\varepsilon, \tag{10.55}$$

and c_t is defined in (9.26).

Further, similarly to (10.46), there is another relation with respect to c_T

$$\sqrt{v_0^2 + c_T^2} = c_T - v = c_T + \hat{c}\varepsilon. \tag{10.56}$$

Comparing this with (10.54) we obtain the strain—vertical speed relation as

$$\frac{\hat{c}}{c}\left(2\varepsilon\sqrt{(1+\varepsilon)\varepsilon} - \frac{\hat{c}}{c}\varepsilon^2\right) = \frac{v_0^2}{c^2}. \tag{10.57}$$

Formulas (10.48), where c_T and ε are defined in (10.54) and (10.57), are valid here.

The asymptotes for small v_0/c are the same as in (10.49), since

$$\hat{c} \sim c \quad (v_0/c \to 0). \tag{10.58}$$

Now consider the case $\lambda > \lambda_0$ where inequality (10.53) does not hold. In this case, the quasifront of the longitudinal wave is divided by two parts. The fore-part propagating faster, is placed ahead of the kink, while the back part, which speeds are below the speed of the kink, propagates behind.

Thus, the above relations with $\lambda = \lambda_0$ are valid for the fore-part, and the speed of the kink (10.54) is

$$c_T = c^0\sqrt{(1+\varepsilon_0)\varepsilon_0} + v^0, \tag{10.59}$$

where

$$c^0 = \sqrt{\frac{E_s(\varepsilon_0)}{\varrho_0}}, \quad v^0 = -\hat{c}_0\varepsilon_0, \quad \hat{c}_0 = \frac{1}{\varepsilon_0}\int_0^{\varepsilon_0} c_t(\varepsilon)\,\mathrm{d}\varepsilon, \tag{10.60}$$

The relation in (10.56) remains valid with the maximal strain on the inclined part of the string. Comparing it with (10.59) we obtain, instead of (10.57), the following relation

$$\frac{\hat{c}}{c^0}\left(2\varepsilon\sqrt{(1+\varepsilon_0)\varepsilon_0} - 2\frac{\hat{c}_0}{c^0}\varepsilon\varepsilon_0 + \frac{\hat{c}}{c^0}\varepsilon^2\right) = \frac{v_0^2}{(c^0)^2}. \tag{10.61}$$

As an example consider the softening type force-strain relation (10.62). In this case

$$F = \frac{E_0\varepsilon}{1+\varepsilon}, \quad E_t = \frac{E_0}{(1+\varepsilon)^2}, \quad c_t = \frac{c_0}{1+\varepsilon} \quad \left(c_0 = \sqrt{\frac{E_0}{\varrho_0}}\right). \tag{10.62}$$

For $\varepsilon > \varepsilon_0$ it follows that

$$\begin{gathered}\varepsilon_0 = 1 \quad (F(\lambda_0)/\lambda_0 = \mathrm{d}F(\lambda_0)/\,\mathrm{d}\lambda), \quad \hat{c} = \frac{c_0}{\varepsilon}\ln(1+\varepsilon), \quad \hat{c}_0 = c_0\ln 2, \\ c^0 = c_0/\sqrt{2}, \quad c_T = c_0(1-\ln 2).\end{gathered} \tag{10.63}$$

Equation (10.61) is reduced to

$$\ln(1+\varepsilon) = \frac{v_0^2}{c_0^2}, \tag{10.64}$$

and it defines the maximal strain as

$$\varepsilon = \exp[\sqrt{v_0^2/c_0^2 + (1-\sqrt{2})^2} - 1 + \ln 2]. \tag{10.65}$$

Comment 10.3 Note that if $\varepsilon > \varepsilon_0$ the kink presence does not influence the distribution of the strain and particle velocities in the string. Besides, the dynamic state in the fore-part of the wave is independent of the tensile force level reached in the back part.

The above-discussed solutions to the nonlinear equations of the string dynamics were constructed using two 'fundamental' dynamic states. One is the longitudinal stress wave propagation and the other is the uniform flow of the string along an arbitrary, uniformly moving contour. It is of interest, however, that there exist not only these more or less trivial problems but solitary and periodic stress waves propagating in a helical string, where both the particle velocities and the tensile force are variable.

This strongly nonlinear dynamic states admit exact analytical solutions (Slepyan et al., 1995a; 1995b, 1998; Krylov et al., 1998). The derivation of relations defining the solitary waves in an inextensible and then in an extensible helicoidal strings are considered below in detail.

10.6 Solitary Waves in an Inextensible Helicoidal String

The helicoidal curve is a unique space curve with uniform geometrical parameters, the curvature, $\varkappa$, and the torsion, χ. In the problem considered below, the string keeps helicoidal shape far away from the 'local' wave. In the formulation, the helix geometry appears only in the representation of the dynamic state as a *propagating-rotating steady-sate wave* with the helicoidal asymptotes for $s \to \pm\infty$. The shape of the wave is defined by the string dynamic nonlinear equations and its deviation from the helix is not small. At a moving point, where the tensile force is maximal, the string crosses the helix axis. Analytical expressions are derived below for the displacements, internal force, axial and angular momenta and energy.

The homogeneous wave Eq. (10.30) is

$$\varrho\ddot{\mathcal{R}}(S,t) - [F(S,t)\mathcal{R}'(S,t)]' = 0, \quad |\mathcal{R}'| = 1, \quad \varrho = \varrho_0, \tag{10.66}$$

where the position vector and coordinate are denoted by $\mathcal{R}$ and S, respectively.

Consider a helix of initial radius R_0 and let γ denote the initial angle between the string and the axis of the helix, x. The helix can be characterized by the orthogonal Frenet triad (10.3) with the initial curvature and torsion as

$$\varkappa = \varkappa_0 = \frac{\sin^2\gamma}{R_0}, \quad \chi = \chi_0 = \frac{\sin\gamma\cos\gamma}{R_0}. \tag{10.67}$$

The lead of the helix is

$$l = 2\pi R_0 \cot\gamma. \tag{10.68}$$

A steady state solution to Eq. (10.66) for a solitary wave in the helix is considered. Note that to an observer moving along the helicoidal string with a constant speed v, and the associated angular velocity about the x-axis, $v\sin\gamma/R_0$ (with an orthogonal triad (10.3) natural to the helix), the initial geometry appears invariant; hence a solitary wave is expected to exist as a steady-state solution in a coordinate system attached to the moving observer.

Letting the position vector $\mathcal{R}(S,t)$ be represented as the sum of the longitudinal vector, $\boldsymbol{R}_x(S,t)$, and a vector $\boldsymbol{R}(S,t)$ lying in the cross-section of the helix, the non-dimensional quantities are introduced as

$$r_x = R_x/R_0, \; r = R/R_0, \quad s = S/R_0 \;\; f = F/\rho v^2, \quad \tau = vt/R_0. \tag{10.69}$$

The vectors r_x and $\boldsymbol{r}$ are expressed as follows:

$$r_x(s,\xi) = [s\cos\gamma + u(\xi)]k_x, \quad \boldsymbol{r} = A(\xi)\mathrm{e}^{\mathrm{i}\alpha s}, \\ \xi = s - \tau, \quad \alpha = \sin\gamma, \tag{10.70}$$

where $u(\xi)$ and $A(\xi)$ are an unknown functions.

Comment 10.4 Note that the *complex representation of a 2D vector* relates to its representation in the polar coordinate system. Let (r, θ) and (x, y) be polar and rectangular coordinate systems, respectively. Then

$$\boldsymbol{r} = r(\boldsymbol{k}_x\cos\theta + \boldsymbol{k}_y\sin\theta) \iff r\mathrm{e}^{\mathrm{i}\theta}, \tag{10.71}$$

where $\boldsymbol{k}_x, \boldsymbol{k}_y$ are the corresponding unit vectors. The imaginary unit, i, separates the components as the unit vectors do. Linear operations, summation, multiplication by a real constant, differentiation with respect to a real variable, are equivalent. Multiplication by the imaginary unit, i, means the rotation of the vector at angle $\pi/2$ as in the complex analysis. A scalar product of two vectors corresponds to that in the complex representation as

$$\boldsymbol{r}_1\boldsymbol{r}_2 \iff \Re(\boldsymbol{r}_1\overline{\boldsymbol{r}}_2) = r_1r_2\cos(\theta_1 - \theta_2). \tag{10.72}$$

For $v > 0$, the solitary wave conditions at infinity must correspond to the initial shape of the helix, i.e.

$$(u,\ u',\ A,\ f) \to 0, \quad A \to 1, \quad (\xi \to +\infty). \tag{10.73}$$

Here and below, primes and dots denote derivatives with respect to the non-dimensional coordinate, s, and time, τ, respectively.

Substitution in Eq. (10.66) then leads to

$$[(\cos\gamma + u')f]' = u'', \quad (1-f)r'' - f'r' - \alpha^2 r - 2\,\mathrm{i}\alpha r' = 0. \tag{10.74}$$

The first equation in (10.74) yields immediately an expression for u', namely

$$u' = \frac{f\cos\gamma}{1-f} \tag{10.75}$$

which satisfies the condition at infinity (10.73). Using the inextensibility condition in (10.6) and the first relation in (10.70), we obtain

$$|r'|^2 = 1 - |r_x'|^2 = 1 - \frac{\cos^2\gamma}{(1-f)^2}. \tag{10.76}$$

Multiplying the latter equation in (10.74) by $\overline{r}'$, the real part leads to

$$(1-f)' - 2f'|r'|^2 - \alpha^2(r^2)' = 0. \tag{10.77}$$

Recall that this corresponds to the scalar product.

As will be shown below, the string crosses the axis of the helix at the point of maximum tension force; this point is denoted to be $\xi = 0$. Furthermore, it is convenient to define: $r = \boldsymbol{r}$ for $\xi \geq 0$ and $r = -|r|$ for $\xi \leq 0$.

Substituting expression (10.76) into (10.77) one obtains $f' = -\alpha^2(r^2)'/2$. This and condition (10.73) yields the relation

$$f = \frac{\alpha^2}{2}(1 - r^2). \tag{10.78}$$

Substituting Eq. (10.78) back into the latter equation in (10.74) and rearranging, leads to

$$r'' - \frac{\alpha^2}{2}[(1 - r^2)r']' - \alpha^2 r - 2\,\mathrm{i}\alpha r' = 0. \tag{10.79}$$

In solving this vector equation, the complex representation of vector $\boldsymbol{r}$ is now used as

$$\boldsymbol{r} = r\mathrm{e}^{\mathrm{i}\varphi}. \tag{10.80}$$

In particular, note that in the initial geometry, $\varphi = \alpha s$. Substituting Eq. (10.80) into Eq. (10.79), separating the real and imaginary parts, we obtain

$$\left[1 - (\alpha^2/2)(1 - r^2)\right]\left[r'' - (\phi')^2 r\right] + \alpha^2 r(r')^2 - \alpha^2 r + 2\alpha\phi' r = 0 \tag{10.81}$$

$$\left[1 - (\alpha^2/2)(1 - r^2)\right](\phi'' r + 2\phi' r') + \alpha^2 r^2 r'\phi' - 2\alpha r' = 0. \tag{10.82}$$

Note that these equations are satisfied by the initial shape of the helix ($r = 1, \varphi' = \alpha s$). However, these equations also lead to a solitary wave solution. To this end we first consider Eq. (10.82) which is linear and of first order in $\psi \equiv \varphi'$. Taking into account the initial helix shape as mentioned above, Eq. (10.82) possesses a solution

$$\psi \equiv \phi' = \alpha\left[1 - (\alpha^2/2)(1 - r^2)\right]^{-1}. \tag{10.83}$$

Substitution of this last result in Eq. (10.81) yields a second-order equation with respect to r; namely

$$r'' + \alpha^2 r\left[1 - (\alpha^2/2)(1 - r^2)\right]^{-2} + \alpha^2 r\left[(r')^2 - 1\right]\left[1 - (\alpha^2/2)(1 - r^2)\right]^{-1} = 0. \tag{10.84}$$

Upon setting $q = r'$ and letting $L(r) = q^2 - 1$, we obtain

$$\frac{dL}{dr} + \frac{2\alpha^2 r}{1-(\alpha^2/2)(1-r^2)} L = -\frac{2\alpha^2 r}{[1-(\alpha^2/2)(1-r^2)]^2}, \tag{10.85}$$

whose solution under the condition $L(\pm 1) = -1$, is

$$L(r) = [\alpha^2(1-r^2) - 1][1-(\alpha^2/2)(1-r^2)]^{-2}. \tag{10.86}$$

Note that using Eq. (10.86), we obtain from Eq. (10.75) and upon integrating Eq. (10.83),

$$u = (r-1)\cos\gamma, \quad \phi = \alpha(r-1+s). \tag{10.87}$$

It can be seen that $L \neq -1$ if $\alpha > 0$ and $|r| < 1$. At the point where $r = 0$, the derivative $q = r'$ has the non-zero value

$$r' = \frac{\alpha^2/2}{1-\alpha^2/2} \quad (r=0), \tag{10.88}$$

thus indicating that the string crosses the x-axis.

Using Eq. (10.86), we obtain, after simple integration, a relation between the coordinate $\xi \equiv s - \tau$ and the deformed helix radius, namely

$$\ln\frac{1+r}{1-r} - \alpha^2 r = \alpha^2\xi. \tag{10.89}$$

From this latter expression, it follows that

$$|r| \sim 1 - 2e^{-\alpha^2(1+|\xi|)} \quad (|\xi| \to \infty) \tag{10.90}$$

that is, the propagating disturbance decays exponentially as $|\xi| \to \infty$.

Thus we conclude that the solution (10.89) corresponds to a solitary wave. It is of interest to note that the shape of the wave depends on the helix angle only and the 'effective' wave length decreases as γ increases.

Finally, we present expressions for the linear and angular momenta p and H, respectively, as well as for the energy, T, of the solitary wave as functions of the helix properties, R_0, γ, and ρ , and the solitary wave parameter, υ. It is worth noting that for a given helix, υ is the sole parameter governing the propagation of the solitary wave. From the first equation in (10.87), it is evident that $\dot{u} = -u' = -r'\cos\gamma$. Then, using Eq. (10.69) one obtains

$$p = 2\rho\upsilon R_0 \int_0^{\infty} \dot{u}\ ds\, k_x = -2\rho\upsilon R_0 \cos\gamma\ \boldsymbol{k}_x, \tag{10.91}$$

$$H = 2\rho v R_0^2 \int_0^\infty r^2 \dot{\phi} \, \mathrm{d}s \, k_x = -\frac{2}{3}\rho v R_0^2 \sin\gamma \, k_x. \tag{10.92}$$

Noting from the latter equation in (10.87) that $\dot{\phi} = -\alpha r'$ and using the representation, Eq. (10.80),

$$T = \rho v^2 R_0 \int_0^\infty (\dot{u}^2 + \dot{r}^2 + r^2\dot{\phi}^2)\, \mathrm{d}s = \frac{2}{3}\rho v^2 R_0 \sin^2\gamma. \tag{10.93}$$

Note that as $\gamma \to 0$ the angular momentum of the wave becomes negligible compared to the linear momentum. In this case, expressions (10.89) and (10.78) yield the asymptotic solution

$$r \sim \tanh \frac{\alpha^2}{2}\xi, \qquad f \sim \frac{\alpha^2}{2}\mathrm{sech}^2\left(\frac{\alpha^2}{2}\xi\right). \tag{10.94}$$

It is of interest that, in this asymptotic case, the force distribution coincides with that for the *soliton* governed by the KdV equation (Korteweg and Vries, 1895; Dodd et al., 1982, 1984) in the form

$$(1 + \alpha^4)\dot{f} + f' + 6\alpha^2 f f' + f''' = 0 \quad (\alpha = \sin\varphi). \tag{10.95}$$

The 'soliton science' is briefly discussed in Chapt. 12.

Conclusion 10.3 Thus, the above-discussed results defined solitary waves in an inextensible string. The deviation from the initial shape of the helix and the tensile force in the wave decay exponentially as the distance tends to infinity. The wave parameters, the effective length, the momenta and energy, are found as functions of the helix angle and the wave speed. Actually, in Eq. (10.93), the wave speed should be considered as a function of the wave energy. This dependence, however, is not valid for too high speed, where extensibility becomes important. Indeed, the wave speed cannot exceed the longitudinal wave speed in an elastic helix. The same problem for the latter is considered in Sect. 10.8. Note that the inextensible string dynamic Eq. (10.30) yield not only solitary waves but a class of periodic waves (Krylov and Slepyan, 1997).

Some numerical results are presented here. In Fig. 1, the phase portrait is shown in $(r, \dot{r})$ space for several values of γ. Results shown in Figs. 2 and 3 are given for $\gamma = \pi/6$. The variation of the helix radius with ξ, based on the relation (19), is shown in Fig. 2. Using the calculated values of r, the force f, evaluated from Eq. (9), is shown as function of $\xi = s - \tau$ in Fig. 3.

A numerical simulation, using a finite discrete system of masses connected by massless inextensible rigid links, was performed to obtain a transient wave response. (A comparison with theoretical results showed the numerical calculations to be highly accurate.) Results of these calculations are presented in Figs. 4 and 5. The shape of the deformed helix with respect to its initial shape is shown in Fig. 4; we note that the disturbance is limited effectively to a finite region of the helix. The spatial distribution of the force with time is presented in Fig. 5 where we observe that the wave travels as a solitary wave.

The collision of two waves propagating in the same and in opposite directions is investigated in Slepyan et al. (1995b). In both cases, the collision is not perfectly elastic and that radiation takes place during the interaction. After collision the main waves continue to propagate as solitary waves correspondingly to the analytical description, but the amplitudes change. In the case of two waves of unequal strength propagating in the same direction, the amplitude of the stronger wave decreases after the collision, while the amplitude of the weaker wave increases. In the case of the collision of two identical waves, which propagate in opposite directions, the amplitudes of both decrease. In both cases, small additional solitary waves are created in the same manner as described above.

Thus, although the solitary wave obtained propagates as a very stable object, the amplitudes of the waves are not maintained under the collision. The reason may be that, in contrast to a 'regular' soliton, a change of the waveguide occurs in the considered system after the wave has passed; namely, a finite shift $u = -2\cos\gamma$ appears in the axial displacement, and $\phi = -2\sin\gamma$ in the rotation (see (10.87) with $r = -1$ for $\xi \to -\infty$). In effect, after the collision the waves must adapt to these new conditions. In spite of such disturbances, the resulting post-collision waves are formed as very stable objects governed by the above-discussed analytical relations.

10.7 Binary Wave in an Inextensible Helicoidal String

In Krylov and Slepyan (1997), a semi-infinite helicoidal string was considered, and a transient wave arising under an axial external force was studied. The latter appeared to be a complex object, which is characterized by the presence of different type waves rotating in opposite directions. The uncommonness of this *binary wave* is a consequence of the waveguide being twisted. This wave is considered below.

The formulation prescribes that the end point of the string ($s = 0$), subjected at $t > 0$ to the action of a constant axial extension force, is held on the axis of the helix (x-axis). It is connected with the helix by an initially small portion of the string, Fig. 10.1. The length of the helix under consideration is chosen to be large enough to eliminate a detectable reflection. The problem is studied numerically using the simplest discrete analogue of the continuous string. It is a chain of masses connected to each other by inextensible, massless links. The finite difference method of the first order with respect to time (the explicit scheme) is used. The accuracy of the scheme is found to be very high.

The wave formed under these conditions is shown in Figs. 10.2 and 10.3. This binary wave consists of two portions: the leading, quasi-periodical, propagating-rotating wave and the rear, standing wave. These waves possess axial rotations in opposite directions: the leading wave untwists the helix, whereas the rear wave twists it. This phenomenon arises for two reasons: there is an initial twist in the waveguide that leads to a wave with rotation, and there is no external angular moment, and therefore the total angular momentum of the binary wave remains to be zero. The wave is characterized by two different velocities which correspond to the leading,

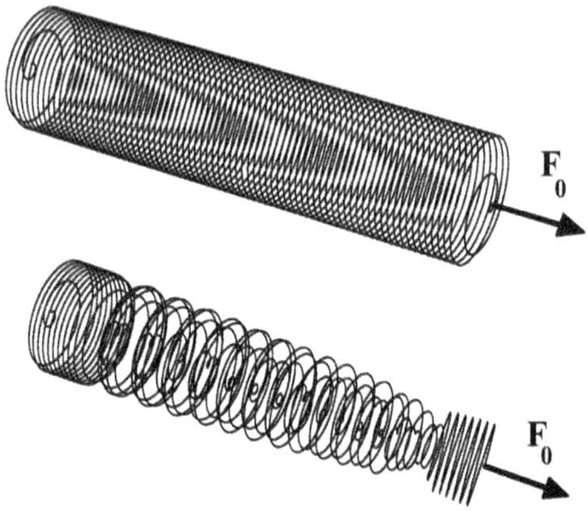

Fig. 10.1 Wave in a helical string

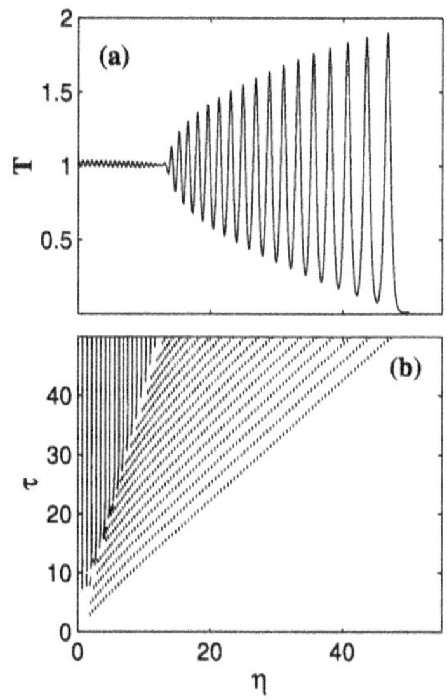

Fig. 10.2 The helicoidal wave parameters 1

high-speed wavefront and the low-speed interface as a short transient region which separates these two different waves. So, the wave as a whole is transient and consists of two totally different parts. At the same time, the structure of each part is in good accordance with the exact solutions for the steady-state waves in the corresponding infinite helix: solitary, periodical and constant-force nonlinear traveling and standing waves.

The numerical results can be compared with analytical relations found for different parts of the wave. First, it is the D'Alembert solution (10.7) valid for the case of a constant tensile force, $F = \varrho_0 c_0^2$, where c_0 is the particle speed relatively the moving wave frame. Further, the relations for the axial displacement (10.87) and the tensile force (10.78) are obtained as

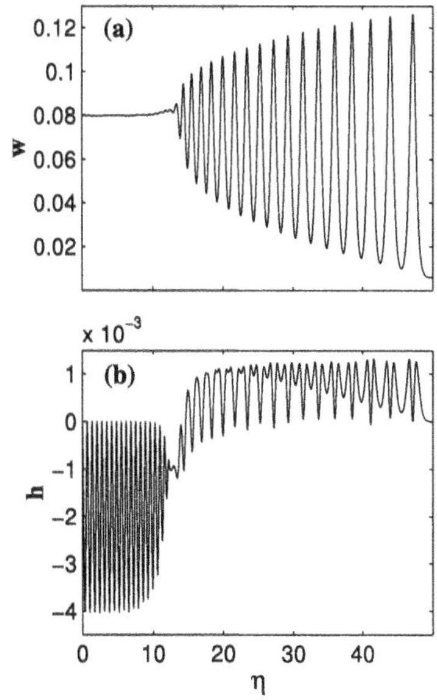

Fig. 10.3 The helicoidal wave parameters 2

$$u = (R - R_0)\cos\gamma, \quad F = \frac{\varrho_0 v^2}{2}\alpha^2\left(1 - \frac{R}{R_0}\right). \tag{10.96}$$

Finally, for the periodic wave in an inextensible helicoidal string (Krylov et al., 1998) it was found, in particular, that the wave length, L, is

$$L = \frac{2\sqrt{2}}{|\alpha|} R_0 \left(1 + \frac{c_0}{v}\right)^{-1} \left(\frac{1 + f_{min}}{\sqrt{f_{min} + f_{max}}} \mathrm{K}(k) - \sqrt{f_{min} + f_{max}}\,\mathrm{E}\,(k)\right) \times \mathrm{sign}(F - \rho v^2), \tag{10.97}$$

where $f = F/(\rho\, v^2)$, $\mathrm{K}(k)$ and $\mathrm{E}(k)$ are complete elliptic integrals of the first and of the second kind, respectively, with the parameter

$$k = \sqrt{\frac{F_{max} - F_{min}}{F_{max} + F_{min}}}. \tag{10.98}$$

We can now compare the structure and velocity of the leading quasi-periodical wave as well as the interface velocity with the relevant analytical results. The wave-front velocity, v_1, as the velocity of the first peak of the wave, is nothing but the velocity of the solitary wave of the same amplitude (see Fig. 10.2 and Eq. (10.96)). The first peak amplitude tends to $2F_0$, and using the latter formula the wave velocity can be expressed in terms of the applied force as follows:

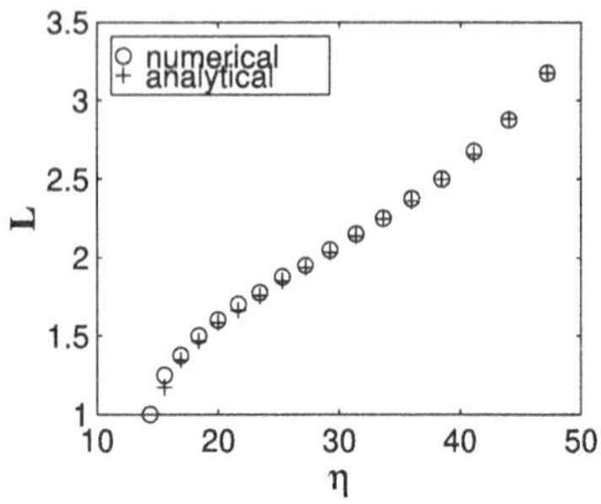

Fig. 10.4 Numerical versus analytical

$$v_1 \sim \frac{2}{\alpha}\sqrt{\frac{F_0}{\rho}}. \tag{10.99}$$

The quasi-periodical leading wave can be described locally by the traveling wave solution: the relation between the velocity of each peak, the minimal and the maximal forces and the distance between two neighboring peaks are in accordance with that defined by Eq. (10.97), Fig. 10.4.

Two waves are separated by a short transient region. Within this region the tensile force, F, remains almost constant, $F \sim F_0$, and the 'interface' propagates with the D'Alembert wave velocity $v_2 \sim c_0 = \sqrt{F_0/\rho}$. Note that the structure of the wave is formed in time, and the above-written relations should be considered in the asymptotic sense. The results presented here differ from the limiting values by 2–3%.

Consider now the rear wave. It consists of standing waves which do not propagate along the string. At the same time, the rear wave as a whole propagates along the string: there is a phase transition from the first, quasi-periodical wave mode to the second, standing wave mode. The rear wave appears to be almost completely polarized in a rotating plane, Fig. 10.1b, c. This wave can also be locally described by an analytical solution.

The standing wave solution can be obtained in the same way as the traveling wave solution. In Eqs. (10.70), (10.80), now in dimensional variables

$$R_x(S,\xi) = [S\cos\gamma + u(\xi)]k_x, \quad \boldsymbol{R} = R(\xi)e^{i\varphi} \quad (\xi = S - t) \tag{10.100}$$

replace the variables to be

$$\boldsymbol{R}_x = R_x(S)\boldsymbol{k}_x, \quad R = R(S), \quad \varphi = \omega t. \tag{10.101}$$

In this case, multiplying Eq. (10.66) by $\boldsymbol{k}_x$ and then taking into account that $|\boldsymbol{R}'| = 1$, we obtain

$$(FR_x')' = 0 \quad \Longrightarrow \quad R_x' = \frac{F_0}{F}, \quad R' = \sqrt{1 - \frac{F_0^2}{F^2}}. \tag{10.102}$$

As a result, Eq. (10.66) yields

$$F = F_{max} - \frac{\rho\,\omega^2}{2} R^2 \quad \text{or} \quad R_{max} = \sqrt{\frac{2(F_{max} - F_0)}{\rho\,\omega^2}}, \tag{10.103}$$

where F_{max} and $F_{min} = F_0$ are minimal and maximal values of the tensile force. The integration of Eqs. (10.102), (10.103) results in the $R - S$ relation in terms of elliptic functions. Calculations show that the numerical results are in good agreement with these expressions.

10.8 Solitary Waves in an Elastic Helicoidal String

A helicoidal elastic string with an arbitrary, stable force-strain relation is considered. A general, analytical, steady-state solution is derived. This solution describes a class of spatial motion of an elastic string which is shown to be a waveguide for subsonic solitary waves. The results confirm that the above-discussed solution for an inextensible string is a low-velocity or low-energy asymptote of the solution presented here. Another type of asymptotic solution derived below corresponds to a high-energy wave. In this case, when the amplitude of the wave increases the wave speed and energy tend to infinity, and the effective wavelength tends to zero.

10.8.1 Formulation

The homogeneous dynamic equation for the extensible string (10.4) is

$$\varrho_0 \ddot{\boldsymbol{R}} - (F\boldsymbol{R}'/\lambda)' = 0, \quad F = F(\lambda)H(\lambda - 1), \quad \lambda = \left|\boldsymbol{R}'\right|. \tag{10.104}$$

Recall that Lagrangian formulation is used. The function $F(\lambda)$ is assumed to be invertible, that is, a positive constant, v_*, is assumed to exist such that for any value of the string stretch in the wave

$$\frac{\mathrm{d}F}{\mathrm{d}\lambda} > \rho_0 c_*^2 \tag{10.105}$$

It is the only restriction on the elasticity law of the string. Also this inequality means stability of the string under extension. Thus a general, geometrically and physically nonlinear, stable-under-extension string is considered. The helix parameters, the initial radius, R_0, and the helix angle, γ, are still as above. Also the above-used representation of the position vector, $\mathcal{R}(S, t)$, as the sum of the longitudinal vector, $\boldsymbol{R}_x(S, t)$, and the normal to the helix axis vector, $R(S, t)$, is used here. Besides, the non-dimensional quantities and the expressions for the vectors, $\boldsymbol{r}_x$ and $\boldsymbol{r}$, are introduced as in (10.69) and (10.70). In addition, note that for the considered uniformly moving-rotating wave the vector $\boldsymbol{r}$ can be represented as

$$\boldsymbol{r} = A(\xi)\mathrm{e}^{\mathrm{i}\alpha s} = B(\xi)\mathrm{e}^{\mathrm{i}\alpha\tau} \quad \left(B = A\mathrm{e}^{\mathrm{i}\alpha\xi}\right). \tag{10.106}$$

10.8.2 Some Preliminary Results

It follows from Eq. (10.106) that the vectors, $\boldsymbol{r}_x$ and $\boldsymbol{r}$, satisfy the following relations:

$$\ddot{\boldsymbol{r}}_x = \boldsymbol{r}_x'', \quad \ddot{\boldsymbol{r}} = \boldsymbol{r}'' - 2\,\mathrm{i}\lambda\boldsymbol{r}' - \lambda^2\boldsymbol{r}. \tag{10.107}$$

Next, a general homogeneous relation between the tensile force and the position vector (10.2) follows immediately from Eq. (10.104), as its projection onto the tangent line yields

$$\lambda(F)F' = \rho_0\ddot{\mathcal{R}}\cdot\mathcal{R}'. \tag{10.108}$$

This allows extension of the string to be expressed in terms of the deformed helix radius. In fact, substitution in Eq. (10.108) leads to

$$\Phi' - \frac{1}{2}(\lambda^2)' = -\frac{\alpha^2}{2}(r^2)', \quad \Phi = \int_0^f \lambda\,\mathrm{d}f. \tag{10.109}$$

Equation (10.109) and condition (10.73) yield the relation with $\lambda = 1$ at infinity, where $r^2 = 1$:

$$U(f) \equiv \Phi - \frac{1}{2}(\lambda^2 - 1) = \frac{\alpha^2}{2}(1 - r^2). \tag{10.110}$$

As follows from (10.105) for the range $0 < v^2 < c_*^2$

$$\frac{\mathrm{d}f}{\mathrm{d}\lambda} > 1 \tag{10.111}$$

and hence, the derivative of U is positive:

$$\frac{\mathrm{d}U(f)}{\mathrm{d}f} = \lambda\left(1 - \frac{\mathrm{d}\lambda}{\mathrm{d}f}\right) > 0 \ (\lambda \geq 1). \tag{10.112}$$

At the same time, local sound velocity in the string

$$c = \sqrt{\frac{1}{\rho_0}\frac{\mathrm{d}F}{\mathrm{d}\lambda}} = v\sqrt{\frac{\mathrm{d}f}{\mathrm{d}\lambda}}. \tag{10.113}$$

It can now be seen that the condition (10.111) means that $v < c$. Thus in the case of a subsonic wave, the derivative, $\mathrm{d}U/\mathrm{d}f$, is positive and hence there is a one-to-one correspondence between the force and the radius of the deformed helix [see (10.110)].

Relation (10.110) can be represented in the form

$$f = \frac{1}{2}(\lambda^2 - 1) - \int_1^{\lambda} (\lambda - 1)\frac{\mathrm{d}f}{\mathrm{d}\lambda}\,\mathrm{d}\lambda + \frac{\alpha^2}{2}(1 - r^2). \tag{10.114}$$

Inequality (10.111) allows us to make the estimation

$$f \leq \lambda - 1 + \frac{\alpha^2}{2} < \lambda. \tag{10.115}$$

Furthermore, under condition (10.111), $U \geq 0$; hence

$$r^2 \leq 1. \tag{10.116}$$

In the partial case of inextensibility when $\lambda \equiv 1$ and $\Phi \equiv f$, Eq. (10.110) yields

$$f = \frac{\alpha^2}{2}(1 - r^2). \tag{10.117}$$

In the case of linear extensibility when

$$\lambda = 1 + M^2 f, \quad \Phi = f + \frac{M^2}{2} f^2 \quad \left(M = \frac{v}{c}\right), \tag{10.118}$$

the non-dimensional internal force is

$$f = M^{-2}\left[\sqrt{1 + \frac{M^2\alpha^2(1 - r^2)}{1 - M^2}} - 1\right]. \tag{10.119}$$

Note that (10.117) follows from (10.119) as a limit of $M \to 0$. In an inextensible helicoidal string, the wave velocity is unbounded. In extensibility is taken into account, the wave velocity is bounded. The upper bound is independent of the helix parameter α in the case of linear extensibility, and it depends on α for a nonlinear elastic law. Consider, for example, the constitutive relation $F = F_0\sqrt{\lambda - 1}$. Then

$$f = \frac{\sqrt{\lambda - 1}}{\omega}, \quad \Phi = f + \frac{1}{3}\omega^2 f^3, \quad c^2 = \frac{F_0}{2\rho_0\sqrt{\lambda - 1}}, \quad \omega = \frac{\rho v^2}{F_0}, \tag{10.120}$$

and the condition, $v^2 < c^2$, leads to the inequality

$$\omega < \frac{1}{2\sqrt{\lambda_{max} - 1}}. \tag{10.121}$$

It now follows from Eq. (10.112) that

$$\lambda^2 + 4\lambda - 5 - 3\alpha^2 < 0, \quad \lambda_{max} = \sqrt{9 + 3\alpha^2} - 2. \tag{10.122}$$

10.8.3 General Solution

Let us return to the general case. For a given α, the internal force, f, and therefore the stretch, λ, depend on r^2 only, and the ratio f/Λ can be denoted as a function of r^2

$$\frac{f}{\lambda} = G(r^2), \quad G \to 0 \ \text{ when } r \to 1 \ \ (s \to \infty). \tag{10.123}$$

This function is defined by a given elastic law, $\lambda(F)$, and the derived Eq. (10.110). It can be seen from (10.115) that $G < 1$.

Substitution of (10.107) in (10.104) [see (10.123)] leads to

$$[G(u' + \cos\gamma)]' = u'', \tag{10.124}$$

and

$$(Gr')' = r'' - 2\mathrm{i}\alpha r' - \alpha^2 r. \tag{10.125}$$

From Eq. (10.124) and condition (10.123), one now has

$$u' = \frac{G\cos\gamma}{1 - G}. \tag{10.126}$$

In solving vector Eq. (10.125), the vector $\boldsymbol{r}$ is now represented as

$$\boldsymbol{r} = r\varepsilon^{\mathrm{i}\varphi}, \tag{10.127}$$

where $r(\xi)$ and $\phi(\xi, s)$ are real functions.

As shown below, as in the case of the inextensible string, the string crosses the axis of the helix at the point of maximum tension force; this point is denoted to be $\xi = 0$. Furthermore, it is convenient to define: $r = |r|$ for $\xi \geq 0$ and $r = -|r|$ for $\xi \leq 0$. In particular, note that in the initial geometry, ϕ is given by the linear relation $\phi = \alpha s$, and under conditions (10.123)

$$r \to 1, \quad \phi \sim \alpha s \ (\xi \to +\infty). \tag{10.128}$$

Substituting representation (10.127) into Eq. (10.125), multiplying by $e^{-i\phi}$ and separating the real and imaginary parts, one obtains

$$(1-G)r\varphi'' + 2\left[(1-G)r' - \frac{\mathrm{d}G}{\mathrm{d}(r^2)}r^2 r'\right]\varphi' = 2\alpha r' \tag{10.129}$$

and

$$[(1-G)r']' - (1-G)r(\varphi')^2 + 2\alpha r\varphi' - \alpha^2 r = 0. \tag{10.130}$$

Equations (10.129) and (10.130) are satisfied by the initial shape of the helix ($r = 1, \varphi' = \alpha$). However, as we now show, these equations also lead to a solitary wave solution. To this end first consider Eq. (10.129) which is linear and of first order in $\psi \equiv \varphi'$. Under the condition at infinity, $\varphi' = \alpha$, Eq. (10.129) possesses a solution

$$\psi \equiv \varphi' = \frac{\alpha}{1-G}. \tag{10.131}$$

Substitution of this last result in Eq. (10.130) yields a second-order equation with respect to r

$$[(1-G)r']' + \frac{\alpha^2 rG}{1-G} = 0. \tag{10.132}$$

Multiplying this equation by $(1-G)r'$ and integrating the resulting expression, one obtains

$$(r')^2 = \frac{\alpha^2}{[1-G(r^2)]^2}\int_{r^2}^{1} G(r^2)\ \mathrm{d}r^2 \quad (r' = 0 \text{ when } r = 1). \tag{10.133}$$

One can see that the derivative, r', is nonnegative at the point $r = 0$ ($r > 0$ if $\xi > 0$)

$$r'(0) = \frac{\alpha}{[1-G(0)]}\left(\int_0^1 G(a)\ \mathrm{d}a\right)^{1/2} > 0. \tag{10.134}$$

Thus the radius of the deformed helix as a function of ξ is defined by the relation

$$\int_0^r \left\{\frac{\alpha^2}{[1-G(r^2)]^2}\int_{r^2}^1 G(a)\ Da\right\}^{-1/2} \mathrm{d}r = \xi \tag{10.135}$$

The axial displacement, u, and the angle, ϕ, can be expressed as follows [see Eqs. (10.126) and (10.131)]

$$u(\xi) = -\int_\xi^\infty \frac{G}{1-G}\ \mathrm{d}\xi \cos\gamma, \quad \phi(\xi) = -\int_\xi^\infty \frac{G}{1-G}\ \mathrm{d}\xi \sin\gamma + s\sin\gamma. \tag{10.136}$$

Note that, as follows from the derived relations, the variables $r(\xi)$, $u(\xi) - u(0)$ and $\theta(\xi) - \theta(0)$ $(\theta = \phi - \alpha s)$ are antisymmetrical, monotonically increasing functions, and the tensile force, $f(\xi)$, is a symmetrical, monotonically decreasing function.

To show that the wave is truly solitary and the integrals are convergent consider the asymptotic expressions for the helix radius and the tension force. Taking into account that

$$f \to 0, \quad \epsilon = \lambda - 1 \sim \frac{F}{\varrho_0 c_0^2} = \frac{f v^2}{c_0^2} \to 0, \quad \Phi = O(f\epsilon) \ (r^2 \to 1), \tag{10.137}$$

the following asymptotic expression of Eq. (10.114) is obtained (c_0 is the sound speed as the limit of c (10.113) at $\varepsilon = 0$)

$$f \sim G \sim \frac{\alpha^2(1-r^2)}{2} + \epsilon \sim \frac{\alpha^2(1-r^2)}{2(1 - v^2/c_0^2)}, \quad c_0 = \lim_{\epsilon \to 0} c. \tag{10.138}$$

Now, as follows from Eq. (10.135)

$$1 - r^2 \sim C \exp\left(- \frac{\alpha^2 |\xi|}{\sqrt{1 - v^2/c_0^2}} \right), \tag{10.139}$$

where C is a constant. Thus the intensity of the wave decreases exponentially when $|\xi| \to \infty$, which is the required result.

10.8.4 Linear Extensibility

Consider the case of the linear extensibility in more detail. In this case, the use of expressions (10.118) and (10.119) in Eq. (10.123) and then in Eq. (10.135) leads to the following relation between the radius of the helix and the coordinate, ξ (recall that $r = |r| \operatorname{sign} \xi$):

$$\ln \frac{(1+r)(\Lambda + r)}{(1-r)(\Lambda - r)} - \frac{2\alpha}{M}\sqrt{1-M^2} \arcsin\left(\frac{\alpha r M}{\sqrt{1 - M^2 + M^2\alpha^2}} \right)$$
$$= \frac{2\alpha^2 \xi}{\sqrt{1-M^2}}, \quad \Lambda = \sqrt{1 + \frac{M^2\alpha^2(1-r^2)}{1-M^2}}, \quad M = \frac{v}{c}. \tag{10.140}$$

Asymptotic representations $[v \to c \ (v < c)]$ follow from Eqs. (10.140) and (10.119) as

$$r \sim \tanh \frac{\alpha^2 \xi}{\sqrt{1-M^2}}, \quad f \sim f_0 \left(\cosh \frac{\alpha^2 \xi}{\sqrt{1-M^2}} \right)^{-1}, \quad f_0 \sim \frac{\alpha}{\sqrt{1-M^2}}, \tag{10.141}$$

where $f_0 = f$ at the peak of the wave ($\xi = r = 0$). As a consequence of Eq. (10.119), the asymptotic expression (10.141) for the force is valid under the condition

$$\frac{\alpha^2(1-r^2)}{1-M^2} >> 1. \tag{10.142}$$

The strain and kinetic energies of the wave are asymptotically the same, and the total energy is

$$W = \frac{\rho v^2}{2} R_0 \int_{-\infty}^{\infty} (M^2 f^2 + \dot{u}^2 + \dot{r}^2 + r^2 \dot{\phi}^2) \sim \frac{2\rho c^2 R_0}{\sqrt{1-M^2}}. \tag{10.143}$$

Thus, with increase of the wave speed, the effective wavelength tends to zero and, at the same time, the energy of the wave increases unboundedly. The ratio f/f_0 for $\alpha = 1/2$ is presented in Fig. 1 for several values of v/c. The dotted curve corresponds to the asymptotic solution (10.141).

The derived results confirm that the solution for an inextensible string is a low-velocity or low-energy asymptote of the solution presented here.

Lastly, consider whether a supersonic steady-state wave can exist. In a general case, when the local sound velocity $c = c(\Lambda)$, and if such a wave does exist, the inequality $v > c$ can be valid within a range of the stretch, $\Lambda_1 < \Lambda < \Lambda_2$, where $\Lambda_1 \geq 1$, $\Lambda_2 \leq \infty$. Within this range, as follows from the relation (10.110), the functions f' and r' coincide in sign. In the case $\Lambda_1 = 1$, at any point with $1 < \Lambda < \Lambda_2$, the tension force $f > 0$, and hence $r > 1 (r = 1$ at$s = +\infty$ where$\Lambda = 1)$. However this conflicts with the expression (10.133) the right hand part of which becomes negative. In the case $\Lambda_1 > 1, r'$ changes sign when Λ crosses the point Λ_1. But this again is inconsistent with (10.133) because r' must be continuous as can be seen in (10.132). Thus, in the considered system, supersonic steady-state waves does not exist.

Chapter 11
Nonlinear Theory of the Elastic Beam Dynamics

11.1 General Formulation

The beam differs from the string by nonzero bending and/or torsion stiffness. This involves, in general, the presence of internal and external forces and moments as 3D vectors. In the classical dynamic equilibrium equations, the beam is considered as a curve with a density and longitudinal, bending and torsion stiffness. In contrast to the string, the cross-section sizes are important here, and we denote the density as ϱ_0 and the density per unit initial length as $\varrho_0 S$, where S is the initial cross-section area. The equation with respect to forces is similar to that for the string (10.2)

$$\varrho_0 S_0 \ddot{\boldsymbol{R}}(s,t) - \boldsymbol{F}'(s,t) = \boldsymbol{q}(s,t)\,; \tag{11.1}$$

however, the force, $\boldsymbol{F}(s,t)$, is here a general 3D vector as well as $\boldsymbol{q}(s,t)$:

$$\begin{aligned} \boldsymbol{F} &= T\boldsymbol{\tau} + N\boldsymbol{n} + B\boldsymbol{b}\,, \\ \boldsymbol{q} &= q_\tau\boldsymbol{\tau} + q_n\boldsymbol{n} + q_b\boldsymbol{b}\,. \end{aligned} \tag{11.2}$$

The moment equation is

$$\boldsymbol{M}' + \tau \wedge F = -\mu \tag{11.3}$$

with

$$M = M_\tau\boldsymbol{\tau} + M_n\boldsymbol{n} + M_b\boldsymbol{b}\,, \tag{11.4}$$

$$\mu = \mu_\tau\tau + \mu_n n + \mu_b b\,. \tag{11.5}$$

Here Frenet triad unit vectors, $\boldsymbol{\tau}, \boldsymbol{n}, \boldsymbol{b}$, are used; the symbol $\wedge$ denotes the cross product. Recall that the Lagrangian formulation Sect. 17.1.2.2 is used.

L. I. Slepyan, *Mechanical Waves: Free, Forced, and Transition*,
https://doi.org/10.1007/978-3-032-15864-2_11

11.2 Plane Problem

In particular, if the symmetry exists, such that the beam axis, being initially a plane curve, remains to be in the plane under in-plane forces

$$\begin{aligned} F &= T\tau + Nn\,, \quad M = M_b b\,, \\ q &= q_\tau \tau + q_n n\,, \quad \mu = \mu_b b\,, \end{aligned} \tag{11.6}$$

a plane problem can be considered. It follows that

$$\varrho_0 S_0 \ddot{\boldsymbol{R}}(s,t) - [T'(s,t) - \lambda\varkappa N(s,t)]\boldsymbol{\tau} - [\lambda\varkappa T(s,t) + N'(s,t)]\boldsymbol{n} = \boldsymbol{q}(s,t) \tag{11.7}$$

with

$$M' = -N - \mu_n \quad (M = M_b) \tag{11.8}$$

and (in the case of an elastic beam)

$$T = T(\lambda)\,, \quad M = M(\varkappa)\,, \quad \lambda = \boldsymbol{R}'\boldsymbol{\tau}\,, \quad \varkappa = \frac{1}{\lambda^2}\left(\boldsymbol{R}'' - \frac{2\lambda'}{\lambda}\boldsymbol{R}'\right)\boldsymbol{n}\,. \tag{11.9}$$

In rectangular axes

$$\boldsymbol{R}(s,t) = \boldsymbol{R}(s,0) + u_x(s,t)\boldsymbol{k}_x + u_y(s,t)\boldsymbol{k}_y\,, \tag{11.10}$$

and Eq. (11.7) yields the system

$$\begin{aligned} \varrho_0 S_0 \ddot{u}_x(s,t) - [T'(s,t) - \lambda\varkappa N(s,t)]\cos\alpha + [\lambda\varkappa T(s,t) + N'(s,t)]\sin\alpha &= q_x(s,t)\,, \\ \varrho_0 \ddot{u}_y(s,t) - [T'(s,t) - \lambda\varkappa N(s,t)]\sin\alpha - [\lambda\varkappa T(s,t) + N'(s,t)]\cos\alpha &= q_y(s,t)\,, \end{aligned} \tag{11.11}$$

where α is the angle between the x-axis and τ:

$$\boldsymbol{\tau}\boldsymbol{k}_x = \boldsymbol{n}\boldsymbol{k}_y = \cos\alpha\,, \quad \boldsymbol{\tau}\boldsymbol{k}_y = -\boldsymbol{n}\boldsymbol{k}_x = \sin\alpha\,. \tag{11.12}$$

For small axial strain, rotations and curvature of an initially straight linearly elastic beam

$$T = ES_0 u_x'\,, \quad M = EI\varkappa\,, \quad \lambda \sim 1\,, \quad \varkappa = w'' \;\; (w \equiv u_y)\,, \quad \alpha \to 0\,, \tag{11.13}$$

and Eq. (11.11) can be linearized (EI is the bending stiffness; I is the moment of inertia of the cross section area with respect to the central axis). For the transversal displacement we come to the *Bernoulli–Euler beam equation*

$$EIw^{IV}(x,t) - Tw''(x,t) + \varrho_0 S_0 \ddot{w}(x,t) = q(x,t) - \mu'(x,t)\,, \tag{11.14}$$

where $q = q_n$, $\mu = \mu_b$.

Comment 11.1 The beam equations considered in this section are based on some assumptions. First of all, the beam is considered as a curve (the centerline) having longitudinal, bending and torsional stiffness. Geometrically this means that a one-dimensional model of the beam is used, where the beam shape and deformations—elongation, bending and torsion—are functions of a single coordinate and time. At the same time, the stiffness is defined based on certain assumptions concerning the stress and strain distribution in the cross-section of the beam. The cross-section is assumed to be flat and normal to the centerline. Note that this assumption is satisfied exactly for a uniform beam having uniform initial and actual curvatures. It is also assumed that there is no normal stresses except the longitudinal ones. The latter assumption is acceptable if the ratio of the beam width to the thickness is small; otherwise, if it is large, width bending strain should be assumed to be zero. In these two cases, the Hooke's law yields a liner stress-strain dependencies as

$$\sigma_{TT} = E\varepsilon_{TT} \quad \text{and} \quad \sigma_{TT} = \frac{E}{1-\nu^2}\varepsilon_{TT}\,, \tag{11.15}$$

respectively. In this model, the beam cross-section area follows the tensile force as

$$S = \frac{S_0}{\lambda} \quad \text{and} \quad S = S_0\left(1 - 2\nu\frac{T}{E}\right) \tag{11.16}$$

for an *incompressible material* and for small strains of an elastic material, respectively. Recall that ν is the Poisson's ratio and λ is the longitudinal stretch.

11.2.1 Steady-State Problem

Consider a steady-state bending wave propagating along an infinite beam with an inextensible axis. We choose the coordinates moving together with the wave, the beam shape is thus assumed to be invariable. Relatively the latter the material axis moves uniformly but the cross-section attached to the axis moves with rotations.

We based on Eqs. (11.7), (11.8) with $\lambda = 1$, $\boldsymbol{q} = 0$, $\mu_n = 0$. The input data are the beam cross-section area, S_0, and the bending stiffness, EI, the speed of the material beam' axis along the fixed frame, v, and the variation of of the angle or an averaged angle between the beam and x-axis

$$\alpha_0 = \alpha(\infty) - \alpha(-\infty) = \int_{-\infty}^{\infty} \varkappa(s)\mathrm{d}s\,. \tag{11.17}$$

Under these conditions there is no acceleration along the beam' axis; hence

$$\ddot{\boldsymbol{R}} = v^2 \varkappa \boldsymbol{n}\,. \tag{11.18}$$

It follows from (11.7) that

$$[T'(s) - \varkappa(s)N(s)]\boldsymbol{\tau} + [\varkappa(s)T(s) + N'(s) - \varrho_0 S_0 v^2 \varkappa(s)]\boldsymbol{n} = 0\,. \quad (11.19)$$

This yields

$$T'(s) + EI\varkappa(s)\varkappa'(s) \quad \Longrightarrow \quad T(s) = T_0 - \frac{1}{2}EI\varkappa^2(s)\,, \quad (11.20)$$

that results in the geometrically exact equation with respect to the beam axis curvature:

$$\varkappa''(s) - \beta^2\varkappa(s) + \frac{1}{2}\varkappa^3(s) = 0\,, \quad \beta = \sqrt{\frac{T_0 - \varrho_0 S_0 v^2}{EI}}\,. \quad (11.21)$$

This nonlinear dynamic equation is equivalent to the static one, where the constant dynamic term, $\varrho_0 S_0 v^2$, plays the role of an axial compressive force. The infinite beam cannot carry such a force, at least it is unstable under compression; hence the integration constant, T_0, must be greater than the 'dynamic term'. (Note that instability of a flexible tube caused by an internal flux is well-known phenomenon.) The problem is thus reduced to that for static states of the beam under extension. This closely relates to the Euler's elastica (see, e.g., Love, 1906; Levien, 2008).

Equation (11.21) admits the first integral as

$$\varkappa'(s) = \pm\sqrt{C + \beta^2\varkappa^2(s) - \varkappa^4(s)/4}\,, \quad (11.22)$$

where C is an integration constant.

11.2.2 A Solitary Wave

It corresponds to $C = 0$. We take $s = 0$ at the critical point, where $\varkappa' = 0$:

$$\varkappa'(s) = 0 \quad \text{at} \quad \varkappa^2(0) = 4\beta^2 \quad \text{and at} \quad \varkappa(0) = 0\,. \quad (11.23)$$

The latter case corresponds to a straight beam: $\varkappa \equiv 0$, whereas the former corresponds to a moving loop, Fig. 11.1. Indeed, from Eq. (11.22) with $C = 0$ it follows that

$$\varkappa(s) = \pm\frac{2\beta}{\cosh \beta s}\,. \quad (11.24)$$

Thus, the curvature vanishes as $s \to \pm\infty$ and the curve straightens, while the directions becomes opposite, $\boldsymbol{\tau}(\infty) = -\boldsymbol{\tau}(-\infty)$, as evidenced by the change in the slope of the curve

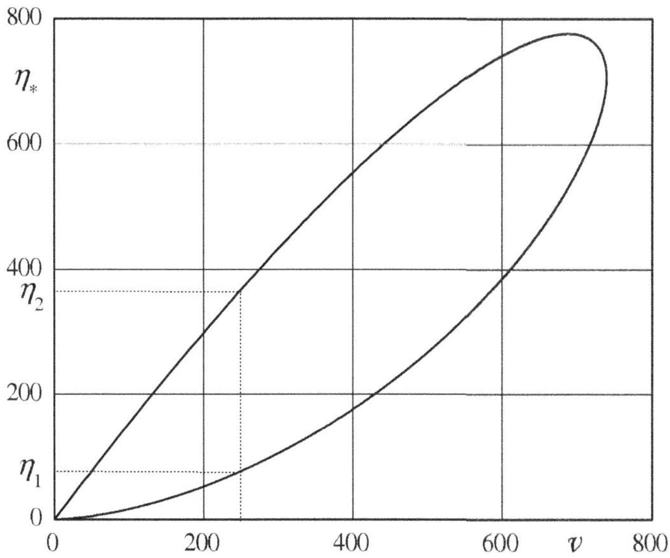

Fig. 11.1 testtttLOOP

$$\alpha_\infty = \int_{-\infty}^{\infty} \varkappa(s)\mathrm{d}s = \pm 2\pi\,. \tag{11.25}$$

Parameter β does not affect the shape but defines the scale. The loop size is inversely proportional to β. By the integration (taken sign '$-$' in the above relations)

$$\begin{gathered} \alpha = \int_0^s \varkappa(s)\mathrm{d}s = \pi - 4\arctan(\exp(\beta s)) \quad (s \geq 0)\,, \\ x = \int_0^s \cos\alpha \mathrm{d}s\,, \quad y = \int_0^s \sin\alpha \mathrm{d}s\,, \end{gathered} \tag{11.26}$$

it is found that the total hight of the wave, h_{total}, the hight of the loop itself, h, and its width, b are

$$\begin{gathered} h_{total} = y(0) - y(\infty) = 2/\beta\,, \\ h = y(0) - y(s_2) \approx 1.423160182/\beta \;\; (s_2 \approx \pm 1.915008048/\beta)\,, \\ b = 2x(s_1) \approx 1.0656799508/\beta \;\; (s_1 \approx \pm 0.8813735870/\beta)\,, \end{gathered} \tag{11.27}$$

where $s = s_2$ is the intersection point. Note that if the wave speed, v, and the tension force far from the loop, $T_0 = T(\pm\infty)$, are given, β is uniquely defined. However, if β is considered as an independent parameter, it defines the wave speed. From (11.21) and (11.26) it follows that

$$
\begin{aligned}
v^2 = \frac{T_0 - EI\beta^2}{\varrho S} &= \frac{T_0 - 4EI/h_{total}^2}{\varrho_0 S_0} \\
&= D^2 - c^2 \frac{4r^2}{h_{total}^2} .
\end{aligned}
\tag{11.28}
$$

Thus, there exists a tension-dependent lower bound for the the loop size

$$
h_{total} > h_{min} = 2\sqrt{\frac{EI}{T_0 S_0}} \tag{11.29}
$$

and the speed increases with the size

$$
v \to \sqrt{\frac{T_0}{\varrho_0 S_0}} \; (h_{total} \to \infty) , \quad v \to 0 \; (h_{total} \to h_{min}) . \tag{11.30}
$$

11.2.3 Rotational Inertia

Note that, in the theory under consideration, the rotational inertia is neglected. If it is taken into account, Eq. (11.8) becomes

$$
M' = -N - \mu_n + \varrho I \ddot{\alpha} \quad (M = M_b) . \tag{11.31}
$$

For the steady-state problem this yields

$$
\ddot{\alpha} = v^2 \alpha'' = v^2 \varkappa' , \quad N = (E - \varrho v^2) I \varkappa' , \tag{11.32}
$$

that is, it results in the substitution of E by $E - \varrho v^2$. Relation (11.28) between the speed and other parameters can now be written in the form

$$
\left(1 - \frac{4r^2}{h_{total}^2}\right) v^2 = D^2 - c^2 \frac{r^2}{4h_{total}^2} \quad \left(D^2 = \frac{T_0}{\varrho_0 S_0} , \quad c^2 = \frac{E}{\varrho_0} , \quad r^2 = \frac{I}{S_0}\right) . \tag{11.33}
$$

In this relation, v is real and $h_{total} > 2rc/D$. For rigid materials $D^2/c^2 = \sigma_{ss}/E \ll 1$, and, in the considered problem, the rotational inertia is really negligible.

Taking a nonzero integration constant, $C \neq 0$, different forms can be obtained. In particular, multiple loop and periodic solutions represent nonlinear waves.

Note that this also follows from the condition of equilibrium.

11.2.4 A Semi-linear Formulation

There exists a way, suggested long ago by V.S. Kalinin, for solving some specific nonlinear problems as if the problem were linear. It is applicable to the Sect. 10.5.2 problem but for a nonzero bending stiffness string, i.e., for an extensible beam model (11.14). This equation is linear as far as the tensile force, T, is given. Otherwise, as in the case of the dynamic problem for an initially unstressed (or weakly stressed) elastic beam, the tensile force arises as a result of the transversal wave, and term Tw'' cannot be considered as linear. So the problem appears to be nonlinear. However, it can be solved using a linear technique if the tensile force may be assumed to be uniform, i.e, x, t-independent, or it can be reduced to a nonlinear ordinary equation if the force is x-independent but depends on t.

For elastic strain of a usual material $\varepsilon_{xx} \ll 1$ and hence $T \ll ES$. It follows that the longitudinal wave speed is much higher than the developed bending wave. This suggests that, in the considered problem, the tensile force can be assumed to depend only on time. Based on the Hooke's law it can be expressed in terms of the transversal displacements as follows:

$$T(t) = \frac{ES}{2ct} \int_{-ct}^{ct} \left(\sqrt{1 + (w'(x,t))^2} - 1 \right) \mathrm{d}x \sim \frac{ES}{4ct} \int_{-ct}^{ct} (w'(x,t))^2 \mathrm{d}x \quad (w'^2 \ll 1)\,. \tag{11.34}$$

Since T is x-independent the Fourier transformation can be used with respect to Eq. (11.14) with $q(x,t) = Q\delta(x)H(t)$, $\mu = 0$. We come to the ordinary equation:

$$\varrho S \ddot{w}^F(k,t) + [EIk^4 + T(t)k^2] w^F(k,t) = QH(t) \tag{11.35}$$

Chapter 12
Solitary Waves of Different Nature

One-dimensional linear wave equation

$$u''(x, t) - \frac{1}{c^2}\ddot{u} = 0 \qquad (12.1)$$

yields waves of an arbitrary shape propagating to the right and left with the constant speed c independently of the wave amplitude

$$u(x, t) = u_1(x - ct) + u_2(x + ct)\,. \qquad (12.2)$$

In particular, $u_{1,2}$ may represent solitary waves, that is, the waves with a finite support or decaying exponentially at infinity. However, it can happen that such a wave equation is only a zero approximation, that is not enough even for a qualitative description of the phenomenon. A simplest adequate equation, having the linear equation as the limit, usually have to contain higher-order derivatives and nonlinearity. There exist such developed wave equations, which, in particular, define solitary waves propagating with a permanent form. Unlike the linear ones a nonlinear solitary wave have a specific form, defined by the corresponding nonlinear equation, and the wave speed does depend on the wave amplitude, i.e., on the wave energy. The above-considered 3D waves in a string (Sect. 10.7) represent an example of the nonlinear solitary waves.

Nonlinear solitary wave can be so stable that it remains unchanged, except for a phase shift, after the collision with another wave propagating with a different speed. For example, the waves can propagate in the opposite directions toward each other, or one wave can propagate faster and pass the other. Such a wave looks like a particle, and it is called *soliton*. Soliton equations are also characterized by the fact that the linearization does not reduce them to the wave equation (12.1). The linearized equation defines waves with dispersion. Thus the soliton is recognized as a dynamic object maintaining a balance between the dispersion and nonlinearity.

L. I. Slepyan, *Mechanical Waves: Free, Forced, and Transition*,
https://doi.org/10.1007/978-3-032-15864-2_12

The ‘soliton science’ has a long history. This phenomenon was first described by John Scott Russell in 1844 (Russell, 1845b). His report is widely cited:

> I was observing the motion of a boat which was rapidly drawn along a narrow channel by a pair of horses, when the boat suddenly stopped – not so the mass of water in the channel which it had put in motion; it accumulated round the prow of the vessel in a state of violent agitation, then suddenly leaving it behind, rolled forward with great velocity, assuming the form of a large solitary elevation, a rounded, smooth and well-defined heap of water, which continued its course along the channel apparently without change of form or diminution of speed. I followed it on horseback, and overtook it still rolling on at a rate of some eight or nine miles an hour, preserving its original figure some thirty feet long and a foot to a foot and a half in height. Its height gradually diminished, and after a chase of one or two miles I lost it in the windings of the channel. Such, in the month of August 1834, was my first chance interview with that singular and beautiful phenomenon which I have called the Wave of Translation.

In more detail, this story and other data with respect to works of John Scott Russell can be found, in particular, in Dodd et al. (1982, 1984). Soliton history can also be fined in Scott et al. (2005).

12.1 Boussinesq and KdV Solitons

12.1.1 Boussinesq Equation

In 1871 and 1872, Boussinesq published the mathematical theory to support Russell’s experimental observation (Boussinesq, 1871; 1872). For a 2D potential flow over a horizontal bed the Boussinesq equation is

$$\ddot{w} - c^2\left(w + \frac{3}{2d}w^2 + \frac{d^2}{3}w''\right)'' = 0\,, \quad c = \sqrt{gd}\,, \tag{12.3}$$

where w is the excess of the wave level above the undisturbed surface, d is the uniform depth and g is the gravity acceleration. It is a weakly nonlinear ‘first approximation’ asymptotic equation for shallow water, that is, for a large ratio of the wave length to the water depth.

A solitary wave solution is sought in the form

$$w = w(\eta)\,, \quad \eta = x - vt\,, \tag{12.4}$$

where v is an unknown wave speed. It follows that

$$w = v^2 w''\,, \tag{12.5}$$

and Eq. (12.3) becomes an ordinary equation

$$\left[\left(\frac{v^2}{c^2}-1\right)w(\eta)-\frac{3}{2d}w^2(\eta)-\frac{d^2}{3}w''(\eta)\right]''=0\,. \tag{12.6}$$

It is also assumed that at infinity the function $w(\eta)$, describing the solitary wave, vanishes together with all its derivatives. Then, Eq. (12.6) can be integrated twice, and now it is of the second order

$$\frac{d^2}{3}w''(\eta)=\left(\frac{v^2}{c^2}-1\right)w(\eta)-\frac{3}{2d}w^2(\eta)\,. \tag{12.7}$$

The regular procedure for solving such an equation is the following. At first, the derivative, $w'(\eta)$, is considered as a function of w as the independent variable, $w'(\eta)=\varphi(w)$. This leads to

$$\frac{d^2}{6}\frac{\mathrm{d}\varphi^2(w)}{\mathrm{d}w}=\left(\frac{v^2}{c^2}-1\right)w-\frac{3}{2d}w^2\,. \tag{12.8}$$

It follows that

$$\varphi(w)\equiv w'(\eta)=\pm\frac{w(\eta)}{d}\sqrt{3\left[\frac{v^2}{c^2}-1-\frac{w(\eta)}{d}\right]}\,, \tag{12.9}$$

where '$\pm$' correspond to the derivative behind the coordinate, where $w=w_{max}$ (let it be $\eta=0$), and ahead of it, respectively. Thus, the derivative, $w'(\eta)$, vanishes at $\eta=\pm\infty$, where $w=0$, and at the wave crest. We have

$$w_{max}=w(0)=\left(\frac{v^2}{c^2}-1\right)d \quad \text{or} \quad v^2=c^2\left(1+\frac{w_{max}}{d}\right); \tag{12.10}$$

hence $v>c$. Integrating Eq. (12.9) yields

$$\begin{aligned}&\int_{w/d}^{w_{max}/d}\frac{\mathrm{d}(w/d)}{(w/d)\sqrt{3\left[v^2/c^2-1-w/d\right]}}\\ &=\frac{2}{\sqrt{3(v^2/c^2-1)}}\operatorname{arctanh}\sqrt{1-\frac{w}{w_{max}}}=\frac{|\eta|}{d}\,.\end{aligned} \tag{12.11}$$

Finally,

$$w(\eta)=\frac{w_{max}}{\cosh^2\psi}\,,\quad \psi=\frac{|\eta|}{2d}\sqrt{\frac{3w_{max}}{d}}\,,\quad v=c\sqrt{1+\frac{w_{max}}{d}}\,. \tag{12.12}$$

12.1.2 KdV Equation

Assuming the wave speed to be close to that following from the linear wave equation, that is assuming that $w_{max} \ll d$, Boussinesq equation (12.3) can be transformed for *waves propagating in only one direction*. To this end represent

$$E \equiv \frac{\partial^2 w}{\partial t^2} - c^2 \frac{\partial^2 w}{\partial x^2} = \left(\frac{\partial}{\partial t} - c\frac{\partial}{\partial x}\right)\left(\frac{\partial w}{\partial t} + c\frac{\partial w}{\partial x}\right). \tag{12.13}$$

Further, assuming that

$$w \sim w(x - vt) \quad \text{and} \quad v \sim c\,, \tag{12.14}$$

we have

$$\frac{\partial}{\partial t} - c\frac{\partial}{\partial x} \sim -2c\frac{\partial}{\partial x} \quad \text{and} \quad E \sim -2c\frac{\partial}{\partial x}\left(\frac{\partial w}{\partial t} + c\frac{\partial w}{\partial x}\right). \tag{12.15}$$

Equation (12.3) can now be integrated once. As a result, we obtain the well-known Korteweg-de Vries equation, the KdV equation (Korteweg and Vries, 1895). For waves propagating to the right

$$\frac{\partial w}{\partial t} + c\frac{\partial}{\partial x}\left(w + \frac{3}{4d}w^2 + \frac{1}{2}\varkappa d^2\frac{\partial^2 w}{\partial x^2}\right) = 0\,, \quad c = \sqrt{gd}\,, \quad \varkappa = \frac{1}{3} - \frac{T}{\varrho g d^2}\,. \tag{12.16}$$

In this equation, not only gravity but the surface tension, T, is also taken into account; ϱ is the density, $\varkappa$ is assumed to be positive (this means that the depth is not too small; numerical estimates are given below). This, as well as the Boussinesq equation, is a weakly nonlinear equation for shallow water, that is, a 'first approximation' asymptotic equation for a large ration of the wave length to the water depth with a small amplitude-to-depth ratio. Note that the 'zero approximation' is the linear PDE equation of the first order. In the coordinate system $\xi = x - ct, t$, the KdV equation looks as

$$\frac{\partial w}{\partial t} + c\frac{\partial}{\partial \xi}\left(\frac{3}{4d}w^2 + \frac{1}{2}\varkappa d^2\frac{\partial^2 w}{\partial x^2}\right) = 0\,, \quad c = \sqrt{gd}\,, \quad \varkappa = \frac{1}{3} - \frac{T}{\varrho g d^2}\,. \tag{12.17}$$

The solitary wave, based on the KdV equation, follows from that, based on the Boussinesq equation, by the substitution $2(v/c - 1)$ instead of $v^2/c^2 - 1$. The wave profile expression in (12.12) remains valid, and the only wave speed expression is changed to

$$v = c\left(1 + \frac{w_{max}}{2d}\right). \tag{12.18}$$

Note the surface tension can be equally introduced in the Boussinesq equation (12.3). With this addition it is

$$\ddot{w} - c^2 \left(w + \frac{3}{2d} w^2 + d^2 \varkappa w'' \right)'' = 0, \quad c = \sqrt{gd}, \tag{12.19}$$

The solitary wave based on the Boussinesq equation with the surface tension taken into account is

$$w(\eta) = \frac{w_{max}}{\cosh^2 \psi}, \quad \psi = \frac{|\eta|}{2d} \sqrt{\frac{w_{max}}{d\varkappa}}, \quad v = c\sqrt{1 + \frac{w_{max}}{d}}, \quad \varkappa = \frac{1}{3} - \frac{T}{\varrho g d^2}. \tag{12.20}$$

To imagine this wave, we estimate the parameters involved in its expression. For water at 20 °C the surface tension $T = 0.0728$ N/m, and $T/(\varrho g) = 0.7420 \cdot 10^{-5}\,\mathrm{m}^2$. It follows that $\varkappa > 0$ if $d > 4.72$ mm and the surface tension is practically negligible if $d > 10$ cm. Judging from the above-quoted story of Russell, the wave velocity was close to $c \approx 4$ m/s, and the depth of the channel was about 1.5 m.

In the case of such a wave, the wavelength is a conditional concept. The 'visible' wave length should correspond to the argument of the hyperbolic cosine $\psi = 2 \div 2.5$ at its borders

$$\frac{l}{2d}\sqrt{\frac{A}{2\varkappa d}} = (2 \div 2.5) \tag{12.21}$$

and with this accuracy

$$l = (4 \div 5) d \sqrt{\frac{2\varkappa d}{A}}. \tag{12.22}$$

Let us take

$$v = 4\mathrm{m/s}, \quad d = 1.5\mathrm{m}, \quad A = 0.4\mathrm{m}, \quad \varkappa = 1/3 \tag{12.23}$$

and the coefficient at the expression for l (12.22) equal to 4. It follows that the theoretical value, $l = 9.5$ m, is rather close to the Scott Russell observation ("some thirty feet long").

Comment 12.1 Is it possible to linearize KdV equation? In accordance with solution (12.12), the tree members in (12.7) relate to each other as

$$\left(\frac{2}{\cosh^2 \psi} - \frac{3}{\cosh^4 \psi} \right) + \frac{3}{\cosh^4 \psi} - \frac{2}{\cosh^2 \psi} \equiv 0. \tag{12.24}$$

Only the second term is negligible at $\psi \to \infty$; it, however is important for not too large ψ. Thus, any member of the equation cannot be ignored, independently of the amplitude-depth ratio, A/d. If Eq. (12.7) is considered without regard to this solution, the nonlinear term can be neglected; however this leads to a spreading wave with dispersion, which cannot be represented as a function of a single variable, $x - vt$. If, moreover, the wavelength-depth ratio is sufficiently large, the dispersion effects are slow and the wave is also spreading slowly but is still spreading. Thus, wave equation (12.1), which follows from KdV equation neglecting the nonlinearity and dispersion, is valid only for a long, low-amplitude wave considered during a short time-interval. At the same time, the wave speed defined by the wave equation, $v = c = \sqrt{\varrho d}$, is the correct limit for the soliton speed as $A/d \to 0$. Thus, the nonlinear term can be arbitrary small; however, a stable solitary wave cannot be obtained without it.

Like in the case of the helicoidal string, the Boussinesq-KdV equations define not only solitary but also periodical waves. The final equations for these waves are derived by introducing of an integrating constant, C, in the integral of (12.8). Now Eq. (12.9) becomes

$$\varphi(w) \equiv w'(\eta) = \pm\sqrt{3\left[\left(\frac{v^2}{c^2} - 1\right)\frac{w^2(\eta)}{d^2} - \frac{w^3(\eta)}{d^3}\right] - C}\,. \tag{12.25}$$

In any C within a certain range of it, there exists an interval of w, where the radicand is positive except the end points, where it is zero, while $w > 0$ in the closed interval. These two end values of w, where $w'(\eta) = 0$, represent the minimal and maximal levels of a periodic wave. Here sign $+(-)$ corresponds to the ranges of η, where the wave level is growing (decreasing).

12.2 Fermi–Pasta–Ulam Soliton in a Chain

The soliton phenomenon was rediscovered by E. Fermi, J. Pasta and S. Ulam in 1953 during computer simulations of a finite nonlinear chain obeying the dynamic equation

$$m\ddot{u}_n(t) = F_n(t) - F_{n-1}(t)\,, \quad F_n = \mu(\varepsilon_n + \alpha\varepsilon_n^\nu)\,, \quad \varepsilon_n = \frac{u_{n+1} - u_n}{a} \tag{12.26}$$

in two versions: $\nu = 2$ and $\nu = 3$ (a is the chain constant, i.e., the distance between neighboring particles). Fermi thought that after many iterations, the system would exhibit thermalization with chaotic oscillations of the particles. However, the system exhibited a complicated quasi-periodic behavior. The results are published in a Los Alamos technical report in 1955 (Fermi et al., 1955a). In fact, they observed the soliton manifestation.

Consider and infinite chain and the problem in a somewhat more general formulation, namely, we do not specify the exponent ν. The conditions are $\nu > 1$, $\varepsilon > 0$ and $\alpha > 0$ (it is assumed of course that $\varepsilon^\nu > 0$). In terms of the strain, Eq. (12.26) becomes

$$\ddot{\varepsilon}_n(t) = \frac{c^2}{a^2}[\varepsilon_{n+1}(t) + \varepsilon_{n-1}(t) - 2\varepsilon_n(t) + \alpha(\varepsilon_{n+1}^\nu(t) + \varepsilon_{n-1}^\nu(t) - 2\varepsilon_n^\nu(t))]\,, \quad c = \sqrt{\frac{\mu a}{m}}\,. \tag{12.27}$$

The Taylor series representation of the finite differences leads to a continuum model as a long-wave asymptotic of the discrete system. Based on the equality

$$\varepsilon_{n+1} \equiv \varepsilon(a(n+1), t) = \sum_{i=0}^{\infty} \frac{a^i}{i!} \varepsilon^{(i)}(an, t), \tag{12.28}$$

where n is now a continuous variable, continuum models of different levels of the approximation can be obtained. If we keep only the first two terms and do not take into account the nonlinearity, the wave equation is obtained as the zero approximation. The first approximation differs by the presence of the fourth order derivative and the 'first order' nonlinear terms, that is

$$\begin{gathered} \varepsilon_{n+1} + \varepsilon_{n-1} - 2\varepsilon_n \sim a^2 \varepsilon''(an) + \frac{a^4}{12} \varepsilon^{IV}(an), \\ \varepsilon_{n+1}^{\nu} + \varepsilon_{n-1}^{\nu} - 2\varepsilon_n^{\nu} \sim \nu a^2 [\varepsilon^{\nu-1} \varepsilon'' + (\nu - 1)\varepsilon^{\nu-2} (\varepsilon')^2] = a^2 (\varepsilon^{\nu})''(an). \end{gathered} \tag{12.29}$$

Substituting this into equation (12.27), we obtain the Boussinesq equation in the form

$$\ddot{\varepsilon} = c^2 \left(\varepsilon + \alpha \varepsilon^{\nu} + \frac{a^2}{12} \varepsilon'' \right)''. \tag{12.30}$$

The following considerations are the same as above. For a solitary wave solution, $\varepsilon = \varepsilon(\eta)$, $\eta = an - vt$, $\varepsilon = \varepsilon' = 0\,(\eta = \pm\infty)$, the latter equation can be integrated twice, and this yields

$$\frac{a^2}{24} \frac{\mathrm{d}(\varepsilon')^2}{\mathrm{d}\varepsilon} = \left(\frac{v^2}{c^2} - 1 \right) \varepsilon - \alpha \varepsilon^{\nu}; \tag{12.31}$$

hence

$$\varepsilon'(\eta) = \pm \frac{2\varepsilon(\eta)}{a} \sqrt{3 \left(\frac{v^2}{c^2} - 1 \right) - \frac{6\alpha}{\nu - 1} \varepsilon^{\nu-1}(\eta)} \quad (\varepsilon > 0,\ \nu > 1). \tag{12.32}$$

It follows that the derivative is zero at

$$\varepsilon = \varepsilon_{max} \equiv \varepsilon_{max} = \left[\frac{\nu - 1}{2\alpha} \left(\frac{v^2}{c^2} - 1 \right) \right]^{1/(\nu-1)} \tag{12.33}$$

and

$$v^2 = c^2 \left(1 + \frac{2\alpha \varepsilon_{max}^{\nu-1}}{\nu - 1} \right), \quad c^2 = \frac{\mu a}{m}. \tag{12.34}$$

Note that the substitutions

$$\varepsilon = \left(\frac{\nu - 1}{2\alpha}\frac{w}{d}\right)^{1/(\nu-1)}, \quad \eta \equiv \eta_\varepsilon = \frac{a}{2(\nu - 1)}\eta_w, \tag{12.35}$$

where η in the Boussinesq equation (12.9) is here denoted by η_w, leads the FPU equation (12.32) to the former equation (12.9). Thus the solution of the FPU equation (12.32) follows from the solution of the Boussinesq equation (12.36) as

$$\varepsilon(\eta) = \frac{\varepsilon_{max}}{(\cosh\psi_\varepsilon)^{2/(\nu-1)}}, \quad \psi_\varepsilon = \sqrt{6\alpha(\nu - 1)\varepsilon_{max}}\,\frac{|\eta|}{a}, \quad v = c\sqrt{1 + \frac{2\alpha\varepsilon_{max}^{\nu-1}}{\nu - 1}}. \tag{12.36}$$

In this section, the solitary wave solutions, based on the Boussinesq, KdV and FPU equations, are discussed. This, however, is only the most straightforward part of the nonlinear field theory of solitary and related periodical waves in various physical applications. In more complicated cases, as in such nonlinear wave formation, evolution and interaction, the KdV *evolution equation* (12.17) plays a crucial role. The subject is discussed in detail, e.g., in the books by Dodd et al. (1982, 1984) and Ablowitz and Clarkson (1991).

Part V
Forced Waves

Chapter 13
The Wave Quasifront

13.1 Wave in a String on an Elastic Foundation

The (non-dimensional) equation for the supported string under (unmoving) oscillating force is

$$\frac{\partial^2 u(x,t)}{\partial t^2} - \frac{\partial^2 u(x,t)}{\partial x^2} + u(x,t) = \delta(x)\mathrm{e}^{\mathrm{i}\omega t}. \tag{13.1}$$

In addition, we have zero initial conditions.

In the following, we base on the Fourier transform on x and the Laplace transform on t, where the latter allows us to satisfy the initial conditions automatically. Here and below we use the parameters k and s, respectively. We have

$$(s^2 + k^2 + 1)u^{LF} = \frac{1}{s - \mathrm{i}\omega}. \tag{13.2}$$

The inverse Laplace transform gives us

$$u^F(k,t) = \frac{1}{k^2 + 1 - \omega^2}\left(\mathrm{e}^{\mathrm{i}\omega t} - \cos(\sqrt{k^2+1}\,t) - \frac{\mathrm{i}\omega\sin(\sqrt{k^2+1}\,t)}{\sqrt{k^2+1}}\right). \tag{13.3}$$

In this expression, the strip $-1 < \omega < 1$ is the band gap, which bounds, $\omega = \pm 1$, correspond to resonant waves, considered below, Chap. 16. So, we assume $\omega > 1$.

Note that although the point $k = k_* = \sqrt{\omega^2 - 1}$ (as well as the symmetric ones) is not a pole but a *quasi-singularity*. Namely, the inverse Fourier transform integral over a small (but nonzero) segment, $-\varepsilon < k - k_* < \varepsilon,\ \varepsilon > 0$, leads to an asymptotic representation of the wave.

L. I. Slepyan, *Mechanical Waves: Free, Forced, and Transition*,
https://doi.org/10.1007/978-3-032-15864-2_13

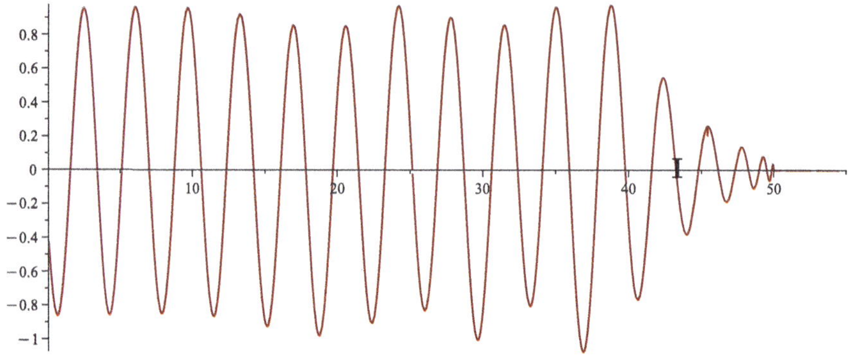

Fig. 13.1 The transient wave, $u(x, 50)$, in the string on an elastic foundation. The point moving with the group velocity $x = v_g t$, $v_g = \sqrt{3}/2$, is marked by the **I**-symbol. The wave is obtained from the expression in (13.4) numerically

Finally

$$u(x,t) = u(-x,t) = \frac{1}{\pi}\int_0^\infty u^F(k,t)\mathrm{e}^{-\mathrm{i}kx}\mathrm{d}k. \tag{13.4}$$

The wave for $\omega = 2$, $t = 50$ is shown in Fig. 13.1.

Two points are remarkable. The first is that the wavefront (now located at $x = 50$) propagates with the same speed as in the free string (in our notations, $c = 1$). That is, since the elastic support, by itself, does not conduct any disturbances, and it does not affect the wave speed in the string but the wave intensity.

Next, we here meet a visible contradiction: the wave penetrates through the boundary (marked by **I**-symbol) moving with the group velocity, $v_g = \sqrt{3}/2$. It looks that energy can propagate faster. As already noted in Sect. 1.2.3, the point is that in the wavefront area, where the wave amplitude is changing, a set of waves of different lengths (and group velocities) takes part in the wave description.

So, of necessity, it appears a slowly expanding *quasifront* instead of point wavefront. At the same time, the energy flux enough behind the latter, where only the exciting single wave propagates, is fixed. It follows that some energy penetrates from the wave behind the v_g-front to an area ahead of it. As a result, we see the quasifront (Fig. 13.2, 13.3).

Thus, the (k, ω)-parameters of the waves are determined. The waves which group speed $v_g > v_0$ place at the right, $\eta = x - v_0 t > 0$, and vice versa. The wave amplitude, A, is defined to satisfy the dynamic equilibrium at $\eta = 0$. Namely, the external force (moment) must be supported by jumps of the internal force (moment).

A resonant wave is excited if at the related point remains valid under the derivative, that is, if the group and phase velocities coincide at that point.

The wave excited by an arbitrary distributed load, $Q(\eta)\exp(\mathrm{i}\omega t)$, is expressed by the convolution. Indeed, if the delta-load acts at $\eta = \xi$, $Q_\eta = Q(\xi)\delta(\eta - \xi)$, then

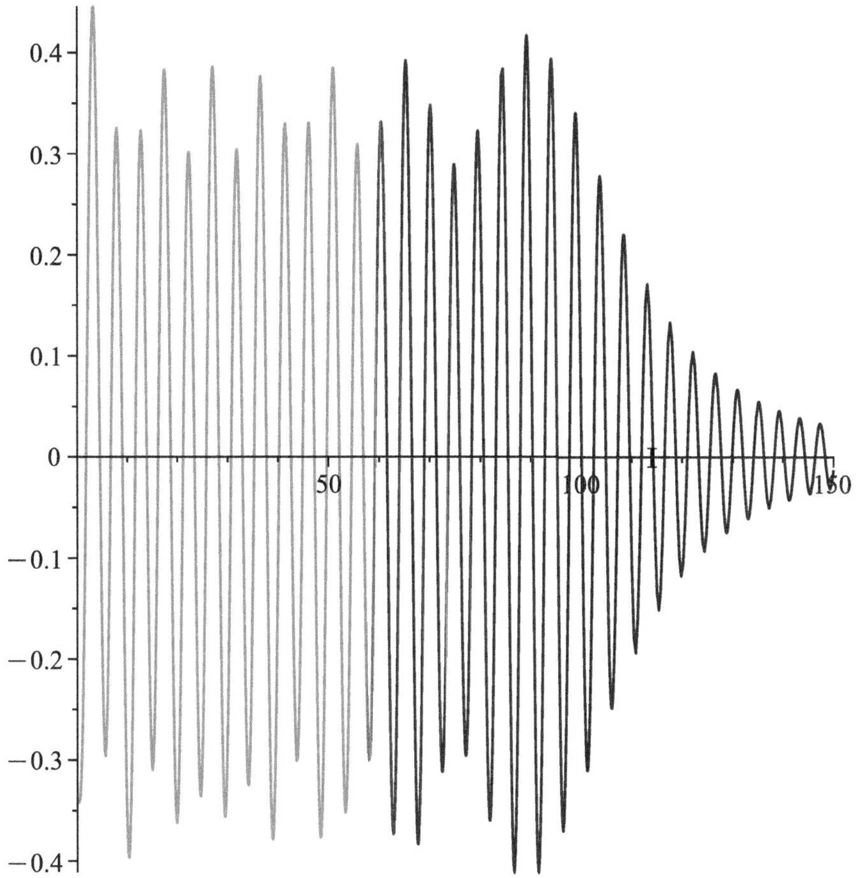

Fig. 13.2 The wave, $u(x, 50)$, in the string on an elastic foundation

$$u_\xi = Q(\xi)u_0(t, x-\xi) = A_+ Q(\xi)\mathrm{e}^{\mathrm{i}(\omega_+ t - k_+(x-\xi))} + A_- Q(\xi)\mathrm{e}^{\mathrm{i}(\omega_- t + k_-(x-\xi))} \quad (13.5)$$

and the superposition principle valid for the uniform linear system considered gives us

$$u = \int_L Q(\xi)u_0(t, x-\xi)\mathrm{d}\xi = A_+ Q^F(k_+)\mathrm{e}^{\mathrm{i}(\omega_+ t - k_+ x)} + A_- Q^F(-k_-)\mathrm{e}^{\mathrm{i}(\omega_+ t - k_- x)}, \quad (13.6)$$

where the $Q^F(k)$ is the Fourier integral transform of $Q(\eta)$, and L is the load support.

To describe the wave in a wave front (quasifront) area, that is in a vicinity of the ray $x = v_g(k_*)t$, we should take into account the dispersion number, $\omega''(k)$, in the asymptotic representation of the dispersion dependency for a vicinity of $k = k_*$

$$\omega(k) - \omega(k_*) \sim v_g(k_*)\xi + \varkappa_n \xi^n, \quad k = k_*, \quad (13.7)$$

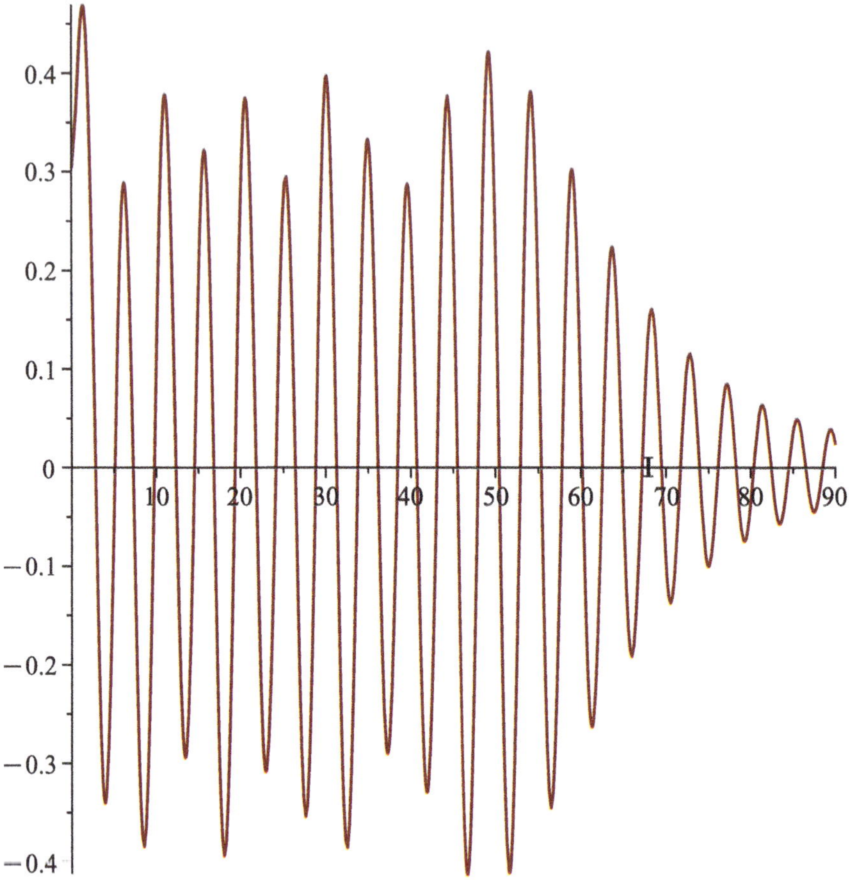

Fig. 13.3 The wave, $u(x, 50)$, in the string on an elastic foundation

where $n \geq 2$ is minimal one for which $\varkappa_n \neq 0$.

The amplitude A_+ (13.6) becomes (we assume here that $v < v_g$)

$$\begin{aligned} A_+ &= \frac{1}{2\pi(v - v_g(k_*))} \int_{-\infty}^{\infty} \frac{\mathrm{e}^{\mathrm{i}((v_g(k_*)t - x)\xi + \varkappa_n \xi^n t)} - \mathrm{e}^{\mathrm{i}(vt-x)\xi}}{\mathrm{i}\xi} \mathrm{d}\xi \\ &= \frac{1}{2\pi(v - v_g(k_*))} \int_{-\infty}^{\infty} \frac{\mathrm{e}^{\mathrm{i}(\zeta_n \xi + \varkappa_n \xi^n)} - \mathrm{e}^{-\mathrm{i}\eta t^{-1/n} \xi}}{\mathrm{i}\xi} \mathrm{d}\xi. \end{aligned} \tag{13.8}$$

Here

$$\zeta_n = \frac{v_g t - x}{t^{1/n}}, \quad \eta = x - vt. \tag{13.9}$$

For $n = 2$ we find

$$A_+ = \frac{1}{2(v - v_g(k_*))}(A_{+1} + A_{+2}),$$
$$A_{+1} = \int_{-\infty}^{\infty} \frac{e^{i\varkappa_2\xi^2}\sin(\zeta_2\xi)}{\pi\xi}d\xi, \quad A_{+2} = \int_{-\infty}^{\infty} \frac{\sin(\xi\eta/\sqrt{t})}{\pi\xi}d\xi. \tag{13.10}$$

Next

$$\Re A_{+1} = \text{FresnelC}(z) + \text{FresnelS}(z),$$
$$\Im A_{+1} = \text{sign}(\varkappa_2)(\text{FresnelC}(z) - \text{FresnelS}(z)), \tag{13.11}$$
$$A_{+2} = 1, \quad z = \frac{\zeta_2}{\sqrt{2\pi|\varkappa_2|}}.$$

With reference to (13.8)–(13.11), we obtain

$$u(t,x) \sim \frac{Q_\eta^{F_\eta}(k_*)}{\Phi^{LF}(i\omega(k_*), k_*)2\omega(k_*)}U(t,x),,$$
$$\Re U(t,x) \sim \frac{1}{2(v_g(k_*) - v)}((1 + \Re A_{+1})\sin(\omega(k_*)t - k_*x)$$
$$+\Im A_{+1}\cos(\omega(k_*)t - k_*x)), \tag{13.12}$$
$$\Im U(t,x) \sim -\frac{1}{2((v_g(k_*) - v)}((1 + \Re A_{+1})\cos(\omega(k_*)t - k_*x)$$
$$-\Im A_{+1}\sin(\omega(k_*)t - k_*x)).$$

To compare the above explicit asymptotic representation (13.12) with that obtained by numerically calculations, consider waves in a bending beam excited by an oscillation load. The equation is

$$\frac{d^2u(t,x)}{dt^2} + \frac{d^4u(t,x)}{dx^4} = Q_0\delta(x)\cos(\omega_0 t), \tag{13.13}$$

where the length and time units are used as the inertia radius, r, of the cross-section area of the beam, and r/c, respectively(c is the sound speed in the beam material). After the Fourier transform on x the equation becomes

$$\frac{d^2u^F(t,k)}{dt^2} + k^4u^F(t,k) = Q_0\cos(t) \quad (\omega_0 = 1). \tag{13.14}$$

It follows

$$u^F(t,k) = Q_0\frac{\sin(k^2t)}{k^2} * \cos(t) = Q_0\frac{\cos(t) - \cos(k^2t)}{k^4 - 1}. \tag{13.15}$$

Thus, we have two expressions

$$u_{num} = \frac{Q_0}{\pi} \int_0^{\infty} \frac{\cos(t) - \cos(k^2 t)}{k^4 - 1} \cos(kx)\mathrm{d}k, \tag{13.16}$$

and the analytical expression (13.12) with

$$k_* = \omega_0 = \omega(k_*) = 1,\ v_g = \varkappa_2 = 1,\ Q = Q_0,\ \Phi^{LF}(\mathrm{i}\omega(k_*), k_*) = 1,\ v = 0. \tag{13.17}$$

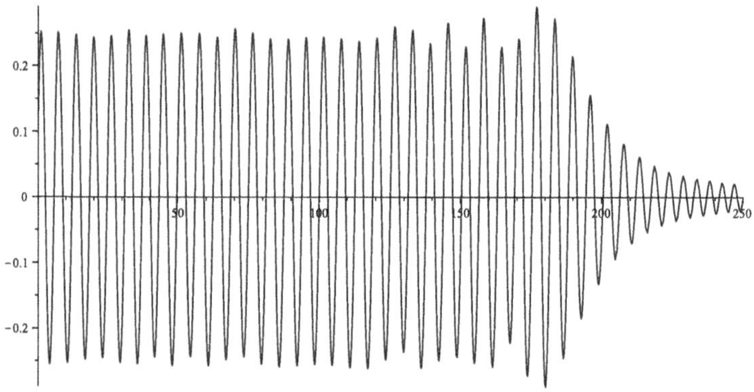

Fig. 13.4 The wave in a bending beam. Asymptotic representation following from the general expressions (13.11), (13.12) with non-dimensional parameters $\omega_0 = k_* = \eta_2 = 1$, $v_g = 2$, $t = 100$

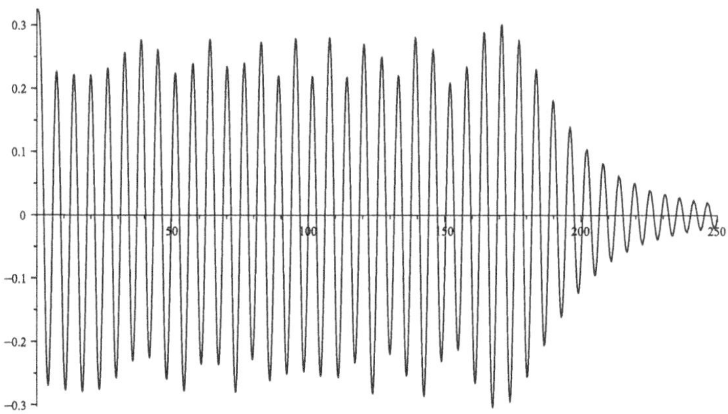

Fig. 13.5 The same function as in Fig. 13.4, but the 'exact' one obtained numerically

Under these conditions

$$u(t, x) = \frac{Q_0}{8}((1 + \Re A_{+1}) \sin(t - x) + \Im A_{+1} \cos(t - x)), \quad \Im u(t, x) = 0. \tag{13.18}$$

The analytical and numerical results for $t = 100$ ($\approx$ 17 periods of oscillations) are shown in Figs. 13.4 and 13.5, respectively.

Chapter 14
Wave Excitation, Steady-State Limit and Causality Principle

14.1 Wave Excitation

14.1.1 Excitation Conditions and the Frequencies

Consider a propagating wave Green's function, $u(t, x)$ excited by a point moving-oscillating load, $Q = Q_0 e^{i\omega_0 t}\delta(\eta)$, $\eta = x - vt$, where v and ω_0 are the load speed and frequency constants, respectively. In the steady-state mode, the over-the-period energy flux from the load to the wave must be positive (otherwise, the wave has no energy to propagate). It follows that the load and the corresponding particle velocity of the wave oscillate with the same frequency (but not necessary of the same phase)

$$\frac{\partial u(t, vt + \eta)}{\partial t} = e^{i(\omega_0 t + \varphi)} w(\eta), \tag{14.1}$$

where the possible phase shift φ is bounded:

$$-\pi/2 < \varphi < \pi/2. \tag{14.2}$$

In these relations, the load-and-wave frequency corresponds to $\eta = \text{const}$, that is, to an observer moving with speed v.

The point lode of such a kind, $Q_0 e^{i\omega_0 t}\delta(\eta)$ can excite a single complex wave, $A e^{i(\omega(k)t - kx)}$ which frequency corresponds (in the form as it presented here) to $x = \text{const}$, that is to an unmoving observer. Here A is a constant amplitude, and $\omega(k)$ is a dispersion dependence. For the moving together with the load, $x = vt + \eta$ it looks as with the load frequency. Thus, the load oscillating with frequency ω_0 and moving with speed v excite a wave with frequency

$$\omega = \omega_0 + kv. \tag{14.3}$$

L. I. Slepyan, *Mechanical Waves: Free, Forced, and Transition*,
https://doi.org/10.1007/978-3-032-15864-2_14

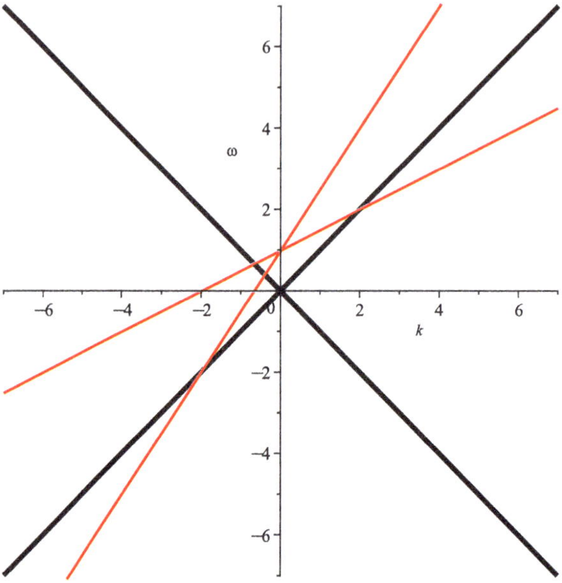

Fig. 14.1 Dispersion dependencies, $\omega = \pm k$ (bold lines) and the load characteristic ray, $\omega = \omega^0 + kv$, $\omega^0 = 1$, $v = 1/2$. The intersection points correspond to the waves excited by this moving-oscillating load. The left point, $\omega = 2/3$, $k = -2/3$, corresponds to the wave placed at the left from the moving load, since its group speed, $v_g = \mathrm{d}\omega/\,\mathrm{d}k = -1 < v$. The right point, $\omega = k = 2$, corresponds to the wave placed at the right due to the opposite relation between these speeds: here $v_g = 1 > v$. Recall that the group velocity is equal to the energy flux velocity in the wave

This is illustrated in Fig. 14.1. Note that the excitation does not take place if the load distributed in such a way that it is orthogonal to the distribution of the particle velocities.

The difference in the frequency of a sinusoidal wave detected by an unmoving observer, ω, and one moving together with the wave source, ω^0, is the well-known *Doppler effect*.

14.2 From a Transient Problem to the Steady-State Limit

14.2.1 Pre-limiting Location of the Singular Points

Consider the corresponding transient problem, where the wave is excited by the force $Q(\eta)e^{i\omega_0 t}H(t)$ under zero initial conditions at $t = 0$. The LF_η-transforms are presented in (18.137)–(18.139). The complex-wave limit, (18.146), if it exists, can be expressed as (see Sect. 8.2.2)

$$\lim_{t\to\infty} u^{F_\eta}(t,k) = \lim_{s\to+0} s u_0^{LF\eta}(s,k) = u^{F_\eta}(+0,k) Q_0^{F\eta}(k). \tag{14.4}$$

The question is whether we can omit the zero symbol in the argument of u^{F_η} or it must be preserved. In this regard, the connection between the pre-limiting location of singular points of the transformed function and the group velocity of the corresponding waves is discussed.

14.2.1.1 Two Examples

First consider the non-dimensional linear dynamics equation for a string

$$\ddot{u}(x,t) - u''(x,t) = Q(x,t). \tag{14.5}$$

The corresponding LF-transform of Green's function is

$$G^{LF}(s,k) = \frac{1}{L(s,k)}, \quad L(s,k) = s^2 + k^2. \tag{14.6}$$

The expression for a (transformed) moving-oscillating load is

$$Q^{LF}(s,k) = \frac{Q_0^{F\eta}(k)}{s - \mathrm{i}(\omega^0 + kv)}, \tag{14.7}$$

and in terms of the transform in the moving coordinate system

$$\begin{gathered} u_0^{LF\eta}(s',k) = \frac{Q_0^{F\eta}(k)}{s' L(s' + \mathrm{i}(kv + \omega^0), k)}, \\ L = (s' + \mathrm{i}(kv + \omega^0 + k))(s' + \mathrm{i}(kv + \omega^0 - k)). \end{gathered} \tag{14.8}$$

If $0 < |\omega^0|$ and $v \neq v_g$ ($v_g = \pm 1$) two zero points of L for $s' = 0$ correspond to the dispersion relations, $\omega = \pm k$:

$$k = k_{1,2} = -\frac{\omega^0}{v \pm 1}, \tag{14.9}$$

where the points are numbered as they are placed in Fig. 14.1: $k_1 < k_2$. Only one zero point corresponds to the case, where $v = \pm 1$ (these cases also can be seen in Fig. 14.1).

Now we consider how these points are shifted as s' takes a small positive value. Noting that the group velocity $v_g = \pm 1$ the corresponding variations can be found from (14.8) as

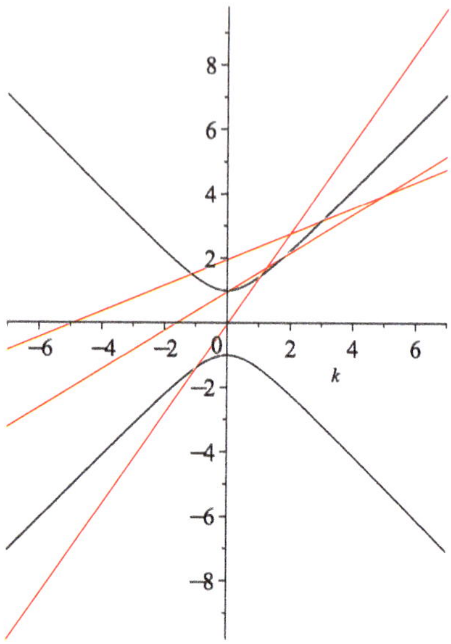

Fig. 14.2 The dispersion dependencies, $\omega = \pm k^2$, and three load related lines $\omega = \omega^0 + kv : \omega = 2, v = 0$ (oscillating unmoving load), $\omega = 0, v = 1/2$ (moving non-oscillating load) and $\omega = 1, v = 1$ (moving-oscillating load). The latter excites four different waves corresponding the four intersection points; two waves are radiated in front of the load, whereas the others are placed behind it

$$\delta k_{1,2} = \frac{\mathrm{i}s'}{v - v_g}. \tag{14.10}$$

Similar problem follows from the Bernoulli-Euler classical dynamic equation for a beam. In a simplest non-dimensional form is reads

$$w^{IV}(x,t) + \ddot{w}(x,t) = q(x,t), \tag{14.11}$$

that corresponds to the above-mentioned dispersion relations, $\omega = \pm k^2$, Fig. 14.2. The corresponding LF-transform of Green's function is

$$G^{LF}(s,k) = \frac{1}{L(s,k)}, \quad L(s,k) = s^2 + k^4. \tag{14.12}$$

In terms of the transform in the moving coordinate system with the same moving-oscillating load (14.7) the displacement is

$$\begin{aligned} w^{LF\eta\omega}(s',k) &= \frac{Q_0^{F\eta}(k)}{s'L((s' + \mathrm{i}(kv + \omega^0), k)}, \\ L &= (s' + \mathrm{i}(kv + \omega^0 + k^2))(s' + \mathrm{i}(kv + \omega^0 - k^2)). \end{aligned} \tag{14.13}$$

If $0 < |\omega^0| < v^2/4$ four zero points of L for $s' = 0$ correspond to the dispersion relations, $\omega = kv + \omega^0 = \pm k^2$:

$$k = k_{1,2} = -\frac{v}{2} \mp \sqrt{\frac{v^2}{4} - \omega^0}, \quad k = k_{3,4} = \frac{v}{2} \mp \sqrt{\frac{v^2}{4} + \omega^0}, \tag{14.14}$$

where the points are numbered as they are placed in Fig. 14.2: $k_1 < k_2 < k_3 < k_4$. Three zero points correspond to the case, where $\omega^0 = 0$, and two ones otherwise (these cases also can be seen in Fig. 14.2).

Now we consider how these points are shifted as s' takes a small positive value. Noting that the group velocity

$$v_g = -2k \ \ (k = k_{1,2}), \quad v_g = 2k \ \ (k = k_{3,4}) \tag{14.15}$$

the corresponding variations can be found from (14.13) as

$$\delta k_{1,2} = \frac{\mathrm{i}s'}{v - v_g}. \tag{14.16}$$

Thus, in both problems, the same expressions, (14.10) and (14.16), are obtained for the prelimiting location of singular points of the solutions. Namely, the corresponding points are in the upper half-plane of the complex k-plane if $v > v_g$ (recall that $s' > 0$), and vice versa. This is not surprising since the wave propagates slower than the load in the former case and faster than the load in the latter case. This implies the obtained location of the singular points.

14.2.2 A More General Consideration

Along with a general expression for the steady-state solution

$$L(s, k) = 1/G(s, k) = L(s' + \mathrm{i}(kv + \omega^0, k)) \quad (s' > 0), \tag{14.17}$$

which for $s' = 0$ defines the wavenumbers, k, of the waves excited by the moving-oscillating load, we consider the original dispersion relation

$$L(\mathrm{i}\omega(k), k) = 0. \tag{14.18}$$

Successive differentiation of the latter with respect to k *along the dispersion curve*, $\omega = \omega(k)$, leads to a sequence of relations:

$$E_1 \equiv \frac{\partial L(\mathrm{i}\omega, k)}{\partial \omega} v_g + \frac{\partial L(\mathrm{i}\omega, k)}{\partial k} = 0,$$

$$E_2 \equiv \frac{\partial^2 L(\mathrm{i}\omega, k)}{\partial \omega^2} v_g^2 + \frac{\partial L(\mathrm{i}\omega, k)}{\partial \omega} \frac{\mathrm{d}v_g}{\mathrm{d}k} \tag{14.19}$$

$$+ 2\frac{\partial^2 L(\mathrm{i}\omega, k)}{\partial \omega \partial k} v_g + \frac{\partial^2 L(\mathrm{i}\omega, k)}{\partial k^2} = 0, \tag{14.20}$$

and so on.

Further, we take the pre-limiting expression for the function $L(s' + \mathrm{i}(kv + \omega^0, k)$ as in (14.17) for a small positive s'. Let $k = k_0$ be a zero of this function for $s' = 0$. A local expansion in a vicinity of $k = k_0$ follows as

$$L(s' + \mathrm{i}(kv + \omega^0, k) = L(s' + \mathrm{i}(kv + \omega^0 + \mathrm{i}v\delta k, k_0 + \delta k) \quad (14.21)$$

$$= \frac{\partial L(\mathrm{i}\omega, k)}{\partial \omega}(v\delta k - \mathrm{i}s') + \frac{\partial L(\mathrm{i}\omega, k)}{\partial k}\delta k + \frac{1}{2}\frac{\partial^2 L(\mathrm{i}\omega, k)}{\partial \omega^2}(v\delta k - \mathrm{i}s')^2 \quad (14.22)$$

$$+ \frac{1}{2}\frac{\partial^2 L(\mathrm{i}\omega, k)}{\partial k^2}\delta k^2 + \frac{\partial^2 L(\mathrm{i}\omega, k)}{\partial \omega \partial k}\delta k(v\delta k - \mathrm{i}s') + \ldots = 0. \quad (14.23)$$

Note that in this variation $v =$ const and the derivatives are taken at the intersection point, k_0, where $k_0 v + \omega^0 = \omega_0(k_0)$.

The pre-limiting location of the zero, $k = k_0 + \delta k$, of function $L(s' + \mathrm{i}(kv + \omega^0, k)$ is now examined. Consider a few particular cases.

(a) Let for $k = k_0$ the following inequalities be valid:

$$\frac{\partial L(\mathrm{i}\omega, k)}{\partial \omega} \neq 0, \quad v \neq v_g. \quad (14.24)$$

In this case, in (14.21), only terms with the derivatives of the first order should be taken into account. Using the first equality in (14.19) we obtain

$$\delta k = \frac{\mathrm{i}s'}{v - v_g} \quad (s' \to +0). \quad (14.25)$$

Thus, under conditions (14.24), the real zero $k = k_0$ is the limit from below, $k = k_0 - \mathrm{i}0$, if $v < v_g$ and from above, $k = k_0 + \mathrm{i}0$, if $v > v_g$, as in the above examples.

(b) Let the second case be

$$\frac{\partial L(\mathrm{i}\omega, k)}{\partial \omega} \neq 0, \quad \frac{\mathrm{d}v_g}{\mathrm{d}k} \neq 0, \quad v = v_g. \quad (14.26)$$

From (14.19) and (14.21) it now follows that the linear in k term is absent, and we have to keep linear in s' and quadratic in k terms. From (14.21) and E_2 in (14.19) we obtain

$$\delta k \sim \pm\sqrt{\frac{2\mathrm{i}s'}{\mathrm{d}v_g/\mathrm{d}k}} \quad (s' \to +0). \quad (14.27)$$

Clearly, $k = k_0$ is a limiting location of two roots; one of them established from above, the other from below. This case may be considered as the limit where two zero points with $v < v_g$ and $v > v_g$ coalesce as $v \to v_g$.

(c) Lastly, consider a four-fold zero point of the function

$$L = (s' + \mathrm{i}(kv + \omega^0))^2 + 2(1 - \cos k)c^2/a^2 \quad (s' \to +0) \tag{14.28}$$

in the case $v = c$, this function corresponds to waves in a regular mass-spring chain, see Sect. 1.2.9. It follows from (14.19) and (14.21) that

$$\frac{a^2c^2}{12}(\delta k)^4 - 2\mathrm{i}s'c\,\delta k - (s')^2 = 0 \tag{14.29}$$

with

$$\delta k_m \sim \left(\frac{24\,s'}{a^2c}\right)^{1/3} \exp\left[\frac{1}{3}\mathrm{i}\pi\left(\frac{1}{2} + 2(m-1)\right)\right], \quad m = 1, 2, 3; \quad \delta k_4 \sim \frac{\mathrm{i}s'}{2c}. \tag{14.30}$$

In this case, only one of the four zeros belongs to the lower half-plane, whereas the three other roots belong to the upper half-plane.

In concluding, note that the above considerations concern one-dimensional and surface/interface waves. In a two-dimensional problem, in general, waves radiated by a moving source are characterized by some segments on the real k-axis corresponding to the curve of intersection of the two-dimensional dispersion surface, $\omega(k_x, k_y)$, with plane $\omega_{s-s} = \omega^0 + k_x v$.

Chapter 15
Phase Shift in Forced Waves

We considered this topic above, in the First level chapter, Sectphsh1. Now we continue to discuss this phenomenon. It is known from long ago; however, problems associated with the phase shift in different areas remain actual nowadays. We here refer to a few of many phase-shift-related publications. In particular, the shift appears in various light systems, even in the simplest case of the reflection at an inclined mirror, Creath and Schmit (2005); Hyun and Zhang (2016). And in optical communications, Jawla and Singh (2013). It is also known in medicine, Deacon and Arendt (1995). The phase shift transformation methods are discussed in Verboomen et al. (2005) and Cano et al. (2017). In this respect, mechanical waves (elastic waves in piezoelectric crystals) were considered in Alshits et al. (1984). This phenomenon is also studied in Slepyan (2022a).

15.1 Preliminary Remarks

Let a moving-oscillating point force

$$Q = Q_0 e^{i\omega t}\delta(x - vt), \tag{15.1}$$

act on an elastic body in y-direction. We here assume that, in a non-resonant case, the loaded point of the body also oscillates periodically (that is the force moves periodically in y-direction). Note that since free oscillations are also excited, this *main* period, $\Theta > \Theta_0 = 2\pi/\omega$. However, the only body frequency equal to the load frequency contributes to a Θ -averaged energy flux. Accordingly, we denote the energy-flux-related body response at that point as

$$w = Ae^{i(\omega t+\varphi)}, \quad -\pi < \varphi \leq \pi, \quad A = \text{const}, \tag{15.2}$$

L. I. Slepyan, *Mechanical Waves: Free, Forced, and Transition*,
https://doi.org/10.1007/978-3-032-15864-2_15

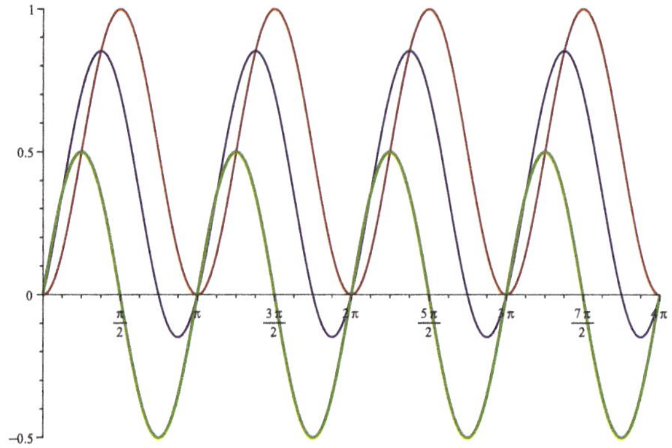

Fig. 15.1 The energy flux rate under the load $Q = \sin(t)$, $N = \sin(t) sin(t + \varphi)$. Top down: $\varphi = 0$, $\varphi = \pi/4$ and $\varphi = \pi/2$). In the latter case, the period-average energy flux is at zero

where w is the speed, A is its amplitude, and φ is the phase shift.

The corresponding rate of the energy flux from the force to the body

$$N = \Re(w(t)\overline{Q}) = \Re(Q\overline{w}(t)) = Q_0 A \cos\varphi. \tag{15.3}$$

For a moving-oscillating force, the frequency, ω, and the speed w correspond to those detected by an observer moving in x-direction together with the load.

In the same way, another energy-related couple, the external moment and the rotation rate, can form the energy flux.

The nonzero phase shift manifests itself by a negative part of the periodic energy flux rate, Fig. 15.1.

For a distributed load, the energy flux rate is equal to the integral of that shown in (15.3) over the load support.

Note that the negative $\cos\varphi$ corresponds to the opposite averaged energy flux, the flux from the body to the force. If this is not the case, the limits of the phase shift narrow

$$-\frac{\pi}{2} \leq \varphi \leq \frac{\pi}{2}. \tag{15.4}$$

Concerning the phase shift, it is clear that the averaged energy flux is at zero in the case of a non-resonant excitation of a finite elastic system (or a waveguide excitation within a bandgap). Hence, in such cases, the phase shifts $\varphi = \pm\pi/2$. So, a nonzero phase shift is not an exception, and the questions can be about its value. In any case, if the force excites propagating waves, the energy flux is nonzero, and $|\varphi| < \pi/2$.

When determining the phase, it is helpful to remember that in the complex representation of the wave, the argument is the phase, regardless of the modulus. In a product, the phase is equal to the sum of the arguments.

Finally, in this section, we note that the above definition, (15.2), leaves the following questions:

(1) What is the role of free oscillations? They do not contribute to the energy flux only *in* Θ*-averaged*. At the same time, their manifestation during the period and the *current phase shift* as that averaged over the load period, can also be of interest.
(2) How much is the main period corresponding to the *whole* motion formed by different-frequency oscillations?
(3) How does the phase shift manifest itself under resonant excitations?
(4) How does dissipation modify the considered dynamic processes?

At first, we will see what happens in this respect in the simplest model: with the harmonic excitation of an oscillator. Illustrations of phenomena associated with this elementary problem can be of interest by themselves and can help as a key introduction.

15.2 An Oscillator Under Harmonic Force

15.2.1 Regular Excitation

From the equation (here without dissipation)

$$\ddot{u}(t) + \omega_0^2 u(t) = \frac{Q_0}{m}\sin(\omega t), \tag{15.5}$$

where m is the oscillator mass, we have the speed of the point, where the force acts on the oscillator massless spring

$$w_{tot}(t) = \frac{Q_0\omega}{\varkappa}\left(\frac{\cos(\Omega\tau) - \cos(\tau)}{1-\Omega^2} + \cos(\Omega\tau)\right), \tag{15.6}$$

where $\Omega = \omega/\omega_0$, $\tau = \omega_0 t$, and $\varkappa$ is the spring stiffness. The load-frequency speed is

$$w(t) = \frac{Q_0\omega}{\varkappa}\frac{2-\Omega^2}{1-\Omega^2}\cos(\Omega\tau)). \tag{15.7}$$

The comparative dependencies of the speed and the energy flux rate

$$N_{tot} = w_{tot}(t)\sin(\omega t), \quad N = w(t)\sin(\omega t), \tag{15.8}$$

are plotted in Figs. 15.2, 15.3, 15.4 and 15.5.

These results show that the associated free oscillations can play an essential role in the energy flux. While it is at zero on average (in the non-resonant case), it can reach considerable value during a part of the main period, where the *current phase shift* is nonzero. The min period grows unboundedly as the load frequency gets closer to the

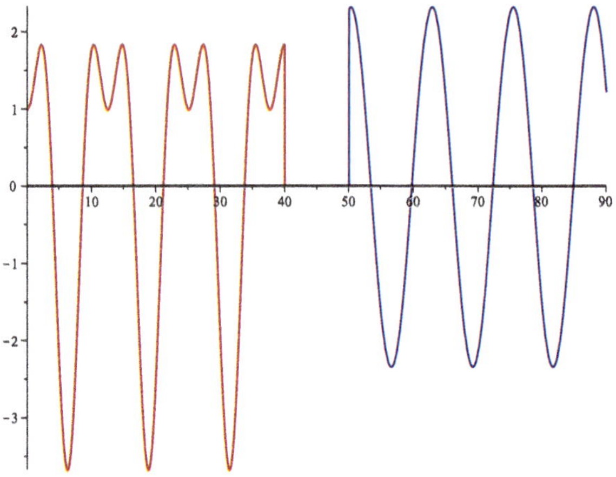

Fig. 15.2 The energy flux-related speed, $\Omega = \omega/\omega_0 = 0.5$. The *whole solution*, $w_{tot}(t)$ (brown, left), and the load-frequency one, $w(t)$ (navy, right)

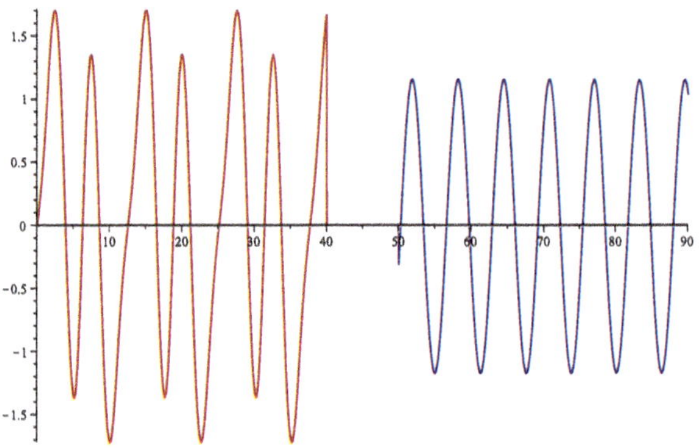

Fig. 15.3 The energy flux rate, $\Omega = \omega/\omega_0 = 0.5$. The *whole solution*, $N_{tot}(t)$ (brown, left), and the load-frequency one, $N(t)$ (navy, right)

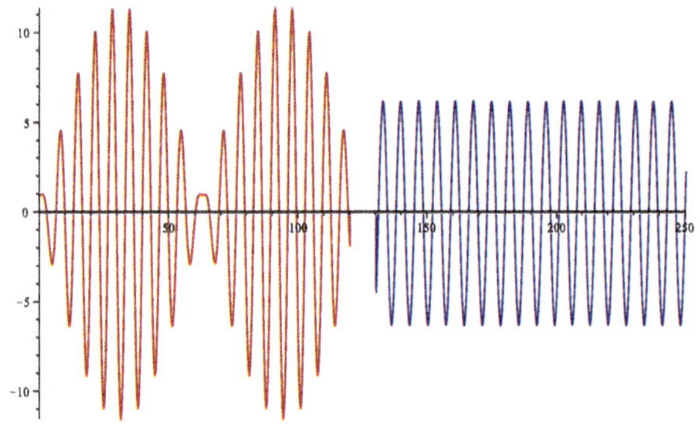

Fig. 15.4 The energy flux-related speed, $\Omega = \omega/\omega_0 = 0.9$. The *whole solution*, $w_{tot}(t)$ (brown, left), and the load-frequency one, $w(t)$ (navy, right)

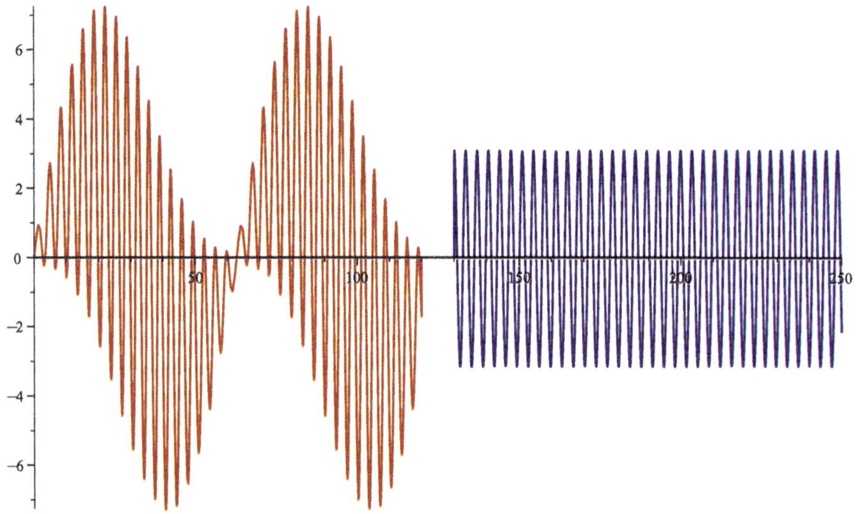

Fig. 15.5 The energy flux rate, $\Omega = \omega/\omega_0 = 0.9$. The *whole solution*, $N_{tot}(t)$ (brown, left), and the load-frequency one, $N(t)$ (navy, right). It is clear that the *current phase shift* (the phase shift averaged over the load period) is positive during a half the main period

oscillator frequency (the resonant case is considered below). At the same time, the phase shift (here $\varphi = \pi/2$) cardinally affects the process. It defines the low-frequency periodicity and the period-averaged quantities.

15.3 Phase Shift in Resonant Waves

Here the load-frequency speed expression (15.7) loses the force, and only the whole one (15.6) should be used. From the latter we see that in the resonant case, $\Omega = 1$, the phase shift quickly disappears (see Fig. 15.6)

$$w_{tot}(t) = \frac{Q_0\omega}{\varkappa}\lambda\left(\frac{1}{2}\tau\sin(\tau) + \cos(\tau)\varrho\right), \tag{15.9}$$

and the energy flux rate linearly grows (see Fig. 15.7)

$$N = \frac{Q_0\omega}{2\varkappa}(\tau\sin^2(\tau) + \sin(2\tau)). \tag{15.10}$$

What is remarkable, there is a discontinuity in the phase shift at $\Omega = 1$:

$$\varphi = \pi/2\ (0 < \Omega < 1), \quad \varphi = 0\ (\Omega = 1). \tag{15.11}$$

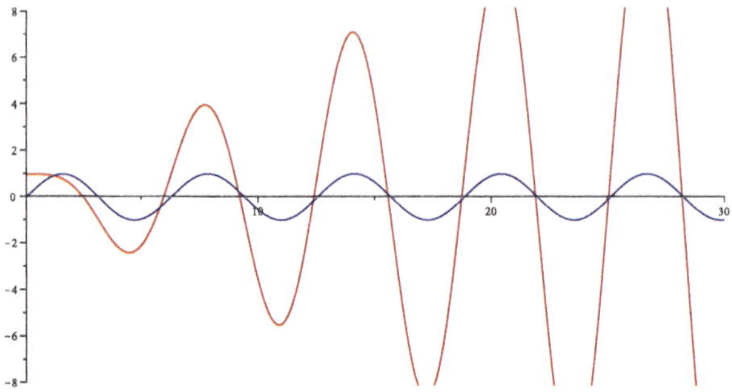

Fig. 15.6 The phase shift under the resonant oscillations. The oscillator growing-amplitude speed (brown) and the sinusoidal load (navy)

Fig. 15.7 The energy flux rate under the resonant excitation

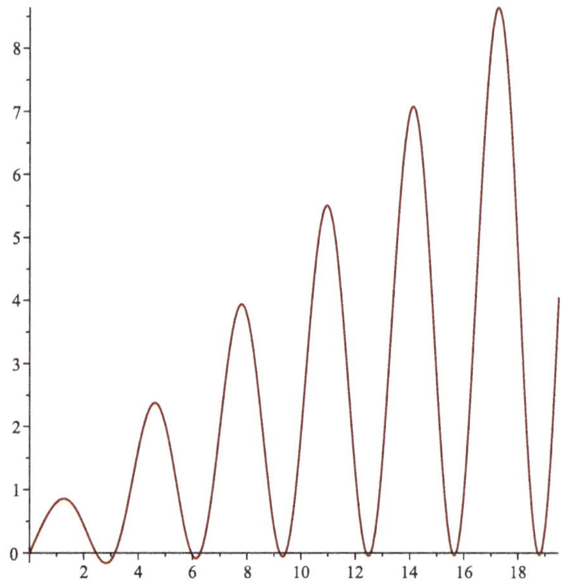

15.3.1 *The Effect of Dissipation*

From the Eq. (15.5) with dissipation

$$\ddot{u}(t) + \omega_0^2 u(t) + \alpha \dot{(}u(t) = \frac{Q_0}{m} \sin(\omega t), \tag{15.12}$$

it can be seen how low dissipation suppresses the free oscillations and getting the two formulations closer, see (Figs. 15.8 and 15.9).

Thus, even under low dissipation, the load-frequency (*partial*) solution quite quickly coincides with the whole one.

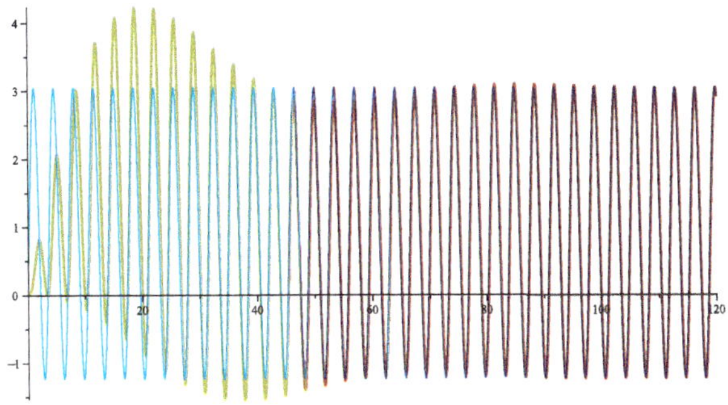

Fig. 15.8 The energy flux rate under the excitation of the oscillator with dissipation. The load frequency $\Omega = 0.9$; the dissipation number $\alpha = 0.1$. The whole solution (brown, variable amplitude) and the load-frequency one (navy, constant amplitude)

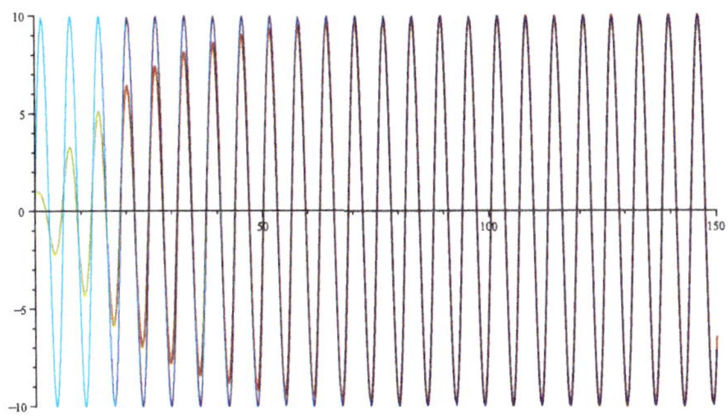

Fig. 15.9 The energy flux rate under the resonant excitation of the oscillator with dissipation. The load frequency $\Omega = 1$; the dissipation number $\alpha = 0.1$. The whole solution (brown, variable amplitude) and the load-frequency one (navy, constant amplitude)

15.4 The Phase Shift in General

At first, consider an unmoving load. Represent the displacement at the action point, $u(t)$, as a convolution of corresponding Green's function, $G(t)$, with the load, $Q = \mathrm{e}^{\mathrm{i}\omega t}$:

$$u(t) = G(t) * \mathrm{e}^{\mathrm{i}\omega t}. \tag{15.13}$$

The Laplace transform (with the parameter s) gives us

$$u^L(s) = G^L(s)\frac{1}{s - \mathrm{i}\omega}, \quad \dot{u}^L(s) = G^L(s)\frac{s}{s - \mathrm{i}\omega}. \tag{15.14}$$

We are looking for the phase shift at the load frequency. So, only the pole at $s = \mathrm{i}\omega$ is meaningful to us. In a non-resonant case, Green's function integral-transformed is regular at this point. We find

$$\dot{u}(t) = \mathrm{i}\omega G^L(\mathrm{i}\omega)\mathrm{e}^{\mathrm{i}\omega t}. \tag{15.15}$$

Thus, for a non-growing Green's function the phase shift is

$$\varphi = \arg(G^L(\mathrm{i}\omega)) + \frac{\pi}{2}. \tag{15.16}$$

15.5 Phase Shift in Forced Waves

15.5.1 General Relations

In an analytical study of forced wave propagating along x-axis, usually the double, Fourier and Laplace transform is used. In this terms,

$$Q^{LF} = \frac{1}{s - \mathrm{i}\omega}, \quad u^L(s,0) = \frac{1}{2\pi}\int_{-\infty}^{\infty} u^{LF}(s,k)\,\mathrm{d}k, \tag{15.17}$$

the expressions in (15.14) become

$$\begin{aligned} u^L(s,0) &= \frac{1}{s-\mathrm{i}\omega}\frac{1}{2\pi}\int_{-\infty}^{\infty} G^{LF}(s,k)\,\mathrm{d}k, \\ \dot{u}^L(s,0) &= \frac{\mathrm{i}\omega}{s-\mathrm{i}\omega}\frac{1}{2\pi}\int_{-\infty}^{\infty} G^{LF}(s,k)\,\mathrm{d}k, \end{aligned} \tag{15.18}$$

and the phase shift can be represented as

$$\varphi = \arg\left(\int_{-\infty}^{\infty} G^{LF}(0+\mathrm{i}\omega)\,\mathrm{d}k\right) + \frac{\pi}{2}, \tag{15.19}$$

where, as usually adopted, $0 + \mathrm{i}\omega = \lim(s' + \mathrm{i}\omega)$ $(0 < s' \to 0)$.

We write such a 'prelimiting' expression, because at $s' = 0$ the integrands can have real poles corresponding to the propagating waves excited by the load. See Fig. 15.10, where the wave frequencies and wavenumbers correspond to crossings of the load line $\omega + kv$ with the dispersion curve. At the same time, the opposite order of the inverse Fourier transform and the substitution $s \to \mathrm{i}\omega$ can be more convenient, especially if the integrals in (15.18) allow us to obtain an explicit expression of $\dot{u}^L(s,0)$

$$\dot{u}^L(s,0) = \frac{s}{s-\mathrm{i}\omega}\frac{1}{2\pi}\int_{-\infty}^{\infty} G^{LF}(s,k)\,\mathrm{d}k = \frac{\Omega(s)}{s-\mathrm{i}\omega}\int_{-\infty}^{\infty} f(k)\,\mathrm{d}k. \tag{15.20}$$

In the case, where $f(k) \geq 0$ and the integral exists, we have no need in the latter, and

$$\varphi = \arg(\Omega(\mathrm{i}\omega)). \tag{15.21}$$

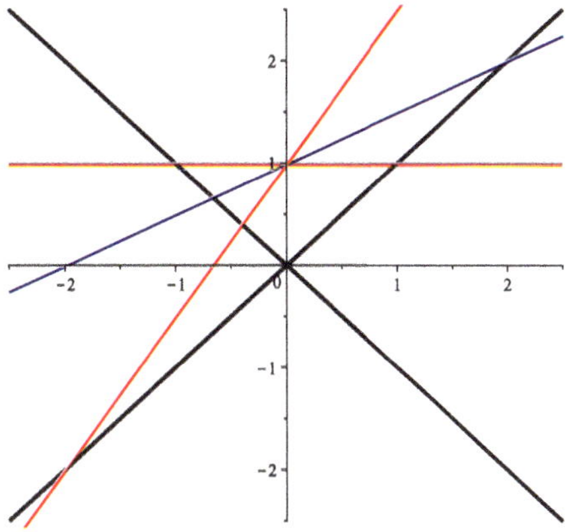

Fig. 15.10 Dispersion dependencies for a string, $\omega_0 = \pm ck$. The load lines, $\omega_0 = \omega + kv$: $v = 0$ (brown), $0 < v < c$ (subsonic, navy) and $v > c$ (supersonic, red) (here $c = \omega = 1$)

15.5.2 Three Examples

From the equation for a *free string* under a unit-amplitude oscillating force

$$\varrho \frac{\mathrm{d}^2 u}{\mathrm{d}t^2} - T \frac{\mathrm{d}^2 u}{\mathrm{d}x^2} = Q_0 \delta(x) \mathrm{e}^{\mathrm{i}\omega_0 t}, \tag{15.22}$$

where ϱ and T are the mass per unit length and the tension force, respectively, we have

$$G^{LF}(s,k) = \frac{1}{\varrho s^2 + Tk^2}, \tag{15.23}$$

and returning to the expressions in (15.20), (15.21) we find

$$\Omega(s) = 1 \quad \Longrightarrow \quad \varphi = 0. \tag{15.24}$$

For a *free bending beam* the equation is

$$\varrho \frac{\mathrm{d}^2 u}{\mathrm{d}t^2} + D \frac{\mathrm{d}^4 u}{\mathrm{d}x^4} = Q_0 \delta(x) \mathrm{e}^{\mathrm{i}\omega_0 t}, \tag{15.25}$$

where D is the bending stiffness. In the same way as above for the string, we find

$$\Omega(s) = s^{-1/2} \quad \Longrightarrow \quad \varphi = -\pi/4. \tag{15.26}$$

Another way allows to demonstrate how the phase shift arises more transparently. The energy flux-related displacement and the corresponding equation are

$$u = U(x)e^{i\omega_0 t}, \quad \frac{d^4U(x)}{dx^4} - \Omega^2 U(x) = \frac{Q_0}{D}\delta(x), \tag{15.27}$$

with

$$\Omega = \sqrt{\varrho/D}\,\omega, \quad U(-x) = U(x), \quad U'(0) = 0, \quad U'''(+0) = Q_0/(2D). \tag{15.28}$$

A non-growing (as $x \to \infty$) solution is

$$u(t,x) = -\frac{Q_0 i}{4D\Omega^{3/2}} e^{i\omega_0 t}\left(e^{-i\sqrt{\Omega}x} - ie^{-\sqrt{\Omega}x}\right) \quad (x > 0), \tag{15.29}$$

and

$$\frac{du(t,0)}{dt} = \frac{Q_0}{4D\sqrt{\Omega}} e^{i\omega_0 t}(1 - i) = \frac{Q_0\sqrt{2}}{8D\sqrt{\Omega}} \exp(i(\omega_0 t - \pi/4)). \tag{15.30}$$

Thus, the phase shift is $\varphi = -\pi/4$, and it arises due to the presence of the localized partial solution.

The phase shift in the case of a bending wave was discovered in Slepyan (2022), where the calculations showed a negative part of the periodic energy flux rate. The phase shift is demonstrated in Fig. 15.11.

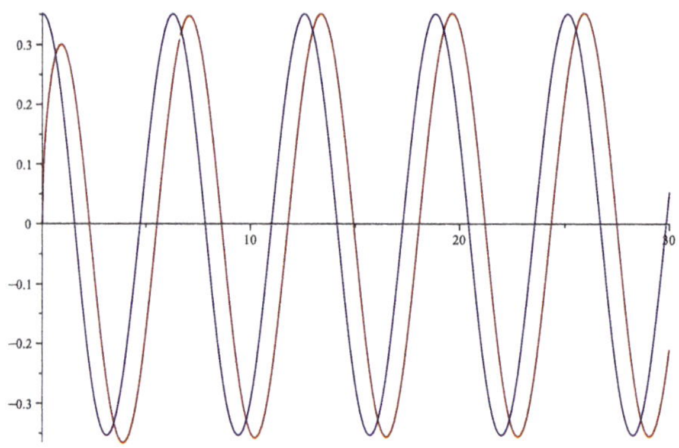

Fig. 15.11 Dispersion dependencies $\omega(k)$ for a string with distributed oscillators. The two branches correspond to the non-dimensional equation $(k^2 - \omega^2)(1 - \omega^2) = \omega^2$ ((15.39) with $\varrho = \omega_{osc} = \varkappa = 1$). The brown branches can be called "The string under the oscillator's influence," and vice versa for the navy one. The gray strips, $1 < \omega < \sqrt{2}$ and symmetric one in the lower half-plane of ω, are the bandgaps

Next, consider the equation of a *beam on an elastic support* ($\varkappa$ is its stiffness)

$$\varrho\frac{\mathrm{d}^2u}{\mathrm{d}t^2}+D\frac{\mathrm{d}^4u}{\mathrm{d}x^4}+\varkappa u=\mathrm{e}^{\mathrm{i}\omega_0 t}. \tag{15.31}$$

The same as above, using the integral transforms, we have

$$\Omega(s)=s\left(s^2+\frac{\varkappa}{\varrho}\right)^{-1/4}. \tag{15.32}$$

It follows

$$\Omega(\mathrm{i}\omega)=\mathrm{i}\omega(\frac{\varkappa}{\varrho}-\omega^2)^{-1/4}. \tag{15.33}$$

Thus,

$$\varphi=\frac{\pi}{2}\ (\omega^2<\varkappa/\varrho),\quad \varphi=-\frac{\pi}{4}\ (\omega^2>\varkappa/\varrho). \tag{15.34}$$

Note that the strip $\omega^2<\varkappa/\varrho$ is the bandgap, oscillations inside which excite no waves. So, in this case, there is no (period-averaged) energy flux, which is just reflected by the first equality in (15.34).

15.5.3 The Role of a Dynamic Structure Attached

The LF-transformed equation with an inertia term 'modified' (Milton & Willis, 2007) by the response of the dynamic (micro) structure (with no interconnections) can be represented as

$$\mathcal{L}^{LF}(s,k)u^{LF}(s,k)=\frac{Q_0}{s-\mathrm{i}\omega_0},\quad \mathcal{L}^{LF}(s,k)=\Omega^L(s)+Ck^{2n}, \tag{15.35}$$

where the inertia term, $\Omega^L(s)>0\,(s>0)$, is formed by both the waveguide inertia and the structure response, $k=1$ for the string, and $n=2$ for the beam, $C>0$ is a constant.

Substituting $k=(\Omega^L(s)/C)^{1/2n}k'$ in the inverse Fourier transform, we obtain the phase shift defining function

$$\mathcal{L}^L(\mathrm{i}\omega_0)=(\Omega^L(\mathrm{i}\omega_0))^{1-1/2n}, \tag{15.36}$$

where also some positive multipliers are omitted.

In particular, if the structure consists of uniformly distributed oscillators,

$$\Omega^L(s) = \varrho s^2 \left(1 + \frac{\varkappa s^2}{\varrho(s^2 + \omega_{osc}^2)}\right), \quad \omega_{osc} = \sqrt{\frac{\varkappa}{m}}, \tag{15.37}$$

where $\varkappa$, m and ω_{osc} are the oscillator spring stiffness, mass (both per unit length) and frequency, respectively. It follows

$$\Omega^L(\mathrm{i}\omega) = -\varrho\omega_0^2 \left(1 + \frac{\varkappa}{\varrho(\omega_{osc}^2 - \omega_0^2)}\right) < 0 \quad \left(\frac{\varkappa}{\varrho(\omega_{osc}^2 - \omega_0^2)} > -1\right). \tag{15.38}$$

So, in this case, the presence of the oscillators do not affect the phase shift angle. Otherwise,

$$\Omega^L(\mathrm{i}\omega) > 0 \quad (0 < \varrho(\omega_0^2 - \omega_{osc}^2) < \varkappa). \tag{15.39}$$

In this case, in accordance with its definition Sect. 15.4, $\varphi = \pi/2$ for any $n > 0$, which means that there is no energy flux to the waveguide.

Thus, in a range of frequencies, such a dynamic structure can block the wave, as it should be. Indeed, the inequalities (15.39) correspond to the bandgaps in the dispersion relations, as can be seen in Fig. 15.12 for the string-with-oscillators waveguide.

A *waveguide on an elastic foundation* is a partial case corresponding to the infinite mass of the oscillator, $\omega_{osc} = 0$, $\varkappa > 0$. For this structure

$$\Omega^L(\mathrm{i}\omega) = \varkappa - \varrho\omega_0^2. \tag{15.40}$$

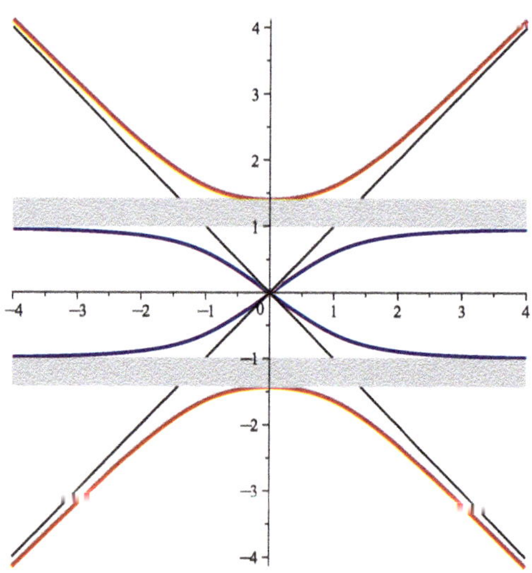

Fig. 15.12 Dispersion dependencies $\omega(k)$ for a string with distributed oscillators. The two branches correspond to the non-dimensional equation $(k^2 - \omega^2)(1 - \omega^2) = \omega^2$ (15.39) with $\varrho = \omega_{osc} = \varkappa = 1$). The brown branches can be called "The string under the oscillator's influence," and vice versa for the navy one. The gray strips, $1 < \omega < \sqrt{2}$ and symmetric one in the lower half-plane of ω, are the bandgaps

Similarly to the above finite-mass oscillators system, the sign of $\Omega^L(\mathrm{i}\omega)$ depends onω

$$\Omega^L(\mathrm{i}\omega) < 0 \quad (\varkappa < \varrho\omega_0^2), \quad \Omega^L(\mathrm{i}\omega) > 0 \quad (\varkappa > \varrho\omega_0^2), \tag{15.41}$$

where the latter inequality corresponds to the bandgap.

15.5.4 *Moving-Oscillating Load*

15.5.4.1 Steady-State Waves

Now the load and displacement are

$$\begin{gathered} Q = \mathrm{e}^{\mathrm{i}\omega_0 t}\delta(x - vt), \\ u^{LF}(s,k) = (u^F(t,k))^L = \frac{G^{LF}(s,k)}{s - \mathrm{i}(\omega_0 + kv)}, \end{gathered} \tag{15.42}$$

and the load frequency-related Laplace inverse transform leads to

$$u^F(t,k) = G^{LF}(0 + \mathrm{i}(\omega_0 + kv), k)\mathrm{e}^{\mathrm{i}(\omega_0 + kv)t}. \tag{15.43}$$

The displacement at the moving load point, $x = vt$, is

$$u(t, vt) = \frac{1}{2\pi}\int_{-\infty}^{\infty} G^{LF}(0 + \mathrm{i}(\omega_0 + kv)\,\mathrm{d}k\,\mathrm{e}^{\mathrm{i}\omega_0 t}, \tag{15.44}$$

and the velocity at that point

$$w \equiv \frac{\mathrm{d}}{\mathrm{d}t}u(t, vt) = \mathrm{i}\omega\frac{1}{2\pi}\int_{-\infty}^{\infty} G^{LF}(0 + \mathrm{i}(\omega_0 + kv)\,\mathrm{d}k\,\mathrm{e}^{\mathrm{i}\omega_0 t}. \tag{15.45}$$

Thus, the phase shift

$$\varphi = \arg\left(\int_{-\infty}^{\infty} G^{LF}(0 + \mathrm{i}(\omega + kv), k)\,\mathrm{d}k\right) + \frac{\pi}{2}. \tag{15.46}$$

15.5.4.2 The Wave in a String

For the *free string* we now have that at $x = vt$

$$G^L(\mathrm{i}\omega_0) = \frac{1}{2\pi\varrho}\lim_{\varepsilon\to 0}\int_{-\infty}^{\infty}\frac{\mathrm{d}k}{(\varepsilon + \mathrm{i}(\omega_0 + kv)^2 + c^2k^2} = \frac{1}{2\mathrm{i}\varrho c\,\omega_0} \quad (\omega_0 > 0, 0 < v < c, \varepsilon > 0). \tag{15.47}$$

So, independently of the (subsonic) speed of the load, $\varphi = 0$ (see 15.43).

15.5.4.3 The Wave in a Beam

The Green's function for a *free beam* is defined as

$$G^{LF} = \frac{1}{\varrho(s + \mathrm{i}kv)^2 + Dk^4}. \tag{15.48}$$

Assuming $s > 0$ we substitute $k = k's$ (the prime is omitted in the following) we represent

$$G^L(s) = \frac{1}{2\pi\varrho s^{3/2}}\int_{-\infty}^{\infty}\frac{\mathrm{d}k}{(1 + \mathrm{i}kvs^{-1/2})^2 + \frac{D}{\varrho}k^4}. \tag{15.49}$$

Next, for $s \to \mathrm{i}\omega_0$

$$\begin{aligned}\mathrm{i}\omega_0 G^{L_\eta}(\mathrm{i}\omega_0) &= \frac{1}{2\pi\varrho\sqrt{\omega_0}}\left(\frac{\varrho}{D}\right)^{1/4}\sqrt{-\mathrm{i}}\,\Psi,\\ \Psi &= \int_{-\infty}^{\infty}\frac{1}{(1 + \sqrt{\mathrm{i}}kV)^2 + k^4}\,\mathrm{d}k,\end{aligned} \tag{15.50}$$

where the non-dimensional speed

$$V = \frac{v}{\sqrt{\omega_0}}\left(\frac{\varrho}{D}\right)^{1/4}. \tag{15.51}$$

Thus, for the bending beam, the phase shift is

$$\varphi = \varphi(V) = \arg(\Psi(V)) - \frac{\pi}{4}. \tag{15.52}$$

This dependence is plotted in Fig. 15.13. Growing from $-\pi/4$ at $V = 0$ it becomes zero at $V = \sqrt{2}$.

The energy flux rates obtained for a cosinusoidal force are plotted in Figs. 15.14 and 15.15 for the load speeds $V = 0$ ($\varphi = -\pi/4$), and $V = \sqrt{2}$ ($\varphi = 0$), respectively.

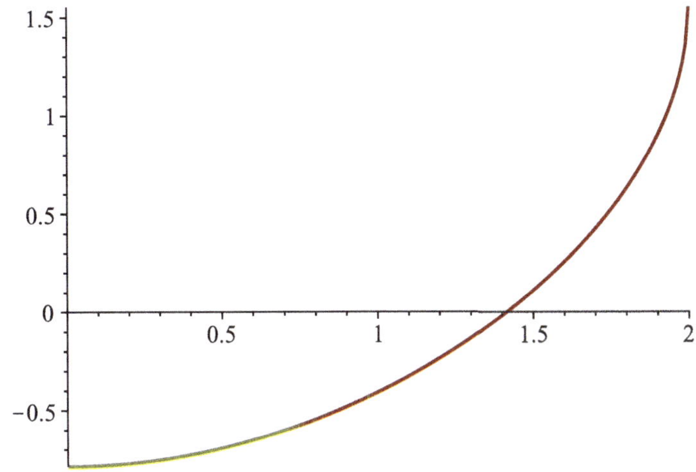

Fig. 15.13 The phase shift in the beam excitation as a function of the (non-dimensional) load speed, $0 \leq \varphi < 2$. In particular, $\varphi = -\pi/4\,(V = 0)$, $\varphi = 0\,(V = \sqrt{2})$, $\varphi \to \pi/2\,(V \to 2)$

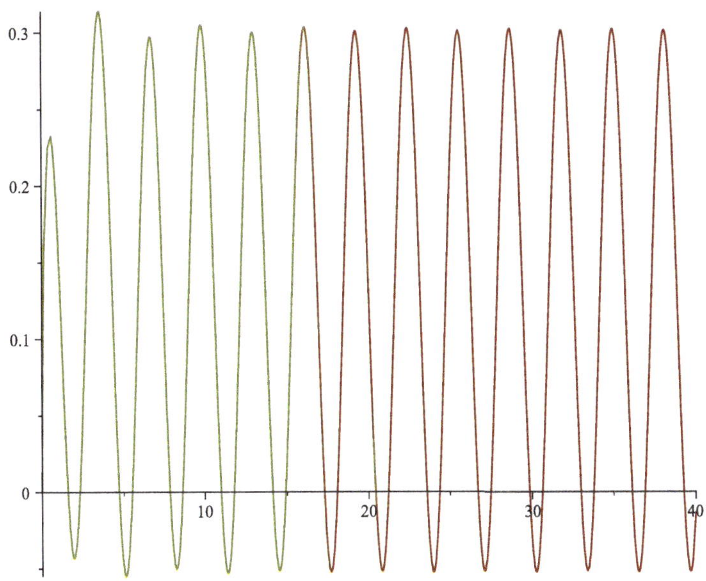

Fig. 15.14 The energy flux rate in the beam excitation for the unmoving force, $V = 0$, where the phase shift $\varphi = -\pi/4$. The partly-negative energy flux reflects the nonzero phase shift

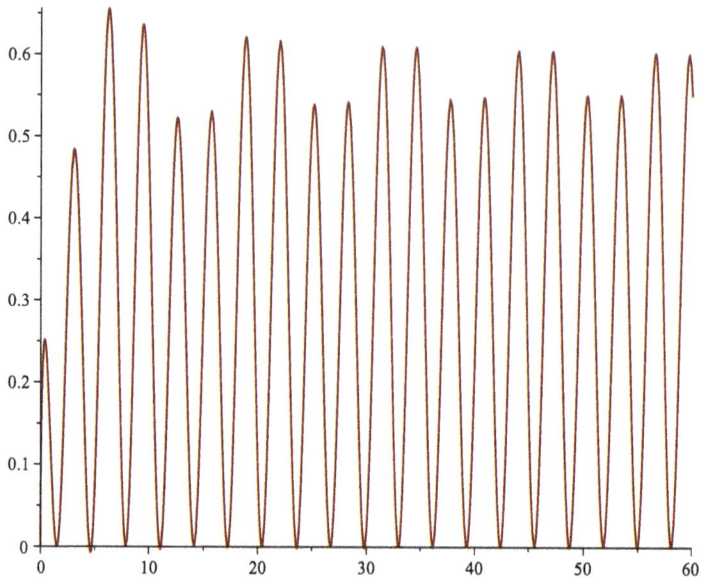

Fig. 15.15 The energy flux rate in the beam excitation for the (non-dimensional) load speed $V = \sqrt{2}$. There is no negative part here that reflects the zero phase shift

15.5.4.4 A Beam on an Elastic Foundation. Sub and Resonant Excitations by a Moving Load

We consider the waves excited by a non-oscillating force moving along the beam with sub-resonant and resonant speeds. The equation is

$$\varrho \frac{\partial^2 u(t,x)}{\partial t^2} + D \frac{\partial^4 u(t,x)}{\partial x^4} + \varkappa u(t,x) = Q_0 \delta(x - vt), \tag{15.53}$$

where $\varkappa$ is the foundation stiffness. In the following, we use non-dimensional variables. We take the values $(D/\varkappa)^{1/4}$ and $\sqrt{\varrho/\varkappa}$ as the length and time units, respectively, and $(Q_0/\sqrt{\varkappa D}$ as the unite external force. Being expressed in these terms, the equation becomes as (Fig. 15.16)

$$\frac{\partial^2 u(t,x)}{\partial t^2} + \frac{\partial^4 u(t,x)}{\partial x^4} + u(t,x) = \delta(x - vt). \tag{15.54}$$

Using the Fourier transform we have

$$\frac{\partial^2 u^F(t,k)}{\partial t^2} + k^4 u^F(t,k) + u^F(t,k) = \mathrm{e}^{\mathrm{i}kvt}. \tag{15.55}$$

From here for zero initial conditions

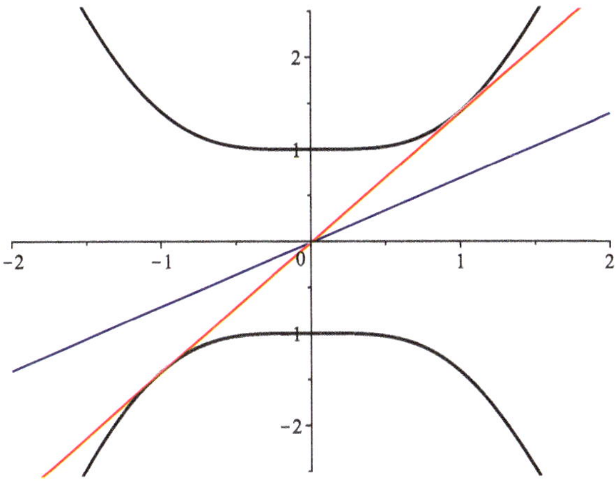

Fig. 15.16 The beam on an elastic foundation. The dispersion curves, $\omega = \pm\sqrt{k^4+1}$ and the moving load lines: the sub-resonant speed, $\omega = 0.7k$ (navy), and the resonant speed, $\omega = \sqrt{2}k$ (red), where the units are used when transforming the Eqs. (15.53)–(15.55)

$$
\begin{aligned}
u^F(t,k) &= \frac{S}{k^4 - v^2k^2 + 1},\\
S &= e^{ikvt} - \cos(\sqrt{k^4+1}\,t) - \frac{ikv}{\sqrt{k^4+1}}\sin(\sqrt{k^4+1}\,t).
\end{aligned}
\tag{15.56}
$$

The displacements at the moving load point

$$
u(t,vt) = \frac{1}{2\pi}\int_{-\infty}^{\infty} \frac{Se^{-ikvt}}{k^4 - v^2k^2 + 1}\,dk,
\tag{15.57}
$$

where for sub-resonant speeds, $v < v_* = \sqrt{2}$, the real and imaginary parts of the function

$$
Se^{-ikvt} = 1 - \left(\cos(\sqrt{k^4+1}\,t) - \frac{ikv}{\sqrt{k^4+1}}\sin(\sqrt{k^4+1}\,t)\right)e^{-ikvt},
\tag{15.58}
$$

are even and odd functions of k, respectively, and $k^4 - v^2k^2 + 1 > 0$. So, $\Im u(t, vt) = 0$. The steady-state solution is the asymptote

$$
u(t, vt+\eta) \sim \frac{1}{2\pi}\int_{-\infty}^{\infty} \frac{\cos(k\eta)}{k^4 - v^2k^2 + 1}\,dk \quad (t \to \infty),
\tag{15.59}
$$

In this case, the displacements shape localized at the load point area moves together with the load, and the latter does not work (the displacements for $v = 1$ are shown in Fig. 15.17). So, for the sub-resonant speed of the load the energy flux is at zero.

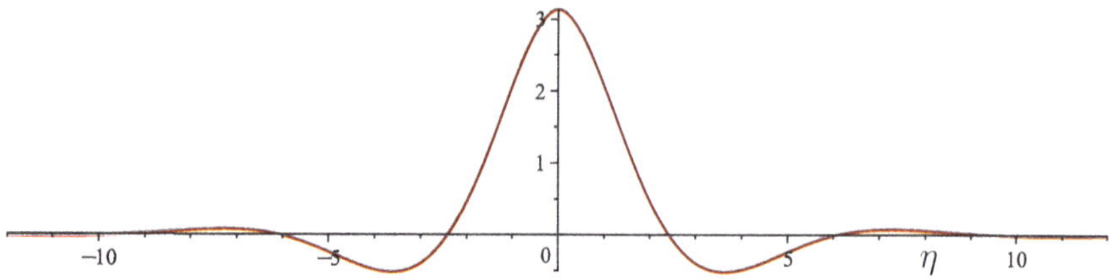

Fig. 15.17 The beam on an elastic foundation. The steady-state displacement under a force moving at a subsonic speed, $v = 0.5$ ($\eta = x - vt$)

Thus

$$\varphi = \frac{\pi}{2} \quad (0 \le v < \sqrt{2}. \tag{15.60}$$

15.5.5 The Resonant Wave

Now

$$u(t, vt) = \frac{1}{2\pi}\int_{-\infty}^{\infty} \frac{R}{(k^2-1)^2} e^{-ikvt}\, dk, \tag{15.61}$$

where $R = S$, (15.56), at $v = \sqrt{2}$.

The above note about the real and imaginary parts of Se^{-ikvt} remains valid for the resonant case too. So, $\Im u(t, vt) = 0$. That also valid for the displacement rate

$$\frac{d}{dt}u(t, vt) = \frac{1}{2\pi}\int_{-\infty}^{\infty} \frac{\sin(\sqrt{k^4+1}t)\cos(\sqrt{2}\,kt)}{\sqrt{k^4+1}}\, dk. \tag{15.62}$$

The resonant displacements and the displacement speed together with of the same amplitude squire root type dependencies are presented in Figs. 15.18 and 15.19. Clearly, it is hard to distinguish differences.

Finally, there is no phase shift, $\varphi = 0$, in the considered excitation of the beam by the non-oscillation load moving at the resonant speed. So, as the load speed reaches the resonant value, $v \to \sqrt{2}$, the pase shift has a jump discontinuity, from $\pi/2$ to 0, similarly as it was in the case of the oscillator considered in Sect. 1.3.7.

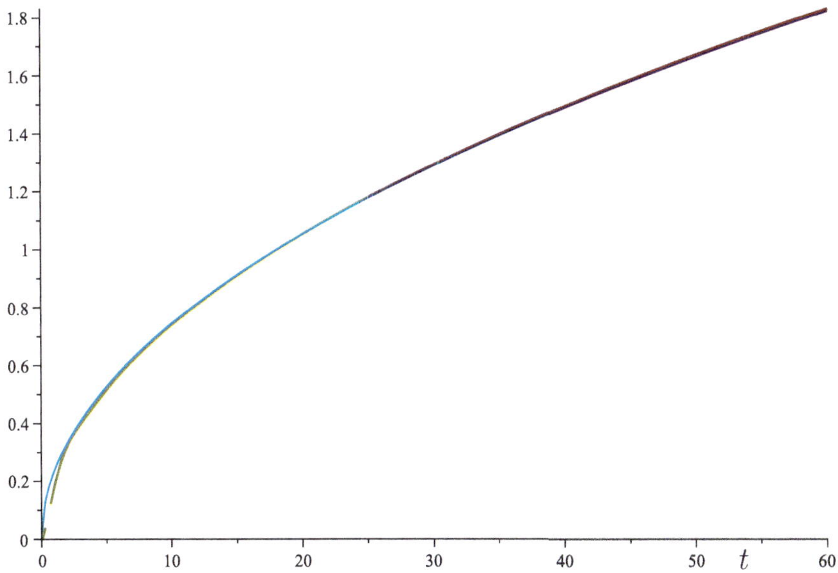

Fig. 15.18 The beam on an elastic foundation. The $\sqrt{t}$-displacement under a force moving at the resonant speed, $v = \sqrt{2}$ (brown), and $0.236\sqrt{t}$ (navy)

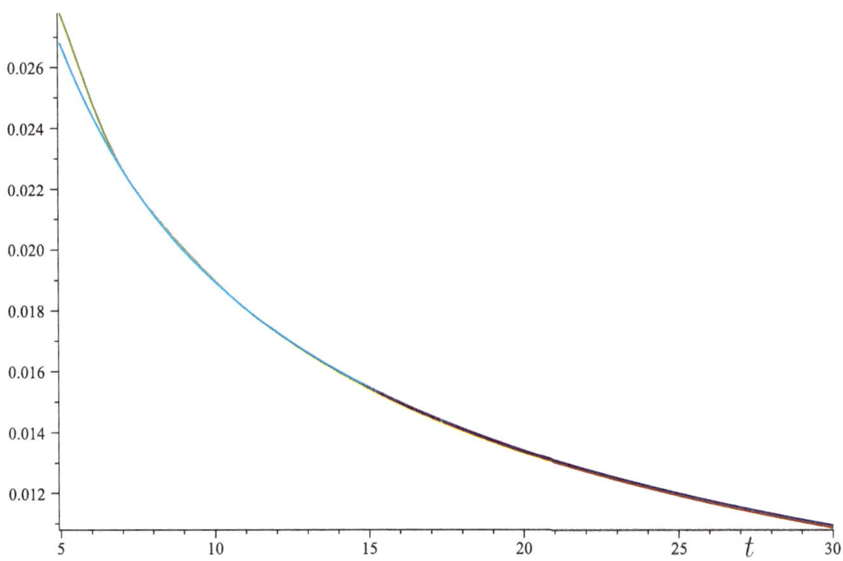

Fig. 15.19 The beam on an elastic foundation. The steady-state displacement speed under a force moving at the resonant speed, $v = \sqrt{2}$ (brown) and $0.060/\sqrt{t}$ (navy)

15.5.6 Conclusions

As the difference between the load and the corresponding velocity phases, the phase shift plays a significant role in mechanical forced oscillations and waves, at least in the energy flux from the load to the waveguide. Under a harmonic excitation, the period-averaged value of the shift is responsible for the same averaged energy flux. At the same time, due to the presence of the free oscillations, the main period can be long (it unboundedly increases as the load frequency gets closer to the resonant point). For this reason, the *current phase shift*, as that averaged over the load oscillation period, also can be considered. Comparative graphs of the load-frequency oscillations and the whole field are shown.

Some general aspects of this topic are presented. In particular, the phase shift following from the Laplace-Fourier-transformed Green's function. The energy flux rates (closely related to the phase shift) are demonstrated. In this way, we considered the forced oscillator and waves in a free string, string with a structure, and in a bending beam. Regular and resonant excitations are considered.

Along with this, the phase shift theme in mechanical waves remains mainly open. Primarily these concerns more complicated structures and the two-three-dimensional problems.

15.6 The Causality Principle for Steady-State Solutions

A steady-state solution corresponding to a given speed v is not unique if there exists a homogeneous solution as a free wave with the same phase speed. The existence of such waves is reflected by real singular points in the $F\eta$-transform of the solution. Indeed, let the steady-state solution obtained as it is, without the above-discussed reference to the underlying transient problem, be defined by the equation as

$$L(\mathrm{i}(kv+\omega^0), k)u_0^{F\eta} = Q_0^{F\eta}(k). \tag{15.63}$$

If there is no zero point of $L(\mathrm{i}(kv+\omega^0), k)$ on the real k-axis we can divide both sides of the equation by L. As a result

$$u_0^{F\eta} = \frac{Q_0^{F\eta}(k)}{L(\mathrm{i}(kv+\omega^0), k)}. \tag{15.64}$$

Otherwise, we have no right to do this. In this latter case, if $L(\mathrm{i}(kv+\omega^0), k)$ has a simple zero at $k = k_0$, there exists a homogeneous solution as a complex wave

$$u_0^{LF\eta}(k) = C\delta(k-k_0), \quad u_0(\eta) = \frac{C}{2\pi}\mathrm{e}^{-\mathrm{i}k_0\eta}, \tag{15.65}$$

where C is a constant. (In the case of the zero of a higher order, the derivatives of the delta function can also present in the $F\eta$-transform, which results in multiplication of the original by $\mathrm{i}\eta$.) However, the complete, inhomogeneous solution appears uncertain.

To address this issue, an additional condition is needed that can be extracted from the physical setting of the problem. In this connection, we assume that the steady state, if exists, is a 'final', limiting or established stage of a transient motion having an origin. The *causality principle* states that *the steady-state solution should be considered as the limit of the corresponding transient solution with zero initial conditions*. In general, if an oscillating factor presents in the transient solution, the limit is assumed to correspond to the solution without this factor as was considered in Sect. 14.2.

It follows from the causality principle as stated in the above *narrow sense causality principle, narrow sense* that the solution should not contain waves carrying energy from infinity or waves exponentially growing to infinity. Indeed, due to zero initial conditions such waves never have time to come from infinity or to be created exponentially growing to infinity. Thus, the narrow-sense causality principle corresponds to the case where no remote energy source is assumed. In a *broad sense,causality principle, broad sense* the causality principle permits all waves whose *remote sources* are prescribed by the problem formulation. If some remote sources exist the corresponding waves carrying energy from infinity are accepted, whereas the other part of the solution should meet the principle in its narrow sense.

The rule following from this principle states that the complex $\mathrm{i}(\omega t - kx)$ appearing in the steady-state solution, where both terms or only one of them are assumed to be nonzero, must be supplemented by $+\mathrm{i}0$, that is it must be considered as the limit from the right. In other words, *the real frequency, ω, must be supplemented by a vanishing negative imaginary part*, that is, ω must be replaced by $\omega - i0$. The latter formulation was suggested by Lighthill (1978). Thus, the rule can be symbolically expressed as

$$\mathrm{i}\omega = \mathrm{i}(kv + \omega^0) \quad \longrightarrow \quad s' + \mathrm{i}\omega \quad \longrightarrow \quad 0 + \mathrm{i}(kv + \omega^0) \quad (s' = +0). \tag{15.66}$$

Considering a prelimiting phase, $s' + \mathrm{i}(kv + \omega^0)$ $(s' \to +0)$, instead of that with $s' = 0$, we make an infinitesimal shift of the singular point from the real axis up or down. This is the same as to indicate a way in the inverse Fourier transform how to round this point. First such rule was posted by Sommerfeld (1949) for a specific equation. The rule is known as *Sommerfeld radiation condition*: "no energy may be radiated from infinity". In his specific case, this means that the phase velocity may not be directed from infinity. Under the *Mandelshtam radiation principle*, Mandelshtam (1972) the group velocity may not be directed from infinity (recall that these velocities can have opposite directions, whereas the energy flux velocity coincides with the latter). The causality principle is also called the *principle of a limiting amplitude* (see Bolotovsky and Stolyarov (1972)). It is still valid when the group velocity is zero.

For an unmoving nonoscillating load the static limit, if exists, can be obtained considering the load having a vanishing imaginary frequency, namely, as follows from the rule (15.66) the frequency must be taken as

$$\omega = -\mathrm{i}\varepsilon, \quad \varepsilon \to +0 \quad \left(\mathrm{e}^{\mathrm{i}\omega t} \quad \longrightarrow \quad \mathrm{e}^{\varepsilon t}, \ \varepsilon \to +0\right). \tag{15.67}$$

15.6.1 *Some Examples of the Steady-State Waves Under a Moving-Oscillating Load*

15.6.1.1 A Free String

Consider the one-dimensional wave equation for a string with the right-hand side corresponding to a moving-oscillating, concentrated external force. It is the first example considered in Sect. 14.2.1.1, but now the complete solution is shown and dimensional values are used. The dynamic equation is

$$\varrho \frac{\partial^2 u(x,t)}{\partial t^2} - T \frac{\partial^2 u(x,t)}{\partial x^2} = Q_0 \delta(\eta) \mathrm{e}^{\mathrm{i}\omega^0 t}, \quad \eta = x - vt, \quad v^2 \neq c^2 = \frac{T}{\varrho}. \tag{15.68}$$

The steady-state solution $u(\eta, t) = u_0(\eta)\mathrm{e}^{\mathrm{i}\omega^0 t}$, in terms of the Fourier transform in the moving coordinate system is

$$u_0^{F\eta}(k) = \frac{Q_0}{T[k^2 + (\mathrm{i}kv + \mathrm{i}\omega^0)^2/c^2]}, \tag{15.69}$$

with singular points $k = \omega^0/(c - v)$ and $k = -\omega^0/(c + v)$ on the real k-axis. At the same time, under rule (15.66) this result must be presented as

$$u_0^{F\eta}(k) = \frac{Q_0}{T[k^2 + (0 + \mathrm{i}kv + \mathrm{i}\omega^0)^2/c^2]}, \tag{15.70}$$

with the singular points

$$k_1 = \frac{\omega^0 - \mathrm{i}0}{c - v}, \quad k_2 = -\frac{\omega^0 - \mathrm{i}0}{c + v}. \tag{15.71}$$

Note that the group velocities are $v_g = \pm c$ for the first and the second expressions in (15.71), respectively.

Let the frequency and the load speed be positive: $\omega^0 > 0$, $v > 0$. The inverse transform is the sum of contributions of the poles. It follows that for a subsonic speed, $v < c$

$$u(\eta, t) = -\frac{\mathrm{i}Q_0 c}{2T\omega^0}\left[\exp\left(-\frac{\mathrm{i}\omega^0\eta}{c-v}\right)H(\eta) + \exp\left(\frac{\mathrm{i}\omega^0\eta}{c+v}\right)H(-\eta)\right]\mathrm{e}^{\mathrm{i}\omega^0 t}. \quad (15.72)$$

$$u'(\eta, t) = \frac{Q_0 c}{2T}\left[-\exp\left(-\frac{\mathrm{i}\omega^0\eta}{c-v}\right)\frac{H(\eta)}{c-v} + \exp\left(\frac{\mathrm{i}\omega^0\eta}{c+v}\right)\frac{H(-\eta)}{c+v}\right]\mathrm{e}^{\mathrm{i}\omega^0 t}. \quad (15.73)$$

The latter expression is also valid for the non-oscillating moving load, $\omega^0 = 0$. It is

$$u'(\eta, t) = \frac{Q_0 c}{2T}\left(-\frac{H(\eta)}{c-v} + \frac{H(-\eta)}{c+v}\right), \quad (15.74)$$

whereas the displacement itself has no steady state limit in this case.

In the case of a supersonic speed, $v > c$, both singular points appear above the real axis, $\Im k_{1,2} = +0$, and the expressions become as

$$u(\eta, t) = \frac{\mathrm{i}Q_0 c}{2T\omega^0}\left[\exp\left(\frac{\mathrm{i}\omega^0\eta}{v-c}\right) - \exp\left(\frac{\mathrm{i}\omega^0\eta}{v+c}\right)\right]H(-\eta)\mathrm{e}^{\mathrm{i}\omega^0 t}. \quad (15.75)$$

$$u'(\eta, t) = \frac{Q_0 c}{2T}\left[-\exp\left(-\frac{\mathrm{i}\omega^0\eta}{v-c}\right)\frac{H(-\eta)}{v-c} + \exp\left(\frac{\mathrm{i}\omega^0\eta}{v+c}\right)\frac{H(-\eta)}{v+c}\right]\mathrm{e}^{\mathrm{i}\omega^0 t}, \quad (15.76)$$

and

$$u'(\eta, t) = -\frac{Q_0 c^2}{T(v^2 - c^2)}H(-\eta) \quad (\omega^0 = 0). \quad (15.77)$$

The corresponding transient solution for the displacements follows from (15.74) and (15.77) as

$$\begin{aligned} u(x, t) &= \frac{Q_0 c}{2T}\left(\frac{ct - x}{c - v}H(\eta) + \frac{ct + x}{c + v}H(-\eta)\right) \quad (v < c), \\ u(x, t) &= -\frac{Q_0 c^2}{v^2 - c^2}\eta H(-\eta) \quad (v > c). \end{aligned} \quad (15.78)$$

The case, where the load propagates with the speed equal to the group velocity of the excited wave, corresponds to a resonant excitation considered in Chap. 15.3.

15.6.1.2 The String on an Elastic Foundation

The presence of the foundation results in an additional term, $\varkappa$, in the relation (15.70), namely, now

$$u_0^{F\eta}(k) = \frac{Q_0}{T[k^2 + (0 + ikv + i\omega^0)^2/c^2] + \varkappa}. \tag{15.79}$$

The corresponding dispersion relation is

$$\omega = \pm\sqrt{c^2k^2 + \varkappa/\varrho} \tag{15.80}$$

The singular points become

$$k_{1,2} = \frac{1}{c^2 - v^2}\left[\omega^0 v \pm c\sqrt{\frac{\varkappa}{\varrho}\left(\frac{v^2}{c^2} + \frac{\varrho}{\varkappa}(\omega^0)^2 - 1\right)}\right]. \tag{15.81}$$

It can be seen that the real poles exist if the couple, v, ω^0, belongs to domain Ω, external to the ellipse

$$\frac{v^2}{c^2} + \frac{\varrho}{\varkappa}(\omega^0)^2 = 1. \tag{15.82}$$

Otherwise, if this couple is inside the ellipse, such a load does not excite waves, and only a localized disturbance exponentially decaying with the distance of the load propagates together with the latter.

15.6.1.3 A Free Elastic Beam

For example, from the equation for an elastic beam, axially compressed by force T and supported by an elastic foundation of stiffness $\varkappa$

$$D\frac{\partial^4 w}{\partial x^4} + T\frac{\partial^2 w}{\partial x^2} + \varkappa w + \varrho S\frac{\partial^2 w}{\partial t^2} = 0 \tag{15.83}$$

the dispersion relation follows

$$\omega = \pm\sqrt{(\varkappa - Tk^2 + Dk^4)/(\varrho S)}. \tag{15.84}$$

Chapter 16
Resonant Waves

Resonant waves have been briefly discussed in Sect. 1.3.5. Now we consider this topic in more detail.

Let us at first mention **resonant oscillations**. A force, oscillating with a frequency equal to one of the frequencies of free oscillations of a finite elastic body, is the resonant one (except a special case where the force is *orthogonal* to the corresponding oscillation form). As the simplest example, consider a single-frequency oscillator. The equation is

$$m\frac{\mathrm{d}^2u(t)}{\mathrm{d}t^2} + \varkappa u(t) = Q\mathrm{e}^{\mathrm{i}\omega t}H(t), \tag{16.1}$$

where $m, \varkappa$ and ω are the oscillator mass, stiffness and frequency, $\omega = \sqrt{\varkappa/m}$, respectively; Q and H are the load amplitude and the Heaviside unit step function, respectively. Using the Laplace transform, Sect. 18.3, we have

$$u^L(s) = \frac{Q}{m(s^2+\omega^2)(s-\mathrm{i}\omega)} = \frac{Q}{m(s+\mathrm{i}\omega)(s-\mathrm{i}\omega)^2)}. \tag{16.2}$$

The growing part of the original corresponds to the second-order pole at $s = \mathrm{i}\omega$. Its contribution is (see Sect. 18.3.1)

$$u(t) \sim \frac{Qt}{2\mathrm{i}\omega m}\mathrm{e}^{\mathrm{i}\omega t}. \tag{16.3}$$

The frequency equality condition is also valid for a resonant wave, but the frequencies must coincide at a point moving with the group velocity. Let v_0 and ω_0 be the load speed and frequency. If this load excites a wave there exists at least a one wavenumber $k = k_*$ where

$$\omega(k_*) - k_*v_0 = \omega_0 \text{ and the group speed } v_g = v_0, \quad (v_g = \mathrm{d}\omega(k)/\mathrm{d}k, \ k = k_*). \tag{16.4}$$

L. I. Slepyan, *Mechanical Waves: Free, Forced, and Transition*,
https://doi.org/10.1007/978-3-032-15864-2_16

These equalities can also be considered as defining the frequency and speed of the resonant load excited the wave with $k = k_*$.

The force-wave frequencies coincidence is not enough for the resonant excitation. The fact is that the energy produced by the load can be carried away by the excited waves. This will not happen, at least concerning the wave in front of the load, if the force moves at the group velocity.

So, in the 1D case considered, the load must satisfy the two conditions: The load speed and frequency, v and ω_0, must coincide with the wave group speed and frequency registered at $x = v_g t$. Generally, these conditions can be satisfied in the following way

Consider a point, $k = k_*$, on the wave dispersion curve, $\omega(k)$. The corresponding group speed $v_g = \mathrm{d}\omega/\mathrm{d}k$ $(k = k_*)$. The complex wave exponent indicator and the frequency at $x = v_g t$, become

$$\mathrm{i}(\omega(k_*)t - k_* x) = \mathrm{i}(\omega_0 t - k_* \eta), \quad \omega_0 = \omega(k_*) - k_* v_g, \quad \eta = \text{const.} \tag{16.5}$$

Thus, the resonant load parameters are

$$\omega_0 = \omega(k_*) - k_* v_g, \quad v = v_g = \frac{\mathrm{d}\omega(k)}{\mathrm{d}k} \; (k = k_*). \tag{16.6}$$

As an **example** consider a wave in a string on an elastic foundation under a concentrated force moving with the resonant speed, starting from $x = 0$ at $t = 0$ (Fig. 16.1). The equation is

$$\varrho\ddot{u}(x,t) - Tu''(x,t) + \varkappa u(x,t) = Q_0 \mathrm{e}^{\mathrm{i}(\omega(k_*) - k_* v_g)t}\delta(x - v_g t), \tag{16.7}$$

and (see (18.137))

$$u^{LF_\eta}(s,k) = u^{LF}(s - \mathrm{i}kv_g, k). \tag{16.8}$$

For a complex wave $\exp(\mathrm{i}\omega t - kx)$

$$\omega(k) = \sqrt{\frac{\varkappa}{\varrho} + c^2k^2}, \quad c = \sqrt{\frac{T}{\varrho}}, \quad v_g(k) = \frac{c^2 k}{\omega}. \tag{16.9}$$

The Laplace transform on time of (16.7) leads to

$$(\varrho^2 s^2 + \varkappa)u^L(x,s) - T(u^L)''(x,s) = \frac{Q}{v_g}\exp\left(-(s - \mathrm{i}\omega_0)\frac{x}{v_g}\right)H(x), \tag{16.10}$$

$$\omega_0 = \omega(k_*) \quad v_g(k_*)k_* \quad \frac{\varkappa}{\varrho\omega(k_*)}$$

Thus,

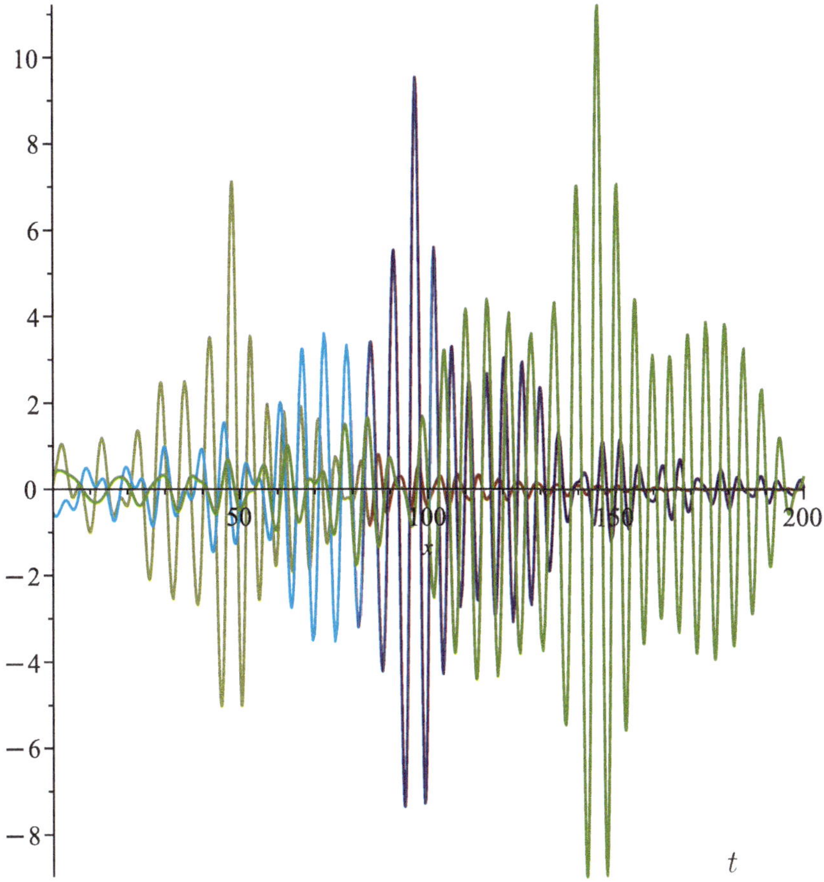

Fig. 16.1 The resonant wave on a beam on an elastic foundation based on. The beam displacement speed under the point force moving at the resonant speed, $v = \sqrt{2}$ (brown) and $0.060/\sqrt{t}$ (navy)

16.1 Phase Shift in Forced Waves

16.2 Resonant Waves in Some Structures

16.3 Load Moving with Periodically Varying Speed

16.3.1 Localized Dynamic Obstacle

Consider a one-dimensional waveguide with a dynamic structure attached at a point. Without loss of generality we can take the latter as the coordinate origin, $x = 0$. Let a sinusoidal wave, excited fare from the zero point, at $x \ll 0$, and propagating alon the structure-free waveguide, manifests itself at $x = 0$ by oscillations as

$$u_0(t, 0) = A_0 \cos(\omega_0 t), \tag{16.11}$$

and let $u_1(t)$ be the additional oscillations of the waveguide caused by the structure response, that is by the force $P(t)$ acting from the structure on the waveguide. We describe the dynamics at this point in terms of the Green's function for waveguide, $G(t)$, and the fundamental solutions for the structure, $R(t)$, such that

$$u_1(t) = G(t) * P(t), \quad P(t) = -R(t) * u(t), \quad u(t) = u_0(t) + u_1(t). \tag{16.12}$$

We assume (for simplicity) that the wave u_0 (16.11) comes to the localized structure at $t = 0$, while the latter and the waveguide at $x \geq 0$ are at rest at $t < 0$. Using the Laplace transform we rewrite the above relations as follows

$$u_0^L(s) = \frac{A_0}{s - \mathrm{i}\omega_0}, \quad u_1^L(s) = G^L(s)P^L(s), \quad P^L(s) = -R^L(s)u^L(s). \tag{16.13}$$

The solution follows from the last two relations

$$\begin{aligned} P^L(s) &= -\frac{R^L(s)}{1 + R^L(s)G^L(s)} \frac{A_0 s}{s^2 + \omega_0^2}, \\ u_1^L(s) &= -\frac{R^L(s)G^L(s)}{1 + R^L(s)G^L(s)} \frac{A_0 s}{s^2 + \omega_0^2}. \end{aligned} \tag{16.14}$$

While $R^L(s)$ is defined separately by the structure, the Green function $G^L(s)$ follows from the waveguide equation. We express the latter as in the previous sections as

$$\begin{aligned} \mathcal{L}^{LF}(s, k)u^{LF}(s, k) &= Q^{LF}(s, k) - R^L(s)u^L(s, 0), \\ u^L(s, 0) &= u_0^{LF} + \frac{1}{2\pi} \int_{-\infty}^{\infty} \frac{P^L(s)u^L(s, 0)}{\mathcal{L}^{LF}(s, k)} \, \mathrm{d}k. \end{aligned} \tag{16.15}$$

(Remain that $Q^{LF}(s, k)$ is the Laplace-Fourier transform of the load, $Q(t, x)$.) It follows that

$$G^L(s) = \frac{1}{2\pi} \int_{-\infty}^{\infty} \frac{1}{\mathcal{L}^{LF}(s, k)} \, \mathrm{d}k. \tag{16.16}$$

The wave $u_1(t, x)$ at $x < 0$ and $x > 0$ is now defined by the Eq. (16.16). For example, in the simplest case of a string with an oscillator,

$$
\begin{aligned}
R^L(s) &= \frac{\varkappa s^2}{s^2 + \omega_{osc}^2}, \\
G^L(s) &= \frac{1}{2\pi}\int_{-\infty}^{\infty} \frac{\mathrm{d}k}{\varrho s^2 + T k^2} = \frac{1}{2\sqrt{\varrho T s^2}},
\end{aligned} \tag{16.17}
$$

where $\omega_{osc} = \sqrt{\varkappa/m}$, and ϱ, T, $\varkappa$ and m are the string's mass per unit length and the tension force, the spring stiffness and mass of the oscillator.

Finally, from the expressions (16.14) and (16.17) we find

$$
\begin{aligned}
P(t) &\sim \frac{2\varkappa\omega_0^2\sqrt{\varrho T}(2\sqrt{\varrho T}(\omega_0^2 - \omega_{osc}^2)\cos(\omega_0 t) - \varkappa\omega_0 \sin(\omega_0 t))}{4\varrho T(\omega_0^2 - \omega_{osc}^2)^2 + \varkappa^2\omega_0^2} A_0, \\
u_1(t) &\sim \frac{\varkappa\omega_0(2\sqrt{\varrho T}(\omega_0^2 - \omega_{osc}^2)\sin(\omega_0 t) + \varkappa\omega_0 \cos(\omega_0 t))}{4\varrho T(\omega_0^2 - \omega_{osc}^2)^2 + \varkappa^2\omega_0^2} A_0, \\
\frac{\mathrm{d}u_1(t)}{\mathrm{d}t} &= \frac{1}{2\sqrt{\varrho T}} P(t), \quad u_1(t, x) = \frac{1}{2\sqrt{\varrho T}} H(ct - |x|) * P(t), \quad c = \sqrt{\frac{T}{\varrho}}.
\end{aligned} \tag{16.18}
$$

Thus, the obstacle induces symmetrically the reflected and refracted waves (as it should). What is defined above, is the wave amplitude, which depends on the waveguide and the oscillator parameters and the incident wave frequency.

Part VI
Some General Relations

Chapter 17
Fundamentals

17.1 Some General Formulations

17.1.1 Invariant Differential Operators

17.1.1.1 Rectangular Coordinates

The invariant differential operators, *gradient* of a scalar function f, and *divergence* and *rotor* of a vector function $\boldsymbol{u}$ have the following expressions in terms of the ∇ operator,

$$\begin{aligned} \operatorname{grad} f &= \nabla f, \\ \operatorname{div} \boldsymbol{u} &= \nabla \boldsymbol{u}, \\ \operatorname{rot} \boldsymbol{u} &= \nabla \times \boldsymbol{u}, \end{aligned} \tag{17.1}$$

where in cartesian coordinates, $\boldsymbol{\nabla} = k_i\, \partial/\partial x_i$ and

$$\begin{aligned} \operatorname{grad} f &= \frac{\partial f}{\partial x_i}\boldsymbol{k}_i, \\ \operatorname{div} \boldsymbol{u} &= \frac{\partial u_i}{\partial x_i}, \\ \operatorname{rot} \boldsymbol{u} &= e_{ijk}\frac{\partial u_k}{\partial x_j}\boldsymbol{k}_i, \end{aligned} \tag{17.2}$$

where k_i are the unit coordinate vectors, the symbol $\times$ denotes the *cross product*, and the *alternating tensor* is

$$e_{ijk} = \begin{cases} +1 & \text{if } ijk \text{ presents an even permutation of } 1, 2, 3 \\ 0 & \text{if any two of } ijk \text{ indices are equal} \\ -1 & \text{if } ijk \text{ presents an odd permutation of } 1, 2, 3. \end{cases} \tag{17.3}$$

L. I. Slepyan, *Mechanical Waves: Free, Forced, and Transition*,
https://doi.org/10.1007/978-3-032-15864-2_17

Here and below summation by repeated indices is assumed unless otherwise stated.

Note that divergence of rotor and rotor of gradient are zero:

$$\nabla(\nabla \times \boldsymbol{u}) = \nabla \times (\nabla f) = 0. \tag{17.4}$$

Also the following vector identity can be useful:

$$\boldsymbol{u}\nabla\boldsymbol{u} = \tfrac{1}{2}u^2 - \boldsymbol{u} \times (\nabla \times \boldsymbol{u}). \tag{17.5}$$

The *Laplace operator*

$$\Delta = \nabla\nabla = \frac{\partial^2}{\partial x_i \partial x_i}. \tag{17.6}$$

The gradient of a scalar function is a vector equal to the maximal directional derivative and pointing in this direction. The divergence of a vector function defines the volume-normalized "flux" of it out of the infinitesimal volume. The rotor (curl) defines the rotation vector field. It follows that the action of these operators and their compositions, for example, of the Laplace operator, $\Delta = \text{div grad} = \nabla^2$, are independent of the coordinate system.

17.1.1.2 Cylindrical Coordinates

In terms of the cylindrical system, r, θ, z, with

$$x_1 = r\cos\theta, \quad x_2 = r\sin\theta, \quad x_3 = z, \quad r = \sqrt{x_1^2 + x_2^2}, \tag{17.7}$$

the invariant differential operators are

$$\begin{gathered}
\nabla f = \frac{\partial f}{\partial r}\boldsymbol{k}_r + \frac{1}{r}\frac{\partial f}{\partial \theta}\boldsymbol{k}_\theta + \frac{\partial f}{\partial z}\boldsymbol{k}_z, \\
\nabla \boldsymbol{u} = \frac{1}{r}\frac{\partial (r u_r)}{\partial r}\boldsymbol{k}_r + \frac{1}{r}\frac{\partial u_\theta}{\partial \theta}\boldsymbol{k}_\theta + \frac{\partial u_z}{\partial z}\boldsymbol{k}_z, \\
\nabla \times \boldsymbol{u} = \left(\frac{1}{r}\frac{\partial u_z}{\partial \theta} - \frac{\partial u_\theta}{\partial z}\right)\boldsymbol{k}_r + \left(\frac{\partial u_r}{\partial z} - \frac{\partial u_z}{\partial r}\right)\boldsymbol{k}_\theta + \frac{1}{r}\left(\frac{\partial (r u_\theta)}{\partial r} - \frac{\partial u_r}{\partial \theta}\right)\boldsymbol{k}_z, \\
\Delta f = \frac{1}{r}\frac{\partial}{\partial r}\left(r\frac{\partial f}{\partial r}\right) + \frac{1}{r^2}\frac{\partial^2 f}{\partial \theta^2} + \frac{\partial^2 f}{\partial z^2}, \\
\Delta \boldsymbol{u} = \left(\Delta u_r - \frac{u_r}{r^2} - \frac{2}{r}\frac{\partial u_\theta}{\partial \theta}\right)\boldsymbol{k}_r + \left(\Delta u_\theta - \frac{u_\theta}{r^2} + \frac{2}{r}\frac{\partial u_r}{\partial \theta}\right)\boldsymbol{k}_\theta + \Delta u_z \boldsymbol{k}_z.
\end{gathered} \tag{17.8}$$

17.1.1.3 Spherical Coordinates

In spherical coordinates, r, α, β, with

$$x_1 = r\sin\alpha\cos\beta,\quad x_2 = r\sin\alpha\sin\beta,\quad x_3 = r\cos\alpha,\quad r = \sqrt{x_1^2 + x_2^2 + x_3^2}, \tag{17.9}$$

the invariant differential operators are

$$\begin{aligned}
\nabla f &= \frac{\partial f}{\partial r}\boldsymbol{k}_r + \frac{1}{r}\frac{\partial f}{\partial \alpha}\boldsymbol{k}_\alpha + \frac{1}{r\sin\alpha}\frac{\partial f}{\partial \beta}\boldsymbol{k}_\beta,\\
\nabla \boldsymbol{u} &= \frac{1}{r^2}\frac{\partial}{\partial r}(r^2 u_r) + \frac{1}{r\sin\alpha}\left[\frac{\partial}{\partial \alpha}(u_\alpha \sin\alpha) + \frac{\partial u_\beta}{\partial \beta}\right],\\
\nabla\times\boldsymbol{u} &= \frac{1}{r\sin\alpha}\left[\frac{\partial}{\partial \alpha}(u_\beta\sin\alpha) - \frac{\partial u_\alpha}{\partial \beta}\right]\boldsymbol{k}_r\\
&+\frac{1}{r}\left[\frac{1}{\sin\alpha}\frac{\partial u_r}{\partial \beta} - \frac{\partial r u_\beta}{\partial r}\right]\boldsymbol{k}_\alpha + \frac{1}{r}\left[\frac{\partial r u_\alpha}{\partial r} - \frac{\partial u_r}{\partial \alpha}\right]\boldsymbol{k}_\beta,\\
\Delta f &= \frac{1}{r^2}\frac{\partial}{\partial r}\left(r^2\frac{\partial f}{\partial r}\right) + \frac{1}{r^2\sin\alpha}\frac{\partial}{\partial \alpha}\left(\frac{\partial f}{\partial \alpha}\sin\alpha\right) + \frac{1}{r^2\sin^2\alpha}\frac{\partial^2 f}{\partial \beta^2},\\
\Delta\boldsymbol{u} &= \left[\Delta u_r - \frac{2u_r}{r^2} - \frac{2}{r^2\sin\alpha}\left(\frac{\partial}{\partial \alpha}(u_\alpha\sin\alpha) + \frac{\partial u_\beta}{\partial \beta}\right)\right]\boldsymbol{k}_r\\
&+\left[\Delta u_\alpha + \frac{2}{r^2}\frac{\partial u_r}{\partial \alpha} - \frac{1}{r^2\sin^2\alpha}\left(u_\alpha + 2\cos_\alpha\frac{\partial u_\beta}{\partial \beta}\right)\right]\boldsymbol{k}_\alpha\\
&+\left[\Delta u_\beta + \frac{2}{r^2\sin\alpha}\frac{\partial u_r}{\partial \beta} + \frac{1}{r^2\sin^2\alpha}\left(2\cos\alpha\frac{\partial u_\alpha}{\partial \beta} - u_\beta\right)\right]\boldsymbol{k}_\beta.
\end{aligned} \tag{17.10}$$

17.1.2 Lagrangian and Eulerian Formulations

There exist different representations for field variables, such as displacements, particle velocities, and stresses. The difference mainly concerns nonlinear formulations. However, even in a linear case, a better physical understanding can be achieved if the formulation type is defined. Also, some relations used in the linear analysis are valid in a more general area, while their mathematical expression does depend on the type of formulation. The main question concerns a function, say, a displacement, expressed as $\boldsymbol{u}(\boldsymbol{x}, t)$. What is $\boldsymbol{x}$? Is it the initial or actual coordinate? The next question concerns the material density. Let an equation be given with inertia term $\varrho\partial^2\boldsymbol{u}/\partial t^2$. Is ϱ the initial density, or is it a function of the coordinates and time as actual density? The answer depends on the type of formulation.

In this connection, two states of the body are distinguished, the *initial state* (this usually means a time-independent unstressed body state) and the *actual state* as a possibly time-dependent stressed body state. The following notations are introduced.

The initial location of a material point is denoted by vector $\boldsymbol{x}$ whose projections onto the Cartesian coordinates are x_i, $i = 1, 2, 3$. The actual location of the same material point is denoted by vector $\boldsymbol{X}$ with the projections X_i. The particle displacement and velocity are

$$\boldsymbol{u} = \boldsymbol{X} - \boldsymbol{x}, \quad \boldsymbol{v} = \frac{\partial \boldsymbol{X}(\boldsymbol{x}, t)}{\partial t}. \tag{17.11}$$

Consider the derivatives

$$\boldsymbol{k}_i = \frac{\partial \boldsymbol{x}}{\partial x_i} \quad \text{and} \quad \boldsymbol{\lambda}_i = \frac{\partial \boldsymbol{X}}{\partial x_i}. \tag{17.12}$$

The first relation represents the unit vector associated with the x_i-coordinate. The vector $\boldsymbol{k}_i \delta l$ represents an oriented material element of length δl in its initial state. The latter derivative in (17.12) multiplied by δl represents this material element in the actual state. So $\lambda_i = |\boldsymbol{\lambda}_i|$ is the *stretch*.

Displacements, velocities and other quantities can be expressed as functions of the coordinates x_i (the *Lagrangian formulation*) or X_i (the *Eulerian formulation*). Quantities of the same physical sense can have different expressions in these formulations.

17.1.2.1 Substantial and Partial Time-Derivatives

In the Lagrangian formulation, the position vector of a particle is a function of $\boldsymbol{x}$ and t, where $\boldsymbol{x}$ is independent of time, the *substantial time-derivative*, as the rate of a value associated with moving particle, is given by the partial time derivative. In contrast, in the Eulerian formulation, where the partial time-derivative is the rate at a fixed point in space, the substantial time-derivative has a different expression. In the Eulerian expression of a scalar or a vector function, $f = f(\boldsymbol{X}, t)$, if it is associated with a particle, the first argument is also a function of time, namely

$$\frac{\mathrm{d}\boldsymbol{X}}{\mathrm{d}t} = \boldsymbol{v}; \tag{17.13}$$

hence the substantial time-derivative usually defined by $\mathrm{d}/\mathrm{d}t$, is

$$\frac{\mathrm{d}f(\boldsymbol{X}, t)}{\mathrm{d}t} = \frac{\mathrm{d}f(\boldsymbol{X}(t), t)}{\mathrm{d}t} = \frac{\partial f(\boldsymbol{X}, t)}{\partial t} + (\boldsymbol{v}\nabla) f(\boldsymbol{X}, t). \tag{17.14}$$

In particular, the *particle velocity and acceleration* in the Lagrangian formulation are

$$\boldsymbol{v} = \frac{\partial \boldsymbol{u}(\boldsymbol{x}, t)}{\partial t}, \quad \boldsymbol{w} = \frac{\partial \boldsymbol{v}(\boldsymbol{x}, t)}{\partial t} = \frac{\partial^2 \boldsymbol{u}(\boldsymbol{x}, t)}{\partial t^2}, \tag{17.15}$$

whereas in the Eulerian formulation the *particle velocity* is defined by the vector equation as

$$\begin{aligned} \boldsymbol{v} = \frac{\mathrm{d}\boldsymbol{u}(\boldsymbol{X},t)}{\mathrm{d}t} = \frac{\mathrm{d}\boldsymbol{u}(\boldsymbol{X}(t),t)}{\mathrm{d}t} = \frac{\partial \boldsymbol{u}(\boldsymbol{X},t)}{\partial t} + (\boldsymbol{v}\nabla)\boldsymbol{u}(\boldsymbol{X},t) \\ \Longrightarrow \left(\delta_{ij} - \frac{\partial u_i}{\partial X_j}\right) v_j = \frac{\partial u_i}{\partial t}, \quad (i,j=1,2,3), \end{aligned} \tag{17.16}$$

and the *particle acceleration* is

$$\boldsymbol{w} = \frac{\partial \boldsymbol{v}(\boldsymbol{X},t)}{\partial t} + (\boldsymbol{v}\nabla)\boldsymbol{v}. \tag{17.17}$$

The latter expression as the substantial derivative is the same as the partial time-derivative in the Lagrangian formulation (where the current state is considered as the initial). In contrast, the partial derivative, $\partial \boldsymbol{v}(\boldsymbol{X},t)/\partial t$ in the Eulerian formulation is the speed variation rate at a fixed point of the space but not the particle acceleration.

The Lagrangian and Eulerian formulations based on the material and space coordinates, respectively, can lead to different definitions of the velocity of a specific immaterial moving point, for example, a point on the wavefront. Let $\boldsymbol{X}(t)$ be the current coordinate of this point in the fixed space frame. Then the velocity $\boldsymbol{c}_a = \mathrm{d}\boldsymbol{X}(t)/\mathrm{d}t$ represents its 'absolute' velocity as the velocity relatively this frame.

Alternatively, let the current coordinate $\boldsymbol{x}(t)$ be the coordinate of the same point in the fixed material frame. Then $\boldsymbol{c}_m = \mathrm{d}\boldsymbol{x}(t)/\mathrm{d}t$ is the corresponding velocity, which can differ from $\boldsymbol{c}_a$ since the particle velocity, $\boldsymbol{v}$, can be nonzero. The 'absolute' speed of this point can be expressed as

$$\boldsymbol{c}_a = \boldsymbol{c}_m + \boldsymbol{v}. \tag{17.18}$$

17.1.2.2 Lagrangian Formulation of Strain

Consider scalar products

$$\boldsymbol{\lambda}_\mathrm{i}\boldsymbol{\lambda}_\mathrm{i} = \lambda_i \lambda_j \cos\gamma_{ij}, \tag{17.19}$$

where $\lambda_i = |\boldsymbol{\lambda}_i|$ and γ_{ij} is the angle between these vectors. Note that $\boldsymbol{k}_i\boldsymbol{k}_j = 0\ (i \neq j)$. Recall that $k_i = |\boldsymbol{k}_i| = 1$ and λ_i is the stretch of the material element initially oriented as axis x_i. The stretch can be expressed as

$$\lambda_i = \sqrt{1 + 2\varepsilon_{ii}}, \tag{17.20}$$

where the strain components, ε_{ij}, are defined by the relation

$$2\varepsilon_{ij} = \boldsymbol{\lambda}_i\boldsymbol{\lambda}_j - \delta_{ij} = \frac{\partial u_i}{\partial x_j} + \frac{\partial u_j}{\partial x_i} + \sum_{m=1}^{3} \frac{\partial u_m}{\partial x_i}\frac{\partial u_m}{\partial x_j}. \tag{17.21}$$

Other geometrical values are defined by the following relations. The shear angle, $\pi/2 - \gamma_{ij}$ $(j \neq i)$, is defined by

$$\cos\gamma_{ij} = \frac{\boldsymbol{\lambda}_i\boldsymbol{\lambda}_j}{\lambda_i\lambda_j} = \frac{2\varepsilon_{ij}}{\sqrt{(1+2\varepsilon_{ii})(1+2\varepsilon_{jj})}} \quad (i \neq j). \tag{17.22}$$

The rotation can be calculated based on the angles, φ_{ij}, between the ˇ$_i$-element and the x_j axis:

$$\cos\varphi_{ij} = \frac{\boldsymbol{\lambda}_i\boldsymbol{k}_j}{\lambda_i} = \frac{\delta_{ij} + \partial u_j/\partial x_i}{\sqrt{1+2\varepsilon_{ii}}}. \tag{17.23}$$

The relative dilatation of a three-dimensional element, Δ, can be expressed by the parallelepipedal product

$$1 + \Delta = |(\boldsymbol{\lambda}_1 \times \boldsymbol{\lambda}_2)\,\boldsymbol{\lambda}_3| = \det\left[\delta_{ij} + \frac{\partial u_i}{\partial x_j}\right] = \sqrt{\det\left[\delta_{ij} + 2\varepsilon_{ij}\right]}. \tag{17.24}$$

If the nonlinear term in (17.21) is neglected, this relation corresponds to linear elasticity. In this case, however, the right-hand side of (17.21) becomes rotation-dependent. For example, if an unstressed body rotates around x_3 by the angle ω, the displacements and the linearized strain components are

$$\begin{aligned} u_1 &= x_1(\cos\omega - 1) - x_2\sin\omega, \\ u_2 &= x_1\sin\omega + x_2(\cos\omega - 1), \\ \varepsilon_{11} = \varepsilon_{22} &= \frac{\partial u_1}{\partial x_1} = \frac{\partial u_2}{\partial x_2} = \cos\omega - 1, \\ \varepsilon_{12} &= \frac{1}{2}\left(\frac{\partial u_1}{\partial x_2} + \frac{\partial u_2}{\partial x_1}\right) = 0. \end{aligned} \tag{17.25}$$

At the same time the exact nonlinear components are rotation-independent as follows directly from the first equality in (17.21).

Note that in one-dimensional cases, where there is no rotations, the exact expression for *relative elongation* in terms of the displacement is linear

$$\lambda - 1 = \frac{du(x)}{dx}. \tag{17.26}$$

Below, in such cases, it is called *strain*. Note that it differs from the corresponding component of the strain tensor defined in (17.21).

17.1.2.3 Eulerian Formulation of Strain

If the actual state is considered as initial and vice versa, the Euler type representation can be obtained in the same way:

$$2\varepsilon_{ij} = \frac{\partial u_i}{\partial X_j} + \frac{\partial u_j}{\partial X_i} - \sum_{m=1}^{3} \frac{\partial u_m}{\partial X_i}\frac{\partial u_m}{\partial X_j},$$

$$\lambda_i = \frac{1}{\sqrt{1-2\varepsilon_{ii}}}, \quad \cos\gamma_{ij}\frac{2\varepsilon_{ij}}{\sqrt{(1-2\varepsilon_{ii})(1-2\varepsilon_{jj})}}, \quad \cos\varphi_{ij} = \frac{\delta_{ij} - \partial u_j/\partial X_i}{\sqrt{1-2\varepsilon_{ii}}}, \tag{17.27}$$

$$1+\Delta = \frac{1}{\det\left[\delta_{ij} - \partial u_i/\partial X_j\right]} = \frac{1}{\sqrt{\det\left[\delta_{ij} - 2\varepsilon_{ij}\right]}}.$$

The Eulerian formulation is commonly used in fluid mechanics where the particle velocities are of interest as functions of time and coordinates and where the stress components are independent of the displacements. This type of field description is also preferable where the boundary conditions are posted on a space surface. In solid mechanics, where the particles are not so mobile and the state does depend on the displacement gradient, the Lagrangian formulation appears, in general, to be more convenient. Note, however, that, in principle, either of these two approaches can be used in the mechanics of both fluid and solid media. For example, Eulerian formulation is preferable in considering the wave momentum.

17.1.3 Dirac Delta Function. Green's Function and Convolution

We now consider the *delta function*, which places a considerable role in the applied analysis. Let us start with a *pre-delta function*

$$\delta(t,a) = \frac{a}{\pi(x^2+a^2)} \quad (a>0). \tag{17.28}$$

The integral over the x-axis of it

$$\int_{-\infty}^{\infty} \delta(t,a)\,\mathrm{d}x = 1 \tag{17.29}$$

independently of the parameter a $(a > 0)$. The Dirac delta function, $\delta(x)$, is a *generalized limit*

$$\delta(x) = \lim_{a\to+0} \delta(x,a). \tag{17.30}$$

So, the delta function is zero outside the zero point, while the integral of it over any segment, containing the zero point inside it, is equal to one

$$\delta(x) = 0 \ (x \neq 0), \quad \int_{x_-}^{x_+} \delta(x)\,\mathrm{d}x = \lim_{a\to+0} \int_{x_-}^{x_+} \delta(t, a)\,\mathrm{d}x = 1 \ (x_- < 0,\ x_+ > 0). \quad (17.31)$$

It follows that the primitive is the Heaviside function, $H(x)$

$$\int_{-\infty}^{x} \delta(x)\,\mathrm{d}x = H(x). \quad (17.32)$$

Thus, the delta function can also be defined as the *generalized derivative*

$$\delta(x) = \frac{\mathrm{d}H(x)}{\mathrm{d}x}. \quad (17.33)$$

Note that different generalized functions also do not exist (in the classical sense) at some values of the argument. The convolution with a regular *test function* is considered instead of the point-by-point definition. Similar to the delta function, two ways of the definition are effective: the generalized limit and a finite order generalized derivative. Usually generalized functions appear as intermediate results in the mathematical analysis of real phenomena, for example, in the integral transforms widely used in the wave theory and applications. This topic is discussed below in Chap. 18.

The *convolution* of two functions, $f(x)$ and $g(x)$, $f(x) * g(x)$, if exists,

$$f(x) * g(x) = \int_{-\infty}^{\infty} f(\xi)g(x-\xi)\,\mathrm{d}\xi = \int_{-\infty}^{\infty} f(x-\xi)g(\xi)\,\mathrm{d}\xi. \quad (17.34)$$

A remarkable fact is that the convolution with the delta function is an identical transformation

$$f(x) * \delta(x) = f(x) \quad (17.35)$$

at any x where $f(x)$ is continuous. It follows from the fact that

$$\delta(x) = H'(x) \quad \Longrightarrow \quad \int_{-a}^{b} \delta(x)\,\mathrm{d}x = 1 \quad (17.36)$$

for any positive a, b. Further, we see that the convolution with the nth derivative of the delta function is the nth derivative

$$f(x) * \delta^{(n)}(x) = \delta^{(n)} f(x), \quad n = 0, 1, 2, \ldots. \quad (17.37)$$

It follows directly from the above definition, (17.34), and the features of the delta function.

Now consider a problem for a system, which displacement $u(x)$ (or any other variable) is defined by a linear operator, L_x, with constant parameters

$$L_x\, u(x) = P(x), \tag{17.38}$$

where $P(x)$ is the external action. At first, let the latter be the delta function. The related displacement is called the *Green's function* (or a *fundamental solution*), $G(x)$

$$L_x\, G(x) = \delta(x). \tag{17.39}$$

It is easy to show that the respond to a general action follows from the latter as a convolution

$$u(x) = G(x) * P(x). \tag{17.40}$$

Indeed, performing the convolution of both sides of the above relation with the action $P(x)$ we find (note that L is independent of x)

$$\begin{aligned}(L_x\, G(x)) * P(x) = L_x \int_{-\infty}^{\infty} G(x-\xi)P(\xi)\,\mathrm{d}\xi = \int_{-\infty}^{\infty} \delta(x-\xi)P(\xi)\,\mathrm{d}\xi \\ = P(x) = L_x\, u(x).\end{aligned} \tag{17.41}$$

The delta function can also be represented as the limit of a sum of two analytical functions. Namely, replace a in pre-delta (17.28) by $y > 0$, and represent

$$\begin{aligned}\delta(x, y) &= \frac{y}{\pi(x^2+y^2)} = \delta_+(x+\mathrm{i}y) + \delta_-(x-\mathrm{i}y), \\ \delta_+ &= -\frac{1}{2\pi\mathrm{i}(x+\mathrm{i}y)}, \quad \delta_- = \frac{1}{2\pi\mathrm{i}(x-\mathrm{i}y)},\end{aligned} \tag{17.42}$$

where δ_+ is a regular analytic function in the upper complex half-plane $z = x + \mathrm{i}y$ $(y > 0)$, and δ_- is such a function in the lower half-plane $z = x - \mathrm{i}y$ $(y > 0)$.

What is important, that such an analytic representation can be simply achieved for any regular function of real variable, $f(x)$. For this goal we can use the convolution with the pre-delta function. Indeed

$$f(x) = \lim_{y\to+0}(f_+ + f_-), \quad f_\pm = \int_{-\infty}^{\infty} f(\xi)\delta_\pm(x-\xi\pm\mathrm{i}y)\,\mathrm{d}\xi. \tag{17.43}$$

Such a decomposition integral is called a *Cauchy-type integral*.

17.2 The Basic Laws

17.2.1 *Hamilton's Principle*

In the classical physics, Hamilton's principle states that there exist a function L called Lagrange function (*Lagrangian*), such that the *action*

$$\int_{t_1}^{t_2} L\,\mathrm{d}t \tag{17.44}$$

is extremal with respect to the variation of independent variables *inside* the integration interval, $t_1 < t < t_2$, where $t_{1,2}$ are arbitrary time points. In mechanics, the lagrangian $L = \mathcal{K} - \mathcal{P}$ ($\mathcal{K}$ and $\mathcal{P}$ are the kinetic and potential energies, respectively). The variation and integration by parts lead to the dynamic equations (*Euler—Lagrange equations*)

$$\frac{\partial L}{\partial u_i} - \frac{\mathrm{d}}{\mathrm{d}t}\frac{\partial L}{\partial \dot{u}_i} = 0, \quad L = L(\boldsymbol{u}, \dot{\boldsymbol{u}}), \quad i = 1, 2, \ldots. \tag{17.45}$$

17.2.2 *Conservation Laws*

17.2.2.1 Momentum Conservation Law

The momentum conservation follows from Hamilton's principle in connection with homogeneity of the space (see e.g. Landau and Livshitz, 1960/1969). Consider spatial region Ω bounded by surface S, which may be moving and variable as in the case of a wave. The momentum variation in Ω is defined by both the action of external forces and the momentum coming into the region with masses crossing its boundary. The law can thus be expresses as

$$\frac{\mathrm{d}}{\mathrm{d}t}\int_\Omega \varrho \boldsymbol{v}\,\mathrm{d}\Omega = \int_\Omega \varrho \boldsymbol{q}\,\mathrm{d}\Omega + \int_S \boldsymbol{\sigma}_n\,\mathrm{d}S + \int_S \varrho[(\boldsymbol{v}_S - \boldsymbol{v}^+)\boldsymbol{n}]\boldsymbol{v}^+\,\mathrm{d}S, \tag{17.46}$$

where ϱ, $\boldsymbol{v}$ and $\boldsymbol{q}$ are the density, particle velocity and distributed body forces per unit mass, respectively; $\boldsymbol{\sigma}_n$ is the stress vector acting on S from outside, $\boldsymbol{v}_S$ is the boundary points velocity, $\boldsymbol{n}$ is the external unit normal to S and $\boldsymbol{v}^+$ is the external limit of the particle velocity on S (it may be discontinuous at S if the latter is the wave front). Similar relation between *angular momentum* and the external moment is valid. Note that for a regular case it follows from the *linear momentum* low (17.46).

Differential forms of this law follow from the integral form (17.46) by contraction the boundary S to an internal regular material point (a boundary point is considered in Sect. 2.2.5). In this case, the last term in the right-hand side of Eq. (17.46) disappears since no matter crosses the boundary. The resulting equations look differently being expresses in Lagrangian or Eulerian formulations.

For a one-dimensional case, in the Lagrangian formulation, the Newton law reads

$$\varrho_0 \frac{\partial v(x,t)}{\partial t} = \frac{\partial \sigma(x,t)}{\partial x} + q(x,t), \tag{17.47}$$

where ϱ_0 is the initial density. In the Eulerian formulation, in accordance with (17.47), it is

$$\varrho \left[\frac{\partial v(X,t)}{\partial t} + v \frac{\partial v(X,t)}{\partial X} \right] = \frac{\partial \sigma(X,t)}{\partial X} + \varrho q(X,t), \tag{17.48}$$

where ϱ is the actual density.

To derive the dynamic equation in Lagrangian formulation (17.47) for a space problem, consider a small rectangular parallelepiped built on material elements $\boldsymbol{k}_1\, \mathrm{d}x_1, \ldots, \boldsymbol{k}_3\, \mathrm{d}x_3$ in the initial state of the body.

In the actual state, whose dynamic equilibrium is considered, this parallelepiped is deformed and rotated, its sides are no longer plane and perpendicular to each other. Consider a side of this deformed parallelepiped built in the initial state on the elements $\mathrm{d}x_m$, $\mathrm{d}x_n$ with $\boldsymbol{k}_i$ as its initial-state external normal (i, m, n are different indices). Denote the force acting on this side from outside by $\tilde{\boldsymbol{\sigma}}_i(\boldsymbol{x}, t)\, \mathrm{d}x_m\, \mathrm{d}x_n$. Its projections onto the rectangular triad $\boldsymbol{k}_j$ are $\tilde{\sigma}_{ij}(\boldsymbol{x}, t)\, \mathrm{d}x_m\, \mathrm{d}x_n$.

Note that, generally speaking, the *Piola-Kirchoff components*, $\tilde{\sigma}_{ij}$, are not true stresses, since the area of the deformed side is not equal to $\mathrm{d}x_m\, \mathrm{d}x_n$. Also, in general, these 'stresses' are not symmetric: $\tilde{\sigma}_{ij}(\boldsymbol{x}) \neq \tilde{\sigma}_{ji}(\boldsymbol{x})$. The dynamic equation has the form

$$\varrho_0 \frac{\partial \boldsymbol{v}(x,t)}{\partial t} = \frac{\partial \tilde{\sigma}_{ij}(\boldsymbol{x},t)}{\partial x_i} \boldsymbol{k}_j + \boldsymbol{q}(x,t), \tag{17.49}$$

where $\tilde{\boldsymbol{q}}$ is the body force per unit volume of the unstrained material.

In the Eulerian formulation, the *Cauchy tensor* of true stresses, σ_{ij}, is used, and the dynamic equation, the *Navier-Stokes* equation is

$$\varrho \left(\frac{\partial \boldsymbol{v}}{\partial t} + (\boldsymbol{v}\nabla)\boldsymbol{v} \right) = -\nabla p + \nabla T + \varrho \boldsymbol{q}. \tag{17.50}$$

Here p is pressure, $p = -\sigma = \frac{1}{3}\sigma_{ii}$, and ∇T is divergence of the tensor deviator of stresses. In the Cartesian coordinates, it is expressed as

$$\nabla T = \frac{\partial}{\partial x_i} e_{ij} \boldsymbol{k}_j, \quad e_{ij} = \sigma_{ij} - \sigma \delta_{ij}. \tag{17.51}$$

17.2.2.2 Energy Conservation and Dissipation in a Perfect Medium

The energy conservation is one of the fundamental physical laws. It is a fundamental observation about nature (Dill and Bromberg, 2003, p. 40). For a closed system having potential energy it follows from Hamilton's Principle of least action as a consequence of homogeneity of time (see, e.g., Landau and Livshitz, 1960/1969). It also constitutes the First Law of thermodynamics.

The total energy density, $\mathcal{E}$, as the energy per unit initial volume of the material element, can vary due to the work of external forces and dissipation. The energy conservation law, in Lagrangian formulation, reads

$$\frac{\partial \mathcal{E}}{\partial t} = -\nabla N + \varrho \boldsymbol{q}\boldsymbol{v} - \varrho D, \tag{17.52}$$

where $\boldsymbol{q}$ is the body forces per unit mass, D is the corresponding dissipation rate and N is the energy flux density

$$N = -\tilde{\sigma}_{ij} v_j \boldsymbol{k}_i. \tag{17.53}$$

In Eulerian formulation for a fixed space element of a unit volume, formula (17.52) is valid with N as a sum of the inductive and convective energy fluxes

$$N = -\sigma_{ij} v_j \boldsymbol{k}_i + \mathcal{E}\boldsymbol{v}. \tag{17.54}$$

It follows from the energy conservation law that dissipation in a material model reflects an energy flux to the microlevel, which does not present in the continuous model of the medium but takes part in the energy conservation. It happens, for example, in friction and a plastic deformation, the resistance to which evidences that sliding activates the microlevel. It also can occur in a perfect continuous medium where under usual conditions energy is conserved. For example, there is dissipation in the front of a shock wave formed in the case of a hardening type of nonlinearity, Sect. 9.1. Also, it occurs in the reflection of a wave at an obstacle moving though the medium, Sect. 1.4.3.

17.2.2.3 Eshelbian Forces

Consider an area, Ω, of a continuous-material model. From the energy point of view, this area is in equilibrium if the potential energy of Ω is minimal,—respective admissible variations of its position. In the case of a regular stress-strain field, the above condition corresponds to the equilibrium of Newtonian forces acting on Ω. However, if Ω contains singular points, lines or surfaces, potential energy may release through a moving singularity,—under the Newtonian forces in the equilibrium. For example, this phenomenon is typical for moving dislocations, cracks, and between-the-phase boundaries. In all these cases, no Newtonian force corresponds to the energy variation, and related forces of a different class is introduced.

Such forces have different names: *the force on a singularity* (Eshelby, 1951), the *material forces* (Maugin, 1993), *configurational forces* (Gurtin, 2000) and the *driving forces*, which exist in thermodynamics too (Dill and Bromberg, 2003).

These forces are defined similarly to the Newtonian forces as the derivative of the corresponding energy; however, in contrast to the latter corresponding to derivatives with respect to the displacement of the loaded material particles, the Eshelbian forces are defined by the derivatives respective the displacement of singularities.

17.2.2.4 Mass Conservation Law

In the Lagrangian formulation, where a fixed material element is under consideration, the mass conservation law is satisfied automatically. In Eulerian formulation, the mass in a fixed space element may vary due to the mass exchange with the surrounding domain. The law thus reads

$$\frac{\mathrm{d}\varrho}{\mathrm{d}t} + \varrho\nabla \boldsymbol{v} = 0 \quad \text{or} \quad \frac{\partial \varrho}{\partial t} + \nabla(\varrho \boldsymbol{v}) = 0. \tag{17.55}$$

17.2.3 *Discrete Versus Continuous Models*

As commonly adopted, the continuous-material model is the long-wave approximation, long comparing with the structure sizes. Nevertheless, in a linear step wave, the continuous model appears adequate despite the wavefront discontinuity: the wave satisfies the dynamic equations and all conservation laws. However, in nonlinear shock waves, dynamic fracture, and other transition waves, such a model contradicts the energy conservation. Also, such singularities appear at a fast-moving elastic contact.[1] Namely, as discussed above, in such waves, there exists a speed-dependent energy release through the moving singularities. Besides, the transition front speed is not defined in the continuum formulation framework.

At the same time, if the wave outside the singularity is smooth enough, the continuous model remains adequate in this (open) area. So, it is reasonable to preserve the model modified with some addition to eliminate the energy imbalance and present the transition criterion needed for its speed determination.

In the shock wave theory, it is assumed that the released energy goes to heat, which creates the *heat pressure*. So, the combined model is a thermomechanical one. Griffith (1921) introduced surface energy as the crack resistance criterion. The combined model (the elastic continuum with the surface energy) resolved the issue; however, there were ignored possible microlevel oscillations. In this sense, it looks

[1] An uncertainty in this problem was resolved in Slepyan and Brun (2012); see Chapter(NN).

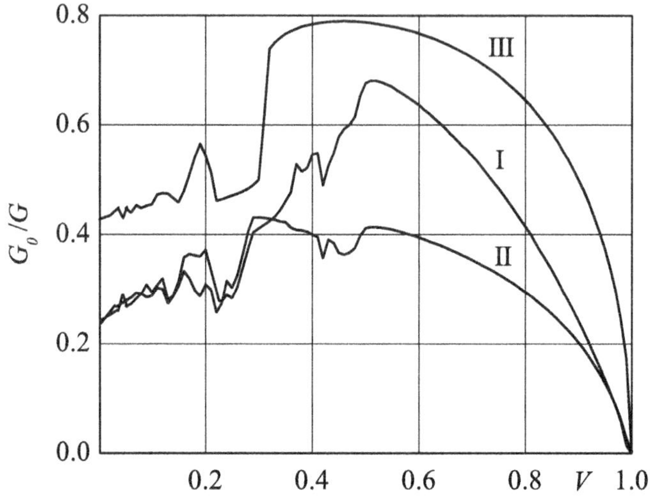

Fig. 17.1 Energy release ratios, G_0/G: I. Mode I ($V = v/c_R$); II. Mode II ($V = v/c_R$); III. Mode III ($V = v/c_2$). Here G is the total energy release, and G_0 is the strain energy of the breaking bonds—the 'surface energy'

as the *cold adiabat* in the shock waves. Note that while waves in a uniform homogeneous medium propagate without dispersion, it is not so in periodic lattices. As a result, moving transition can excite sinusoidal waves carrying the energy release away from the transition front. This phenomenon leads to an essential (speed-dependent) increase in crack resistance energy. Crack dynamics analysis based on some discrete lattices, Slepyan (1981, 2002), Kulakhmetova et al. (1984), showed that the energy of the oscillations is considerable even in a quasi-static crack development. It unboundedly grows as the crack speed is approaching the shear, c_2 (mode III) or Rayleigh, c_R modes I and II) wave speeds, Fig. 17.1.

Finally, an adequate combined model is a discrete-continuous one. The continuous medium serves to describe the macrolevel field, and the crack resistance is determined based on the lattice model. What is remarkable, since the microlevel length/time units are small, one can consider a macrolevel transient problem with variable transition front speed based on the steady-state lattice solution corresponding to a constant speed.

Chapter 18
Generalized Functions and Integral Transforms

18.1 Generalized Functions

18.1.1 Little on Generalized Functions

Generalized functions (also called *distributions*) were first introduced by Sergei Lvovich Sobolev (1935), and then independently by Laurent Schwartz in late 1940s. L. Schwartz developed a comprehensive theory of the distributions defined on the set of infinitely differentiable, finite-support *test functions* usually denoted by $\varphi(x)$. An example of a test function with support $[-a, a]$ is

$$\varphi(x) = \exp\left(-\frac{a^2}{a^2 - x^2}\right) H(a^2 - x^2),\ a = \text{const.} \tag{18.1}$$

Such test functions are valid for any distribution. In special cases, the requirements concerning the differentiability and locality can be relaxed.

For example, the delta function considered in Sect. 17.1.3 is defined as such that its convolution with any test function (in this case, any continuous function) is an identical transform, Eq. (17.35)

A generalized function can also be defined as a generalized limit or a generalized derivative. For example, the delta function, which is the general derivative of the Heaviside function Sect. 17.1.3.

18.1.2 Generalized Function as the Limit

Consider *prelimiting* function $f(\varepsilon, x)$ locally integrable for $\varepsilon > 0$, such that $f(+0, x) = f(x)$ at regular points of $f(x)$, if such points exist, and extend the definition of $f(x)$ on the whole of the x-axis as follows:

L. I. Slepyan, *Mechanical Waves: Free, Forced, and Transition*,
https://doi.org/10.1007/978-3-032-15864-2_18

$$\langle f(x), \varphi(x) \rangle = \lim_{\varepsilon \to +0} \int_{-\infty}^{\infty} f(\varepsilon, x)\varphi(x)\,\mathrm{d}x. \tag{18.2}$$

Thus, the limit of the integral is considered instead of the integral of the limit. In the case of a usual function, the change of the order does not affect the result. In contrast, the latter order has no sense for the singular case, while the limit of the integral can exist. In this sense, the expression of a generalized function, $f(+0, x)$, means not a nonexisting limit of $f(\varepsilon, x)$ but the limit in (18.2). If it exists for any test function from the class mentioned above, the prelimiting function is such, that is, the relation in (18.2) does define a generalized function. It is also true concerning a set of prelimiting functions, which lead to the same result for any given test function (generally, different for different test functions). We preserve notation $f(x)$ for the generalized function bearing in mind that it is defined as the functional (18.2). Note that, in some cases, the integral symbol is used instead of the functional notation – with the same sense as in (18.2)

$$\int_{-\infty}^{\infty} f(x)\varphi(x)\,\mathrm{d}x = \lim_{\varepsilon \to +0} \int_{-\infty}^{\infty} f(\varepsilon, x)\varphi(x)\,\mathrm{d}x. \tag{18.3}$$

The discussed technique looks as follows in the application to the generalized Fourier transform (18.86). The corresponding functional is

$$\begin{aligned}\langle f^F(k), \varphi(k) \rangle &= \lim_{\varepsilon \to 0} \int_{-\infty}^{\infty} \int_{-\infty}^{\infty} f(x)\mathrm{e}^{-\varepsilon|x|+ikx}\,\mathrm{d}x\,\varphi(k)\,\mathrm{d}k \\ = \lim_{\varepsilon \to 0} \int_{-\infty}^{\infty} \int_{-\infty}^{\infty} & f(x)\mathrm{e}^{-\varepsilon|x|+ikx}\varphi(k)\,\mathrm{d}k\,\mathrm{d}x = \int_{-\infty}^{\infty} f(x)g(x)\,\mathrm{d}x,\end{aligned} \tag{18.4}$$

where the change of the integration order is justified by Fubini's theorem and

$$g(x) = \int_{-\infty}^{\infty} \mathrm{e}^{ikx}\varphi(k)\,\mathrm{d}k \tag{18.5}$$

is an infinitely differentiable (analytical) *fast decreasing function* of x: for any number n

$$g(x) = o(1/x^n)\ (x \to \pm\infty). \tag{18.6}$$

Despite the infinite support, it can also play the test function's role concerning the class of slow-growing functions.

Note that analytical functions can represent prelimiting functions as in the case of the generalized Fourier transform (18.86). Such a representation is called the *analytical representation* of the generalized function. Along with this, any given generalized function can be represented by different prelimiting functions, not necessarily analytical.

Also, note that in contrast to any analytical function, which is uniquely continued from its part in a regular domain (for example, in any segment of the real axis), generalized functions do not possess this feature. For example, function $f(x) = 1/x$ given for $x > 0$ is not defined for $x \leq 0$.

18.1.3 Null Functions

Let $f_1(\varepsilon, x)$ and $f_2(\varepsilon, x)$ be different prelimiting function for a given generalized function. In this case, the difference, $f_1(\varepsilon, x) - f_2(\varepsilon, x)$ tends, as $\varepsilon \to 0$, to a generalized function equal to zero. The limit, however, not necessarily exists or is zero in the usual sense. This also applies to a prelimiting function multiplied by a vanishing quantity. For example

$$
\begin{aligned}
f_1(\varepsilon, x) = \frac{\varepsilon}{x^2+\varepsilon^2},\ f_2(\varepsilon, x) = \frac{2\varepsilon^3}{(x^2+\varepsilon^2)^2} = \langle f_1(+0, x), \\
\varphi(x)\rangle = \langle f_2(+0, x), \varphi(x)\rangle = \varphi(0) = \langle f_1(+0, x) - f_2(+0, x), \\
\varphi(x)\rangle = 0,\ \lim_{\varepsilon \to +0} \int_{-\infty}^{\infty} \sqrt{\varepsilon} f_1(\varepsilon, x)\varphi(x)\, \mathrm{d}x = 0,
\end{aligned}
\tag{18.7}
$$

whereas

$$
f_1(\varepsilon, 0) - f_2(\varepsilon, 0) = -1/\varepsilon \to -\infty,\ \sqrt{\varepsilon} f_1(\varepsilon, 0) = 1/\sqrt{\varepsilon} \to \infty\ (\varepsilon \to +0).
\tag{18.8}
$$

Such zero generalized functions are called *null functions*. It can be seen below that they play an important role in some considerations, especially when solving the homogeneous problems (see Sect. 18.1.10). Note that although, in general, the product of generalized functions is not defined (see Sect. 18.1), the product of the null functions may exist as a nonzero generalized function. For example

$$
(\sqrt{\varepsilon} f_1(\varepsilon, x))^2 = \frac{1}{2} f_2(\varepsilon, x) \to \frac{1}{2} f_2(+0, x).
\tag{18.9}
$$

18.1.4 Generalized Function as the Derivative

Next important way is the representation of generalized functions as *generalized derivatives*. This is based on the usual rule of the integration by parts

$$
\langle f^{(n)}(x), \varphi(x)\rangle = -\langle f^{(n-1)}(x), \varphi'(x)\rangle = (-1)^n \langle f(x), \varphi^{(n)}(x)\rangle
\tag{18.10}
$$

which is valid for generalized functions since it is valid for the prelimiting functions. Using this rule generalized function $f(x)$ can be defined as a *generalized derivative* of a regular function. Namely, any generalized function can be represented as a finite-order generalized derivative of a regular antiderivative, $F(x)$, as

$$\begin{aligned}\langle f(x), \varphi(x)\rangle &= \langle F^{(n)}(x), \varphi(x)\rangle = (-1)^n \langle F(x), \varphi^{(n)}(x)\rangle \\ &= (-1)^n \int_{-\infty}^{\infty} F(x)\varphi^{(n)}(x)\,\mathrm{d}x.\end{aligned} \tag{18.11}$$

It can be seen that, in this representation, the non-uniqueness of the antiderivative does not matter. Note that this and the previous definitions do not contradict to each other.

18.1.5 Generalized Power Functions

Functions $x_\pm^\lambda$

For $\lambda > -1$ regular functions $x_\pm^\lambda$ are defined as

$$x_\pm = (\pm x)^\lambda H(\pm x). \tag{18.12}$$

For fractional $\lambda < -1$ ($\lambda = -\beta - n, 0 < \beta < 1, n = 1, 2, \ldots$) these functions are defined as generalized derivatives

$$x_\pm^\lambda = \frac{(\mp 1)^n}{\beta(\beta+1)\ldots(\beta+n-1)}(x_\pm^{-\beta})^{(n)} = \frac{(\mp 1)^n \Gamma(\beta)}{\Gamma(\beta+n)}(x_\pm^{-\beta})^{(n)}. \tag{18.13}$$

They are regular for $x \neq 0$

$$\begin{aligned} x_+^\lambda &= \begin{cases} x^\lambda & (x > 0) \\ 0 & (x < 0), \end{cases} \\ x_-^\lambda &= \begin{cases} 0 & (x > 0) \\ (-x)^\lambda & (x < 0). \end{cases}\end{aligned} \tag{18.14}$$

We have

$$\langle x_\pm^\lambda, \varphi(x)\rangle = \frac{(\pm 1)^n \Gamma(\beta)}{\Gamma(-\lambda)} \langle x_\pm^{-\beta}, \varphi^{(n)}\rangle = \frac{(\pm 1)^n \Gamma(\beta)}{\Gamma(-\lambda)} \int_{+0}^{\infty} x_\pm^{-\beta}, \varphi^{(n)}\,\mathrm{d}x. \tag{18.15}$$

This functional is an analytic function of λ with poles at $\lambda = -1, -2, \ldots$. From the definition it follows, in particular, that the product

$$x_\pm^\lambda x = \pm x_\pm^{\lambda+1}. \tag{18.16}$$

Regularization

In the representation (18.15), the singular point, $x = 0$, is outside the integration domain that allows us to integrate by parts n times again to express the functional $\langle x_+^\lambda, \varphi(x)\rangle$ explicitly. At each step of the integration, we choose the antiderivative such that preserves the integrant in (18.15) to be integrable. This results in subtraction from the test function first n terms of the Taylor series of it. So we introduce function $\varphi_n(x)$ as

$$\varphi_n(x) = \varphi(x) - \sum_{m=0}^{n-1} \frac{x^m}{\Gamma(m+1)} \varphi^{(m)}(0). \tag{18.17}$$

Clearly, this partial sum of the series does not influence the integrant in (18.15), and the latter is still integrable ($0 < x < \infty$) at each step of the integration by parts. In this way, the *regularization* of the divergent integral is obtained

$$\langle x_+^\lambda, \varphi(x)\rangle = \int_{+0}^{\infty} x^\lambda \varphi_n(x)\,\mathrm{d}x. \tag{18.18}$$

In the same way, it can be found that

$$\begin{aligned}\langle x_-^\lambda, \varphi(x)\rangle &= \frac{(-1)^n \Gamma(\beta)}{\Gamma(-\lambda)} \int_{-\infty}^{-0} (-x)^{-\beta} \varphi^{(n)}(x)\,\mathrm{d}x \\ &= \int_{+0}^{\infty} x^\lambda \left[\varphi(-x) - \sum_{m=0}^{n-1} \frac{(-x)^m}{\Gamma(m+1)} \varphi^{(m)}(0) \right] \mathrm{d}x.\end{aligned} \tag{18.19}$$

Functions $(x \pm \mathrm{i}0)^\lambda$

This functions are defined as

$$(x \pm \mathrm{i}0)^\lambda = x_+^\lambda + \mathrm{e}^{\pm \mathrm{i}\pi\lambda} x_-^\lambda, \tag{18.20}$$

and this is in agreement with the generalized limits of the corresponding analytical functions

$$(x \pm \mathrm{i}0)^\lambda = \lim_{\varepsilon \to +0} (x \pm \mathrm{i}\varepsilon)^\lambda. \tag{18.21}$$

For fractional λ the inverse relations

$$x_{+}^{\lambda} = \frac{(x - \mathrm{i}0)^{\lambda} \exp(\mathrm{i}\pi\lambda) - (x + \mathrm{i}0)^{\lambda} \exp(-\mathrm{i}\pi\lambda)}{2\mathrm{i}\sin\pi\lambda},$$
$$x_{-}^{\lambda} = \frac{(x + \mathrm{i}0)^{\lambda} - (x - \mathrm{i}0)^{\lambda}}{2\mathrm{i}\sin\pi\lambda} \tag{18.22}$$

give us the analytical representations for $x_{\pm}^{\lambda}$, whereas the definition in (18.20) together with relations (18.17)–(18.19) defines the regularization of $\langle (x \pm \mathrm{i}0)^{\lambda}, \varphi(x)\rangle$.

Thus, the considered power functions are generated by the analytic function z^{λ} with fractional λ. In the this case, there are two branch points: $z = 0$ and $z = \infty$. We assume that the branch cut is drown along the negative x-axis, and

$$z^{\lambda} = x^{\lambda} \ (y = 0,\ x > 0). \tag{18.23}$$

The analytic continuation from the positive half-axis leads to

$$z^{\lambda} = \begin{cases} (-x)^{\lambda} \exp(\mathrm{i}\pi\lambda) & (x < 0,\ y = +0) \\ (-x)^{\lambda} \exp(-\mathrm{i}\pi\lambda) & (x < 0,\ y = -0), \end{cases} \tag{18.24}$$

and this is in agreement with (18.21).

18.1.6 The Principal Value and the Derivatives of the Delta Function

Analytic function $1/z^{n}$ ($z = x + \mathrm{i}y$) generates two generalized functions

$$(x \pm \mathrm{i}0)^{-n} = \lim_{y \to \pm 0} z^{-n}. \tag{18.25}$$

For $n = 1$

$$\lim_{y \to \pm 0} \frac{1}{z} = \lim_{y \to \pm 0} \frac{x - \mathrm{i}y}{x^2 + y^2} = \mathcal{P}\left(x^{-1}\right) \mp \mathrm{i}\pi\delta(x), \tag{18.26}$$

where $\mathcal{P}$ means the *principal value*:

$$\left\langle \mathcal{P}\left(x^{-1}\right), \varphi(x)\right\rangle = V.p. \int_{-\infty}^{\infty} \frac{\varphi(x)}{x}\,\mathrm{d}x = \lim_{\varepsilon \to 0}\left[\int_{-\infty}^{-\varepsilon} \frac{\varphi(x)}{x}\,\mathrm{d}x + \int_{\varepsilon}^{\infty} \frac{\varphi(x)}{x}\,\mathrm{d}x\right]$$
$$= \int_{+0}^{\infty} \frac{\varphi(x) - \varphi(-x)}{x}\,\mathrm{d}x. \tag{18.27}$$

For a general positive integer n consider linear combinations of the functions $(x \pm \mathrm{i}0)^{-n}$. One half of the sum is called the *principal value* of x^{-n} denoted as

$$\mathcal{P}\left(x^{-n}\right) = \frac{1}{2}\left[\frac{1}{(x+\mathrm{i}0)^n} + \frac{1}{(x-\mathrm{i}0)^n}\right]. \tag{18.28}$$

The difference results in the *derivatives of the delta function*

$$\delta^{(n-1)}(x) = \frac{(-1)^n \Gamma(n)}{2\pi \mathrm{i}}\left[\frac{1}{(x+\mathrm{i}0)^n} - \frac{1}{(x-\mathrm{i}0)^n}\right] \; (n = 1, 2, \ldots). \tag{18.29}$$

The inverse relation is

$$(x \pm \mathrm{i}0)^{-n} = \mathcal{P}\left(x^{-n}\right) \pm \frac{(-1)^n \pi \mathrm{i}}{\Gamma(n)} \delta^{(n-1)}(x). \tag{18.30}$$

Thus

$$\langle \mathcal{P}(x^{-1}), \varphi(x) \rangle = \int_{+0}^{\infty} \frac{\varphi(x) - \varphi(-x)}{x} \, \mathrm{d}x \tag{18.31}$$

and

$$\begin{aligned} \langle (x \pm \mathrm{i}0)^{-n}, \varphi(x) \rangle &= \frac{1}{\Gamma(n)} \langle (x \pm \mathrm{i}0)^{-1}, \varphi^{(n-1)}(x) \rangle \\ = \mp \frac{\pi \mathrm{i}}{\Gamma(n)} \varphi^{(n-1)}(0) &+ \frac{1}{\Gamma(n)} \int_{+0}^{\infty} \frac{\varphi^{(n-1)}(x) - \varphi^{(n-1)}(-x)}{x} \, \mathrm{d}x. \end{aligned} \tag{18.32}$$

It follows that

$$\langle \mathcal{P}(x^{-n}), \varphi(x) \rangle = \frac{1}{\Gamma(n)} \int_{+0}^{\infty} \frac{\varphi^{(n-1)}(x) - \varphi^{(n-1)}(-x)}{x} \, \mathrm{d}x. \tag{18.33}$$

Note that a generalized function $\mathcal{G}_p$ with a point support at $x = x_*$ *point-support generalized function* is a finite sum of the delta function and its derivative

$$\mathcal{G}_p = \sum_{m=0}^{n} c_m \delta^{(m)}(x - x_*), \tag{18.34}$$

where a_m are constants.

18.1.7 Fourier Transform of the Power Functions

For $\lambda > -1$ the generalized Fourier transform of $x_{\pm}^{\lambda}$ results in

$$\begin{aligned}\left(x_{+}^{\lambda}\right)^{F} &= \int_{0}^{\infty} x_{+}^{\lambda} e^{ikx}\, dx = \frac{\Gamma(\lambda+1)}{(0-ik)^{\lambda+1}} = \frac{\Gamma(\lambda+1)}{(k+i0)^{\lambda+1}} e^{i\pi(\lambda+1)/2},\\ \left(x_{-}^{\lambda}\right)^{F} &= \int_{-\infty}^{0} x_{-}^{\lambda} e^{ikx}\, dx = \frac{\Gamma(\lambda+1)}{(0+ik)^{\lambda+1}} = \frac{\Gamma(\lambda+1)}{(k-i0)^{\lambda+1}} e^{-i\pi(\lambda+1)/2}.\end{aligned} \tag{18.35}$$

Thus the generalized Fourier transform of $x_{\pm}^{\lambda}$ for $\lambda > -1$ is proportional to the generalized functions $(k \pm i0)^{-\lambda-1}$, respectively. By the integration by parts it can be shown that these formulas are still valid for any $\lambda \neq -1, -2, \ldots$.

Now from (18.20) it follows that

$$\left[(0 \pm ix)^{\lambda}\right]^{F} = \frac{2\pi}{\Gamma(-\lambda)} k_{\mp}^{-\lambda-1}, \tag{18.36}$$

where identity

$$\Gamma(\lambda+1)\Gamma(-\lambda) = -\frac{\pi}{\sin \pi\lambda} \tag{18.37}$$

was used in the determination of the latter transforms.

Further, referring to (18.28)

$$\left[\mathcal{P}(x^{-n})\right]^{F} = \frac{\pi i (ik)^{n-1}}{\Gamma(n)} \quad (n = 1, 2, \ldots) \tag{18.38}$$

and as follows directly from (18.35)

$$(x^{n})^{F} = (-i)^{n} 2\pi \delta^{(n)}(k) \; (n = 0, 1, \ldots), \tag{18.39}$$

while the transforms of the delta function and its derivatives are

$$[\delta(x)]^{F} = 1\,, \; \left[\delta^{(n)}(x)\right]^{F} = (-ik)^{n}. \tag{18.40}$$

Generalized function x^{-n}, $n = 1, 2, \ldots$, can be considered as

$$x^{-n} = \frac{(-1)^{n-1}}{\Gamma(n)} \frac{d^{(n)} \ln(|x|)}{dx^{(n)}} \text{ or } x^{-n} = \frac{1}{2}\left((x+i0)^{-n} + (x-i0)^{-n}\right). \tag{18.41}$$

For $n = 1$ it is sufficient to take into account that it is an odd function. It follows from this as well, as from (18.41), that

$$\left(x^{-1}\right)^F(k) = \int_{-\infty}^{\infty} \frac{\mathrm{i}\sin(kx)}{x}\,\mathrm{d}x = \pi\mathrm{i}\,\mathrm{sign}(k) \tag{18.42}$$

and

$$\left(x^{-n}\right)^F(k) = \frac{\pi\mathrm{i}^n}{\Gamma(n)} k^{n-1}\,\mathrm{sign}(k). \tag{18.43}$$

In particular

$$\left(x^{-2}\right)^F(k) = -\pi|k|. \tag{18.44}$$

Now consider $1/|x|$. Is it a generalized function? The answer depends on the definition. It can be defined as a generalized derivative as

$$\frac{1}{|x|} = f_1(x) = \frac{\mathrm{d}\,(\mathrm{sign}\,x\,\ln|x|)}{\mathrm{d}x}, \tag{18.45}$$

which is true outside the singular point, $x = 0$. In this case, integrating by parts we obtain

$$\begin{aligned}\left(\frac{1}{|x|}\right)^F(k) &= 2\lim_{\varepsilon\to 0}\int_0^{\infty} \ln(x)[\varepsilon\cos(kx) + k\sin(kx)]\mathrm{e}^{-\varepsilon x}\,\mathrm{d}x \\ &= -2\gamma - \lim_{\varepsilon\to 0}\ln(k^2+\varepsilon^2) = -2(\gamma + \ln(|k|)),\end{aligned} \tag{18.46}$$

and this evidences that under the definition in (18.45) it really is a distribution. However, it also can be defined as

$$\frac{1}{|x|} = \lim_{\varepsilon\to 0} f_2(\varepsilon, x),\ f_2(\varepsilon, x) = \frac{1}{\sqrt{x^2+\varepsilon^2}}, \tag{18.47}$$

that also coincides with $1/|x|$ outside the singular point. In this case

$$\begin{aligned}f_2^F(\varepsilon, k) &= \int_0^{\infty} \frac{\cos(kx)}{\sqrt{x^2+\varepsilon^2}}\,\mathrm{d}x = K_0(\varepsilon|k|) \\ &= -2(\gamma + \ln(|k|) + 2\ln(2/\varepsilon) + \mathrm{O}(\varepsilon)\ (\varepsilon \to +0),\end{aligned} \tag{18.48}$$

where K_0 is the Bessel function. This differs from the previous result by a k-independent term infinitely growing as $\varepsilon \to 0$. With refer to (18.40) it follows from this that the function

$$f_2(\varepsilon, x) - 2\ln(2/\varepsilon)\delta(\varepsilon, x) \to f_1(x)\ (\varepsilon \to +0) \tag{18.49}$$

is a representation of the generalized function in (18.45). In Sect. 18.1.8, this relation is discussed in terms of a mechanical problem. Note that a general definition of the Fourier transform of generalized functions is presented in Sect. 18.1.7.

18.1.8 On Generalized Functions in Physics

Delta Function and Saint-Venant's Principle

Generalized functions can arise in different ways. They can be introduced as an idealization of an external action, or appear in the course of mathematical conversions as intermediate results, or arise as the final results. In particular, the delta function can be used as the source of Green's function (Green's tensor) or as a unite concentrated force expressed in terms of stresses. In the former case, this formalism results in the convolution related to a given regular load as

$$f(x) = \int_{(\Omega)} G(x, \xi)Q(\xi)\,\mathrm{d}\xi \text{ or } f(x) = \int_{(\Omega)} G(x - \xi)Q(\xi)\,\mathrm{d}\xi, \qquad (18.50)$$

where $f(x)$ is the solution, Ω is the support of the external load, $Q(x)$, whereas $G(x, \xi)$ or $G(x)$ is the Green's function (the latter type of the Green's function and the convolution corresponds to a homogeneous system). In this case, if the load is represented by a regular function, the result, as a rule, is also regular, and no question arises concerning acceptability of the singular delta and the corresponding Green's function. In contrast, in the case where the delta is related to a concentrated force and the corresponding Green's function is the final result, the question is not so transparent.

Generalized function $f(x)$ can be characterized by a set of regular pre-limiting functions, $f(\varepsilon, x)$, each of which tends to $f(x) = f(+0, x)$, in the sense of (18.2). The main question is whether the state of the considered physical system depends only on this limit independently of the prelimiting function. If so we may use this formalism; otherwise, it cannot be used without regard to its origin. Consider the Dirac delta function, $\delta(x - \xi)$, as an external action at the upper half-plane boundary, $x = \xi$, $y = 0$, and the corresponding Green's function, $G(\xi, x, y)$. Denote the latter as the response to a prelimiting delta by $G(\varepsilon, \xi, x, y)$. The question is whether the equality

$$G(\xi, x, y) = \lim_{\varepsilon \to 0} G(\varepsilon, \xi, x, y) \; ((x - \xi)^2 + y^2 > 0, \varepsilon > 0), \qquad (18.51)$$

is the same for any pre-delta. The issue is closely related to Saint-Venant's principle [SV1] stated that the strains that can be produced in a body by the application, to a small part of its surface, of a system of forces statically equivalent to zero force and zero couple, are of negligible magnitude at distances which are large compared with the linear dimensions of the part. Since the support of any pre-delta or at least of any 'essential part' of is tightening to a point as $\varepsilon \to 0$, any given nonzero distance from it is becoming infinitely large being compared with the size of the support. The principle thus states that the limit in (18.51) is really the same for any pre-delta, and hence, if it is valid, the use of the delta function with the corresponding Green's function is justified. In addition, in the case of an infinite homogeneous body, the concept of a zero limit by itself becomes meaningless from the physical point of view since there is no any natural size. In this case, the limit can only be applied to the ratio of the support to the given distance from it, just as in Saint-Venant's principle. Thus, Saint-Venant's work of 1855 [SV1] can be mentioned as the first one in the Dirac delta function history.

Note that the principle is valid at least for a standard situation in homogeneous elasticity. However, in some models, for example for Winkler foundation, it fails.

18.1.9 Energy Flux from a Moving Force

The problem of multiplication of generalized functions arises, in particular, in determining the energy flux from the concentrated force. Consider the problem by an example of a string under a moving force. The linear wave equation for a string under a moving concentrated force applied at $t = 0$ is

$$\varrho\ddot{u}(x,t) - Tu''(x,t) = Q\delta(x - vt)H(t), \tag{18.52}$$

where ϱ, T and Q are the density per unit length, the tensile force and the external force, respectively, and $v = \text{const}$ ($|v| \neq c = \sqrt{T/\varrho}$) is the speed of the latter along the string. Using the integral transforms it can be found that the particle velocities of the string in front and behind the force are

$$\begin{aligned} \dot{u}(x,t) &= \frac{Q}{2\varrho(c-v)}H(x - vt)H(ct - x), \\ \dot{u}(x,t) &= \frac{Q}{2\varrho(c+v)}H(vt - x)H(ct + x), \end{aligned} \tag{18.53}$$

respectively. This is the Green's function multiplied by Q.

By the definition, the energy flux from the force is

$$N = Q\int_{-\infty}^{\infty} \delta(x - vt)\dot{u}(x,t)\,\mathrm{d}x. \tag{18.54}$$

However, if $v \neq 0$ the particle velocities are discontinuous at the point where the force acts. There exist different ways to overcome this difficulty. One way is to calculate energy fluxes in front and behind the force application point. The result is

$$N = \frac{Q}{2}[\dot{u}(vt+0,t) + \dot{u}(vt-0,t)] = \frac{Q^2 c}{2\varrho(c^2 - v^2)} \quad (|v| < c),$$
$$N = \frac{Q}{2}\dot{u}(vt-0,t) = \frac{Q^2 |v|}{2\varrho(v^2 - c^2)} \quad (|v| > c). \tag{18.55}$$

(The resonant case, $v = c$, is considered in Chap. 16). The result coincides with that following from (18.54) with the use a pre-delta function assuming it is symmetric.

We now show that, with respect to the relation (18.54), the pre-delta can be considered as if it where always symmetric. Moreover, with respect to a Green's function for a homogeneous system this statement is quite general. The latter follows from the fact that the pre-delta is not only the multiplier but also is the source of the prelimiting Green's function.

Denote the Green's function, which can be discontinuous, the pre-delta function and the response of the linear system to the latter by $G(x)$, $\delta_\varepsilon(x)$ and $G_\varepsilon(x)$, respectively. The *superposition principle* following from the linearity and uniformity of the system states

$$G_\varepsilon(x) = G(x) * \delta_\varepsilon(x) \equiv \int_{-\infty}^{\infty} G(\xi)\delta_\varepsilon(x-\xi)\,\mathrm{d}\xi. \tag{18.56}$$

It follows that, in the prelimiting relation, the product of the prelimiting functions, $G_\varepsilon(x)\delta_\varepsilon(x)$, is equal to the product of the Green's function, $G(x)$, and the symmetric convolution of the pre-deltas, $\Delta_\varepsilon(x)$:

$$\int_{-\infty}^{\infty} G_\varepsilon(x)\delta_\varepsilon(x)\,\mathrm{d}x = \int_{-\infty}^{\infty} G(\xi)\Delta_\varepsilon(\xi)\,\mathrm{d}\xi,$$
$$\Delta_\varepsilon(\xi) = \int_{-\infty}^{\infty} \delta_\varepsilon(x-\xi)\delta_\varepsilon(x)\,\mathrm{d}x = \delta_\varepsilon(\xi) * \delta_\varepsilon(-\xi), \tag{18.57}$$

where the symmetric function, $\Delta_\varepsilon(x)$, is also a pre-delta since

$$\delta_\varepsilon(x) \to \delta(-x) \to \delta(x)\ (\delta(-x) = \delta(x)) \text{ and } \delta(x) * \delta(x) = \delta(x). \tag{18.58}$$

Note that the convolution in (18.57) shows its symmetry explicitly: $\Delta_\varepsilon(-\xi) = \Delta_\varepsilon(\xi)$. We now may write

$$\int_{-\infty}^{\infty} G(x)\delta(x)\,\mathrm{d}x = \lim_{\varepsilon\to 0}\left[\int_{-\infty}^{0} G(\xi)\Delta_\varepsilon(\xi)\,\mathrm{d}\xi + \int_{0}^{\infty} G(\xi)\Delta_\varepsilon(\xi)\,\mathrm{d}\xi\right]$$
$$= \frac{1}{2}[G(-0) + G(+0)]. \tag{18.59}$$

18.1.10 Null Functions and Homogeneous Solutions

In physics, such functions are important because they can, in particular, reflect *remote actions*. The fact that the corresponding function tends to a null generalized function does not necessarily mean that the influence of this action vanishes. On the contrary, such a function can give rise a nontrivial homogeneous solution. Consider, for example, a two-dimensional stationary problem for the two-dimensional wave equation

$$\frac{\partial^2 u(x,y,t)}{\partial x^2} + \frac{\partial^2 u(x,y,t)}{\partial y^2} - \frac{1}{c^2}\frac{\partial^2 u(x,y,t)}{\partial t^2} = 0 \tag{18.60}$$

defined in the upper half-plane, $-\infty < x < \infty, 0 < y < \infty$, with mixed boundary conditions at $y = 0$: the function, u, is given at $x > vt,\ 0 \le v = \text{const} < c$, and the normal derivative, $\partial u/\partial y$, is given at $x < vt$. In the considered stationary problem, $u = u(\eta, y),\ \eta = x - vt$. Denoting $u(\eta) = u(\eta, 0),\ \sigma(\eta) = \mu\partial u(\eta, y)/\partial y\ (y = 0)$, and using the Fourier transform on η with the causality principle (see Sect. 15.6) the relation is found as

$$\begin{gathered}\sigma^F(\varepsilon,k) + \mu\sqrt{1-v^2/c^2}L(\varepsilon,k)u^F(\varepsilon,k) = 0,\\ L(\varepsilon,\varepsilon,k) = L_+(\varepsilon,k)L_-(\varepsilon,k),\ L_\pm(\varepsilon,k) = \sqrt{\varepsilon/(c\mp v)\mp ik}\ (0<\varepsilon\to 0).\end{gathered} \tag{18.61}$$

To solve this equation represent (see Sect. 18.5 where the Wiener-Hopf technique is discussed in more detail)

$$\sigma^F(\varepsilon,k) = \sigma_+(\varepsilon,k) + \sigma_-(\varepsilon,k),\ u^F(\varepsilon,k) = u_+(\varepsilon,k) + u_-(\varepsilon,k), \tag{18.62}$$

where the functions with subscripts $\pm$ correspond to the generalized Fourier transform on the intervals $(0, \infty)$ and $(-\infty, 0)$, respectively. For instance

$$\begin{aligned}\sigma_+(\varepsilon,k) &= \int_0^\infty \sigma(\eta)e^{ik\eta}\,d\eta\ (\Im k = \varepsilon > 0),\\ \sigma_+(\varepsilon,k) &= \int_{-\infty}^0 \sigma(\eta)e^{ik\eta}\,d\eta\ (\Im k = -\varepsilon < 0).\end{aligned} \tag{18.63}$$

As far as $\varepsilon > 0$ the plus (minus)-functions are regular in the upper (lower) half-planes of k, respectively, including the real axis (the corresponding original functions marked by $\pm$ superscripts are assumed slowly growing). We here consider the homogeneous problem assuming $\sigma^-(\eta) = u^+(\eta) = 0$.

Physically, any nontrivial solution to such a homogeneous problem must have a source located at infinity or at the moving separating point, $\eta = 0$. Allowing only an energy release through this point, we introduce the load corresponding to a source at infinity, $\eta \to -\infty$. Consider the load as

$$\sigma^{-}(\varepsilon_1, x) = -A'\varepsilon_1^{n+1/2}(-x)^n e^{\varepsilon_1 x} H(-x),\ \sigma_{-}(\varepsilon, k) = -A' \frac{\varepsilon_1^{n+1/2}\Gamma(n+1)}{(\varepsilon_1 + \mathrm{i}k)^{n+1}},$$
$$A' = \text{const},\ \varepsilon_1 = \varepsilon/(c-v),\ n = 0, 1, \ldots. \tag{18.64}$$

This expression for any nonnegative integer n satisfies the following conditions: (a) It is a regular function of x tending to zero as $\varepsilon \to +0$ (maximum of the load is equal to $\sqrt{\varepsilon_1}\, n^n e^{-n}$ $(x = -n/\varepsilon_1)$), (b) The Fourier transform of it (it is the left-sided transform) is a null generalized function (it corresponds to a generalized function multiplied by a vanishing parameter), and (c) The product with a given right-sided transform, namely, $\sigma_{-}(\varepsilon, k)(\varepsilon_1 - \mathrm{i}k)^{-1/2} \equiv \sigma_{-}(\varepsilon, k)/L_{+}(\varepsilon, k)$, tends to $C\delta(k)$ as $\varepsilon \to +0$, where $C = \text{const} \neq 0$. Recall that the product of a null function with a nonzero generalized function has the support only at the singular points; in the considered case, it can be the delta function only. It follows that the condition (c) can be replaced by the following

$$\int_{-\infty}^{\infty} \frac{\sigma_{-}(\varepsilon, x)}{\sqrt{\varepsilon_1 - \mathrm{i}k}} \mathrm{d}k \to C\ (\varepsilon \to +0). \tag{18.65}$$

In the considered case

$$C = -A'\pi 2^{1/2-2n}[1 + 2(n-1)]!!\ ((-1)!! = 1). \tag{18.66}$$

Any one of these null functions can be used to reflect the remote load for the considered homogeneous equation. The Eq. (18.61) takes the form

$$\sigma_{+}(\varepsilon, k) + \mu\sqrt{1 - v^2/c^2}\, L(\varepsilon, k) u_{-}(\varepsilon, k) = A' \frac{\varepsilon_1^{n+1/2}\Gamma(n+1)}{(\varepsilon_1 + \mathrm{i}k)^{n+1}}. \tag{18.67}$$

Clearly, it satisfies the boundary condition ($\sigma = 0$ at $x < 0$) at $\varepsilon = +0$. Next we separate the plas/mines parts in the left-hand side, and based on the above considerations and the analytical representation of the delta function (18.29), make of the same sense separation in right-hand side

$$\frac{\sigma_{+}(\varepsilon, k)}{\mu L_{+}(\varepsilon, k)} + \sqrt{1 - v^2/c^2}\, L_{-}(\varepsilon, k) u_{-}(\varepsilon, k) = A' \frac{\varepsilon_1^{n+1/2}\Gamma(n+1)}{\mu(\varepsilon_1 + \mathrm{i}k)^{n+1}(\varepsilon_1 - \mathrm{i}k)^{1/2}}$$
$$= A\left(\frac{1}{\varepsilon_1 - \mathrm{i}k} + \frac{1}{\varepsilon_1 + \mathrm{i}k}\right),\ A = \text{const}. \tag{18.68}$$

In the following, we disregard the prelimiting difference between ε and $\varepsilon/(c \pm v)$ (in fact, we could do this from the very beginning). It follows that in the limit, $\varepsilon = +0$, see Sect. 18.4

$$\sigma_+(k) = A\mu(0 - \mathrm{i}k)^{-1/2},\ u_-(k) = \frac{A}{\sqrt{1 - v^2/c^2}}(0 + \mathrm{i}k)^{-3/2},$$
$$\sigma^+(\eta) = \frac{A\mu H(\eta)}{\sqrt{\pi\eta}},\ u^-(\eta) = \frac{2A}{\sqrt{1 - v^2/c^2}}\sqrt{\frac{\eta}{\pi}}. \tag{18.69}$$

Thus, in the case where $L_+(k) = (0 - ik)^{\alpha}$, $\alpha = 1/2$, to reflect a remote source the delta function, $\delta(k)$, can be introduced in the right-hand side of the homogeneous equation (18.61) divided by $L_+(k)$. It can be shown that this is true for any positive exponent, $0 < \alpha$. In parallel, derivatives of the delta can be introduced as $\delta'(k)$ if $\alpha > 1$ and $\delta^{(n)}(k)$ if $\alpha > n$. It is in agreement with the fact that $\delta^{(n)}(k)k^{\alpha} = 0$ if $\alpha > n$, that is, in this case, the original equation (18.61) remains to be homogeneous

$$\frac{\sigma_+(k)}{(0 - \mathrm{i}k)^{\alpha}} + \mathrm{const} \times L_-(k)u_-(k) = 2\pi A\delta^{(m)}(k)$$
$$\Longrightarrow \sigma_+(k) + \mathrm{const} \times (0 - \mathrm{i}k)^{\alpha} L_-(k)u_-(k) = 2\pi A\delta^{(m)}(k)(0 - \mathrm{i}k)^{\alpha} = 0, \tag{18.70}$$

where m and n are nonnegative integer and $m \le n < \alpha$. Besides, in this sense, $\delta^{(m)}(k - k_0)$ corresponds to $L_+(k) = (0 - \mathrm{i}(k - k_0))^{\alpha}$.

18.1.11 A Singular Green's Function and the 'Dark Energy'

Consider the dynamic Green's function for the two-dimensional wave equation (18.60) defined in the upper half-plane, $y > 0$. The initial and boundary conditions are

$$u(x, y, t) = 0\ (t < 0),\ \sigma_y(x, 0, t) = \mu\frac{\partial u}{\partial y} = -\delta(t)\delta(x). \tag{18.71}$$

The function u represents displacements, in the case of the anti-plane strain of an elastic body, or the potential of velocities in a compressible fluid. Using the integral transforms, the Laplace transform on t and the Fourier transform on x we obtain equation

$$\frac{\mathrm{d}^2}{\mathrm{d}y^2}u^{LF}(k, y, s) - (k^2 + s^2/c^2)u^{LF}(k, y, s) = 0 \tag{18.72}$$

which solution vanishing at $y = \infty$ is

$$u^{LF}(k, y, s) = G^{LF}(k, y, s) = \frac{1}{\mu\sqrt{k^2 + s^2/c^2}}\mathrm{e}^{-\sqrt{k^2+s^2/c^2}\,y}, \tag{18.73}$$

where a positive value of the square root for real k, s is assumed. Such a uniform double transform can be inverted directly (see Sect. 18.3.2). In the considered case, the original represents a cylindrical wave as

$$u(x, y, t) = G(x, y, t) = \frac{c}{\pi\mu}(c^2t^2 - r^2)_+^{-1/2} \tag{18.74}$$

This generalized function can be considered as product of two functions of the type discussed above in Sect. 18.1.5

$$(c^2t^2 - r^2)_+^{-1/2} = (ct + r)_+(ct - r)_+, \tag{18.75}$$

which, however, have common singular point $t = r = 0$. It also can be considered as the generalized time derivative of the below locally integrable function (18.77) with a logarithmic singularity at $r = 0$, or as a generalized function of two variables (defined on the set of test functions of the same features as considered above but for two variables), or as a generalized function defined by the locally integrable *LF* transform (18.73)). Note that the latter expression for $k = 0$ represents a step wave of the displacement.

The wave induced by a load of a general view, $q(x, t)H(t)$, applied at the boundary, $y = 0$, is represented by the double convolution on x and t

$$u(x, y, t) = \frac{c}{\pi\mu} q(x, t) * *[c^2t^2 - x^2 - y^2]_+^{-1/2}. \tag{18.76}$$

In particular, for an invariable load concentrated at $x = 0$, $q(x, t) = Q_0\delta(x)$

$$\begin{aligned} u(x, y, t) &= Q_0 G_x(x, y, t) \\ &= \frac{Q_0}{\pi\mu} \ln\left(\frac{ct}{r} + \sqrt{\left(\frac{ct}{r}\right)^2 - 1}\right) H(ct - r), \end{aligned} \tag{18.77}$$

where $r = \sqrt{x^2 + y^2}$. This can also be represented as a sum of two terms

$$\begin{aligned} u(x, y, t) &= u_{qst}(x, y, t) + u_{dyn}(x, y, t), \\ u_{qst}(x, y, t) &= -\frac{Q_0}{\pi\mu} \ln(r)H(ct - r), \\ u_{dyn}(x, y, t) &= \frac{Q_0}{\pi\mu} \ln\left(ct + \sqrt{(ct)^2 - r^2}\right) H(ct - r). \end{aligned} \tag{18.78}$$

The first, quasistatic term is infinite at $r = 0$. This means that the concept of a concentrated force is not applicable to such a continuous medium, and the considered solution can be used only as Green's function, that is, as an intermediate result useful in obtaining a solution for the action of a distributed load. The latter, dynamic term is regular; however it tends to infinity as $t \to \infty$ at any given r, that is, there is no static limit in this case. This is due to the fact that, in a plane static state, the infinite medium cannot carry any load of a nonzero main vector, that is, the formulation based on the assumption of infinity cannot be accepted in statics. Note that the latter remark is valid not only for the two-dimensional wave equation but also for the plane elasticity (see Part IV), whereas the displacement at $r = 0$, where the concentrated force is applied, is infinite in tree-dimensional elasticity as well.

The fact is important that since, in the solution, the displacement ar $r = 0$ is infinite, the energy of the body $Q_0 u(0, 0, t) = \infty$. The is still valid for any small nonzero load, and this suggests the conclusion that a zero load can produce a nonzero energy. In this end, consider the load uniformly distributed at a segment, such that the prelimiting delta $\delta_\varepsilon = H(x^2 - \varepsilon^2)/(2\varepsilon)$. In accordance with (18.57) the energy is

$$\begin{gathered}\mathcal{E} = Q^2 \int_{-\infty}^{\infty} G_\varepsilon(x, 0, t)\delta_\varepsilon(x)\,\mathrm{d}x = Q^2 \int_{-\infty}^{\infty} G_x(x, 0, t)\Delta_\varepsilon(x)\,\mathrm{d}x, \\ \Delta_\varepsilon(x) = \delta_\varepsilon(x) * \delta_\varepsilon(-x),\end{gathered} \tag{18.79}$$

The symmetric pre-delta, $\Delta_\varepsilon(x)$, is found in (18.57). Now it is

$$\Delta_\varepsilon(x) = \frac{1}{2\varepsilon}\left(1 - \frac{|x|}{2\varepsilon}\right) H\left(1 - \frac{|x|}{2\varepsilon}\right) \tag{18.80}$$

and with refer to (18.77)

$$\mathcal{E} = \frac{Q_0^2}{\pi\varepsilon\mu} \int_0^{2\varepsilon} \ln\left(\frac{ct}{x} + \sqrt{\left(\frac{ct}{x}\right)^2 - 1}\right)\left(1 - \frac{x}{2\varepsilon}\right) \mathrm{d}x. \tag{18.81}$$

It follows that for $ct \le 2\varepsilon$

$$\mathcal{E} = \frac{Q_0^2}{\pi\mu} \frac{ct(2\pi\varepsilon - ct)}{4\varepsilon^2}, \tag{18.82}$$

whereas for $ct \ge 2\varepsilon$

$$\begin{aligned}\mathcal{E} &= \frac{Q_0^2}{\pi\mu}\left[\ln\left(\frac{ct}{2\varepsilon} + \sqrt{\left(\frac{ct}{2\varepsilon}\right)^2 - 1}\right) + \frac{ct(\sqrt{c^2t^2 - 4\varepsilon^2} - ct + 4\varepsilon \arcsin(2\varepsilon/(ct))}{4\varepsilon^2}\right] \\ &\sim \frac{Q_0^2}{\pi\mu}\left(\ln\frac{ct}{\varepsilon} + \frac{3}{2}\right) \quad (t > 0,\ \varepsilon \to 0).\end{aligned} \tag{18.83}$$

Thus, if the force is fixed the energy increases unboundedly as the support of the load tends to a point. This, of course, also follows directly from the fact that, in the framework of the continuous medium model, the displacement under the concentrated force is infinite. Now consider the prelimiting solution (18.83) with $Q_0 = Q_{00}/|\ln\varepsilon|^\nu$. We find that as $\varepsilon \to 0$

$$\mathcal{E} \to 0\ (\nu > 1/2),\ \mathcal{E} \to \frac{Q_{00}^2}{\pi\mu}\ln ct\ (\nu = 1/2),\ \mathcal{E} \to \infty\ (\nu < 1/2). \quad (18.84)$$

At the same time $Q_0 \to 0$ if $\nu > 0$. We come to the following conclusion. In a continuous medium, zero force as a null-function action can produce finite or even infinite energy. Added to this is that, in this case, all the energy is stored in the singular point, $r = 0$. Indeed, the area outside this point, $0 < r \le ct$, is finite at any finite time, while the energy density in this domain tends to zero as $\varepsilon \to 0$.

Mechanics of elastic continuum gives different examples of loss of energy. One of them is considered above, the other is the energy release in fracture. It appears in a moving singular stress-strain field. (Griffith had founded fracture mechanics (Griffith, 1921) by the fracture criterion stated that the energy release, by which a quasi-statically growing crack is accompanied, should be equal to the surface energy, that is the energy released from the bulk of the body is converted into the energy of new surfaces. The surface energy, however, is outside the framework of the elasticity.) In the case of a crack on an elastic body, the crack tip singularities are of the square root type; however, in beam/plate classical model, there exist an antisymmetrical fracture mode with the stresses in front of the crack described by a generalized function $x_+^{-3/2}$. Of course, this model is not acceptable for the estimation of the stresses; however, it gives a correct result concerning the energy release rate. Well-known Barenblatt's cohesive forces introduced in a vicinity of the crack tip to eliminate the singularity (Barenblatt, 1959; 1959b) accumulate the energy released from the body thus 'restoring' the energy conservation low . Under this condition the total cohesive force vanishes as the support of the cohesive forces is tighten to the crack tip. Thus, in the limit, the zero force produces a nonzero energy (see Slepyan, 2002).

Such paradoxical phenomena are inherent to the continuous models.

18.2 Fourier Transform

With its modifications and consequences, the Fourier transform is the primary mathematical tool in the analytical study of linear waves in homogeneous, discrete and continuous-discrete periodic structures.

18.2.1 The Transform

The Fourier transform, $f^F(k)$, of a regular (*locally integrable*) function, $f(x)$, $-\infty < x < \infty$, is

$$f^F(k) = \int_{-\infty}^{\infty} f(x)e^{ikx}\, dx, \quad -\infty < k < \infty, \tag{18.85}$$

where $f(x)$ is called the *original function*, more briefly the *original*. The transform exists if the integral converges. This, however, imposes too restrictive conditions on the original. For example, the integral does not converge if the original is a constant. An essential extension of the applicability of the transform is achieved by the *generalized Fourier transform*

$$f_\alpha^F = f_+(k) + f_-(k), \tag{18.86}$$

with the one-sided transforms

$$f_+(k) = \int_0^{\infty} f(x)e^{ik_+x}\, dx,\ f_-(k) = \int_{-\infty}^{0} f(x)e^{ik_-x}\, dx,\ k_\pm = k \pm i\alpha, \tag{18.87}$$

where k is real, $-\infty < k < \infty$, and $\alpha > 0$. The one-sided transforms are applicable for constants, polynomials and any locally integrable *slowly growing function* (the latter term means that there exists a number, n, such that $|f(x)| < |x|^n$ as $x \to \pm\infty$). Moreover, in this case, the one-sided transforms, $f_\pm(k)$, are analytical functions with no singular points in the upper and lower half-planes of k, respectively.

The functions $f_\pm(k)$ analytic in the (open) upper and lower half-planes of the complex variable k can be analytically continued to the lower and upper (closed) half-planes, respectively. Each of these functions, however, may have singularities outside the open half-planes of regularity, in particular, on the real k-axis.

As follows from (18.87), as $\alpha \to +0$, k_+-axis goes down to the real axis, while k_- goes up to it. This means the in the Fourier transform Sect. 18.2, the real singular points of $f_\pm^F(k)$ are placed at $\Im k = \mp 0$, respectively. This conclusion is important for the inverse transform on the complex plane, see Sect. 18.2.2.

Some Useful Relations

The transform of a derivative. The transform of a derivative follows by integration by parts

$$(u'(x))^F = -iku^F(k),\ (\delta'(x))^F = -ik. \tag{18.88}$$

The transform of an integral. In the same way, the transform of integrals

$$\int_0^{\infty}\left(\int_0^x u(\xi)\,\mathrm{d}\xi\right)\mathrm{e}^{\mathrm{i}(k+\mathrm{i}0)}\,\mathrm{d}x = \frac{1}{0-\mathrm{i}k}u_+(k),$$
$$\int_{-\infty}^{0}\left(\int_x^0 u(\xi)\,\mathrm{d}\xi\right)\mathrm{e}^{\mathrm{i}(k-\mathrm{i}0)}\,\mathrm{d}x = \frac{1}{0+\mathrm{i}k}u_-(k). \tag{18.89}$$

The Transform of Power Functions

$$(H(x))^F = \frac{1}{0-\mathrm{i}k}\quad (H'(x)=\delta(x)),$$
$$(x^{\nu}H(x))^F = \frac{\Gamma(\nu+1)}{(0-\mathrm{i}k)^{\nu+1}}\quad (\nu\neq -1,-2,\ldots). \tag{18.90}$$

The transform of a convolution. The convolution is defined in (17.34)[1] If the change of the integration order is justified, the transform of it is expressed as the product of the transforms:

$$(f(x)*g(x))^F = \int_{-\infty}^{\infty}\int_{-\infty}^{\infty} f(\xi)\mathrm{e}^{\mathrm{i}k\xi}g(x-\xi)\mathrm{e}^{\mathrm{i}k(x-\xi)}\,\mathrm{d}\xi\,\mathrm{d}x$$
$$= \int_{-\infty}^{\infty} f(\xi)\mathrm{e}^{\mathrm{i}k\xi}\,\mathrm{d}\xi\int_{-\infty}^{\infty} g(x-\xi)\mathrm{e}^{\mathrm{i}k(x-\xi)}\,\mathrm{d}x = f^F(k)g^F(k). \tag{18.91}$$

If the functions do not satisfy the above condition, a normalization can be used, for example, by multiplication by a fast enough decreasing function $\varphi(x,\alpha)$, such that $\varphi(x,\alpha)\to 1$ as $\alpha\to 0$.

18.2.2 The Inverse Transform

In the inverse transform, we make the below integral convergent in the same way as in the direct transform. We introduce a function of $z = x\pm \mathrm{i}\beta$ as follows

$$f(x,\alpha,\beta) = f^+ + f^-,$$
$$f^+ = \frac{1}{2\pi}\int_0^{\infty} f_{\alpha}^F\mathrm{e}^{-(\beta+\mathrm{i}x)k}\,\mathrm{d}k = \frac{1}{2\pi}\int_0^{\infty}\int_{-\infty}^{\infty} f(\xi)\mathrm{e}^{-\alpha|\xi|-(\beta+\mathrm{i}(x-\xi))k}\,\mathrm{d}\xi\,\mathrm{d}k, \tag{18.92}$$
$$f^- = \frac{1}{2\pi}\int_{-\infty}^{0} f_{\alpha}^F\mathrm{e}^{(\beta-\mathrm{i}x)k}\,\mathrm{d}k = \frac{1}{2\pi}\int_{-\infty}^{0}\int_{-\infty}^{\infty} f(\xi)\mathrm{e}^{-\alpha|\xi|+(\beta-\mathrm{i}(x-\xi))k}\,\mathrm{d}\xi\,\mathrm{d}k,$$

[1] A uniform system response to an action expressed by a convolution of Green's function with an action was considered in Sect. 17.1.3.

where both, α and β are positive. Under this condition, we can change the integration order. We obtain (see Sect. 17.1.3)

$$
\begin{aligned}
f(x,\alpha,\beta) &= \int_{-\infty}^{\infty} f(\xi)\mathrm{e}^{-\alpha|\xi|}\delta(x-\xi,\beta)\,\mathrm{d}\xi, \\
\delta(x-\xi,\beta) &= \frac{\beta}{\pi((x-\xi)^2+\beta^2)}, \ \delta(x-\xi,+0)=\delta(x-\xi).
\end{aligned}
\tag{18.93}
$$

Note that if the original is such that there is no need in the analytical extensions, and you can do under $\alpha=\beta=0$, the direct and inverse transforms are

$$
f^F(k) = \int_{-\infty}^{\infty} f(x)\mathrm{e}^{\mathrm{i}kx}\,\mathrm{d}x, \ f(x) = \frac{1}{2\pi}\int_{-\infty}^{\infty} f^F(k)\mathrm{e}^{-\mathrm{i}kx}\,\mathrm{d}k. \tag{18.94}
$$

The inverse transform of a test function

$$
\varphi(x) = \frac{1}{2\pi}\int_{-\infty}^{\infty}\varphi^F(k)\mathrm{e}^{-\mathrm{i}kx}\,\mathrm{d}k = \frac{1}{2\pi}\int_{-\infty}^{\infty}\int_{-\infty}^{\infty}\varphi(\xi)\mathrm{e}^{\mathrm{i}k(\xi-x)}\,\mathrm{d}\xi\,\mathrm{d}k. \tag{18.95}
$$

The delta function follows as

$$
\frac{1}{2\pi}\int_{-\infty}^{\infty}\mathrm{e}^{\mathrm{i}kx}\,\mathrm{d}k = \delta(x). \tag{18.96}
$$

Note that the Fourier transform of the (infinitely differentiable, finite support) test function is an analytic *fast-decreasing function*, that is, a function decreasing (as $k\to\pm\infty$) faster than $|k|^{-n}$ for any n.

Cauchy Type Integral

Let us represent the relation (18.96) as the limit

$$
\begin{aligned}
\delta(x) &= \frac{1}{2\pi}\int_{-\infty}^{\infty}\mathrm{e}^{\mathrm{i}kx}\,\mathrm{d}k = \lim_{y\to 0}(\delta_+(z)+\delta_-(z)), \\
\delta_+(z) &= \frac{1}{2\pi}\int_{0}^{\infty}\mathrm{e}^{\mathrm{i}kz}\,\mathrm{d}k = -\frac{1}{2\pi\mathrm{i}z} \ (z=x+\mathrm{i}y, y>0), \\
\delta_-(z) &= \frac{1}{2\pi}\int_{-\infty}^{0}\mathrm{e}^{\mathrm{i}kz}\,\mathrm{d}k = \frac{1}{2\pi\mathrm{i}z} \ (z=x+\mathrm{i}y, y<0).
\end{aligned}
\tag{18.97}
$$

It follows that the Cauchy type integral in the form

$$
f(z) = f_+(z)+f_-(z), \ f_\pm(z) = \mp f(x)*\delta_\pm(z), \tag{18.98}
$$

where $f(x)$ gives us a representation of is a slowly growing function, $f(x)$ by two functions analytical in the upper $(f_+(z))$ and lower $(f_-(z))$ half-planes. In sum, these functions tends, $y \to 0$, to $f(x)$ (at the point of continuity)

$$f(x) = \lim_{y \to 0}(f_+(z) + f_-(z))$$
$$= \frac{1}{2\pi}\left(\int_0^{\infty} \frac{f(\xi)}{\xi - x - \mathrm{i}0} - \int_{-\infty}^{0} \frac{f(\xi)}{\xi - x + \mathrm{i}0}\right). \quad (18.99)$$

Thus, as was already discussed in Sect. 17.1.3, the Cauchy type integral gives analytical representation (18.98) for functions of a real variable. The Cauchy type integral is used, in particular, in the Wiener-Hopf technique (see Sect. 18.5).

The Inverse Transform on the Complex Plane

The inverse transform path can shifted on the k-plane without crossing singular points and branch cuts in the lower or upper half-plane for $f_{\pm}$, respectively, as shown in Fig. 18.1 for the former.

In this way, we can make such a *normalization* of slow growing transforms that the integration over the the semicircles, $R = \text{const}$, tends to zero as $R \to \infty$. This allows us to represent the original, $g(x)$, as a sum of contributions of the singularities.

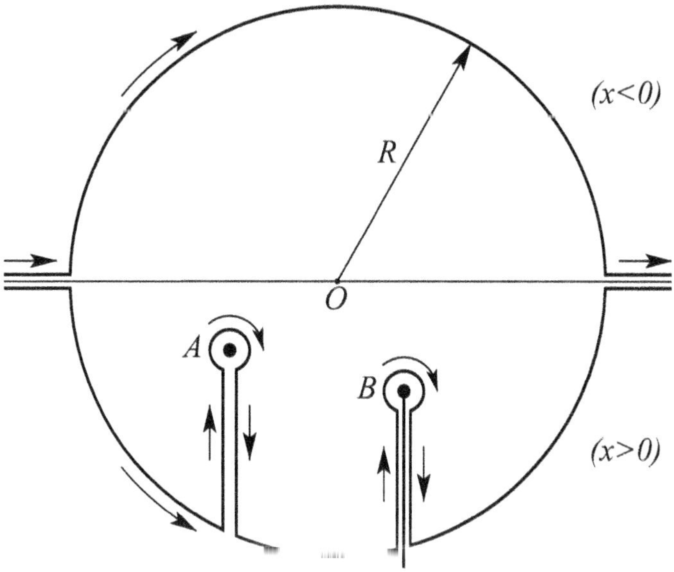

Fig. 18.1 Integration contours in inverse transform of $f_+(k)$. For the poles the contributions of the integration over the cut banks disappears. The upper path for $x < 0$ gives zero, as it should be

Finally, in accordance with (18.89)

$$f(x) = \frac{\mathrm{d}^n}{\mathrm{d}x} g(x). \tag{18.100}$$

Consider the right-side Fourier transform f_+ and its normalised version

$$g_+ = \frac{f_+}{(0 - \mathrm{i}k)^n}, \tag{18.101}$$

where the integer number n is minimal one under which

$$|g_+| \to 0 \ (R \to \infty). \tag{18.102}$$

Under the condition (18.102), the *Jordan's lemma* states that the integral taken over the lower semicircle, R = const, tends to zero as $R \to \infty$, and only contributions of poles and cuts (in the case of branch points) remains.

Remaind that for a singular point of $f_\pm$ on the real axes the integration contour should be placed above or below it, respectively (see Sect. 18.2.1). For example, the transform of the Heaviside function (18.90) has the pole at $k = -\mathrm{i}0$.

The contribution of a pole at $k = k_0$,

$$u^F(k) \sim \frac{1}{k - k_0} \ (k \to k_0), \tag{18.103}$$

given by the integration over a small enough circumstance (no singularities, except the pole, to be in the closed domain inside the integration contour, r = const) is

$$u_0(x) = \int_0^{2\pi} f^F(k) r \mathrm{i} \mathrm{e}^{\mathrm{i}\theta}\, \mathrm{d}\theta = 2\pi \mathrm{i}, \ (k - k_0 = r\mathrm{e}^{\mathrm{i}\theta}). \tag{18.104}$$

In the case of the Heaviside function, $H(x)$, the pole is $1/(-\mathrm{i}k)$, $f^F = f_+ \sim f_0^F = 1/(0 - \mathrm{i}k) = \mathrm{i}\mathrm{e}^{-\mathrm{i}\theta}/r$, the integration direction is opposite (clockwise), and there is the multiplier $1/(2\pi)$. So, we return to the original. Namely,

$$f(x) = \frac{1}{2\pi} \int_{-\infty}^{\infty} f_+ \mathrm{e}^{-\mathrm{i}kx}\, \mathrm{d}k = \frac{1}{2\pi} \int_{3\pi/2}^{-\pi/2} f_0^F r \mathrm{i} \mathrm{e}^{\mathrm{i}\theta}\, \mathrm{d}\theta = H(x). \tag{18.105}$$

Cosine and Sine Transforms

If a function is symmetric, $f(-x) = f(x)$, or antisymmetric, $f(-x) = -f(x)$, the Fourier transform is reduced to the cosine or sine transform, respectively

$$
\begin{aligned}
f^F(k) &= \int_{-\infty}^{\infty} f(x)e^{ikx}\,dx = 2f^C(k) = 2\int_0^{\infty} f(x)\cos(kx)\,dx, \\
f(x) &= \frac{1}{2\pi}\int_{-\infty}^{\infty} f^F(k)e^{-ikx}\,dk = \frac{2}{\pi}\int_0^{\infty} f^C(k)\cos(kx)\,dk
\end{aligned}
\tag{18.106}
$$

and

$$
\begin{aligned}
f^F(k) &= \int_{-\infty}^{\infty} f(x)e^{ikx}\,dx = 2if^S(k) = 2i\int_0^{\infty} f(x)\sin(kx)\,dx, \\
f(x) &= \frac{1}{2\pi}\int_{-\infty}^{\infty} f^F(k)e^{-ikx}\,dk = \frac{2}{\pi}\int_0^{\infty} f^S(k)\sin(kx)\,dk.
\end{aligned}
\tag{18.107}
$$

The transforms of the derivatives are obtained by the integration by parts

$$
(f'(x))^S(k) = -kf^C(k), \; (f'(x))^C(k) = -f(0) + kf^S(k). \tag{18.108}
$$

These transforms are useful for functions defined on a semi-infinite region and analytically continued as the symmetric or antisymmetric ones. The continuations are valid if the even or odd derivatives, respectively, at $x = 0$ are zero.

For example, for $f^{IY}(x)$, if in the first case, $f'(0)$ and $f'''(0)$ are given, we can assume the function such that the odd derivatives at $x = 0$ to be zero, and to use the cosine transform. Otherwise, if $f(0)$ and $f''(0)$ are given, then sine transform is applicable.

18.3 Laplace Transform and LF Double Transform

18.3.1 The Transforms

The *one-sided Laplace transform Laplace transform one-sided* or simply the *Laplace transform* is the right-sided Fourier transform with $k = is$. The latter rename means that the s-complex plate is the k-plane rotate by an angle $-\pi/2$, that is clockwise.

$$
f^L(s) = f_+(is) = \int_{-0}^{\infty} f(t)e^{-st}\,dt, \tag{18.109}
$$

where the real part of s is assumed large enough, namely, if $f(t)$ is locally integrable and $f(t) = O[\exp(\alpha_* t)]$, $t \to \infty$, then $\Re s > \alpha_*$. It is assumed here that α_* is a minimal value suitable for this estimate. The integration limit (-0) denotes that the functions of the point support at $t = 0$ are also incorporated. We will not repeat this minus, but keep this remark in mind.

The above "useful relations" become The transform of a derivative

$$(u'(t))^L = su^L(s), \ (\delta'(t))^L = s. \tag{18.110}$$

The transform of an integral

$$\int_0^\infty \left(\int_0^t u(\eta)\, \mathrm{d}\eta \right) \mathrm{e}^{-st}\, \mathrm{d}t = \frac{1}{s} u^L(s). \tag{18.111}$$

The transform of power functions

$$\begin{aligned} (H(t))^L &= \frac{1}{s} \quad (\mathrm{d}H(t)/\,\mathrm{d}t = \delta(t)), \\ (t^\nu H(t))^L &= \frac{\Gamma(\nu+1)}{s^{\nu+1}} \quad (\nu \neq -1, -2, \ldots). \end{aligned} \tag{18.112}$$

The transform of a convolution (18.91) remains valid

$$(f(t) * g(t))^L = f^L(s) g^L(s). \tag{18.113}$$

In addition, the Laplace transform of a function (if exists) multiplied by an exponent, is

$$\left(f(t)\mathrm{e}^{s_0 t}\right)^L = f^L(s - s_0). \tag{18.114}$$

The delay theorem states that the Laplace transform of a disturbance starting as $t \geq \eta > 0$ is

$$\begin{aligned} (f(t-\tau))^L = \int_\tau^\infty f(t-\tau)\mathrm{e}^{-st}\, \mathrm{d}t = \mathrm{e}^{-s\tau} \int_0^\infty f(t)\mathrm{e}^{-st}\, \mathrm{d}t = \mathrm{e}^{-s\tau} f^L(s), \\ f(t) = 0 \ \ (t < 0). \end{aligned} \tag{18.115}$$

The rename $k = \mathrm{i}s$ means that the s-complex plane is the k-plane rotated by angle $-\pi/2$ (that is rotated clockwise). The inverse transform formula follows as

$$f(t) = \frac{1}{2\pi \mathrm{i}} \int_{\alpha_0 - \mathrm{i}\infty}^{\alpha_0 + \mathrm{i}\infty} f^L(s)\mathrm{e}^{st}\, \mathrm{d}s \ \ (\alpha_0 > \alpha_*). \tag{18.116}$$

The only difference is that here we admit exponentially growing originals. So, the inverse transform path places at a proper distance from the axis (now it is the imaginary one), $\Re s = \alpha_0 > \alpha_*$. For $t < 0$ it defines the original equal to zero since there are no singular points in the right, at $\Re s \geq \alpha_0$.

Remarkable that similarly to the Fourier inverse transform the original function, $f(t)$. can be determined as the contribution of the singularities places at the left, $\Re s < \alpha_0$.

18.3.2 *Double, Laplace and Fourier Transform (LF Transform)*

In general, the inverse transform is defined as a double integral, as the Fourier and Laplace inverse transforms, one after another. This general method of inversion can be greatly simplified in the case of a homogeneous structure of the transform.

To this end consider a version of the *Cagniard-de Hoop method* applicable in obtaining the original of a certain type of the double transform, the Laplace transform on t and the Fourier transform on x. Let the LF transform be a uniform function of s and k, such that after the substitution $k = (s/c)q$ the double transforms

$$f^{LF}(s/c, k) = u_0^{LF}(s/c, k)\mathrm{e}^{-w(s/c,k)} \tag{18.117}$$

becomes

$$f^{LF}(s/c, qs/c) = (s/c)^{-\nu}u_0^{LF}(1, q)\mathrm{e}^{-(s/c)w(1,q)}, \tag{18.118}$$

where w is either zero or, otherwise

$$\begin{gathered} w(1, -q) = w(1, q) > 0\ (\Im q = 0), \\ \frac{\mathrm{d}w(1, q)}{\mathrm{d}q} > 0\ (q > 0),\ w_0 = w(1, 0) \geq 0,\ w(1, q) \to \infty\ (q \to \infty), \end{gathered} \tag{18.119}$$

where ν is a constant. Such a transform corresponds to a Green's function for a system possessing no length and time natural units.

In the Cagniard-de Hoop method, Cagniard (1962), the Fourier inversion formula is rearranged into the form as the Laplace transform. As a result, the original can be found without any integration. However, this requires the integration path to be deformed, that makes the method more complicated. In the considered version of the method, Slepyan (1972), there is no need for the deformation of the integration path. Instead analytical functions delivered by one-side transforms are considered.

In this way, we represent the Laplace transform f^L as the inverse Fourier transform of f^{LF} as follows

$$\begin{aligned} f^L(s/c, (s/c)q) &= \frac{(s/c)^{1-\nu}}{2\pi}u^L,\ u^L = u_+ + u_-, \\ u_+ &= \int_0^\infty u_0^{LF}(1, q)\exp(-(s/c)(w(1, q) + \mathrm{i}xq)\,\mathrm{d}q, \\ u_- &= \int_{-\infty}^0 u_0^{LF}(1, q)\exp(-(s/c)(w(1, q) + \mathrm{i}xq)\,\mathrm{d}q \\ &= \int_0^\infty u_0^{LF}(1, -q)\exp(-(s/c)(w(1, -q) - \mathrm{i}xq)\,\mathrm{d}q. \end{aligned} \tag{18.120}$$

Next, we rename

$$x \longrightarrow \pm \mathrm{i} z_{\pm}, \ (z_{\pm} = \beta \pm \mathrm{i}x, 0 < \beta \to 0), \tag{18.121}$$

for $u_{\pm}$, respectively. Note that $u_{\pm}$ are analytic functions in the upper/lower half-planes of the complex variable x, and we can consider $z_+ = z_- = \beta > 0$ with the analytic continuation $\beta \longrightarrow \mp \mathrm{i}x + 0$ at the final (or any other) step of the considerations.

Now

$$\begin{aligned} u_+ &= \int_0^\infty u_0^{LF}(1, q) \exp(-(s/c)(w(1, q) - z_+ q)\, \mathrm{d}q, \\ u_- &= \int_0^\infty u_0^{LF}(1, -q) \exp(-(s/c)(w(1, -q) - z_- q)\, \mathrm{d}q. \end{aligned} \tag{18.122}$$

In the first case, $w \equiv 0$, we put $q = ct/z_{\pm}$ for $u_{\pm}$, respectively. We find

$$u^L = \int_0^\infty \left(\frac{1}{z_+} u_0^{LF}(1, ct/z_+) + \frac{1}{z_-} u_0^{LF}(1, ct/z_-) \right) \mathrm{e}^{-st}\, \mathrm{d}t. \tag{18.123}$$

Thus, in this case,

$$\begin{aligned} u(x, t) &= \left(\frac{1}{z_+} u_0^{LF}(1, ct/z_+) + \frac{1}{z_-} u_0^{LF}(1, ct/z_-) \right). \\ f(x, t) &= \frac{1}{2\pi \Gamma(\nu - 1)} t^{\nu - 2} * u(x, t). \end{aligned} \tag{18.124}$$

An example. Consider the function

$$\begin{aligned} f^{LF}(s/c, k) &= \frac{2}{(s/c)^2 + k^2} \Longrightarrow \frac{2}{(s/c)^2(1 + q^2)} \ (k = (s/c)q), \\ f^L &= \frac{2}{(s/c)2\pi} \int_{-\infty}^\infty \frac{1}{1 + q^2} \mathrm{e}^{-\mathrm{i}(s/c)xq}\, \mathrm{d}q \\ \Longrightarrow & \frac{1}{(s/c)\pi} \left(\int_0^\infty \frac{1}{1 + q^2} \mathrm{e}^{-\mathrm{i}(s/c)xq}\, \mathrm{d}q + \int_0^\infty \frac{1}{1 + q^2} \mathrm{e}^{\mathrm{i}(s/c)xq}\, \mathrm{d}q \right) \mathrm{e}^{-\mathrm{i}(s/c)xq}\, \mathrm{d}q. \end{aligned} \tag{18.125}$$

Next, substituting $x = \mp \mathrm{i} z_{\pm}$, $q = ct/z_{\pm}$ we come to the Laplace transform

$$f^L = \frac{c^2}{s\pi} \int_0^\infty \left(\frac{z_+}{z_+^2 + (ct)^2} + \frac{z_-}{z_-^2 + (ct)^2} \right) \mathrm{e}^{-st}\, \mathrm{d}t. \tag{18.126}$$

From here we obtain the original

$$f(x,t) = \frac{c^2}{\pi}\int_0^t \left(\frac{z_+}{z_+^2 + (ct)^2} + \frac{z_-}{z_-^2 + (ct)^2}\right) \mathrm{d}t$$

$$= \frac{c^2}{2\pi}\int_0^t \left(\frac{1}{ct - x + \mathrm{i}\beta} + \frac{1}{ct - x - \mathrm{i}\beta} + \frac{1}{ct + x + \mathrm{i}\beta} - \frac{1}{ct + x - \mathrm{i}\beta}\right) \mathrm{d}t \tag{18.127}$$

$$= \frac{c^2}{\pi}\int_0^t \left(\frac{\beta}{(ct - x)^2 + \beta^2} + \frac{\beta}{(ct + x)^2 + \beta^2}\right) \mathrm{d}t$$

$$\to c^2 \int_0^t (\delta(ct - x) + \delta(ct + x))\, \mathrm{d}t = cH((ct)^2 - x^2),$$

where the limit corresponds to $\beta \to +0$. Concerning the prelimiting expression of the delta function see Sect. 17.1.3.

In the next, more general case, $w \neq 0$, to represent the inverse Fourier transform as the direct Laplace one, we put (in this connection, see (18.120))

$$w(1,q) = v(q) + w(1,0),\ v(q) + zq = ct,\ z > 0 \text{ (finally } z \mp \mathrm{i}x + 0 \text{ for } u_\pm),$$

$$\mathrm{d}q = \frac{\mathrm{d}q}{\mathrm{d}t}\,\mathrm{d}t = \frac{c\,\mathrm{d}t}{\mathrm{d}v(q)/\,\mathrm{d}q + z} = \frac{c\,\mathrm{d}t}{\mathrm{d}w(q)/\,\mathrm{d}q + z}. \tag{18.128}$$

Under the conditions in (18.119) the solution of this equation, $q = q(t, z)$ exists, and as follows from (18.128), $q = 0$ $(t = 0)$ $(q = \infty$ $(t = \infty$, and the integration interval under the change $q \longrightarrow t$ is not changed. The inverse transform becomes

$$u^{L}_{\pm}(z,s) = \int_0^\infty u_0^{LF}(1, \pm q(z,t))\frac{\partial q(z,t)}{\partial t}\mathrm{e}^{-st}\,\mathrm{d}t. \tag{18.129}$$

Hence

$$u_\pm(z,t) = u_0^{LF}(1, \pm q(z,t))\frac{\partial q(z,t)}{\partial t}. \tag{18.130}$$

Next example. Consider the LF transform

$$f^{LF}(k,y,s) = \frac{1}{\sqrt{k^2 + (s/c)^2}}\mathrm{e}^{-|y|\sqrt{k^2+(s/c)^2}},\ y > 0. \tag{18.131}$$

The relation (18.130) becomes

$$w = |y|\sqrt{q^2+1},\ u_\pm(x,y,t) = \frac{1}{\sqrt{q^2+1}}\frac{c}{\mathrm{d}w(q)/\mathrm{d}q + z}$$
$$= \frac{c}{|y|q + z\sqrt{q^2+1}}, \tag{18.132}$$

where q satisfies the equation (in this connection, see (18.128))

$$|y|\sqrt{q^2+1} + zq = |y| + c(t-\tau) = ct, \tag{18.133}$$

where the time delay $\tau = |y|/c$ corresponds to the multiplier $\exp(-(|y|/c)s)$ in the Laplace transform. It follows

$$q = \frac{-ctz + y\sqrt{c^2t^2 - y^2 + z^2}}{y^2 - z^2},\ \sqrt{q^2+1} = \frac{ct - zq}{|y|}. \tag{18.134}$$

In the considered case, $\nu = 1$, $u_+ = u_-$, and finally we obtain

$$f(x,y,t) = \frac{1}{2\pi}(u_+ + u_-) = \frac{H(t^2 - r^2/c^2)}{\pi\sqrt{t^2 - r^2/c^2}},$$
$$r = \sqrt{x^2+y^2}\ \ (z^2 = -x^2). \tag{18.135}$$

18.3.3 Integral Transforms in Unmoving and Moving Coordinates

Let in the unmoving coordinate, the wave satisfy a linear equation

$$\mathcal{L}(x,t) ** u(x,t) = Q(\eta,t) = Q_0(\eta)\mathrm{e}^{\mathrm{i}\omega_0 t},\ \eta = x - vt. \tag{18.136}$$

where $\mathcal{L}$ is a uniform linear (x,t)-operator.

We here consider two type of the integral transforms related to the wave excited by a moving (and oscillating) load. Namely, the *LF* transform, as the Laplace transform on t, and the Fourier transform on x, where x is the unmoving coordinate, and the LF_η transform on t and $\eta = x - vt$, $v = \text{const}$, the moving coordinate.

The connection between the *LF* and LF_η transforms of a function is as follows

$$f^{LF}(s,k) = \int_{-\infty}^{\infty}\int_0^{\infty} f(x,t)\mathrm{e}^{-st+\mathrm{i}kx}\,\mathrm{d}t\,\mathrm{d}x$$
$$= \int_{-\infty}^{\infty}\int_0^{\infty} f(vt+\eta,t)\mathrm{e}^{-(s-\mathrm{i}kv)t+\mathrm{i}k\eta}\,\mathrm{d}t\,\mathrm{d}\eta = f^{LF_\eta}(s - \mathrm{i}kv, k). \tag{18.137}$$

Accordingly

$$f^{LF_\eta}(s,k) = f^{LF}(s + \mathrm{i}kv, k). \tag{18.138}$$

The load (the right-hand part of (18.136) LF_η transform

$$Q^{LF_\eta} = \frac{Q_0^{F_\eta}(k)}{s - \mathrm{i}\omega_0}. \tag{18.139}$$

The LF transform of the Eq. (18.136) can be represented as

$$\mathcal{L}^{LF}(s,k)u^{LF}(s,k) = \frac{Q_0^{F_\eta}(k)}{(s - \mathrm{i}(\omega_0 + kv))}, \tag{18.140}$$

and (with account of (18.137))

$$u^{LF}(s,k) = u^{LF_\eta}(s - \mathrm{i}kv, k) = \frac{Q_0^{F_\eta}(k)}{\mathcal{L}^{LF}(s,k)(s - \mathrm{i}(\omega_0 + kv))}, \tag{18.141}$$

Note that it is convenient to preserve the LF-transform of $\mathcal{L}(x,t)$.
Let us now put $s = \mathrm{i}(\omega_0 + kv) + s'$, $s' \to 0$. We have

$$u^{LF_\eta}(s' + \mathrm{i}\omega_0) = \frac{Q_0^{F_\eta}(k)}{s'\mathcal{L}^{LF}(s' + \mathrm{i}(\omega_0 + kv), k)}, \tag{18.142}$$

The limit at $t \to \infty$, if exists, is equal to

$$\lim_{s' \to 0} u^{LF_\eta}(s' + \mathrm{i}\omega_0) = \frac{Q_0^{F_\eta}(k)}{s'\mathcal{L}^{LF}(s' + \mathrm{i}(\omega_0 + kv), k)}. \tag{18.143}$$

In the resonance case, $v_g = v$, the lower of the nonzero derivative, $\mathrm{d}^{(n)}\omega/\,\mathrm{d}k^n, n = 2, 3, \ldots$, plays the decisive role.

An Example

Consider an elastic string based on an elastic foundation and subjected by a normally directed moving and oscillating load. Namely, we consider the following equation

$$\varrho\ddot{u}(x,t) - Tu''(x,t) + \varkappa u(x,t) = Q_0(x - vt)\mathrm{e}^{\mathrm{i}\omega_0 t}, \tag{18.144}$$

where u, ϱ, T, $\varkappa$, Q, v, ω_0 are the string displacement, mass per unit length, tension force, foundation stiffness, load, its speed, and the load frequency, respectively. As discussed above, Sect. 14.1.1, the propagating wave frequency at the moving load coincides with the load frequency. Based on Eqs. (18.137) and (14.3) we can express the wave frequency and wavenumber as follows

$$\omega = \sqrt{\frac{\varkappa}{\varrho} + c^2k^2} = \omega_0 + kv, \; c = \sqrt{\frac{T}{\varrho}},$$
$$k = k_\pm = \frac{1}{c^2 - v^2}\left(\omega_0 v \pm \sqrt{\omega_0^2 c^2 - \frac{\varkappa}{\varrho}(c^2 - v^2)}\right). \tag{18.145}$$

Note that the load excites a propagation wave only if k in this expression appears real, that is if the load frequency is in the wave frequencies domain (the frequencies at $x = v_0 t$).

Consider η-related Green function, that is the wave excited by the moving concentrated force, $Q_0(\eta) = Q^0\delta(\eta)$, $Q^0 = \text{const}$. The continuous displacement is

$$u = A\left(\mathrm{e}^{\mathrm{i}(\omega_0 t - k_+\eta)} H(\eta) + \mathrm{e}^{\mathrm{i}(\omega_0 t - k_-\eta)} H(-\eta)\right). \tag{18.146}$$

The equilibrium at $\eta = 0$ gives us

$$T(u'_{\eta=-0} - u'_{\eta=+0}) = AT\mathrm{i}(k_+ - k_-) = Q^0$$
$$\Longrightarrow \; A = -\frac{\mathrm{i}Q^0}{T(k_+ - k_-)}. \tag{18.147}$$

Such wave configurations are valid as far as $k_+ \neq k_-$. The equality can occur, in particular, as the frequency $\omega_0 = 1$ of the middle line in Fig. 18.2, or $\omega_0 = 2$ upper line, decreases, under $v = \text{const}$, until it becomes a tangent to the dispersion curve. Note that in this way, the phase and group speeds of the excited wave coincide, that corresponds to the resonant excitation. Resonant waves are considered in Sect. 15.3.

Similar results correspond to another waveguides, although some detail can be different. For example, in a supported bending beam, the resonant wave can also be excited by a non-oscillation moving load, see Fig. 1.19.

18.3.4 Multidimensional Fourier Transform and Hankel Transform

Consider the Fourier multidimensional transform for a function of r

$$f^{F_n}(\boldsymbol{k}) = \overbrace{\int_{-\infty}^{\infty} \cdots \int_{-\infty}^{\infty}}^{n} f(r) \exp[\mathrm{i}\boldsymbol{k}\boldsymbol{x}]\, \mathrm{d}x_1 \ldots \mathrm{d}x_n, \tag{18.148}$$

with

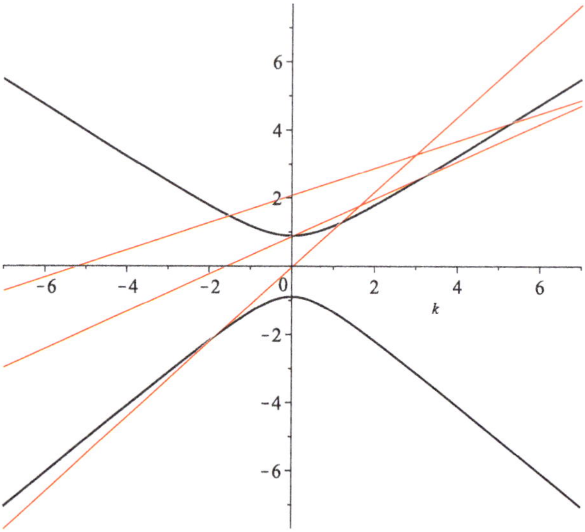

Fig. 18.2 Waves excited by a moving load in a string on an elastic foundation. The dispersion dependencies, $\omega = \pm\sqrt{1+k^2}$, and three load related lines: $\omega_0 = 0$, $v = 1.4\,k$; $\omega_0 = 1$, $v = 0.6\,k$; $\omega_0 = 2$, $v = 0.4\,k$. The corresponding wavenumbers are $(k_\pm = \pm 1.0206)$, $(k_+ = 1.875, k_- = 0)$, $(k_+ = 3.0686, k_- = -1.6294)$

$$\begin{gathered}\boldsymbol{x} = r\boldsymbol{\mu} = x_1\boldsymbol{\mu_1} + \cdots + x_n\boldsymbol{\mu_n},\ \boldsymbol{k} = k\boldsymbol{\nu} = k_1\boldsymbol{\nu_1} + \cdots + k_n\boldsymbol{\nu_n},\\ \sqrt{x_1^2 + \cdots + x_n^2} = r,\ \sqrt{k_1^2 + \cdots + k_n^2} = k,\\ x_i = r\alpha_i,\ k_i = k\beta_i,\ \sum_{i=1}^{n}\alpha_i^2 = \sum_{i=1}^{n}\beta_i^2 = 1,\ \alpha_i = \boldsymbol{\mu\mu_i},\ \beta_i = \boldsymbol{\nu\nu_i},\end{gathered} \tag{18.149}$$

where $\boldsymbol{\mu}, \boldsymbol{\mu_i}, \boldsymbol{\nu}, \boldsymbol{\nu_i}$ are unit vectors, $i = 1, \ldots, n$.

Following Bochner (1959), to simplify the scalar product is (18.148), we introduce in x-space a new rectangular coordinate system, y_i, with the same origin. In this system, the coordinate y_1 is oriented relative to the coordinates x_i as vector $\boldsymbol{k}$ is oriented in the k-space. This means that $x_i = y_1\beta_i$ and this is the only requirement for the y-system orientation. Substituting the latter relation and the corresponding relation for k_i from (18.149) in exponent in (18.148) we obtain

$$\mathrm{e}^{\mathrm{i}\boldsymbol{kx}} = \mathrm{e}^{\mathrm{i}ky_1} \text{ and } r = \sqrt{y_1^2 + \varrho^2},\ \varrho = \sqrt{y_2^2 + \cdots + y_n^2}. \tag{18.150}$$

Further, we can express the space element in terms of the y-system: $\mathrm{d}x_1 \ldots \mathrm{d}x_n \longrightarrow \mathrm{d}y_1 \ldots \mathrm{d}y_n$. Next we note that the multidimensional integral of a function, which depends on the radial coordinate, ϱ, but not on the coordinates y_i separately $(i = 2, \ldots, n)$ is $(n > 1)$

$$\int_{-\infty}^{\infty} \cdots \int_{-\infty}^{\infty} f\left(\sqrt{y_1^2+\varrho^2}\right) \mathrm{d}y_2 \ldots \mathrm{d}y_n = \frac{2\pi^{(n-1)/2}}{\Gamma[(n-1)/2]} \int_0^{\infty} f\left(\sqrt{y_1^2+\varrho^2}\right) \varrho^{n-2}\, \mathrm{d}\varrho. \tag{18.151}$$

Thus

$$f^{F_n}(\boldsymbol{k}) = \frac{2\pi^{(n-1)/2}}{\Gamma[(n-1)/2]} \int_{-\infty}^{\infty} \int_0^{\infty} f\left(\sqrt{y_1^2+\varrho^2}\right) \varrho^{n-2} \mathrm{e}^{\mathrm{i}ky_1}\, \mathrm{d}\varrho\, \mathrm{d}y_1. \tag{18.152}$$

Finally, we put $y_1 = r\cos\varphi$, $\varrho = r\sin\varphi$ $(0 \le \varphi \le \pi)$ and obtain

$$\begin{aligned} f^{F_n}(\boldsymbol{k}) &= \frac{2\pi^{(n-1)/2}}{\Gamma[(n-1)/2]} \int_0^{\infty} f(r) r^{n-1} \int_{-\infty}^{\infty} \mathrm{e}^{\mathrm{i}kr\cos\varphi} (\sin\varphi)^{n-2}\, \mathrm{d}\varphi\, \mathrm{d}r \\ &= \int_0^{\infty} f(r) L(r,k)\, \mathrm{d}r,\ L(r,k) = \frac{(2\pi)^{n/2}}{k^{n/2-1}} r^{n/2} J_{n/2-1}(kr). \end{aligned} \tag{18.153}$$

As usual, $\Gamma(x)$ is the Gamma function and $J_\nu(x)$ is Bessel function.
This result is also valid for $n = 1$. In partial cases

$$L(r,k) = \begin{cases} 2\cos kr & (n=1) \\ 2\pi r J_0(kr) & (n=2) \\ 4\pi(r/k)\sin kr & (n=3). \end{cases} \tag{18.154}$$

The first kernel, coincides with that in the Fourier one-dimensional transform of an even function, it relates to the cosine transform. The second one relates to the Hankel transform

$$\begin{aligned} f^{H_\nu}(k) &= \int_0^{\infty} f(x) x J_\nu(kx)\, \mathrm{d}x \ (0 < k < \infty, \nu > -1/2), \\ f(x) &= \int_0^{\infty} f^{H_\nu}(k) k J_\nu(kx)\, \mathrm{d}k. \end{aligned} \tag{18.155}$$

We have shown how this transform for $\nu = 0$ is connected with the 2D Fourier transform. These relations, however, are valid for any $\nu > -1/2$, since

$$\begin{aligned} \int_0^{\infty} J_\nu(kx) J_\nu(k'x) kx\, \mathrm{d}x &= \delta(k-k'), \\ \int_0^{\infty} J_\nu(kx) J_\nu(kx') kx\, \mathrm{d}k &= \delta(x-x')\ (\nu > -1/2). \end{aligned} \tag{18.156}$$

The last kernel ($n = 3$) is a 3D-modified sine transform

$$f^{S_3}(k) = \int_0^\infty f(x)(x/k)\sin(kx)\,\mathrm{d}x,$$
$$f(x) = \frac{2}{\pi}\int_0^\infty f^{S_3}(k)(k/x)\sin(kx)\,\mathrm{d}k. \tag{18.157}$$

Now consider function

$$f(r,\theta) = g(r)\mathrm{e}^{-in\theta},\ n = 0, 1, \ldots, \tag{18.158}$$

where r, θ are the polar coordinates

$$x = r\cos\theta,\ y = \sin\theta,\ r = \sqrt{x^2 + y^2}. \tag{18.159}$$

In the double Fourier transform

$$f^{FF}(k_x, k_y) = \int_{-\infty}^\infty\int_{-\infty}^\infty g(r)\exp[-\mathrm{i}(n\theta - k_x x - k_y y)]\ \mathrm{d}x\ \mathrm{d}y, \tag{18.160}$$

put

$$k_x = k\sin\alpha,\ k_y = k\cos\alpha,\ \mathrm{d}x\ \mathrm{d}y = r\ \mathrm{d}r\ \mathrm{d}\theta. \tag{18.161}$$

We have

$$f^{FF}(k_x, k_y) = \int_0^\infty g(r)r\int_0^{2\pi}\exp[-\mathrm{i}(n\theta - kr\sin(\theta+\alpha))]\,\mathrm{d}\theta\,\mathrm{d}r$$
$$\int_0^\infty g(r)r\mathrm{e}^{\mathrm{i}n\alpha}\int_0^{2\pi}\cos(n\varphi - kr\sin\varphi)\,\mathrm{d}\varphi\,\mathrm{d}r = 2\pi\mathrm{e}^{\mathrm{i}n\alpha}\int_0^\infty g(r)rJ_n(kr), \tag{18.162}$$

that is

$$f^{FF}(k\sin\alpha, k\cos\alpha) = 2\pi\mathrm{e}^{\mathrm{i}n\alpha}g^{H_n}(k). \tag{18.163}$$

With regard to a function of a general dependence on both variables, $f(x, y)$, Hankel transform can be applied for the corresponding *Fourier series* as

$$f(x, y) = \sum_{n=-\infty}^{\infty} g_n(r)\mathrm{e}^{\mathrm{i}n\theta}. \tag{18.164}$$

Thus two-dimensional Fourier transform can be expressed as a *double Fourier transform*

$$f^{FF}(k_x, k_y) = \int_{-\infty}^{\infty} \int_{-\infty}^{\infty} f(x, y) \mathrm{e}^{\mathrm{i}(k_x x + k_y y)} \, \mathrm{d}x \, \mathrm{d}y \tag{18.165}$$

or as a series on Hankel transforms, a general term of which is presented in (18.163). This also concerns the Fourier inverse 2D transform. Along with the direct formulae

$$f(x, y) = \frac{1}{4\pi^2} \int_{-\infty}^{\infty} \int_{-\infty}^{\infty} f^{FF}(k_x, k_y) \exp[-\mathrm{i}(k_x x + k_y y)] \, \mathrm{d}k_x \, \mathrm{d}k_y, \tag{18.166}$$

Hankel inverse transform (18.157) can be used with respect to each Fourier coefficient, $g^{H_n}(k)$.

18.4 Limiting Relations and Asymptotic Representations

18.4.1 Limiting Relations

Consider locally integrable, exponentially growing function $f(x)$ and the one-sided Fourier transform, $f_{\pm}(\pm \mathrm{i}s)$ (the Laplace transform, $f^L(\pm s)$)

$$\begin{aligned} f_+(\mathrm{i}s) = f^L(s) &= \int_0^{\infty} f(x) \mathrm{e}^{-sx} \, \mathrm{d}x, \\ f_-(-\mathrm{i}s) = f^L(-s) &= \int_{-\infty}^{0} f(x) \mathrm{e}^{sx} \, \mathrm{d}x, \end{aligned} \tag{18.167}$$

where $\Re s$ is assumed to be large enough, such that the above considered integrals converge. The following limiting relations are valid: If the corresponding limit of the original function exists, then

$$\begin{aligned} \lim_{x \to +0} f(x) &= \lim_{s \to \infty} s f_+(is) = \lim_{s \to \infty} s f^L(s), \\ \lim_{x \to -0} f(x) &= \lim_{s \to \infty} s f_-(-is) = \lim_{s \to \infty} s f^L(-s), \\ \lim_{x \to \infty} f(x) &= \lim_{s \to 0} s f_+(is) = \lim_{s \to +0} s f^L(s), \\ \lim_{x \to -\infty} f(x) &= \lim_{s \to 0} s f_-(-is) = \lim_{s \to +0} s f^L(-s). \end{aligned} \tag{18.168}$$

These relations can be obtained in the following way

$$\begin{aligned} & s \int_0^{\infty} f(x) \mathrm{e}^{-sx} \, \mathrm{d}x = \int_0^{\infty} f\left(\frac{x}{s}\right) \mathrm{e}^{-x} \, \mathrm{d}x \\ & \to \begin{cases} f(+0) \int_0^{\infty} \mathrm{e}^{-x} \, \mathrm{d}x = f(+0) & (s \to \infty) \\ f(+\infty) \int_0^{\infty} \mathrm{e}^{-x} \, \mathrm{d}x = f(+\infty) & (s \to +0). \end{cases} \end{aligned} \tag{18.169}$$

The same considerations are valid for f_-. Relations (18.168) remain valid if the original function, $f(x)$, is replaced by its average

$$\langle f(x) \rangle = \frac{1}{x} \int_0^x f(\xi)\, \mathrm{d}\xi. \tag{18.170}$$

The limiting relations (18.168) with respect to the averaged values can be obtained in the same way as used in (18.169). Note that the limit of $\langle f(x) \rangle$ is equal to the corresponding limit of $f(x)$ if the latter exists. However, the former may also exist in the case when $f(x)$ has no limit. For example, $f(x) = 1 + \sin x$ has no limit at $x = \infty$, while $\lim \langle f(x) \rangle = 1$.

Similar relationships concerning asymptotic representations for power type singularities are considered in Sect. 18.4.3.

18.4.2 Some Asymptotic Representations of Inverse Transforms

Function $f_1(x)$ is called the asymptotic representation of function $f(x)$ for $x \to x_0$ if

$$\frac{f_1(x)}{f(x)} \to 1 \ (x \to x_0). \tag{18.171}$$

Such an asymptotic equality is denoted as $f(x) \sim f_1(x)\,(x \to x_0)$. Often $x_0 = \infty$; however, the definition is still valid for any real or complex limiting point. It follows from the definition that if $f \sim f_1$ then $f_1 \sim f$; however, usually an asymptotic representation of a function is introduced as a function which is simpler then the original function itself. For example, an elementary function can asymptotically represent a special function.

In addition, asymptotic series are considered as

$$f(x) \sim \sum_{n=1} f_n(x) \ (x \to x_0) \tag{18.172}$$

which means that

$$f(x) - \sum_{n=1}^{N} f_n(x) \sim f_{N+1}(x) \ (x \to x_0, N = 1, 2, \ldots). \tag{18.173}$$

Such a series is not required to be convergent. Instead, any finite segment of it, $\sum_{n=1}^{n=N} f_n$, must be an asymptotic representation of $f(x)$ and satisfy (18.173).

Asymptotic representation of an integral with a large (infinitely growing) parameter is usually based on the following. (a) On the path of integration the point is determined, such that the contribution of integration outside an arbitrarily small vicinity of it is asymptotically negligible (if there is no such point, then we can try to change the integration path to get this point). (b) The integrand is replaced by a simpler function, asymptotically equivalent to the original function in that vicinity without violation of condition (a). (c) Also to simplify the result the integration path is extended to infinity. Some important specific methods are discussed below.

18.4.3 Asymptotic Contribution of a Singular Point

Consider the Fourier inverse transform as a slowly growing analytical function of k with separate singular points. Some of these points may lie on the real k-axis. In this case, the causality principle gives us the rule how these points must be round (see Sect. 15.6). Using the above-discussed transformation of the integration path, Sect. 18.2.2, consider the asymptotic contribution of a singular point, $k = k_0$. Due to the presence of the fast decreasing factor, $\mathrm{e}^{-k'x}$ $(x \to +\infty, 0 < k' < \infty)$, the integration over the branch banks outside an arbitrarily small vicinity of the singular point is asymptotically negligible since its contribution is exponentially small

$$\begin{aligned} &\int_{\varepsilon}^{\infty} f^F(k_0 - \mathrm{i}k')\mathrm{e}^{-(\mathrm{i}k_0+k')x}\,\mathrm{d}k' \\ &= \mathrm{e}^{-\mathrm{i}k_0-\varepsilon x}\int_0^{\infty} f^F(k_0 - \mathrm{i}\varepsilon - \mathrm{i}k'')\mathrm{e}^{-k''x}\,\mathrm{d}k'' = \mathrm{O}\left(\mathrm{e}^{-\varepsilon x}\right). \end{aligned} \tag{18.174}$$

This allows us to use the *Laplace method*, namely, first we can use an asymptotic representation of $f^F(k)$ $(k \to k_0)$ instead of this function itself. Then, by the same reason, we can restore the integration path $(k_0, k_0 - \mathrm{i}\infty)$ for the asymptotic representation, that is, we can take instead $f^F(k)$ any slow growing function, $f_0^F(k)$, defined on the semi-infinite straight line $k_0, k_0 - \mathrm{i}\infty$, such that $f^F(k) \sim f_0^F(k)$ $(k \to k_0)$

$$\int_{\Gamma} f^F(k)\mathrm{e}^{-\mathrm{i}kx}\,\mathrm{d}k \sim \int_{\Gamma} f_0^F(k)\mathrm{e}^{-\mathrm{i}kx}\,\mathrm{d}k, \tag{18.175}$$

where Γ is the integration path paved around the singular point and along the branch banks as shown in Fig. 18.1. Note that singular points below the real axis give an exponentially decreasing contribution $(x \to \pm\infty)$, which is negligible if such points on the real axis exist. Otherwise, in the asymptotic representation, the points with minimal imaginary part should be taken into account.

It follows that if the Fourier transform has a singular point of the power type, expression (18.90) is still valid as an asymptote. Namely, if

$$f^F(k) \sim C[0 - \mathrm{i}(k - k_0)]^{\nu} \ (k \to k_0, \Im k_0 \le 0, C = \mathrm{const}), \tag{18.176}$$

the asymptotic contribution of this singular point is

$$f_0(x) \sim \frac{Cx^{-\nu-1}}{\Gamma(-\nu)} \mathrm{e}^{-\mathrm{i}k_0 x} \ (x \to +\infty). \tag{18.177}$$

Similarly, if

$$f^F(k) \sim C[0 + \mathrm{i}(k - k_0)]^\nu \ (k \to k_0) \tag{18.178}$$

then

$$f_0(x) \sim \frac{C(-x)^{-\nu-1}}{\Gamma(-\nu)} \mathrm{e}^{-\mathrm{i}k_0 x} \ (x \to -\infty). \tag{18.179}$$

The above relations (18.176)–(18.179), as connections between the asymptotic representation of the transform and that of the original function, represent a nontrivial generalization of the limiting relations in Sect. 18.4.1.

18.4.4 The Methods of Stationary Point

The Method of Stationary Phase

Above we derived an asymptote of the Fourier inverse transform with a single large parameter, x, as a factor in the exponent, $\mathrm{e}^{-\mathrm{i}kx}$. Now consider the Fourier transform comprised of complex sinusoidal waves (see Sect. 3.4.3), where the corresponding inverse transform contains two parameters, x and t. Consider a one-dimensional waveguide, which initial state is characterized by displacements $u(x, 0)$ and particle velocities $\dot{u}(x, t)$, and denote dispersion dependencies for waves in the waveguide by $\pm\omega(k)$. It follows that

$$\begin{aligned} u^F(k, t) &= g_+(k)\mathrm{e}^{\mathrm{i}\omega(k)t} + g_-(k)\mathrm{e}^{-\mathrm{i}\omega(k)t}, \\ g_\pm(k) &= \frac{1}{2}[u^F(k, 0) \mp \mathrm{i}\dot{u}^F(k, 0)/\omega(k)]. \end{aligned} \tag{18.180}$$

The inverse transform for the first term looks as

$$\begin{aligned} u(x, t) &= \frac{1}{2\pi} \int_a^b g(k)\mathrm{e}^{\mathrm{i}S(k,x,t)} \,\mathrm{d}k, \\ S(k, x, t) &= \omega(k)t - kx, \ (g = g_+), \end{aligned} \tag{18.181}$$

where the interval corresponds to *propagating waves*, only to those waves which can be detected far from the source, that is it corresponds to real ω. Often $a = -\infty, b = \infty$; however, this interval for a dispersion relation can also be finite (see, e.g., Mishuris et al. 2009). Besides, an expression similar to that in (18.181) corresponds to a two-dimensional steady-state waves radiated by a moving action. In the latter case, the *phase* looks as $S(k, x, y) = \psi(k)y - kx$, where $\psi(k)$ is a function, and y is the coordinate normal to x as usual. Note that the results discussed below are completely valid for the second term as well. We assume $g(k)$ and $\omega(k)$ to be infinitely differentiable within the interval.

Substitute $x = r\cos\theta$, $t = r\sin\theta$ and consider an asymptote of $f(x, t)$ for $r \to \infty$. If angle θ is such that the group velocity $c_g = \omega'(k) \neq \cot\theta = x/t$, the integration by parts

$$2\pi u(x,t) \sim \int_a^b g(k)\mathrm{e}^{\mathrm{i}(\omega(k)t-kx)}\,\mathrm{d}k$$
$$= \left[\frac{g(k)}{\mathrm{i}r(c_g(k)\sin\theta - \cos\theta)}\mathrm{e}^{\mathrm{i}r(\omega(k)\sin\theta - k\cos\theta)}\right]\Bigg|_a^b \tag{18.182}$$
$$-\int_a^b \frac{\mathrm{d}}{\mathrm{d}k}\left[\frac{g(k)}{\mathrm{i}r(c_g(k)\sin\theta - \cos\theta)}\right]\mathrm{e}^{\mathrm{i}r(\omega(k)\sin\theta - k\cos\theta)}\,\mathrm{d}k$$

leads to the conclusion that the main contribution to the asymptotic representation is given by the end points of the integration interval, and the main asymptotic term is of the order of $1/r$, whereas the integration along the inner domain results in a fast decreasing function ($r \to \infty$).

A more significant asymptotic representation arises due to the presence of a *stationary phase* point, that is the point $k = k_*$ where

$$\frac{\partial S(k, x, t)}{\partial k} = c_g(k)t - x = 0, \tag{18.183}$$

within the interval or at its end point(s). This reflects the fact that the energy in sinusoidal waves propagates with the *group velocity*. By the same reason, if the stationary point exists, the only arbitrarily small vicinity of it gives the asymptotic representation, and the latter is defined by an asymptotic representation of the Fourier transform for $k \to k_*$.

First consider the case $\partial^2 S(k_*, x, t)/\partial k^2 \neq 0$. We have

$$S(k, x, t) - S(k_*, x, t) \sim \frac{\partial^2 S(k_*, x, t)}{\partial k^2}\frac{(k-k_*)^2}{2} \quad (k \to k_*). \tag{18.184}$$

It follows that

$$u(x,t) \sim \mathrm{e}^{\mathrm{i}(\omega(k_*)t - k_* x)} g(k_*)\Phi_2(t),$$
$$\Phi_2(t) = \frac{1}{2\pi}\int_{-\infty}^{\infty} \mathrm{e}^{\mathrm{i}\omega''(k_*)t(k-k_*)^2/2}\,\mathrm{d}(k-k_*) \tag{18.185}$$
$$= \frac{1}{\sqrt{2\pi|\omega''(k_*)|t}}\mathrm{e}^{\pm\mathrm{i}\pi/4} \quad (\pm\omega''(k_*) > 0).$$

This expression corresponds to the case $a < x_* < b$. If $x_* = a$ or $x_* = b$, then Φ must be replaced by $\Phi/2$ since it is now defined by the integral over the right or the left half-axis. Recall that this integral limits are is taken as $-\infty, \infty$, or 0∞, or $-\infty, 0$ instead of $-\varepsilon, \varepsilon$, or $0, \varepsilon$, or $-\varepsilon, 0$, respectively, for simplicity. In any case the asymptotic representation ($t \to \infty$) remains valid.

Now consider the case $\partial^2 S(k_*, x, t)/\partial k^2 = \omega''(k_*) = 0$. The latter equality along with that in (18.183) can be satisfied only on a separate ray in the x, t-plane, $x = x_*(t)$, since k_* is completely defined by the equality $\omega''(k_*) = 0$. This is in contrast to the previous case in which, for a given t, the stationary point is a function of x: $k_* = k_*(x)$, where x can vary. To determine the asymptotic representation in a vicinity of this ray we use the following series expansion valid in a vicinity of $k = k_*$

$$S(k, x, t) - S(k_*, x_*, t) \sim \frac{\partial S}{\partial k}(k - k_*) + \frac{\partial S}{\partial x}(x - x_*) + \frac{\partial^2 S}{\partial k \partial x}(k - k_*)(x - x_*)$$
$$+\frac{1}{6}\frac{\partial^3 S}{\partial k^3}(k - k_*)^3 = -k_*(x - x_*) - (k - k_*)(x - x_*) + \frac{1}{6}\omega'''(k_*)t(k - k_*)^3, \tag{18.186}$$

or

$$S(k, x, t) \sim \omega(k_*)t - k_* x - (k - k_*)(x - x_*)$$
$$+ \frac{1}{6}\omega'''(k_*)t(k - k_*)^3. \tag{18.187}$$

It follows that

$$u(x, t) \sim e^{i(\omega(k_*)t - k_* x)} g(k_*)\Phi_3(x, t),$$
$$\Phi_3(x, t) = \frac{1}{2\pi}\int_{-\infty}^{\infty} \exp\{i[\omega'''(k_*)t(k - k_*)^3/6 - (k - k_*)(x - x_*)]\}\, d(k - k_*) \tag{18.188}$$
$$= \left(\frac{2}{|\omega'''(k_*)|t}\right)^{1/3} \mathrm{Ai}\left(-\left(\frac{2}{|\omega'''(k_*)|t}\right)^{1/3}(x - x_*)\operatorname{sign}\omega'''(k_*)\right),$$

where the *Airy wave function*, $Ai(x)$, is defined as

$$\mathrm{Ai}(x) = \frac{1}{\pi}\int_0^{\infty} \cos\left(\frac{\xi^3}{3} + x\xi\right) d\xi,\ \ \mathrm{Ai}(0) = 3^{-2/3}/\Gamma(2/3). \tag{18.189}$$

The Saddle Point Method

It is also called the *steepest descent approximation* and the *method of pass*. Consider again the integral in (18.181) assuming the functions, $g(k)$ and $\omega(k)$, are analytical with no singular points in a vicinity of the stationary point, $k = k_*$. The asymptotic representation of $iS(k, x, t)$ ($k \to k_*$) can by represented as

$$
\begin{aligned}
Q &= \mathrm{i}[S(k,x,t) - S(k_*,x,t)] \sim \mathrm{i}\,t\frac{\mathrm{d}^n\omega(k_*)}{\mathrm{d}k^n}\frac{(k-k_*)^n}{\Gamma(n+1)} \\
&= t\left|\frac{\mathrm{d}^n\omega(k_*)}{\mathrm{d}k^n}\right|\frac{|k-k_*|^n}{\Gamma(n+1)}[\cos(n\theta+\theta_0) + \mathrm{i}\sin(n\theta+\theta_0)],
\end{aligned}
\tag{18.190}
$$

where n is a minimal integer corresponding to a nonzero derivative: $n \geq 2$, and the angle θ corresponds to representation of $k - k_*$ as a complex variable, $\theta_0 = \text{const}$. It can be seen that there are $4n$ rays from the stationary point, separating equal sectors

$$
\theta = -\frac{\theta_0}{n} + \frac{\pi m}{2n}, \quad m = 0, 1, \ldots, 4n-1, \tag{18.191}
$$

such that for $|k - k_*| > 0$

$$
\begin{aligned}
&\Re Q > 0,\ \Im Q = 0\ \ (m=0),\ \Re Q = 0,\ \Im Q > 0\ \ (m=1),\\
&\Re Q < 0,\ \Im Q = 0\ \ (m=2),\ \Re Q = 0,\ \Im Q < 0\ \ (m=3),
\end{aligned}
\tag{18.192}
$$

and this sequence repeats in the following n times. where the real or (in fact, the separating lines can be considered as straight rays only in a small vicinity of the stationary point since only an asymptotic representation of S is considered). For $\Re Q$ or $|\exp(Q)|$ the sectors between the 'imaginary Q'-rays represent 'ridges' $[(m = 4n-1, m = 1)$ $((m = 3, m = 5)$, ..., $(m = 4n-5, m = 4n-3)]$, and 'valleys' $[(m = 1, m = 3)$, $(m = 5, m = 7)$, ..., $(m = 4n-3, m = 4n-1)]$. With $t \to \infty$ the ascent and descent are increasingly steep, especially at the middle rays: $m = 0, 4, \ldots, 4n-4$ (the steepest ascent) and $m = 2, 6, \ldots, 4n-2$ (the steepest descent). The existence of a fast decreasing factor, allows the integration path in a vicinity of the stationary point alon the real k-segments, $0 < k < a$ and $-a < k < 0$, to be changed to that on the nearest steepest descent rays, and vice versa. Besides, this factor allows the asymptotic representation to be obtained by the integration over an arbitrarily small vicinity of the stationary point, the same as in the stationary phase technique.

Thus the stationary phase and the steepest descent results are asymptotically equivalent. It can be seen that the contribution of the integration over an arc, connected the considered finite rays, is of the order of $1/t$. The equivalency is true for any couple of rays in any 'valley' as well. This also follows from the fact that the integral over the corresponding semi-infinite ray

$$
\int_0^\infty \exp\left(-x^n \mathrm{e}^{\mathrm{i}n\theta}\right)\mathrm{e}^{\mathrm{i}\theta}\,\mathrm{d}x = \frac{\Gamma(1/n)}{n}\ \left(-\frac{\pi}{2n} \leq \theta \leq \frac{\pi}{2n}\right), \tag{18.193}
$$

is independent of θ in the above-mentioned range. Note that this path transformation is closely related to that in the inverse Fourier transform discussed in Sect. 18.2.2.

Thus, the integral along the steepest descent, Γ, corresponds to the negative exponent, $Q = -R > 0$, in (18.190) as

$$\int_{\Gamma} g(k_*) e^{-R(k,x,t)}\, \mathrm{d}\Gamma,$$
$$R \sim= -t \left| \frac{\mathrm{d}^n \omega(k_*)}{\mathrm{d}k^n} \right| \frac{|k - k_*|^n}{\Gamma(n+1)} \quad (n \geq 2). \tag{18.194}$$

Let contour Γ contain the stationary point inside, and let the integrant be asymptotically even function in a vicinity of the stationary point. In this case, the contribution of the end points is exponentially small. As a result not only the above-considered asymptotic representation but an asymptotic series can be obtained. To this end a Taylor series of $g(k) = g(k_*) + g'(k_*)(k - k_*) + \cdots$ should be used, each term of which results in the corresponding term of the asymptotic series. Further, if the conditions are satisfied as (a) contour Γ can be extended to infinity, as well as a nearest 'imaginary' path ($Q = \pm \mathrm{i}R$, $R > 0$), (b) there are no singular points between these pathes and (c) function g is slow growing, then using considerations similar to those in Sect. 18.2.2 we can conclude that the integrals over Γ and over the semi-infinite imaginary path are equal to each other. It follows that the same asymptotic series corresponds to the method of stationary phase.

Thus, in the considered cases, the method of stationary phase and the saddle point method are equivalent, and both could be called the *method of stationary point*. The path transformation is, however, possible if the corresponding functions are analytical (or can be, in principle, properly approximated by analytical functions in a vicinity of the stationary point). Along with this, the method of stationary phase or the steepest descent method as they are, without the integration path transformation, can be used for functions of a more general view.

In addition, note that the stationary point can be outside the given integration contour. In this case, if the functions are analytical, it may be possible to transform the contour to include this point and to obtain an asymptotic representation using the above-considered technique.

18.4.5 Quasi-Singularity and Asymptotic Representation

Expression

$$F(k,t) = \frac{\sin(k^n t)}{k}, \; n = 1, 3, \ldots, \tag{18.195}$$

is regular, and the integral

$$\int_{-\infty}^{\infty} F(k,t)\, \mathrm{d}k = \frac{\pi}{n} \tag{18.196}$$

is independent of t for $t > 0$.

This trivial fact, however, ignores a *nontrivial* time-dependence of the $F(k,t)$-distribution on k-axis. Namely,

$$F(k,t) \sim \frac{\pi}{n}\delta(k) \ (t \to \infty). \tag{18.197}$$

Indeed, consider the above integral, Eq. (18.196), over a finite segment, $-\varepsilon < k < \varepsilon$. Substituting $k = t^{-1/n} p$, we have that for any small $\varepsilon > 0$ the integral is the same as in (18.196)

$$\int_{-\varepsilon}^{\varepsilon} F(k,t)\,\mathrm{d}k = \int_{-\varepsilon t^{1/n}}^{\varepsilon t^{1/n}} \frac{\sin(p^n)}{p}\,\mathrm{d}p \sim \frac{\pi}{n} \ (t \to \infty). \tag{18.198}$$

So, the point $k = 0$ can be called *quasi-singular*.

A more general quasi-singularity is (Fig. 18.3)

$$F_{m,n}(k,t) = \frac{\sin(k^m t)}{k^n},\ 0 < n \le m, \tag{18.199}$$

with ($\varepsilon > 0$)

$$\int_{-\varepsilon}^{\varepsilon} F_{m,n}(k,t)\,\mathrm{d}k = t^{t(n-m)/m} \int_{-\varepsilon t^{1/m}}^{\varepsilon t^{1/m}} \frac{\sin(p^m)}{p^n}\,\mathrm{d}p. \tag{18.200}$$

So

$$\int_{-\varepsilon}^{\varepsilon} F_{m,n}(k,t)\,\mathrm{d}k \sim t^{t(n-m)/m} \int_{-\infty}^{\infty} \frac{\sin(p^m)}{p^n}\,\mathrm{d}p \ (t \to \infty). \tag{18.201}$$

That is

$$F_{m,n}(k,t) \sim t^{t(n-m)/m} \frac{\sin(p^m)}{p^n}\,\delta(k) \ (t \to \infty). \tag{18.202}$$

In particular, for m odd integer

$$\int_{-\infty}^{\infty} F_{m,1}(k,t)\,\mathrm{d}k = \frac{\pi}{m}. \tag{18.203}$$

Consider the inverse Fourier transform as

$$u(t,x) = \frac{1}{2\pi} \int_{-\infty}^{\infty} \frac{1 - \exp(\mathrm{i}k^n t)}{k^m} \mathrm{e}^{-\mathrm{i}kx}\,\mathrm{d}k,\ 1 \le m \le n. \tag{18.204}$$

For any t the point $k = 0$ is regular, but can be called *quasi-singular*. The point is that

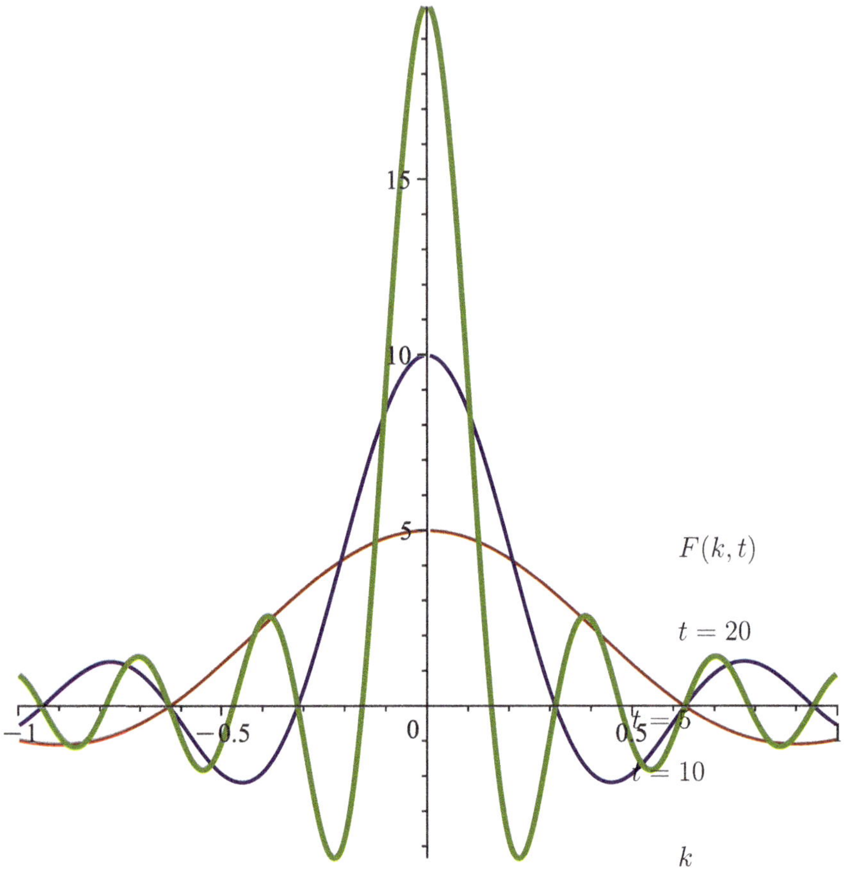

Fig. 18.3 The positive part of the dispersion branches, Ω_1, Ω_2, for a free string equipped by distributed oscillators, $c = \varkappa = \varrho_0 = \varrho = 1$. The dependence for the nonstructural string, $\varrho_0 = 0$, is also shown, Ω_{1a}. Also the asymptote of ω_2 ($k \to \pm\infty$) is presented, Ω_{2a}. The negative part differs only by the sign, Eq. (1.43)

18.5 Wiener-Hopf Technique

18.5.1 The Equation

Consider an equation containing the one-sided *LF* transforms, $f_+(k, s)$ and $f_-(k, s)$, separately:

$$f_+(k, s) + L(k, s) f_-(k, s) = C(k, s), \tag{18.205}$$

where $C(k, s)$ is a given function and $L(k, s)$ is the transform of the dynamic Green's function, $\mathcal{L}(x, t)$, related to this equation. Really, if $C(k, s) \equiv 0$ and $f_- = -1$ [this means that the original is $-\delta(x)\delta(t)]$, then $f_+(k,s) = L(k,s)$. Such equation arises in *mixed problems* where the boundary conditions defined at $-\infty < x < 0$ and $0 \le x < \infty$ are different. In this case, $x = 0$ is the *separation point*.

In the case of a moving separation point, $x = vt$, we can use the moving coordinate system, $\eta = x - vt, t$. Referring to we obtain

$$f_+(k, s' + \mathrm{i}kv) + L(k, s' + \mathrm{i}kv) f_-(k, s' + \mathrm{i}kv) = C(k, s' + \mathrm{i}kv). \quad (18.206)$$

Recall that if $\Re s' > 0$ then no singular points exist on the real k-axis.

The problem is how to find both functions, $f_\pm$, from this single equation. It appears that, under certain conditions, it can be done using the Wiener-Hopf technique, Wiener and Hopf (1931), considered below.

18.5.2 Factorization

In this method, the first step is the *factorization* of $L(k)$ (for short we here omit the second argument of the functions), that is the representation

$$L(k) = L_+(k)L_-(k), \quad (18.207)$$

where $L_+(k)$ has neither singular nor zero points in the upper half-plane including the real axis, while $L_-(k)$ has no such points in the lower half-plane with the real axes. Note that this means the representation of the original Green's function, $\mathcal{L}(x)$, as a convolution

$$\mathcal{L}(x) = L^+(x) * L^-(x), \quad (18.208)$$

where $L^+(x)$ is zero at $x < 0$ and $L^-(x) = 0$ at $x > 0$. If there exists a generalized function with a point support at $x = 0$ then we have to decide to which function it is included, to $L^-(x)$ or $L^+(x)$. In the latter case, $L^+(x)$ is zero at $x < 0$ and $L^-(x) = 0$ at $x \le 0$.

In some cases, Green's function Fourier transform can be factorized straightforwardly. For example, if $L(k) = (k^2 + s^2)^\nu$ then $L_\pm(k) = (s \mp \mathrm{i}k)^\nu$. In more complicated cases, this can be done by the use of the Cauchy-type integral, Sect. 18.6.4, in its application to $\ln L(k)$, namely

$$L_\pm(k) = \exp\left[\pm\frac{1}{2\pi\mathrm{i}}\int_{-\infty}^{\infty}\frac{\ln L(\xi)}{\xi - k}\,\mathrm{d}\xi\right], \quad (18.209)$$

where $\Im k > 0$ for $L_+(k)$, while $\Im k < 0$ for $L_-(k)$. Indeed, the integral represents the functions regular in the corresponding half-planes and hence $L_+(k)$ and $L_-(k)$ have neither singular nor zero points at $\Im k \ge 0$ and at $\Im k \le 0$, respectively. Separating half-residue at $\xi = k$ for continuity points we have

$$L_{\pm}(k) = \sqrt{L(k)} \exp\left[\pm\frac{1}{2\pi \mathrm{i}} V.p. \int_{-\infty}^{\infty} \frac{\ln L(\xi)}{\xi - k}\, \mathrm{d}\xi\right]. \qquad (18.210)$$

Recall that for real k, $\ln|L(k)|$ is a symmetric function and $\Im L(k)$ is an antisymmetric function. The latter implies that if we take Arg $(0) = 0$, without loss of generality, $\mathrm{Arg}(k)$ will be an antisymmetric function. In this way, it follows that for $k = 0$, if this is a regular point

$$L_{\pm}(0) = \sqrt{L(0)} \exp\left[\pm\frac{1}{\pi} V.p. \int_{0}^{\infty} \frac{\mathrm{Arg}\, L(\xi)}{\xi}\, \mathrm{d}\xi\right]. \qquad (18.211)$$

It may happen, however, that the origin is a singular point for $\ln L(k)$, for instance if $|L(k)| \sim L_0(k) = C|k|^{\nu}$ $(k \to 0,\, C > 0)$ with a real $\nu \neq 0$. In this case, Cauchy type integral yields $(0 < \varepsilon' \to 0)$

$$\begin{aligned} &\pm\frac{1}{2\pi \mathrm{i}} \int_{-\infty}^{\infty} \frac{\ln|L(\xi)|}{\xi \mp \mathrm{i}\varepsilon'}\, \mathrm{d}\xi = \frac{1}{\pi} \int_{0}^{\infty} \frac{\ln|L(\xi)|\varepsilon'}{\xi^2 + (\varepsilon')^2}\, \mathrm{d}\xi \\ &= \frac{1}{\pi} \int_{0}^{\infty} \frac{\ln L_0(\varepsilon'\zeta)}{\zeta^2 + 1}\, \mathrm{d}\zeta + \frac{1}{\pi} \int_{0}^{\infty} \frac{\ln[|L(\varepsilon'\zeta)|/L_0(k)]}{\zeta^2 + 1}\, \mathrm{d}\zeta. \end{aligned} \qquad (18.212)$$

The latter integral tends to zero as $\varepsilon' \to +0$ and

$$\int_{0}^{\infty} \frac{\ln \zeta}{\zeta^2 + 1}\, \mathrm{d}\zeta = 0. \qquad (18.213)$$

As a result, with analytical continuation, $\varepsilon' = \varepsilon \mp ik$, where $0 < \varepsilon \to 0$, $0 < \varepsilon' \to 0$, $k \to 0$, we obtain a 'symmetrical' factorization for $k \to 0$

$$L_{\pm}(0) \sim \sqrt{C}(\varepsilon \mp \mathrm{i}k)^{\nu/2} \quad \to \quad \sqrt{C}(0 \mp \mathrm{i}k)^{\nu/2}. \qquad (18.214)$$

Note that factorization is not unique. For example, we can consider CL_+, L_-/C instead of L_+, L_- $(C \neq 0)$, and this does not affect the mixed problem solution. However, Cauchy type integral gives us a unique factorization.

The Index

The Cauchy type integral (18.209) exists if $\ln L(x)$ is locally integrable and tends to zero when $x \to \pm\infty$. The latter implies that the following should be true:

$$\begin{aligned} &|L(k)| \to 1 \ (k \to \pm\infty), \\ &\mathrm{Ind}\, L(k) \equiv \frac{1}{2\pi}\left[\mathrm{Arg}L(\infty) - \mathrm{Arg}L(-\infty)\right] = 0. \end{aligned} \qquad (18.215)$$

Argument

$$\mathrm{Arg}L(k) = \Im \ln L(k) \;\; (L(k) = |L(k)| \exp(\mathrm{i}\mathrm{Arg}L(k)) \tag{18.216}$$

is a multiple-valued function: $\mathrm{Arg}(..) = \arg(..) + 2\pi n$, $-\pi < \arg(..) \le \pi$, $n = 0$, $\pm 1, \ldots$. In contrast, the index is a single-valued quantity. It is determined by considering how the Argument continuously varies as variable k runs along the real axis, where $L(k)$ is assumed regular. In this way, a specific 'initial' value of the Argument at a point, that is a specific value of integer n, does not matter.

However, if $L(k)$ has singular points or/and zeros on the real axis, which are singular points of $\ln L(k)$), without some additional information it is impossible to trace the argument variation. For example, let $L(k) = k$ or $L(k) = 1/k$. Take $\mathrm{Arg}L(-\infty) = \pi$. We have $\mathrm{Arg}L(k) = -\pi$ $(k < 0)$; however, we do not know what is its value for $k > 0$: 0 or 2π or, for example, -6π. This non-uniqueness does not matter for $L(k)$ itself but matters for $\ln L(k)$. Thus the index can be defined only in the case where such special points of $L(k)$, zero points or singular points, are placed outside the real axis; otherwise, we need a rule how to round these points or, equivalently, to shift them from the integration path. Such rule (15.66) following from the *causality principle* is discussed in detail in Chap. 14.

Now consider special point $k = k_0$ with $\Im k_0 \neq 0$. Let $L(k) = k - k_0$. Then the graph of $L(k)$ is a straight line parallel to the real axis, below it if $\Im k_0 > 0$ and above it if $\Im k_0 < 0$. Clearly, as k runs from $-\infty$ to ∞, $\mathrm{Arg}L(k)$ gains increment π in the former case and $-\pi$ in the latter case. Thus, $\mathrm{Ind}L(k) = \pm 1/2$, respectively. It follows that

$$\mathrm{Ind}(k - k_0)^\nu = \mathrm{Ind}(\Im k_0 + \mathrm{i}(k - \Re k_0))^\nu = \frac{\nu}{2}\mathrm{sign}\Im k_0 \tag{18.217}$$

for any real, positive or negative ν. Next, suppose that $L(k)$ can be represented as

$$L(k) = L^0(k) \prod_{n=1}^{N} (k - k_n)^{\nu_n}. \tag{18.218}$$

Then

$$\mathrm{Ind}L(k) = \mathrm{Ind}L^0(k) + \frac{1}{2}\sum_{n=1}^{N} \nu_n \mathrm{sign}\Im k_n \;\; (\Im k_n \neq 0). \tag{18.219}$$

Note that if in representation (18.218), $\mathrm{Arg}L^0$ is constant, for example, if $L(k)$ is positive $(-\infty < k < \infty)$, then its index is zero, and only the contribution of the special points remains in (18.219). In more complicated cases, the index can be determined by considering the graph of $L(k)$ on the complex k-plane. While the index in equal to the total polar angle divided by 2π, its integer part is equal to the number of revolutions around the origin with the corresponding sign.

Normalization

If the index is not zero, it also may be possible to use the Cauchy-type integral, but for a normalized function. To this end we first note that the Fourier transform modulus of Green's function is a symmetric function since Green's function itself is real. With this in mind we consider $L(k)$ satisfying the following conditions

$$|CL(k)k^{\mu}| \to 1 \ (k \to \pm\infty), \ \mathrm{Ind}\, L(k) = \nu. \tag{18.220}$$

This suggests the normalization as

$$\begin{gathered} L^0(k) = L(k)l(k), \ l(k) = l_+(k)l_-(k), \\ l_+(k) = C(1 - \mathrm{i}k)^{\nu+\mu/2}, \ l_-(k) = (1 + \mathrm{i}k)^{-\nu+\mu/2}, \end{gathered} \tag{18.221}$$

where $L^0(k)$ satisfies the above requirements (18.220) and can be factorized using the Cauchy type integral, while $l(k)$ is already factorized directly. We obtain

$$L_+(k) = L_+^0(k)/l_+(k), \ L_-(k) = L_-^0(k)/l_-(k). \tag{18.222}$$

18.5.3 Solution in Terms of the Transform

Decomposition of the Right-Hand Side

Recall that functions $L_+(k)$ and $L_-(k)$ have no singular and zero points at $\Im k \geq 0$ and $\Im k < 0$, respectively, It follows that functions $P_\pm(k) = 1/L_\pm(k)$ have no such special points in the corresponding half-planes, and ones the factorization is done, we can represent (18.206) as

$$P_+(k)f_+(k) + L_-(k)f_-(k) + = P_+(k)C(k), \tag{18.223}$$

where the first term in the left-hand side of this equation is regular in the upper half-plane including the real axis, while the second is regular in the lower half-plane.

The next step is to achieve such a decomposition on the right-hand side. This can be done in different ways, in particular, by means of the use of the Cauchy-type integral. There exists an easy case where $C(k) = C_0/(k - k_0)$ with $\Im k_0 > 0$. In this case, $C(k)$ has a pole in the upper half-plane, while $P_+(k)$ has singular points in the lower half-plane. Hence, a decomposition is required. It can be done as follows:

$$\begin{gathered} P_+(k)C(k) = C_+(k) + C_-(k), \\ C_-(k) = P_+(k_0)C(k), \\ C_+(k) = C(k)\left(P_+(k) - P_+(k_0)\right), \end{gathered} \tag{18.224}$$

where the point $k = k_0$ is regular for $C_+(k)$ and hence this function is really regular in the lower half-plane.

A straightforward decomposition can also be performed by means of a procedure leading to a homogeneous problem with $C(k) = 0$ in the limit. As discussed in Sect. 18.1.10, $C(k)$ as a null generalized function can be generated by a remote load. In this case, the right-hand side of Eq. (18.223) can be represented as the delta function and/or its derivatives depending on the order of zero of $L_+(k)$ at a point, $k = k_0$, on the real k-axis. Recall that the required decomposition of such point-support functions into plus/minus terms is given by the corresponding analytical representation (18.29). For example, the below function $L_+(k)$ and the corresponding null function can be expressed as

$$L_+(k) = [s - i(k - k_0)]^\nu,\ C(k) = \frac{(2s)^\nu}{s + \mathrm{i}(k - k_0)} \to 0\ \ (s \to 0). \tag{18.225}$$

This results in

$$P_+(k)C(k) \to 2\pi\delta(k - k_0) = \frac{1}{0 + \mathrm{i}(k - k_0)} + \frac{1}{0 - \mathrm{i}(k - k_0)}\ \ (s \to +0), \tag{18.226}$$

where the required separation is achieved by the analytical representation of the delta function. It can be checked that the ratio, $C(k)/[2\pi L_+(k)]$, satisfies the sufficient conditions for the delta function listed in Sect. 17.1.3. Note that the parameter s reflects the Laplace transform.

Solution

If the decomposition of the right-hand side in (18.223) is made, we have the equality as

$$P_+(k) f_+(k) - C_+(k) = C_-(k) - L_-(k) f_-(k). \tag{18.227}$$

Because the functions marked by the subscript '+' are regular in the upper half-plane with the real axis, while the subscript '−' means the regularity in the lower half-plane with the real axis, both the left and right sides of Eq. (18.227) are regular in the whole of the complex plane. Such a function can be represented by a power series; denote it $\mathcal{P}(k)$. Thus,

$$\begin{aligned} f_+(k) &= [C_+(k) + \mathcal{P}(k)]L_+(k),\\ f_-(k) &= [C_-(k) - \mathcal{P}(k)]P_-(k). \end{aligned} \tag{18.228}$$

Usually function $\mathcal{P}(k)$ can be determined based on some additional conditions, in particular, geometrical and energy restrictions should be satisfied. For example, if the condition is $f^-(-0) = 0$, while $L_\pm(\pm i\infty) = 1$ as in (18.220) and $C_-(k) \to 0$ $(k \to i\infty)$, then one must put $\mathcal{P}(k) = 0$.

In this way, the solution in terms of the Fourier transforms is, in principle, determined. In some cases, however, in particular in the nonuniform separation point movement, the below considerations 'on the original plane' are preferable.

18.5.4 Considerations in Terms of the Original Functions

Each function in the left-hand side of (18.227) corresponds to an original function with the support at $x \geq 0$, while the support of the original functions in the right-hand side is the left half-axis, $x \leq 0$. The product of functions marked by the subscript '+' ('−') is also regular in the upper (lower) half-plane, and hence the corresponding convolutions satisfy the inequalities

$$P^+(x) * f^+(x) = 0 \ (x < 0), \ L^-(x) * f^-(x) = 0 \ (x > 0), \tag{18.229}$$

and this concerns $C^+(x)$ and $C^-(x)$, respectively. Recall that the replacement of a subscript by the corresponding superscript means the replacement of a one-sided transform by the corresponding original function. Thus, in the original relation corresponding to (18.227),

$$L^+(x) * f^+(x) - C^+(x) = C^-(x) - f^-(x) * P^-(x), \tag{18.230}$$

the support of the left-hand side is $x \geq 0$, while it is $x \leq 0$ for the right-hand side. So, the supports have a single common point, $x = 0$. A generalized function with the point support can be presented as a finite sum

$$\mathcal{P}_0(x) = \sum_{n=0} a_n \frac{\mathrm{d}^n \delta(x)}{\mathrm{d}x^n}, \ \mathcal{P}(k) = \mathcal{P}_0^F(k) = \sum_{n=0} a_n(-\mathrm{i}k)^n. \tag{18.231}$$

Both the left-hand side and the right-hand side of (18.230) are zero outside of $x = 0$ since their supports are different, and they are equal to the same generalized function at $x = 0$. So,

$$\begin{aligned} L^+(x) * f^+(x) &= C^+(x) + \mathcal{P}_0(x), \\ P^-(x) * f^-(x) &= C^-(x) - \mathcal{P}_0(x). \end{aligned} \tag{18.232}$$

We now note that equation of the type

$$a(x) * f(x) = b(x) \tag{18.233}$$

can be easily solved using the direct and inverse Fourier transforms:

$$f^F(k) = b^F(k)c^F(k) \; (c^F(k) = 1/a^F(k))$$
$$\longrightarrow \; f(x) = b(x) * c(x). \tag{18.234}$$

It follows that

$$\begin{aligned} f^+(x) &= \left[C^+(x) + \mathcal{P}_0(x)\right] * P^+(x), \\ f^-(x) &= \left[C^-(x) - \mathcal{P}_0(x)\right] * L^-(x), \end{aligned} \tag{18.235}$$

which is the solution in (18.228) expressed in the originals. The use of this technique in a dynamic problem for a nonuniform movement of the separation point is presented in Sect. 16.3.

18.6 Discrete Fourier Transform

18.6.1 The Transform

Definition

Consider a slowly increasing function, $f(x)$, defined on a set of discrete points, $x = am, m = 0, \pm 1, \ldots$. Introduce the one-sided discrete transforms as

$$\begin{aligned} f_+(k + \mathrm{i}\varepsilon) &= \sum_{m=0}^{\infty} f_m \mathrm{e}^{\mathrm{i}(k+\mathrm{i}\varepsilon)am}, \; f_m = f(am), \\ f_-(k - \mathrm{i}\varepsilon) &= \sum_{m=-\infty}^{-1} f(am)\mathrm{e}^{\mathrm{i}(k-\mathrm{i}\varepsilon)am} \quad (\Im k = 0). \end{aligned} \tag{18.236}$$

Note that sometimes f_0 is included by equal parts of it to both f_+ and f_-, or completely to f_-. These functions, as in the homogeneous case, are regular in the corresponding half-planes. The discrete Fourier transform is

$$\begin{aligned} f^F(k) &= \lim_{\varepsilon \to 0} \left[f_+(k + \mathrm{i}\varepsilon) + f_-(k - \mathrm{i}\varepsilon)\right] \\ &= f_+(k + \mathrm{i}0) + f_-(k - \mathrm{i}0). \end{aligned} \tag{18.237}$$

As follows from definition (18.236), the transform is a periodic function

$$f^F(k + 2\pi/a) = f^F(k). \tag{18.238}$$

With account taken of the analytical continuation of $f_\pm(k)$ to the lower/upper half-planes, respectively, the transform, $f^F(k)$, is an analytic function in the k-plane; however, $f_\pm(k)$ may have singular points in the lower/upper half-planes, respectively.

18.6.2 Inverse Transform

The original, $f(x)$, can be obtained by means of the inverse transform

$$f(x) = \frac{a}{2\pi} \int_{-\pi/a}^{\pi/a} f^F(k) \mathrm{e}^{-\mathrm{i}kam} \, \mathrm{d}k. \tag{18.239}$$

Indeed, substituting the limit (18.237) we obtain

$$\begin{aligned} f(x) &= \sum_{n=-\infty}^{\infty} f(\xi)\delta_D(\xi - x), \; \xi = 0, \pm an, \ldots, \\ \delta_D(\xi - x) &= \frac{a}{2\pi} \int_{-\pi/a}^{\pi/a} \exp\left[\mathrm{i}k(\xi - x)\right] \mathrm{d}k = \frac{a \sin[\pi(\xi - x)/a]}{\pi(\xi - x)} = \delta_{mn}. \end{aligned} \tag{18.240}$$

The latter symbol denotes the *Kroniker delta*

$$\delta_{mn} = \begin{cases} 1 & (m = n) \\ 0 & (m \neq n). \end{cases} \tag{18.241}$$

Note that the inverse transform results in a periodic function of x as a continuous variable, which, the function, coincides with the original function of discrete argument at $x = am$.

It is clear that the discrete Fourier transform is similar to the representation of a function by a Fourier series. In this sense, the original function of the discrete argument is the set of Fourier coefficients, while the transform is the function represented by the series.

Inverse Transform of $f_{\pm}(k)$

This can also be expressed as a recurrence. From (18.236) it follows that

$$\begin{aligned} f(0) &= \lim_{k \to \mathrm{i}\infty} f_+(k), \\ f(a) &= \lim_{k \to \mathrm{i}\infty} \left[f_+(k) - f(0)\right] \mathrm{e}^{-\mathrm{i}ka}, \\ f(-a) &= \lim_{k \to -\mathrm{i}\infty} f_-(k)\mathrm{e}^{-\mathrm{i}ka}. \end{aligned} \tag{18.242}$$

Further, for $n = 1, 2, \ldots$

$$f(na) = \lim_{k \to \mathrm{i}\infty} \left[f_+(k) - \sum_{0}^{n-1} f(ma)\mathrm{e}^{\mathrm{i}kam} \right] \mathrm{e}^{-\mathrm{i}kan}, \tag{18.243}$$

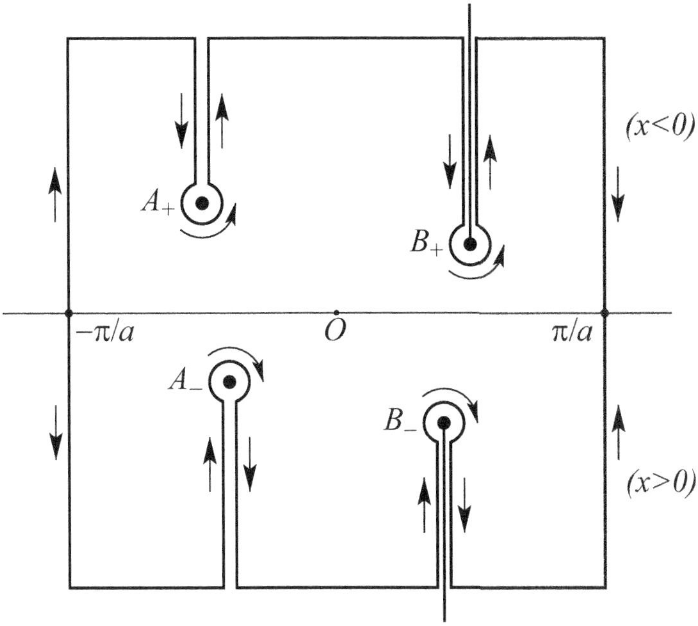

Fig. 18.4 Integration contour in inverse transform. Discrete case

and for $n = -2, -3, \ldots$

$$f(na) = \lim_{k \to -i\infty} \left[f_-(k) - \sum_{n+1}^{-1} f(ma) e^{ikam} \right] e^{-ikan}. \tag{18.244}$$

Inverse Transform on the Complex Plane

It is similar to that for the continuous transform; however, due to the periodicity the integration path on the complex half-planes of k is now located in a strip as shown in Fig. 18.4. If there are no singular points on the real k-axis, as in the case of the generalized transform (18.236) with $\varepsilon > 0$, the integration path over the segment $-\pi/a, \pi/a$ can be drawn upward for $x < 0$ and downward for $x \geq 0$.

Since $f^F(k) \exp(-ikam)$ is a periodic function (18.238), the integration over the left and the right vertical lines, $\Re k = \pm\pi/a$, cancel each other. If, in addition, $f^F(k) \exp(-ikam) \to 0$ ($\Im k \to \pm\infty$ for $m \leq -1$ and $m \geq 0$, respectively) then the integration over the horizontal segments vanishes when $\Im k \to \pm\infty$, and only the contributions of the singular points remain. In this case, the inverse transform on the complex plane leads to the representation of the original function as a sum of such contributions.

Consider function $f(x) = e^{-\varepsilon|x|}$, $\varepsilon > 0$, as an example. In this case

$$f_+(k) = \left[1 - \exp(ika - \varepsilon a)\right]^{-1} \ (\Im k \geq 0),$$
$$f_-(k) = \exp(-ika - \varepsilon a)\left[1 - \exp(-ika - \varepsilon a)\right]^{-1} \ (\Im k \leq 0), \quad (18.245)$$
$$f^F(k) = [1 + \exp(-2\varepsilon) - 2\exp(-\varepsilon)\cos k]^{-1} \ (\Im k = 0).$$

These relations can be considered as analytically continued on the whole of the complex plane k except the poles at $k = \mp \mathrm{i}\varepsilon + 2\pi n/a$ $(n = 0, \pm 1, \dots)$. It can be seen that $f^F(k) \to 0$ $(\Im k \to \pm\infty)$. In the inverse transform (18.240) with the contour similar to that shown in Fig. 18.4, only singular points $k = \pm \mathrm{i}\varepsilon$ remain and the residues at these points yield the original function.

18.6.3 Selective Discrete Transform

Along with the regular discrete Fourier transform

$$f^F(k) = \sum_{m=-\infty}^{\infty} f(m)\mathrm{e}^{\mathrm{i}km} \quad (18.246)$$

consider a *selective discrete transform*

$$f^F_{n\nu}(k) = \sum_{\mu=-\infty}^{\infty} f(\nu + \mu n)\mathrm{e}^{\mathrm{i}k(\nu+\mu n)}, \quad (18.247)$$

where ν and n are integer, $0 \leq \nu \leq n - 1$, $n \geq 1$.

The statement is true that the latter can be expressed in terms of the former as follows

$$f^F_{n\nu}(k) = \frac{1}{n}\sum_{m=0}^{n-1} \exp\left(\frac{2\pi \mathrm{i} m\nu}{n}\right) f^F\left(k - \frac{2\pi m}{n}\right). \quad (18.248)$$

Indeed, for any integer ν', $0 \leq \nu' \leq n - 1$, we find that the original function corresponding to this transform is

$$\begin{aligned} f_{n\nu}(\nu' + \mu n) &= \frac{1}{n}\sum_{m=0}^{n-1} \exp\left(\frac{2\pi \mathrm{i} m\nu}{n}\right) \frac{1}{2\pi}\int_{-\pi}^{\pi} f^F\left(k - \frac{2\pi m}{n}\right) \mathrm{e}^{-\mathrm{i}k(\nu'+\mu n)}\,\mathrm{d}k \\ &= \frac{1}{2\pi}\int_{-\pi}^{\pi} f^F\left(k - \frac{2\pi m}{n}\right) \exp\left[-\mathrm{i}\left(k - \frac{2\pi m}{n}\right)(\nu' + \mu n)\right] \mathrm{d}k \\ &\quad \times \frac{1}{n}\sum_{m=0}^{n-1} \exp\left[\frac{2\pi \mathrm{i} m}{n}(\nu - \nu')\right] \\ &= f(\nu' + \mu n)\frac{1}{n}\sum_{m=0}^{n-1} \exp\left[\frac{2\pi \mathrm{i} m}{n}(\nu - \nu')\right]. \end{aligned} \tag{18.249}$$

Thus

$$\begin{aligned} f_{n\nu}(\nu' + \mu n) &= f(\nu + \mu n)\ (\nu' = \nu), \\ f_{n\nu}(\nu' + \mu n) &= f(\nu + \mu n)\frac{1}{n}\frac{1 - \exp[2\pi \mathrm{i}(\nu - \nu')]}{1 - \exp[2\pi \mathrm{i}(\nu - \nu')/n]} = 0\ \ (\nu' \neq \nu). \end{aligned} \tag{18.250}$$

Note that we here based on the fact that the integrant is a 2π-periodic function of k.

18.6.4 Some Related Topics

Cauchy Type Integral

The Cauchy type integral splits $f^F(k)$ into two functions, $f_+(k)$ and $f_-(k)$, regular in the upper and lower half-planes, respectively, as discussed in Sect. 18.2.2. Now it is for a periodic function. Recall that $f_\pm(k) = f^\pm(x)$, where for the discrete variable, x,

$$\begin{aligned} &f^+(x) = 0\ (x = -a, -2a, \ldots), \\ &f^-(x) = 0\ \ (x = 0, a, \ldots),\ f^+(x) + f^-(x) = f(x). \end{aligned} \tag{18.251}$$

To find the corresponding kernels we obtain the corresponding transforms starting from $f^F(k)$

$$
\begin{aligned}
f_+(k) &= \sum_{m=0}^{\infty} \frac{a}{2\pi} \int_{-\pi/a}^{\pi/a} f^F(p) \mathrm{e}^{-\mathrm{i}pam} \,\mathrm{d}p \, \mathrm{e}^{\mathrm{i}kam} \\
&= \frac{a}{2\pi} \int_{-\pi/a}^{\pi/a} f^F(p) \delta_{D+}(p-k) \,\mathrm{d}p \ (\Im k > 0), \\
f_-(k) &= \sum_{m=-\infty}^{-1} \frac{a}{2\pi} \int_{-\pi/a}^{\pi/a} f^F(p) \mathrm{e}^{-\mathrm{i}pam} \,\mathrm{d}p \, \mathrm{e}^{\mathrm{i}kam} \\
&= \frac{a}{2\pi} \int_{-\pi/a}^{\pi/a} f^F(p) \delta_{D-}(p-k) \,\mathrm{d}p \ (\Im k < 0),
\end{aligned}
\tag{18.252}
$$

where

$$
\begin{aligned}
\delta_{D+} &= \frac{a}{2\pi} \sum_{m=0}^{\infty} \mathrm{e}^{-\mathrm{i}(p-k)am} = \frac{a}{2\pi} \left(1 - \mathrm{e}^{-\mathrm{i}(p-k)a}\right)^{-1} \ (\Im k > 0), \\
\delta_{D-} &= \frac{a}{2\pi} \sum_{m=-\infty}^{-1} \mathrm{e}^{-\mathrm{i}(p-k)am} = -\frac{a}{2\pi} \left(1 - \mathrm{e}^{-\mathrm{i}(p-k)a}\right)^{-1} \ (\Im k < 0).
\end{aligned}
\tag{18.253}
$$

It can be seen that the functions $\delta_{D\pm}(p-k)$ are regular in the corresponding half-planes, and this is true for $f_\pm(k)$ as well. Besides, $\delta_{D+}(p-k) \to 1$ $(\Im k \to \infty)$ and $\delta_{D-}(p-k) \to 0$ $(\Im k \to -\infty)$.

This analogue of the *Cauchy-type integral for a periodic function* appearing in the discrete Fourier transform was mentioned by Slepyan (1982). The same integral, but for another $f_\pm$-decomposition,

$$
\begin{aligned}
f_+(k) &= \frac{f(0)}{2} + \sum_{k=1}^{\infty} f(an) \mathrm{e}^{\mathrm{i}kan}, \\
f_-(k) &= \frac{f(0)}{2} + \sum_{k=-\infty}^{-1} f(an) \mathrm{e}^{\mathrm{i}kan},
\end{aligned}
\tag{18.254}
$$

was used by Eatwell and Willis (1982). In this case

$$
\delta_{D\pm} = \pm \frac{a}{4\pi \mathrm{i}} \cot \frac{(p-k)a}{2} \ (\pm \Im k > 0). \tag{18.255}
$$

Note that both versions formally go over to the classical kernels (18.97) as $a \to 0$.

Wiener-Hopf Technique for the Discrete Transform

The Wiener-Hopf technique, in its application to the discrete Fourier transform used in the case of periodic structures, is basically similar to that introduced for the continuous transform. The main dissimilarity is the Cauchy-type integral which differs from the original version intended for the continuous case.

Convolution

The convolution of two functions of a discrete argument is defined as

$$\mathcal{C}(am) = f(am) * g(am) = \sum_{n=-\infty}^{\infty} f(an)\, g[a(m-n)]. \tag{18.256}$$

It follows that, as in the case of continuous functions,

$$\begin{aligned} \mathcal{C}^F(k) &= \sum_{n=-\infty}^{\infty} \sum_{m=-\infty}^{\infty} f(am)\mathrm{e}^{\mathrm{i}kam} g[a(n-m)]\mathrm{e}^{\mathrm{i}ka(n-m)} \\ &= f^F(k) g^F(k). \end{aligned} \tag{18.257}$$

The Parseval Identity

From the definition of the discrete transform it follows that

$$g(-am)^F(k) = g^F(-k), \quad \left[\overline{g(-am)}\right]^F(k) = \overline{g^F(k)}. \tag{18.258}$$

The latter identity and Eq. (18.257) give us

$$\begin{aligned} \left[f(am) * \overline{g(-am)}\right]^F(k) &= \left[\sum_{n=-\infty}^{\infty} f(an)\overline{g(a(n-m))}\right]^F(k) \\ &= f^F(k)\overline{g^F(k)}. \end{aligned} \tag{18.259}$$

Now, using the inverse-transform formula (18.239) for $m = 0$, we obtain the Parseval identity

$$\sum_{n=-\infty}^{\infty} f(an)\overline{g(an)} = \frac{a}{2\pi} \int_{-\pi/a}^{\pi/a} f^F(k)\overline{g^F(k)}\, \mathrm{d}k. \tag{18.260}$$

Steady-State Regime and Duality in the Independent Variable

The discrete transform can be used not only for analysis of a regular discrete system but for a periodic structure of a general view. In this way, the Floquet wave approach is used (see Sects. 3.4.9 and 3.4.13).

As in the continuous case, the steady-state regime is characterized by variable $\eta = x - vt$ instead of x and t separately. Now x is a discrete variable, $x = am$. So η is a discrete variable for a given t or a continuous variable for a given m. However, in the use of the Fourier transform, η should be considered as a single variable. There are two possibilities.

18.6.5 *Some Asymptotes*

Limiting Relations and Power Asymptotes

It follows directly from the definition of the discrete transform (18.236) that

$$f(0) = \lim_{k \to \mathrm{i}\infty} f_+(k), \tag{18.261}$$

and another expression follows from (18.239):

$$f(0) = \frac{a}{2\pi} \int_{-\pi/a}^{\pi/a} f^F(k)\, \mathrm{d}k. \tag{18.262}$$

Note that these relations differ from the corresponding relation for the continuous transform (18.168). At the same time, relations for the limits of $f(am)$, $m \to \pm\infty$, if the limits exist, are similar to (18.168), namely

$$f(\infty) = \lim_{k \to 0}[(-\mathrm{i}ka) f_+(k)], \; f(-\infty) = \lim_{k \to 0}[(\mathrm{i}ka) f_-(k)]. \tag{18.263}$$

Indeed, take $N \to \infty, kN \to 0$. Then

$$\begin{aligned}\lim_{k \to 0}\left[(-\mathrm{i}ka) f_+^F(k)\right] &= \lim_{k \to 0}\left[f(\infty)(-\mathrm{i}ka) \sum_{m=N}^{\infty} \mathrm{e}^{\mathrm{i}(k+\mathrm{i}0)am}\right] = f(\infty),\\ \lim_{k \to 0}\left[(\mathrm{i}ka) f_+^F(k)\right] &= \lim_{k \to 0}\left[f(-\infty)(-\mathrm{i}ka) \sum_{m=-\infty}^{-N} \mathrm{e}^{\mathrm{i}(k+\mathrm{i}0)am}\right] = f(-\infty).\end{aligned} \tag{18.264}$$

In the discrete case, the asymptotes for $x \to \pm\infty$, (18.177) and (18.179), which are determined considering inverse transform on the complex plane, are still true as well (see Fig. 18.4).

The Related Continuum

The asymptotic contribution of the zero point, $k_0 = 0$, presents a long-wave approximation of the result corresponding to the discrete system. Indeed, a discrete system differs from the related continuum by the fact that finite differences are present in the corresponding equations instead of the derivatives. However, they coincide with each other if $k \to 0$:

$$\begin{aligned} \mathrm{d}_1 &= \left[\frac{u(x+a)-u(x)}{a}\right]^F = \frac{\mathrm{e}^{-\mathrm{i}ka}-1}{a}u^F \to -\mathrm{i}ku^F, \\ \mathrm{d}_2 &= \left[\frac{u(x+a)-2u(x)+u(x-a)}{a^2}\right]^F = \frac{2[\cos(ka)-1]}{a^2}u^F \to -k^2u^F, \end{aligned} \tag{18.265}$$

and so on. In this context, note that an original function defined by the inverse Fourier transform can be considered as a function of the continual variable, x, as a parameter of the integrand. This is in spite of the fact that x was considered as a discrete variable in the direct transform. Thus, the discrete transform leads to a *quasi-continuum* (Kunin, 1982, 1983).

18.7 Structural Discontinuity and Fourier Transform

We study waves in a segmented structure, possibly connected with a continuous medium. The discontinuities are considered as localized strains induced by self-equilibrated localized actions, Slepyan (2018). Under this formulation augmented by interface conditions, the linearized formulation remains valid. This approach eliminates the need for examining separate sections with subsequent conjugation. Only conditions related to the discontinuities should be satisfied, while the continuity in other respects preserves itself automatically. No obstacle remains for the continuous Fourier transform. For a uniform partitioning, the discrete transform is used together with the continuous one. We demonstrate the technique by obtaining the Floquet wave dispersion relations with their dependence upon interface stiffness. To this end, we briefly consider the flexural wave in the segmented beam on Winkler's foundation, the gravity wave in a plate (also segmented) on deep water and the Floquet-Rayleigh wave in such a plate on an elastic half-space. Besides, we present the wave equations developed for an elastic medium with discontinuities.

We show how the continuous Fourier transform, an effective method in the linear analysis commonly used for a system homogeneous in the respective coordinate, can be extended to a segmented structure possibly interacting with a continuous medium.

Let a homogeneous system have point contacts with some other structures. For example, it could be an elastic beam supported by deformable rods. We can carry out the Fourier transform with accounting for the (unknown) contact forces and obtaining then the latter based on respective force-displacement relations. Note that in the case

(a)

(b)

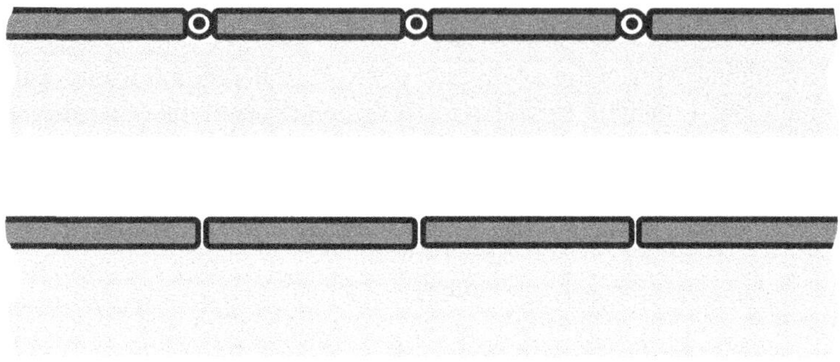

Fig. 18.5 The segmented elastic beam in contact with a 2D medium homogeneous in the longitudinal coordinate. The discontinuities allowing rotations under the bending moment (**a**) and entirely separated parts of the beam, which can interact with each other through the medium (**b**)—as Professor Gennady Mishuris sees it

of a distributed contact, we end up with an integral equation on the corresponding domain. In the case of discrete supports distributed uniformly, the discrete Fourier transform leads to a single ratio. Gueorguiev et al. (2000) examined such a system as a waveguide with several periodical arrays of attachments.

It may seem less evident that similar considerations are applicable not only for the contact inhomogeneities but local structural discontinuities as well. Note that structures with discontinuities present in different areas. An array of surface cracks in a material (in this connection, see Joglekar & Mitran, 2016) and a cracked sea ice cover can serve as examples. Segmented constructions can also be mentioned such as stationary or pontoon bridges. Note that in the ice cover case, in-plane compressive forces and the crack closure effect strongly affect the 'effective' hinge stiffness manifested in angle interactions of the adjacent sections. In this connection, see Rice and Levy (1972), Dempsey et al. (1995), Slepyan et al. (1995a). (The in-plane forces also play a nontrivial role in the analysis developed below, see Sect. 18.7.3.)

We consider an analytical technique that permits the Fourier integral transform in studying the static deformations, oscillations, and waves in discrete-continuous structures. As illustrative examples, we examine a homogeneous elastic beam (or a plate) divided into sections and coupled with a continuous medium. We show two possible configurations schematically in Fig. 18.5 (the parts can rotate and shift transversely relative to each other under the bending moment and transverse force, respectively).

For the analytical study of such systems we use the approach earlier discussed in fracture mechanics where a crack was represented as the result of the action of generalized forces on the continuous medium, Slepyan (1981, 2002, Sect. 5.7).

We use Lagrangian formulation considering the discontinuities as generalized strains created in an initially homogeneous structure by localized, self-equilibrated, generalized loads. Mathematically, this approach differs by using some higher order 'deltas,' the derivatives of the Dirac delta function, higher than those related to real external localized forces.

For example, the functions $\delta(x)$ and $\delta'(x) = \mathrm{d}\delta(x)/\mathrm{d}x$, corresponding to jump discontinuities in the transverse force and bending moment in the beam, make the internal force and moment discontinuous. These facts are well-known. At the same time, no one prevents us from continuing with the derivatives of higher orders taking generalized loads $\delta''(x)$ and $\delta'''(x)$. The latter corresponds to the structural discontinuities, the jumps in the inclination angle of the beam $w'(x)$ and its displacement $w(x)$, respectively. Concerning the above considerations, we note that the delta function can be defined as the derivative of a unite jump of discontinuity (see the related topics, e.g., in Bremermann, 1965).

Formally, the 'loads' creating structural discontinuities do not affect the regular strain outside of the singular points but result in supersingular strain at these points. The latter, however, have no connection with the actual interface interaction, for which we introduce relations between the discontinuities and the corresponding force factors independently.

In doing so, we eliminate the obstacle for the use of the integral Fourier transform for the segmented structure with an adjoining medium, which reflects itself by Green's function. In the case of a uniform discrete distribution of the discontinuities, the coupled discrete and continuous transforms present in the analysis.

18.7.1 General Formulation

So, we distinguish two formulations, physical, which we must eventually follow, and mathematical, which we use in solving the problems. In the former, we have the segmented structure, which parts under internal forces can have limited turn and shift relative to each other. In contrast, in the mathematical formulation, which is preferable for the analysis, we have the same but homogeneous structure, where the same as above discontinuities arise due to the action of self-equilibrated point-localized loads.

Since outside the singular points, the equations and boundary conditions adopted in these formulations coincide, the difference concerns only these points. In the physical formulation, we assume that there are relations between the structural discontinuities and related forces, whereas in the mathematical formulation we have supersingular strain there. Fortunately, we can merely eliminate this difference. We ignore the singularity replacing it by the (physical) finite relations. Thereby, we make the mathematical solution entirely satisfying the physical formulation.

18.7.2 Discontinuities and Delta Loads

We consider a uniform linear differential equation of a general view including that for a 1D system coupled with a 2D homogeneous medium (homogeneous in the longitudinal direction)

$$\sum_{n=1}^{N} a_n u^{(n)}(x) + \mathcal{G}(x) * u(x) = q(x), \tag{18.266}$$

where a_n are constants, $C(x)$ is (inverse) Green's function as the response of the medium to $w(x) = \delta(x)$, $*$ means the convolution, and $q(x)$ is a regular function. We assume that, under reasonable boundary conditions, this equation is satisfied by a regular function $u(x)$. Note that

$$\mathcal{G}(x) * u(x) \quad \longrightarrow \quad a_0 u(x) \text{ if } \mathcal{G}(x) = a_0\delta(x) \ (a_0 = \text{const}). \tag{18.267}$$

Also note that the equation can be one of a system. Therefore it can contain other functions. We, however, do not show it explicitly, since it does not affect the connections between the discontinuities and related deltas.

In dynamics, the functions u, $\mathcal{G}$ and q also depend on time, and $*$ means the convolution on both the coordinate and time. We will show this dependence explicitly in due time. In the current sections, it does not affect our considerations.

Now, let the jth order derivative have a discontinuity at $x = x_m$

$$\Delta u^{(j)}(x_m) = u^{(j)}(x_m + 0) - u^{(j)}(x_m - 0). \tag{18.268}$$

We find

$$\begin{aligned} &\sum_{n=1}^{N} a_n u^{(n)}(x)|_{x \neq x_m} + \mathcal{G}(x) * u(x) \\ &+\Delta u^{(j)}(x_m) \sum_{n=j+1}^{N} a_n \delta^{(n-j-1)}(x - x_m) = q(x). \end{aligned} \tag{18.269}$$

Outside of the singular point $x = x_m$ this equation is satisfied by a regular function $u(x)$, but the equality does not hold at $x = x_m$ due to the presence of the singular term (the last sum). To compensate the singularity we introduce the same term as an additional discontinuity-associated load. The equation becomes

$$\sum_{n=1}^{N} a_n u^{(n)}(x) + \mathcal{G}(x) * u(x) = q(x)$$
$$+\Delta u^{(j)}(x_m) \sum_{n=j+1}^{N} a_n \delta^{(n-j-1)}(x - x_m). \tag{18.270}$$

The Fourier transform leads to the solution

$$u^F(k) = \left(q^F(k) + \Delta u^{(j)}(x_m) \sum_{n=j+1}^{N} a_n (-\mathrm{i}k)^{n-j-1} \right) \mathrm{e}^{\mathrm{i}kx_m}$$
$$\times \left(\sum_{n=1}^{N} a_n (-\mathrm{i}k)^n u^F(k) + \mathcal{G}^F(k) \right)^{-1}. \tag{18.271}$$

It is not difficult to see that this solution does correspond to the discontinuity. Up to a term, the inverse-transform integral of which converges uniformly (hence defining a continuous function)

$$\left(u^{(j)}\right)^F = (-\mathrm{i}k)^j u^F(k) = \Delta u^{(j)}(x_m) \frac{1}{\pi} \int_0^{\infty} \frac{\sin[k(x - x_m)]}{k} \mathrm{d}k$$
$$= \frac{1}{2} \Delta u^{(j)}(x_m) \mathrm{sign}(x - x_m). \tag{18.272}$$

The above result corresponds to the prescribed discontinuities. However, in some cases, as shown below, the structural discontinuities break the continuity of the internal forces (as if a non-self-equilibrated action is present), which affects the analysis.

18.7.3 Localized Loads: Forces and Self-equilibrated Actions

It may happen that a single discontinuity associates not only with a self-equilibrated action but a force or a moment. In this case, another discontinuity should exist, such that the non-equilibrated effects cancel out. It follows from the equilibrium and the fact that only a self-equilibrated localized load produces the discontinuity.

Consider this issue by an example of the same elastic beam as above (see Eq. 15.1) but the beam under a tensile force, T. The equation for the continuous beam is

$$D\frac{\mathrm{d}^4 w(x)}{\mathrm{d}x^4} - T\frac{\mathrm{d}^2 w(x)}{\mathrm{d}x^2} + \varkappa w(x) = q(x), \quad -\infty < x < \infty. \tag{18.273}$$

Let both geometrical values $w(x)$ and $w'(x)$ have discontinuities at $x = 0$. Under these (and only these) discontinuities the equation becomes

$$D\frac{d^4w(x)}{dx^4} - T\frac{d^2w(x)}{dx^2} + \varkappa w(x) = q(x)$$
$$-D[\alpha\delta''(x) + \beta\delta'''(x)] + T[\alpha\delta(x) + \beta\delta'(x)]. \tag{18.274}$$

Recall that $\alpha = \Delta w'(0)$ and $\beta = \Delta w(0)$.

In this case, due to the presence of the tensile force, the localized external force $T\alpha$ and moment $T\beta$ appear related to the discontinuity of $w'(0)$ and $w(0)$, respectively. These forces, however, do not produce the discontinuities, because the latter are the results of self-equilibrated actions. The forces are the matter of equilibrium, and they must cancel out in the equation of equilibrium. It follows that the prescribed irregularities cause discontinuities in $w'''(0)$ and $w''(0)$, respectively, such that

$$\Delta\mathcal{Q} = -D\Delta w'''(0) = -T\alpha, \ \Delta\mathcal{M} = -D\Delta w''(0) = -T\beta. \tag{18.275}$$

So, in the presence of the tensile force, the structure discontinuities cause the discontinuities in the bending moment and transverse forces in the beam.

Thus, the displacement (the Fourier transform of it) is

$$w^F(k) = \frac{q^F(k) - Dk^2(\alpha - \mathrm{i}k\beta)}{Dk^4 + Tk^2 + \varkappa}. \tag{18.276}$$

18.7.4 The Problem in the Classical Framework

For the reader who would prefer traditional analysis another way of deriving the solution, Eq. (18.271), is below presented, where we avoid consequent differentiation of the discontinuity and the necessity to use deltas at all.

Let a discrete set of the singular points, $x = x_m$, be associated with the equation (18.266). The Fourier transform of a single term over a single segment *between* the neighboring singular points, $x_m < x < x_{m+1}$, with the integration by parts is

$$\int_{x_m+0}^{x_{m+1}-0} u^{(n)}(x)e^{\mathrm{i}kx}\,dx = (-\mathrm{i}k)^n u^{F_m}(k) + A_- + A_+,$$
$$A_- = -\sum_{\nu=0}^{n-1}(-\mathrm{i}k)^\nu u^{(n-1-\nu)}(x_m+0)e^{\mathrm{i}kx_m}, \tag{18.277}$$
$$A_+ = \sum_{\nu=0}^{n-1}(-\mathrm{i}k)^\nu u^{(n-1-\nu)}(x_{m+1}-0)e^{\mathrm{i}kx_{m+1}}.$$

Note that the inverse transform results in a zero outside this segment.

The transform on the previous segment, $x_{m-1} < x < x_m$ gives us the same integrated terms at $x = x_m$ as A_+ but with $x = x_m$. Similarly, this action on the next segment, $x_{m+1} < x < x_{m+2}$ gives us at $x = x_{m+1}$ the same integrated terms as A_-

but with $x = x_{m+1}$. It follows that the integrated terms at $x = x_m$ and $x = x_{m+1}$ annihilate if the corresponding derivatives at these points are continuous. (Indeed, the continuous Fourier transform does not contain integrated terms.) If the left (right) integration limit goes to $\mp\infty$, it brings no integrated terms if the generalized Fourier transform is assumed (Bremermann, 1965).

Now, let there be the discontinuity, $\Delta u^{(j)}(x_m)$. In this case, we obtain that its m-contribution to the right-hand side of the equation

$$\Delta u^{(j)}(x_m)a_n(-ik)^{n-j-1}e^{ikx_m} \tag{18.278}$$

coincide with that reflected in (18.271). Making the transform for all n and combining the finite-support transforms we obtain the solution (18.271).

After these 'classical' considerations we, however, should then discuss what happens at the discontinuity points (as we already did in the previous sections).

18.7.5 *Periodic Array of the Discontinuities*

In the case of a uniform distribution of a single discontinuity (as considered above in (18.268)–(18.271))

$$\Delta u^{(j)}(x) = \Delta u^{(j)}(x_m),\ x_m = am,\ a = \text{const}, \tag{18.279}$$

we have the following right-hand part of the equation

$$S(x) = q(x) + \sum_{m=-\infty}^{\infty} \Delta u^{(j)}(x_m)a_n(m)\delta^{(n-j-1)}(x - x_m), \tag{18.280}$$

which in terms of the discrete Fourier transform becomes

$$\begin{aligned} S^F(k) &= q^F(k) + S^{Fd}(k), \\ S^{Fd}(k) &= \sum_{m=-\infty}^{\infty} \Delta u^{(j)}(x_m)a_n(m)(-ik)^{n-j-1}e^{ikx_m}. \end{aligned} \tag{18.281}$$

So, in this case, we have both the continuous and discrete Fourier transform coupled. With the aim to use the interface relations like those in (18.299), we have to rearrange the coupled transform to the discrete one.

18.7.6 The Fourier Transform: From Continuous-Discrete Coupled to Discrete

Consider a function $f^F(k)$, which can be treated as the Fourier transform of a function $f(x)$

$$f^F(k) = \int_{-\infty}^{\infty} f(x) \mathrm{e}^{\mathrm{i}kx} \, \mathrm{d}x. \tag{18.282}$$

Apparently, the corresponding discrete transform

$$f^{Fd}(k) = \sum_{n=-\infty}^{\infty} f(an) \mathrm{e}^{\mathrm{i}kan}, \; a > 0, \tag{18.283}$$

can be expressed from the continuous one in the following way

$$\begin{aligned}
f^{Fd}(k) = \frac{1}{2\pi} \int_{-\infty}^{\infty} f^F(k') \sum_{n=-\infty}^{\infty} \mathrm{e}^{\mathrm{i}(k-k')an} \, \mathrm{d}k' &= \int_{-\infty}^{\infty} f^F(k') \Phi(k-k') \, \mathrm{d}k', \\
\Phi(k-k') &= \Phi_+(k-k') + \Phi_-(k-k'), \\
\Phi_+(k-k') &= \frac{1}{2\pi} \frac{1}{1 - \exp[\mathrm{i}(k-k'+\mathrm{i}0)a]}, \\
\Phi_-(k-k') &= \frac{1}{2\pi} \frac{\exp[-\mathrm{i}(k-k'-\mathrm{i}0)a]}{1 - \exp[-\mathrm{i}(k-k'-\mathrm{i}0)a]}, \\
\Phi_+(k-k') + \Phi_-(k-k') &= \sum_{n=-\infty}^{\infty} \delta[a(k-k') + 2\pi n].
\end{aligned} \tag{18.284}$$

Thus, the expression is

$$f^{Fd}(k) = \frac{1}{a} \sum_{n=-\infty}^{\infty} f^F\left(k + \frac{2\pi n}{a}\right) \tag{18.285}$$

It follows that the discrete transform defined by the last relation is a $2\pi/a$ periodic function of k, as well as the functions $\Phi_\pm$, as they should be, and the latter represent functions regular in the upper/lower complex half-planes of k, respectively. So, the functions $\Phi_\pm$ are the kernels of the Cauchy-type integral (for the discrete transform)

$$\begin{aligned}
f_\pm^{Fd}(k) &= \int_{-\infty}^{\infty} f^F(k') \Phi_\pm(k-k') \, \mathrm{d}k' \\
&= a \int_{-\pi/a}^{\pi/a} f^{Fd}(k') \Phi_\pm(k-k') \, \mathrm{d}k'.
\end{aligned} \tag{18.286}$$

In the same manner, it can be seen that similar results correspond to a product of the continuous and discrete Fourier transforms. Due to the periodicity of the latter

$$h^{Fd}(ak + 2\pi n) = h^{Fd}(k) \ \left(f^F(k) = g^F(k)h^{Fd}(k)\right), \tag{18.287}$$

the corresponding discrete transform is

$$f^{Fd}(k) = \frac{1}{a} h^{Fd}(k) \sum_{n=-\infty}^{\infty} g^F\left(k + \frac{2\pi n}{a}\right). \tag{18.288}$$

Concerning the summation, it should be noted that each part of the sum, $\sum_{n=-\infty}^{-1}$ and $\sum_{n=0}^{\infty}$, separately can diverge as well as the total sum as it is presented in (18.288), while its another form, where the two sums are summed sequentially

$$\begin{aligned}\sum_{n=-\infty}^{\infty} g^F\left(k + \frac{2\pi n}{a}\right) &= g^F(k) \\ &+ \sum_{n=1}^{\infty}\left[g^F\left(k + \frac{2\pi n}{a}\right) + g^F\left(k - \frac{2\pi n}{a}\right)\right] \\ &= \frac{1}{2}\sum_{n=-\infty}^{\infty}\left[g^F\left(k + \frac{2\pi n}{a}\right) + g^F\left(k - \frac{2\pi n}{a}\right)\right],\end{aligned} \tag{18.289}$$

converges. This can happen for an asymptotically odd function of n where

$$g^F\left(k - \frac{2\pi n}{a}\right) \sim -g^F\left(k + \frac{2\pi n}{a}\right) \quad (n \to \infty). \tag{18.290}$$

Note that if the sum $\sum_{n=-\infty}^{\infty} g^F(k + 2\pi/a)$ converges, then

$$\sum_{n=-\infty}^{\infty} g^F\left(k + \frac{2\pi n}{a}\right) = \sum_{n=-\infty}^{\infty} g^F\left(k - \frac{2\pi n}{a}\right), \tag{18.291}$$

and the representation (18.289) is identical to that in (18.288). Otherwise, only the rule (18.289) should be accepted, and we use it for $Q_q^F(k)$ in the beam related section.

Note that due to the presence of $(k + 2\pi m/a)^4$ term, the infinite sums in (18.288), at least in the form in (18.289), converge fast.

The Beam-Plate Problems

We now consider the beam with two types of the discontinuity, the jumps of both the displacement derivative and the displacement itself

$$\alpha_n = w'(x_n + 0) - w'(x_n - 0),\ \beta_n = w(x_n + 0) - w(x_n - 0),$$
$$n = 1, 2, \dots, N. \tag{18.292}$$

Here we do not specify the medium with which the beam is coupled. In the same way as in the above example, we get the equation

$$D\frac{\mathrm{d}^4 w(x)}{\mathrm{d}x^4} + \mathcal{G}(x) * w(x)$$
$$= q(x) + D\sum_n \alpha_n \delta''(x - x_n) + \beta_n \delta'''(x - x_n), \tag{18.293}$$

where $\mathcal{G}(x)$ is (inverse) Green's function as the response to the displacement $w(x) = \delta(x)$, and $*$ is the convolution symbol. In dynamics, the discontinuity parameters α and β become functions of time as well as $w = w(x,t)$, $\mathcal{G} = \mathcal{G}(x,t)$, and the convolution also includes the beam inertia. However, in this section, we do not show the time-dependence explicitly, since it does not affect our reasoning.

The Fourier transform results in

$$\left[Dk^4 + \mathcal{G}^F(k)\right] w^F(k) = q^F(k) - Dk^2 \sum_m (\alpha_m - \mathrm{i}k\beta_m)\mathrm{e}^{\mathrm{i}kx_m}. \tag{18.294}$$

The dependencies for the displacement and its derivatives for the singular interface located at $x = 0$ are as follows (the relations for other points differ only by the exponential multipliers as in (18.294))

$$w^F(k) = \frac{q^F(k) - Dk^2(\alpha - \mathrm{i}k\beta)}{Dk^4 + \mathcal{G}^F(k)},$$
$$(w')^F(k) = \underline{\beta} - \frac{\mathrm{i}kq^F(k) + \beta\mathcal{G}^F(k) - D\alpha\mathrm{i}k^3}{Dk^4 + \mathcal{G}^F(k)},$$
$$(w'')^F(k) = \underline{\alpha - \mathrm{i}k\beta} + (w''_{reg})^F(k),$$
$$(w''')^F(k) = \underline{-\mathrm{i}k\alpha - k^2\beta} + (w'''_{reg})^F(k), \tag{18.295}$$

with

$$(w''_{reg})^F(k) = -\frac{k^2q^F(k) + (\alpha - \mathrm{i}k\beta)\mathcal{G}^F(k)}{Dk^4 + \mathcal{G}^F(k)},$$
$$(w'''_{reg})^F(k) = \frac{\mathrm{i}k^3q^F(k) + (\mathrm{i}k\alpha + k^2\beta)\mathcal{G}^F(k)}{Dk^4 + \mathcal{G}^F(k)}, \tag{18.296}$$

where the underlined terms correspond to the deltas at $x = 0$, which do not affect the forces and should be ignored as discussed in Sect. 18.7.1.

Expressions for the total bending moment and the transverse force due to the external force and the discontinuities at $x = x_m$, $m = 1, \dots, N$, follow from these relations as

$$\mathcal{M}^F(k) = M_q^F(k) + \sum_{m=1}^{N} \left[\Lambda_\alpha^F(k)\alpha_m + \Lambda_{\alpha\beta}^F(k)\beta_m\right] \mathrm{e}^{\mathrm{i}x_m},$$
$$\mathcal{Q}^F(k) = Q_q^F(k) + \sum_{m=1}^{N} \left[\Lambda_{\alpha\beta}^F(k)\alpha_m + \Lambda_\beta^F(k)\beta_m\right] \mathrm{e}^{\mathrm{i}x_m}, \quad (18.297)$$

with

$$M_q^F(k) = \frac{Dk^2 q^F(k)}{Dk^4 + \mathcal{G}^F(k)}, \quad Q_q^F(k) = -\frac{D\mathrm{i}k^3 q^F(k)}{Dk^4 + \mathcal{G}^F(k)},$$
$$\Lambda_\alpha^F(k) = \frac{D\mathcal{G}^F(k)}{Dk^4 + \mathcal{G}^F(k)}, \quad \Lambda_\beta^F(k) = -\frac{Dk^2\mathcal{G}^F(k)}{Dk^4 + \mathcal{G}^F(k)}, \quad (18.298)$$
$$\Lambda_{\alpha\beta}^F(k) = -\frac{D\mathrm{i}k\mathcal{G}^F(k)}{Dk^4 + \mathcal{G}^F(k)}.$$

The relations between the discontinuities α and β on the one hand, and the corresponding forces, $\mathcal{M}$ and $\mathcal{Q}$ on the other hand are

$$\mathcal{M}(x_n) = -\varkappa_{Mn}\alpha_n, \; \mathcal{Q}(x_n) = \varkappa_{Qn}\alpha_n. \quad (18.299)$$

Thus, we have $2N$ interface conditions (18.299), which allow us to determine the discontinuity parameters α_n and β_n. After this, the expression in (18.294) uniquely defines the beam displacement.

Uniformly Distributed Discontinuities

Here, we reduce the formulation to the case of uniformly distributed discontinuities: $x_n = an, n = 0, \pm 1, \ldots, a = \text{const} > 0$ with $\varkappa_{(M,Q)n} = \varkappa_{(M,Q)}$. In this case, the sums on m in (18.294) and (18.297) become the Fourier discrete transforms

$$\left[Dk^4 + \mathcal{G}^F(k)\right] w^F(k) = q^F(k) - Dk^2\left(\alpha^{Fd}(k) - \mathrm{i}k\beta^{Fd}(k)\right), \quad (18.300)$$

where for a function $f(n)$

$$f^{Fd}(k) = \sum_{n=-\infty}^{\infty} f(n)\mathrm{e}^{\mathrm{i}an}. \quad (18.301)$$

The expressions for the bending moment and transverse force (18.297) become

$$\begin{aligned}\mathcal{M}^F(k) &= M_q^F(k) + \Lambda_\alpha^F(k)\alpha^{Fd}(k) + \Lambda_{\alpha\beta}^F(k)\beta^{Fd}(k),\\ \mathcal{Q}^F(k) &= Q_q^F(k) + \Lambda_{\alpha\beta}^F(k)\alpha^{Fd}(k) + \Lambda_\beta^F(k)\beta^{Fd}(k),\end{aligned} \tag{18.302}$$

where the coefficients $\Lambda^F(k)$ are defined in (18.298). Also, as follows from (18.299)

$$\mathcal{M}^{Fd}(k) = -\varkappa_M \alpha^{Fd}(k),\ \mathcal{Q}^{Fd}(k) = \varkappa_Q \alpha^{Fd}(k). \tag{18.303}$$

Now, referring to (18.303) and (18.285), we get the following equations concerning the discontinuity parameters α and β

$$\begin{aligned}\mathcal{M}^{Fd}(k) &= M_q^{Fd}(k) + \Lambda_\alpha^{Fd}(k)\alpha^{Fd}(k) + \Lambda_{\alpha\beta}^{Fd}(k)\beta^{Fd}(k) = -\varkappa_M \alpha^{Fd}(k),\\ \mathcal{Q}^{Fd}(k) &= Q_q^{Fd}(k) + \Lambda_{\alpha\beta}^{Fd}(k)\alpha^{Fd}(k) + \Lambda_\beta^{Fd}(k)\beta^{Fd}(k) = \varkappa_Q \beta^{Fd}(k),\end{aligned} \tag{18.304}$$

where

$$\Lambda_{\alpha,\alpha\beta,\beta}^{Fd}(k) = \frac{1}{a}\sum_{n=-\infty}^{\infty} \Lambda_{\alpha,\alpha\beta,\beta}^{F}\left(k + \frac{2\pi n}{a}\right). \tag{18.305}$$

As the discontinuity parameters are determined from these two equations, we can go back to the beam displacement defined in (18.300). However, the homogeneous solution (the dispersion relation in the case of waves) follows from (18.304) directly. If $q = 0$ a nontrivial solution satisfies the relation as

$$(\Lambda_\alpha^{Fd}(k) + \varkappa_M)(\Lambda_\beta^{Fd}(k) - \varkappa_Q) - (\Lambda_{\alpha\beta}^{Fd}(k))^2 = 0, \tag{18.306}$$

and in the case where $\beta^{Fd} = 0$ ($\varkappa_Q = \infty$), it reduces to

$$\Lambda_\alpha^{Fd}(k) = -\varkappa_M. \tag{18.307}$$

At the limit $\varkappa_M = \infty, \varkappa_Q = \infty$, we return to the continuous beam with the Floquet-type dispersion relation

$$\prod_{n=-\infty}^{\infty}\left[D\left(k + \frac{2\pi n}{a}\right)^4 + \mathcal{G}^F\left(k + \frac{2\pi n}{a}\right)\right] = 0. \tag{18.308}$$

This expression represents the set of modes indistinguishable at the discrete set of points $x = an$, where the Floquet wave is defined. So at this limit, the set is identical to the single dispersion relation

$$Dk^4 + \mathcal{G}^F(k) = 0, \quad -\pi < ka \le \pi. \tag{18.309}$$

Summarising, the general procedure is as follows. Assuming Green's function $\mathcal{G}^F(k)$ is given, first the Λ^{Fd}-functions should be determined based on (18.298) and

(18.285)

$$
\begin{aligned}
\Lambda_{\alpha}^{Fd}(k) &= \frac{1}{a}\sum_{n=-\infty}^{\infty}\frac{D\mathcal{G}^{F}(k')}{D(k')^4+\mathcal{G}^{F}(k')},\\
\Lambda_{\beta}^{Fd}(k) &= -\frac{1}{a}\sum_{n=-\infty}^{\infty}\frac{D(k')^2\mathcal{G}^{F}(k')}{D(k')^4+\mathcal{G}^{F}(k')},\\
\Lambda_{\alpha\beta}^{Fd}(k) &= -\frac{1}{a}\sum_{n=-\infty}^{\infty}\frac{D\mathrm{i}k'\mathcal{G}^{F}(k')}{D(k')^4+\mathcal{G}^{F}(k')},\quad k'=k+\frac{2\pi n}{a}.
\end{aligned} \tag{18.310}
$$

Then, if the complete solution is required, similar transformation is to be applied for the load functions in (18.298)

$$
M_q^F(k) \longrightarrow M_q^{Fd}(k),\; Q_q^F(k) \longrightarrow Q_q^{Fd}(k) \tag{18.311}
$$

based on (18.285) and (18.289), respectively.

As a result, for the case $q=0$ we obtain the 'dispersion relation' (18.306) defining the Floquet wavenumber k. Otherwise, the discontinuity parameters $\alpha^{Fd}(k)$ and $\beta^{Fd}(k)$ can be found from (18.304).

Finally, the beam displacement and internal forces are defined in the form of (continuous) Fourier transform in (18.300) and (18.302).

Below we present some illustrative examples assuming that the beam (or plate) has only angle discontinuities periodically distributed, whereas the displacement is continuous. In this case, the system (18.304) reduces to a single equation as

$$
M_q^{Fd}(k) + (\Lambda_{\alpha}^{Fd}(k) + \varkappa_M)\alpha^{Fd}(k) = 0. \tag{18.312}
$$

18.7.7 The Periodically Segmented Beam on Elastic Foundation

General Dependencies

In this case, $\mathcal{G}^F(k) = \varkappa = \text{const}$. We assume that the beam is under the force $q = P\delta(x)$, $q^F(k) = P$. The functions in (18.297) become

$$
\begin{aligned}
M_q^{Fd}(k) = aPS_2,\; \Lambda_{\alpha}^{Fd}(k) = \frac{DS_0}{a},\; \Lambda_{\alpha\beta}^{Fd}(k) = -\frac{\mathrm{i}DS_1}{a^2},\\
Q_q^{Fd}(k) = -\mathrm{i}PS_3,\; \Lambda_{\beta}^{Fd}(k) = -\frac{DS_2}{a^3},
\end{aligned} \tag{18.313}
$$

with

$$S_0 = \sum_{n=-\infty}^{\infty} \frac{\varkappa}{D(k+2\pi n/a)^4+\varkappa} = \frac{1}{2}\Re(\hat{\varkappa}^{1/4}\Psi),$$

$$S_1 = \sum_{n=-\infty}^{\infty} \frac{a(k+2\pi n/a)\varkappa}{D(k+2\pi n/a)^4+\varkappa}$$

$$= \frac{\sqrt{\hat{\varkappa}}}{2}\frac{\sinh(\tau)\sin(\tau)\sin(ak)}{(\cosh(\tau)\cos(\tau)-\cos(ak))^2+(\sinh(\tau)\sin(\tau))^2},$$

$$S_2 = \sum_{n=-\infty}^{\infty} \frac{(k+2\pi n/a)^2\varkappa}{a^2(D(k+2\pi n/a)^4+\varkappa)} = \frac{1}{2}\Im(\hat{\varkappa}^{-1/4}\Psi), \tag{18.314}$$

$$S_3 = \sum_{n=-\infty}^{\infty} \frac{1}{2a}\left(\frac{(k+2\pi n/a)^3 D}{D(k+2\pi n/a)^4+\varkappa} - \frac{(-k+2\pi n/a)^3 D}{D(-k+2\pi n/a)^4+\varkappa}\right)$$

$$= \frac{1}{2}\Re\frac{\sin(ak)}{\cosh(\sqrt{\mathrm{i}}\hat{\varkappa}^{1/4})-\cos(ak)},$$

$$\Psi = \frac{\sqrt{\mathrm{i}}\,\sinh(\sqrt{\mathrm{i}}\,\hat{\varkappa}^{1/4})}{\cosh(\sqrt{\mathrm{i}}\,\hat{\varkappa}^{1/4})-\cos(ak)},\ \hat{\varkappa} = \frac{\varkappa a^4}{D},\ \tau = \frac{\sqrt{2}}{2}\hat{\varkappa}^{1/4}.$$

The Beam in Statics

The first equation in (18.304) becomes

$$aPS_2 + \left(\frac{DS_0}{a} + \varkappa_M\right)\alpha^{Fd} = 0, \implies \alpha^{Fd} = -\frac{\hat{P}S_2}{S_0+\hat{\varkappa}_M}, \tag{18.315}$$

and

$$w^F(k) = \frac{P - Dk^2\alpha^{Fd}}{Dk^4+\varkappa} = \frac{\hat{P}a^2}{(ak)^4+\hat{\varkappa}}\left(1+\frac{(ak)^2 S_2}{S_0+\hat{\varkappa}_M}\right) \tag{18.316}$$

with

$$\hat{P} = \frac{a^2 P}{D},\ \hat{\varkappa} = \frac{a^4\varkappa}{D},\ \hat{\varkappa}_M = \frac{a\varkappa_M}{D}. \tag{18.317}$$

The beam displacement corresponding to $\hat{P} = 1$ and some values of $\hat{\varkappa}_M$ for $\hat{\varkappa} = 1$ and $\hat{\varkappa} = 100$ is presented in Figs. 18.6 and 18.7, respectively.

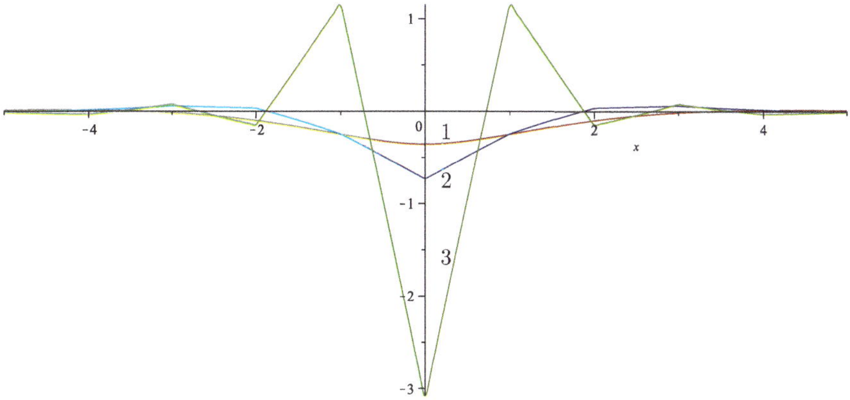

Fig. 18.6 The beam displacement for $\hat{P} = \hat{\varkappa} = 1$ and different values of the hinge stiffness: $\varkappa_M = \infty$ (1), $\varkappa_M = 1/2$ (2) and $\varkappa_M = 0$ (3)

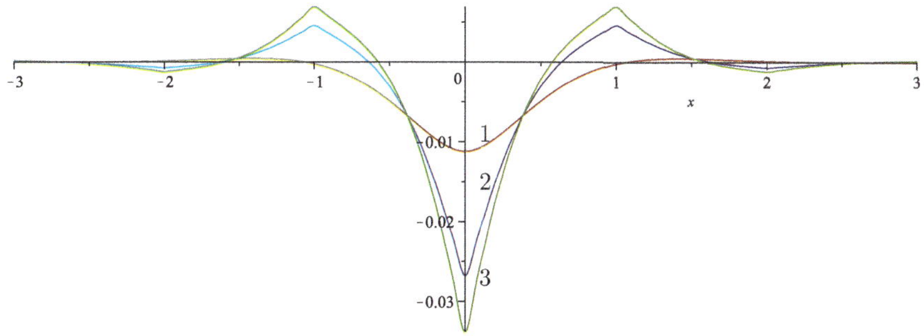

Fig. 18.7 The beam displacement for $\hat{P} = \hat{\varkappa} = 100$: and different values of the hinge stiffness: $\varkappa_M = \infty$ (1), $\varkappa_M = 1/2$ (2) and $\varkappa_M = 0$ (3)

The Beam in Oscillations

In a general dynamic case, we replace

$$\mathcal{G}^F(k) \quad \longrightarrow \quad \varrho S s^2 + \mathcal{G}^{LF}(s, k), \tag{18.318}$$

where ϱ, S and s a are the beam material density and the cross-section area of the beam and the parameter of the Laplace transform on time t, respectively, and $\mathcal{G}(t, x)$ corresponds to the medium dynamic response to $w(t, x) = \delta(t)\delta(x)$.

For the case of sinusoidal oscillations where the functions contain multiplier $\exp(\mathrm{i}\omega t)$, omitting the latter we return to the previous formulation with the change

$$\varkappa \quad \longrightarrow \quad \varrho S(0 + \mathrm{i}\omega)^2 + \varkappa, \; \hat{\varkappa} \quad \longrightarrow \quad (0 + \mathrm{i}\hat{\omega})^2 + \hat{\varkappa}, \; \hat{\omega} = \frac{a^2\omega}{rc}, \tag{18.319}$$

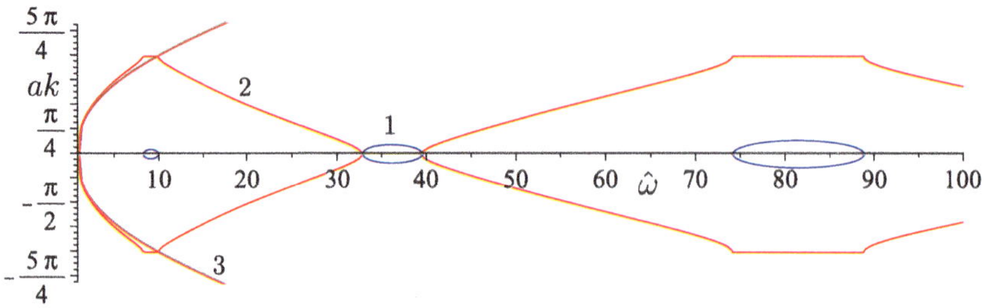

Fig. 18.8 The beam on the elastic foundation. The Floquet normalized wavenumber ak for $\hat{\varkappa}_M = 10$ and $\hat{\varkappa} = 1$: (1) $a\Re k$ (red), (2) $a\Im k$ (blue) and (3) the regular normalized wavenumber for the intact beam, $\hat{\varkappa}_M = \infty$ (brown)

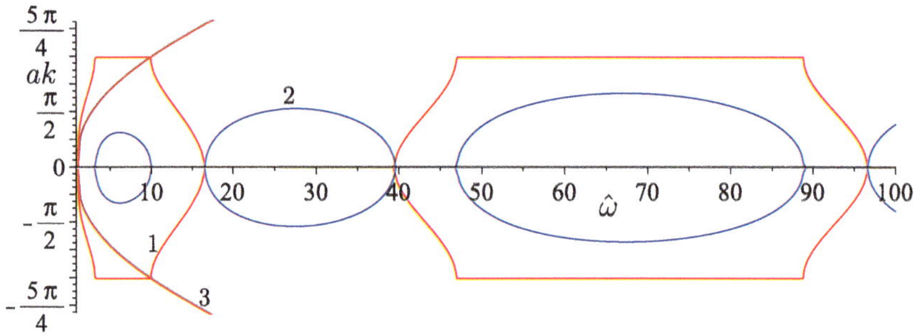

Fig. 18.9 The beam on the elastic foundation. The Floquet normalized wavenumber ak for $\hat{\varkappa}_M = 1$ and $\hat{\varkappa} = 1$: (1) $a\Re k$ (red), (2) $a\Im k$ (blue) and (3) the regular normalized wavenumber for the intact beam, $\hat{\varkappa}_M = \infty$ (brown)

where r is the radius of inertia of the beam cross-section area and $c = \sqrt{E/\varrho}$ is the longitudinal wave speed in the corresponding elastic rod. The homogeneous relation (18.307) yields

$$ak = \arccos\left(\cos\theta - \frac{\theta\sin\theta}{2\hat{\varkappa}_M}\right),\ \theta = (\hat{\omega}^2 - \hat{\varkappa})^{1/4}. \tag{18.320}$$

The dependence $ak(\hat{\omega})$ for $\hat{\varkappa} = 1$ and several values of the hinge stiffness $\hat{\varkappa}_M$ is plotted in Figs. 18.8, 18.9 and 18.10, where the dependence for the intact beam, $ak = (\hat{\omega}^2 - \hat{\varkappa})^{1/4}$ ($\hat{\varkappa}_M = \infty$) is shown too.

Can be seen that the band gaps grow as $\hat{\varkappa}_M$ decreases and as $\hat{\varkappa}_M \to 0$ cover almost whole domain, except narrowing vicinities of

$$\hat{\omega} = \sqrt{(m\pi)^4 + 1},\ m = 1, 2, \ldots. \tag{18.321}$$

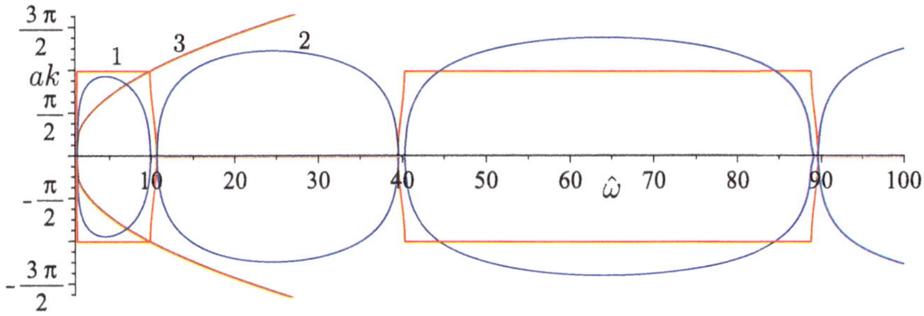

Fig. 18.10 The beam on the elastic foundation. The Floquet normalized wavenumber ak for $\hat{\varkappa}_M = 0.1$ and $\hat{\varkappa} = 1$: (1) $a\Re k$ (red), (2) $a\Im k$ (blue) and (3) the regular normalized wavenumber for the intact beam, $\hat{\varkappa}_M = \infty$ (brown)

18.7.8 A Periodically Segmented Elastic Plate on Deep Water

Gravity Waves

We consider a floating elastic plate rested on deep incompressible water, $z < 0$. The formulation for the continuous plate coincides with that for capillary waves (ripples), see, e.g., Lighthill (1978), if the pressure by the surface tension term is replaced by the one corresponding to the plate: $Tk^2 \longrightarrow Dk^4$. So, we have

$$\mathcal{G}^F(k) = \varrho_0 \left(g - \frac{\omega^2}{\sqrt{0 + k^2}} \right), \tag{18.322}$$

where γ_0 and g are the water density and the acceleration due to gravity, respectively. The frequency in the case of the intact plate is defined as

$$\omega^2 = \left(g + \frac{Dk^4}{\varrho_0} \right) \sqrt{0 + k^2}. \tag{18.323}$$

Consider the homogeneous problem ($q = 0$). The function $\Lambda_\alpha^{Fd}(k)$ (18.310) follows as

$$\Lambda_\alpha^{Fd}(k) = \frac{D}{a} \sum_{n=-\infty}^{\infty} \frac{\hat{\mathcal{G}}^F(\zeta)}{\zeta^4 + \hat{\mathcal{G}}^F(\zeta)}, \; \hat{\mathcal{G}}^F(\zeta) = C_g - \frac{12h\varrho_0}{a\varrho\hat{\gamma}}\hat{\omega}^2,$$
$$C_g = \frac{a^4 \varrho_0 g}{D}, \; \hat{\gamma} = \sqrt{0 + \zeta^2} \; \hat{\omega} = \frac{a^3\omega}{h^2 c}, \; \zeta = ak + 2\pi n. \tag{18.324}$$

We come to the dispersion relation (18.307). In the non-dimensional form, it is

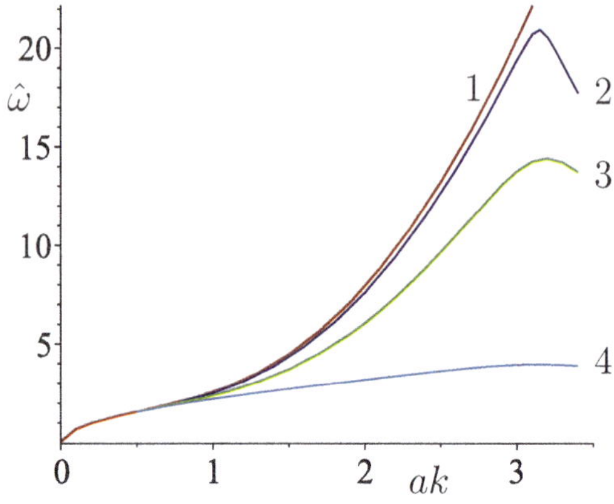

Fig. 18.11 The 2π-periodic Floquet dispersive dependence $\hat{\omega}(-ak) = \hat{\omega}(ak)$ for the gravity wave in a fluid covered by an intact elastic plate. The regular dependence corresponding to the intact plate, $\varkappa_Q = \varkappa_M = \infty$, coincides with the plotted one (1) and continues for greater k in accordance with (18.326). The dependencies for finite values of $\varkappa_M = \infty$ ($\varkappa_Q = \infty$) plotted for $\varkappa_M = 10$ (2), $\varkappa_M = 1$ (3) and $\varkappa_M = 0$ (4)

$$\sum_{n=-\infty}^{\infty} \frac{\hat{\mathcal{G}}^F(\zeta)}{\zeta^4 + \hat{\mathcal{G}}^F(\zeta)}) = -\hat{\varkappa}_M, \ \hat{\varkappa}_M = \frac{a\varkappa_M}{D}, \tag{18.325}$$

The relation for the intact plate, $\hat{\varkappa}_M = \infty$, follows as

$$\hat{\omega} = \sqrt{\frac{\varrho a}{12\varrho_0 h} \left((ak)^4 + C_g\right) \sqrt{0 + (ak)^2}}. \tag{18.326}$$

The dispersion dependencies for several values of the hinge stiffness are shown in Fig. 18.11. It can be seen how the frequency maximal value decreases from $\hat{\omega}_{max} \approx 23$ for the intact plate ($\varkappa_M = \infty$) to $\hat{\omega}_{max} \approx 4$ for the case of the perfect hinges, $\hat{\varkappa}_M = 0$. The plots correspond to $C_g = 3$ and $12h\varrho_0 = 0.6a\varrho$. Also, the discontinuities entail the band gap extending as $\varkappa_M$ decreases. Its upper bound is

$$\omega^*_{max} = \sqrt{\frac{\pi g}{a} + \frac{\pi^5 D}{\varrho_0 a^5}} \tag{18.327}$$

(it is equal to ω_{max} at $\varkappa_M = \infty$), whereas ω^*_{min} is the frequency at $k = \pi/a$ corresponding to the stiffness $\varkappa_M$ under consideration. Graphs of the bounds are plotted in Fig. 18.12. Note that ω^*_{max} independency of λ_M follows from the fact that for $k = \pi/a$ the terms $n = 0$ and $n = -1$ (in the sum by n) coincide.

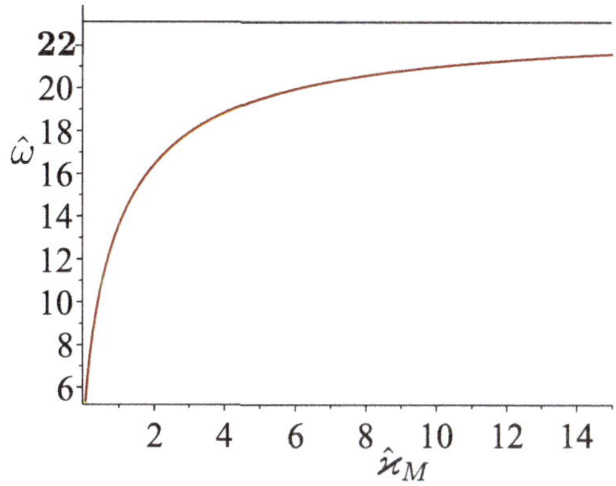

Fig. 18.12 The 2π-periodic Floquet dispersive dependence $\hat{\omega}(-ak) = \hat{\omega}(ak)$ for the gravity wave in a fluid covered by an elastic plate with the periodically placed discontinuities. The band gap upper and lower bounds

18.7.9 The Plate in Sliding Contact with an Elastic Medium

The waves in the homogeneous flexible plate contacting with the elastic half-space where studied in Achenbach and Epstein (1967) and Ozisik and Akbarov (2003). Below we consider this problem for the segmented plate as in the previous section, but for an elastic half-plane with the gravity neglected.

Green's function for the corresponding plane problem can be found, e.g., in Slepyan (2002, Sect. 9.1.3). For our case where $w = u_2 = \delta(x)\delta(t),\ w^{FL}(s,k) = 1$

$$\mathcal{G}_0^{LF}(s,k) = \sigma_{22}^{LF} = -\frac{1}{S_{22}^{LF}} = \frac{\mu c_2^2 R}{s^2 n_1}\ (\sigma_{12} = 0),$$
$$R = (n_2^2 + k^2)^2 - 4k^2 n_1 n_2,\ n_{1,2} = \sqrt{k^2 + \frac{s^2}{c_{1,2}^2}},\ c_1^2 = \frac{\lambda + 2\mu}{\varrho_1},\ c_2^2 = \frac{\mu}{\varrho_1}, \tag{18.328}$$

where λ and μ are the Lamé elastic constants and ϱ_1 is the medium density.

For the complex wave with dependence on time as $e^{i\omega t}$, the above relations remain valid as $s \longrightarrow 0 + i\omega$. We come to the same expression for $\Lambda_\alpha^{Fd}(k)$ as in (18.324), but with the following function $\mathcal{G}^F(\zeta)$

$$\hat{\mathcal{G}}^F(\zeta) = -A\hat{\omega}^2 - \frac{B[(2\zeta^2 - \hat{\omega}^2)^2 - 4\zeta^2\sqrt{\zeta^2 - \hat{\omega}^2/4}\sqrt{\zeta^2 - \hat{\omega}^2/4}]}{\hat{\omega}^2\sqrt{\zeta^2 - \hat{\omega}^2/4}},$$
$$A = \frac{12c_2^2 a^2}{c^2 h^2},\ B = \frac{12\mu(1-\nu^2)a^3}{Eh^3},\ \zeta = ak + 2\pi n, \tag{18.329}$$

where the A-term corresponds to the plate inertia.

The results of calculation of the dispersion relation for some values of $\hat{\varkappa}$ for $A = 30$, $B = 15$ are presented in Fig. 18.13. Note that lowering of the frequency

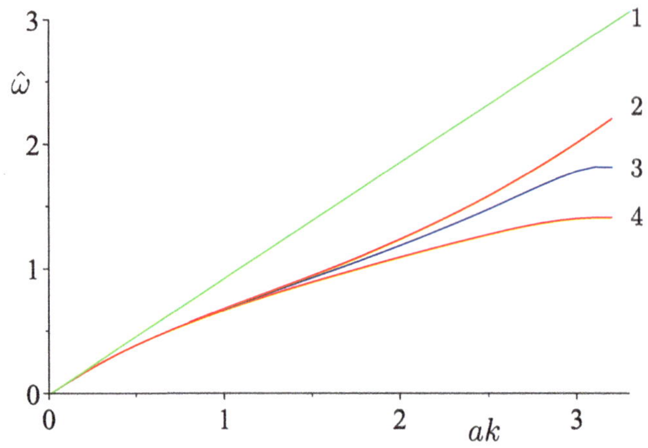

Fig. 18.13 The half-space elastic medium covered by the plate: the Rayleigh wave for the free surface) (1); the Rayleigh wave for the intact plate (with the sliding contact), $\varkappa_M = \infty$ (2); the Rayleigh- Floquet wave for $\varkappa_M = 1/2$ (3) and for $\varkappa_M = 0$ (4)

with the decrease of the hinge stiffness, which occurs at shorter waves, is caused by the decrease of the effective bending stiffness of the plate.

18.7.10 Wave Equations for Elastic Body with Discontinuities

The wave equations (which can be seen, e.g., in Slepyan, 2002, p. 103) is

$$\Delta\varphi - \frac{1}{c_1^2}\frac{\partial^2\varphi}{\partial t^2} = -\frac{f}{\lambda+2\mu}, \; c_1 = \sqrt{\frac{\lambda+2\mu}{Gr}},$$
$$\Delta\psi - \frac{1}{c_2^2}\frac{\partial^2\psi}{\partial t^2} = -\frac{\boldsymbol{p}}{\mu}, \; c_2 = \sqrt{\frac{\mu}{\varrho}}, \tag{18.330}$$

where φ, ψ and f, $\boldsymbol{p}$ are the (scalar and vector) potentials of the displacement $\boldsymbol{u}$ and body force $\boldsymbol{q}$, respectively, such that

$$\boldsymbol{u} = \nabla\varphi + \nabla\wedge\psi, \; \boldsymbol{q} = \nabla f + \nabla\wedge\boldsymbol{p}. \tag{18.331}$$

We now consider the (fundamental) x, y strain plane problem. The task is to determine the potentials f and $\boldsymbol{p}$ for the point-localized generalized forces corresponding to the delta-discontinuity. The latter places at the same point as the discontinuity (here it is $x = y = 0$), where no localized mass is assumed. It follows that the force-discontinuity relations are the same as in statics, where the forces (clearly presented in Slepyan, 1981/1990) depend on time as a parameter. So we have no need showing them explicitly. For mode I discontinuity $u_y(x, \pm 0) = \pm\delta(x)$ (with the shear stresses $\sigma_{xy}(x, 0) = 0$) the forces are

$$q_x = 2\mu\frac{1-4\nu}{3-4\nu}\delta'(x)\delta(y) = \frac{\partial f}{\partial x} + \frac{\partial p}{\partial y},$$
$$q_y = 2\mu\frac{5-4\nu}{3-4\nu}\delta(x)\delta'(y) = \frac{\partial f}{\partial y} - \frac{\partial p}{\partial x}, \qquad (18.332)$$

where $p = p_z$ and ν is Poisson's ratio. Note that only the normal-to-the-plane components of $\boldsymbol{\psi}$ and $\boldsymbol{p}$ remain in the plane problem.

Using the (generalized) Fourier transform it can be found from these equations that

$$f^F = \frac{2\mu}{k_x^2+k_y^2}\left(\frac{1-4\nu}{3-4\nu}k_x^2 + \frac{5-4\nu}{3-4\nu}k_y^2\right),$$
$$p^F = -\frac{8\mu}{3-4\nu}\frac{k_x k_y}{k_x^2+k_y^2}. \qquad (18.333)$$

It follows that

$$f = 2\mu\delta(x)\delta(y) - \frac{4\mu}{\pi(3-4\nu)}\frac{x^2-y^2}{(x^2+y^2)^2},$$
$$p = -\frac{8\mu}{\pi(3-4\nu)}\frac{xy}{(x^2+y^2)^2}, \qquad (18.334)$$

where the expressions apart from the delta-term are also singular at $x = y = 0$. To show all the singularities explicitly, we rearrange the f^F expression as follows

$$f^F = f_{11}^F + f_{12}^F,\ f_{11}^F = 2\mu\frac{1-4\nu}{3-4\nu},\ f_{12}^F = \frac{8\mu}{3-4\nu}\frac{k_y^2}{k_x^2+k_y^2},$$
$$f^F = f_{21}^F + f_{22}^F,\ f_{21}^F = 2\mu\frac{5-4\nu}{3-4\nu},\ f_{22}^F = -\frac{8\mu}{3-4\nu}\frac{k_x^2}{k_x^2+k_y^2}. \qquad (18.335)$$

The original functions

$$f_{11} = 2\mu\frac{1-4\nu}{3-4\nu}\delta(x)\delta(y),\ f_{21} = 2\mu\frac{5-4\nu}{3-4\nu}\delta(x)\delta(y), \qquad (18.336)$$

fully define the forces (18.332), respectively, whereas the other terms f_{12}^F and ψ, and f_{22}^F and ψ cancel out

$$\begin{aligned}\frac{\partial f_{12}}{\partial x}+\frac{\partial \psi}{\partial y}&=0\ (-\mathrm{i}k_x f_{12}^F=\mathrm{i}k_y\psi),\\ \frac{\partial f_{22}}{\partial y}-\frac{\partial \psi}{\partial x}&=0\ (-\mathrm{i}k_y f_{22}^F=-\mathrm{i}k_x\psi).\end{aligned}\tag{18.337}$$

For mode II with the jump $u_x(x,\pm 0)=\pm\delta(x)$ (under the condition $\sigma_{yy}(x,0)=0$), based on the relations as

$$q_x=2\mu\delta(x)\delta'(y)=\frac{\partial f}{\partial x}+\frac{\partial p}{\partial y},\ q_y=2\mu\delta'(x)\delta(y)=\frac{\partial f}{\partial y}-\frac{\partial p}{\partial x},\tag{18.338}$$

we find

$$f^F=\frac{4\mu k_x k_y}{k_x^2+k_y^2},\ p^F=\frac{2\mu(k_x^2-k_y^2)}{k_x^2+k_y^2},\tag{18.339}$$

and outside of $x=y=0$, where the below expressions are singular

$$f=\frac{4\mu}{\pi}\frac{xy}{(x^2+y^2)^2},\ p=-\frac{2\mu}{\pi}\frac{x^2-y^2}{(x^2+y^2)^2}.\tag{18.340}$$

Similarly to mode I, we can represent

$$p^F=2\mu-\frac{4\mu k_y^2}{k_x^2+k_y^2}=-2\mu+\frac{4\mu k_x^2}{k_x^2+k_y^2}.\tag{18.341}$$

Now the first terms of the potential p represent the forces q_x and q_y, respectively, whereas the other terms annihilate together with f.

Thus, with the above remarks, the force potentials are found as in (18.333) and (18.334) (mode I) and (18.339) and (18.340) (mode II).

Finally, for mode III we have a single equation

$$\frac{\partial^2 u_z}{\partial x^2}+\frac{\partial^2 u_z}{\partial y^2}-\frac{1}{c_2^2}\frac{\partial^2 u_z}{\partial t^2}=-\frac{q_z}{\mu}=2\delta(x)\delta'(y).\tag{18.342}$$

References

Abbott, B. P., et al. (2016). Observation of Gravitational Waves from a Binary Black Hole Merger. *Physical Review Letters,116*, Article 061102.

Ablowitz, M. J., & Clarkson, P. A. (1991). *Solitons, nonlinear evolution equations and inverse scattering*. Cambridge: Cambridge University Press.

Achenbach, J. D., & Epstein, H. I. (1967). Dynamic Interaction of a Layer and a Half-Space. *Journal of the Engineering Mechanics Division, ASCE,93*, 27–42.

Adams, S. D. M., Craster, R. V., & Williams, D. P. (2007). Rayleigh waves guided by topography. *Proceedings of the Royal Society A, 463*, 531–550.

Airy, G. B. (1841). Tides and Waves. In *Encyclopaedia Metropolitana*, (Vol. III). Mixed Sciences.

Alshits, V. I. et al. (1984). The phase shift for reflection of elastic waves in hexagonal piezoelectric crystals. *Wave Motion,6*, 459–472.

Anderson, D. L. (1961). Elastic Wave Propagation in Layered Anisotropic Media. *Journal of Geophysical Research, 66*(9), 2953–63.

Antenucci, J. P., & Imburger, J. (2001). Energetics of long internal gravity waves in large lakes. *Limnology and Oceanography, 46*(7), 1760–1773.

Ayzenberg-Stepanenko, M. V., Mishuris, G. S., & Slepyan, L. I. (2014). Brittle fracture in a periodic structure with internal potential energy. Spontaneous crack propagation. *Proceedings of the Royal Society A, 470*, 20140121.

Babich, V. M., & Bilyi, Y. I. (1979). Wave-guide properties of an underwater ridge. *Fluid Dynamics,14*(3), Article 449451.

Balk, A. M., Cherkaev, A. V., & Slepyan, L. I. (2001). Dynamics of chains with non-monotone stress-strain relations. I. Model and numerical experiments. *Journal of the Mechanics and Physics of Solids, 49*, 131–148.

Balk, A. M., Cherkaev, A. V., & Slepyan, L. I. (2001). Dynamics of chains with non-monotone stress-strain relations. II. Nonlinear waves and waves of phase transition. *Journal of the Mechanics and Physics of Solids, 49*, 149–171.

Barenblatt, G. I. (1959a). On equilibrium cracks formed in brittle fracture. *Doklady Akademii Nauk SSSR,127*(1), 47–50.

Barenblatt, G. I. (1959b). Concerning equilibrium cracks forming during brittle fracture: The stability of isolated cracks. Relationships with energetic theories. *Journal of Applied Mathematics and Mechanics,23*(5), 1273–1282.

Barenblatt, G. I., & Cherepanov, G. P. (1960). On the Wedging of a Brittle Body. *Prikladnaya Matematika i Mekhanika, 24*(4), 666–682.

L. I. Slepyan, *Mechanical Waves: Free, Forced, and Transition*,
https://doi.org/10.1007/978-3-032-15864-2

Barenblatt, G. I., & Goldstein, R. V. (1972). Wedging of an elastic body by a slender wedge moving with a constant super-Rayleigh subsonic velocity. *International Journal of Fracture, 8*(4), 427–434.

Barenblatt, G. I., & Sivashinsky, G. I. (1970). Self-similar solutions of the second kind in the problem of propagation of intense shock waves. *PMM / Journal of Applied Mathematics and Mechanics, 34*(4), 655–662.

Barros, A. C., & Gendron, P. J. (2019). A computational bayesian approach for localizing an acoustic scatterer in a stratified ocean environment. *Journal of the Acoustical Society of America, 146* (3), EL245–EL250.

Bažant, Z. P., & Jirásek, M. (2008). Nonlocal integral formulations of plasticity and damage: survey and benchmark tests. *Journal of Engineering Mechanics (ASCE),134*(1), 1–16.

Bažant, Z. P., Caner, F. C., Kim, J.-K., & Le, J.-L. (2007). Unified size effect law for quasibrittle and ductile fracture: its application to notched and unnotched structures. *Engineering Mechanics, 134*(9), 918–930.

Bigoni, D. (2012). *Nonlinear solid mechanics: Bifurcation theory and material instability*. Cambridge University Press.

Bitzek, E., Kermode, G. R., & Gumbsch, P. (2015). Atomistic aspects of fracture. *International Journal of Fracture, 191*, 13–30.

Biot, M. A. (1937). Bending of an infinite beam on an elastic foundation. *Journal of Applied Physics, 12*, 155–164.

Bloch, F. (1928). Über die Quantenmechanik der Elektronen in Kristallgittern. *Zeitschrift für Physik, 52*, 555–600.

Bochner, S. (1959). *Lectures on fourier integrals*. Princeton, NJ: Princeton University Press.

Bok, E., Park, J. J., Choi, H., Han, C. K., Wright, O. B., & Lee, S. H. (2018). Metasurface for water-to-air sound transmission. *Physical Review Letters, 120*, Article 044302.

Bolotovsky, B. M., & Stolyarov, S. N. (1972). On radiation principles for a medium with dispersion. In *Problems fort Theoretical physics* (pp. 267–280). Moscow: Nauka.

Boucher, J. P., Clanet, C., Quéré, D., & Chevy, F. (2017). Popsicle-stick cobra wave. *Physical Review Letters, 119*(8), Article 084301.

Boussinesq, J. (1871). Théorie de l'intumescence liquide, appelée onde solitaire ou de translation, se propageant dans un canal rectangulaire. *Comptes Rendus de l'Académie des Sciences,72*, 755–759.

Boussinesq, J. (1872). Théorie des ondes et des remous qui se propagent le long d'un canal rectangulaire horizontal. *Journal de Mathématiques Pures et Appliquées, Série,2*(17), 55–108.

Bremermann, H. J. (1965). *Distributions, complex variables, and fourier transforms*. Addison-Wesley.

Brillouin, L. (1925). Sur les tensions de radiation. *Annalen der Physik, 4*, 528–586.

Brillouin, L. (1960). *Wave Propagation in Periodic Structures: Electric Filters and Crystal Lattices* (2nd ed.). New York: Dover.

Broberg, K. B. (1999). *Cracks and Fracture*. San Diego: Academic Press.

Brock, L. M. (2002). Sliding of a cup-shaped die on a thermoelastic half-space. *International Journal of Solids and Structures, 39*(15), 4031–4046.

Brun, M., Movchan, A. B., & Slepyan, L. I. (2013). Transition wave in a supported heavy beam. *Journal of the Mechanics and Physics of Solids, 61*(10), 2067–2085.

Brun, M., Guenneau, S., Movchan, A. B., & Bigoni, D. (2010). Filtering and focusing effects for elastic waves in periodic structured media. *Journal of the Mechanics and Physics of Solids, 58*, 1212–1224.

Brun, M., Giaccu, G. F., Movchan, A. B., & Slepyan, L. I. (2014). Transition wave in the collapse of the San Saba Bridge. *Frontiers in Materials, 1*, 12.

Cagniard, L. (1939). *Réflexion et réfraction des ondes séismiques progressives*. Gauthier-Villars.

Cagniard, L. (1962). *Reflection and refraction of progressive seismic waves*. New York: McGraw-Hill.

Cai, C. W., Cheung, Y. K., & Chan, H. C. (1988). Dynamic response of infinite continuous beams subjected to a moving force—An exact method. *Journal of Sound and Vibration, 123*(3), 461–472.
Cano Rodríguez, J. M., Mojumdar, M. R. R., Norniella, J. G., & Orcajo, G. A. (2017). Phase shifting transformer model for direct approach power flow studies. *International Journal of Electrical Power & Energy Systems,91*, 71–79.
Cardozo, T. M., & Nascimento, M. A. C. (2009). Energy partitioning for generalized product functions. *The Journal of Chemical Physics,130*, Article 104102.
Cherkaev, A., Cherkaev, E., Slepyan, L. (2005). Transition waves in bistable structures. I. Delocalization of damage. *Journal of the Mechanics and Physics of Solids, 53*(2), 383–405.
Comninou, M. (1977). The Interface Crack. *Journal of Applied Mechanics, 44*, 631–636.
Comninou, M. (1978). The interface crack: The case of a non-homogeneous bonding material. *Journal of Applied Mechanics, 45*, 287–290.
Craggs, J. W., & Roberts, A. M. (1967). On the motion of a heavy cylinder over the surface of an elastic half-space. *Journal of Applied Mechanics, 34*, 207–209.
Creath, K. & Schmit, J. (2005). Errors in phase-measuring interferometry. In P. K. Rastogi & D. Inaudi (Eds.), *Trends in optical non-destructive testing and inspection* (pp. 67–83). Amsterdam, Netherlands: Elsevier.
Darrigol, O. (2002). *Between hydrodynamics and elasticity: The first five births of the navier–stokes equations*. Archive for History of Exact Sciences.
De Hoop, A. T. (1958/1959). *Representation theorems for the displacement in an elastic solid and their application to elastodynamic diffraction theory* (doctoral thesis, Delft).
Deacon, S., & Arendt, J. (1995). Melatonin-induced temperature suppression and its acute phase-shifting effects correlate in a dose-dependent manner in humans. *Brain Research,688*, 77–85.
Dempsey, J. P., Slepyan, L. I., & Shekhtman, I. I. (1995). Radial cracking with closure. *International Journal of Fracture,73*(3), 233–261.
Derrick, G. (1964). Comments on nonlinear wave equations as models for elementary particles. *Journal of Mathematics and Physics, 5*, 1252–1254.
Dill, K. A., & Bromberg, S. (2003). *Molecular driving forces: Statistical thermodynamics in chemistry and biology*. New York: Garland Science.
Dodd, R. K., Eilbeck, J. C., Gibbon, J. D., & Morris, H. C. (1982). *Solitons and nonlinear wave equations*. London/New York: Academic Press.
Dodd, R. K., Eilbeck, J. C., Gibbon, J. D., & Morris, H. C. (1984). *Solitons and nonlinear wave equations*. NY: Academic.
Eatwell, G. P., & Willis, J. R. (1982). The excitation of a fluid-loaded plate stiffened by a semi-infinite array of beams. *IMA Journal of Applied Mathematics,29*(3), 247–270.
Engelbrecht, J. (2020). *Nonlinear waves in continuous media. In Encyclopedia of Continuum Mechanics*. Berlin/Heidelberg: Springer.
Eshelby, J. D. (1951). The force on an elastic singularity. *Philosophical Transactions A, 244*, 87–112.
Eshelby, J. D. (1956). The continuum theory of lattice defects. *Solid State Physics*, (Vol. 3). pp. 79–144.
Evans, D. V., & Porter, R. (2007). Penetration of flexural waves through a periodically constrained thin elastic plate in vacuo and floating on water. *Journal of Engineering Mathematics,58*(14), Article 317337.
Falkovich, G. (2011). *Fluid Mechanics (A short course for physicists)*. Cambridge University Press.
Falnes, J. (2007). A review of wave-energy extraction. *Marine Structures, 20*, 185–201.
Fermi, E., Pasta, J., & Ulam, S. (1955a). Studies of nonlinear problems. I. Los Alamos report LA-1940.
Fermi, E., Pasta, J., Ulam, S., & Tsingou, M. (1955b). *Studies of Nonlinear Problems*. Los Alamos Scientific Laboratory Report LA-1940.
Floquet, G. (1883). Sur les équations différentielles linéaires à coefficients périodiques. *Annales scientifiques de l'École Normale Supérieure, Série, 2*(12), 47–88.
Freund, L. B. (1990). *Dynamic Fracture Mechanics*. Cambridge: Cambridge University Press.

Friba, L. (1999). *Vibration of solids and structures under moving loads (Third addition)*. USA: ASCE Press.

Fu, Y. B., Rogerson, G. A., & Wang, W. F. (2013). Surface waves guided by topography in an anisotropic elastic half-space. *Proceedings of the Royal Society A, 469*, 20120371.

Galin, L. A., & Cherepanov, G. P. (1966). Self-sustaining failure of a stressed brittle body. *Soviet Physics - Doklady, 11*(3), 267–269.

Galin, L. A., & Gladwell, G. M. L. (2008). *Contact problems: The legacy of L. A. Galin*. Dordrecht: Springer.

Gao, H., Abraham, F. F., & coauthors,. (2001). Continuum and atomistic studies of intersonic crack propagation. *Journal of the Mechanics and Physics of Solids, 49*, 2113–2132.

Garipov, R. M. (1965). Nonsteady waves above an underwater ridge. *Soviet Physics - Doklady, 10*, 194–196.

Garipov, S. N. (2000). Transition through the Critical Velocity for a Moving Load in an Elastic Waveguide. *Technical Physics, 45*(4), 515–518.

Georgiadis, H. G., & Barber, J. R. (1993). On the super-rayleigh subsonic regime of velocities in dynamic rayleigh-wave propagation. *Journal of Applied Mechanics, 60*(2), 443–447.

Geubelle, P. H., & Kubair, D. V. (2001). Intersonic crack propagation in homogeneous media under shear-dominated loading: Numerical analysis. *Journal of the Mechanics and Physics of Solids, 49*(3), 571–587.

Griffith, A. A. (1921). The Phenomena of Rupture and Flow in Solids. *Philosophical Transactions of the Royal Society A,221*, 163–198.

Grigoryan, S. S. (1967). Some problems of the mathematical theory of deformation and fracture of hard rocks. *PMM, 31*(4), 667–686.

Gueorguiev, D., McDaniel, J. G., Dupont, P., & Felsen, L. B. (2000). Analysis of Floquet wave generation and propagation in a plate with multiple arrays of line attachments. *Journal of Sound and Vibration,234*(5), 819–840.

Gurtin, M. E. (2000). *Configurational forces as basic concepts of continuum physics*. New York: Springer.

Hamilton, W. R. (1839). On a General Method in Dynamics Philosophical Transactions of the Royal Society of London. *Part, II*, 247–308.

Haslinger, S. G., Movchan, N. V., Movchan, A. B., & Jones, I. S. (2016). Dynamic interfacial trapping of flexural waves in structured plates. *Proceedings of the Royal Society A, 472*, 20150658.

Haslinger, S. G., Movchan, N. V., Movchan, A. B., Jones, I. S., & Craster, R. V. (2017). Controlling flexural waves in semi-infinite platonic crystals with resonator-type scatterers. *The Quarterly Journal of Mechanics and Applied Mathematics, 70*(3), 216–247.

Hayes, M. (1977). A note on group velocity. *Proceedings of the Royal Society of London, Series A: Mathematical and Physical Sciences, 354*, 533–535.

Huang, J. H., & Liu, H.-K. (1998). On a flat ellipsoidal inclusion or crack in three-dimensional anisotropic media. *International Journal of Engineering Science, 36*(2), 143–155.

Hyun, S., & Zhang, S. (2016). Enhanced phase-scanning profilometry for high-speed 3D shape measurement. *Applied Optics*, 55(21), 5481–5487.

Jawla, S., & Singh, R. K. (2013). Different modulation formats used in optical communication system. *IOSR Journal of Electronics and Communication Engineering*, *8*(4), 84–90.

Joglekar, M., & Mitran, S. (2016). The effect of surface crack arrays on the effective properties of materials. *Journal of the Mechanics and Physics of Solids,94*, 24–47.

Kachanov, M., & Sevostianov. I. (2018). *Micromechanics of materials, with applications*. Springer Nature.

Kaina, N., Causier, A., Bourlier, Y., Fink, M., Berthelot, T., & Lerosey, G. (2017). Slow waves in locally resonant metamaterials line defect waveguides. *Science and Reports, 7*, 15105.

Kapitza, P. L. (1951). Dynamic stability of a pendulum when its point of suspension vibrates. *Soviet Physics JETP, 21*, 588–592.

Kapitza, P. L. (1951). Pendulum with a vibrating suspension. *Uspekhi Fizicheskikh Nauk, 44*, 7–15.

Kaplunov, Y. D., & Muravskii, G. B. (1986). *Mekhanika Tverdogo Tela, 1*, 155–160.

Korteweg, D. J., & de Vries, G. (1895). On the change of form of long waves advancing in a rectangular canal and a new type of long stationary waves. *Philosophical Magazine Series, 39*, 422–443.
Korycansky, D. G. (2013). Energy conservation and partition in CTH impact simulations. Lunar and Planetary Science Conference, p. 1370.
Krylov, V., & Slepyan, L. (1997). Binary wave in a helical fiber. *Physical Review B, 55*(21), 14067–14070.
Krylov, V. I., Parnes, R., & Slepyan, L. I. (1998). Nonlinear waves in an inextensible flexible helix. *Wave Motion, 27*, 117–136.
Kulakhmetova, S. A., Saraikin, V. A., & Slepyan, L. I. (1984). Crack dynamics in a lattice. *Mekhanika Tverdogo Tela,19*(3), 112–118.
Kunin, I. A. (1975). *Elastic Media with Microstructure*. Science, Moscow: Nonlocal theory of elasticity.
Kunin, I. A. (1982). *Elastic media with microstructure*. Springer.
Kunin, I. A. (1983). *Elastic media with microstructure II*. Springer.
Landau, L. D., & Lifshitz, E. M. (1960). *Theory of elasticity*. Oxford: Pergamon Press.
Landau, L. D., & Lifshitz, E. M. (1969). *Mechanics*. Pergamon Press.
Landau, L. D., Liftshitz, E. M., & Pitaevskii, L. P. (1984). *Electrodynamics of continuous media*. New York: Pergamon Press.
Larose, E., Margerin, L., Van Tiggelen, B. A., & Campillo, M. (2004). Weak localization of seismic waves. *Physical Review Letters, 93*(4), 048501–4.
Le, J.-L., Bažant, Z. P., & Bazant, M. Z. (2011). Unified nano-mechanics based probabilistic theory of quasibrittle and brittle structures: I. Strength, crack growth, lifetime and scaling. *Journal of the Mechanics and Physics of Solids, 59*, 1291–1321.
Lighthill, J. (1978). *Waves in fluids*. Cambridge University Press.
Mandelshtam, L. I. (1972). *Lectures of optics, theory of relativity and quantum mechanics*. Moscow: Nauka.
Marder, M., & Gross, S. (1995). Origin of crack tip instabilities. *Journal of the Mechanics and Physics of Solids, 43*, 1–48.
Martinsson, P. G., & Movchan, A. B. (2003). Vibrations of lattice structures and phononic band gaps. *The Quarterly Journal of Mechanics and Applied Mathematics, 56*(1), 45–64.
Mask, L. R., & Jay, B. E. (1967). The partition of energy in standing gravity waves of finite amplitude. *Journal of Geophysical Research, 72*, 573–581.
Maugin, G. A. (1993). *Material inhomogeneities in elasticity*. London: Chapman & Hall.
Maugin, G. A. (2010). *Configurational forces*. Chapman and Hall/CRC Press.
Maugin, G. A., & Rousseau, M. (2015). *Wave momentum and quasi-particles in physical acoustics*. World Scientific.
McIntyre, M. E. (1981). On the "wave momentumâŁž myth. *Journal of Fluid Mechanics, 106*, 331–347.
Milton, G. W., & Willis, J. R. (2007). On modifications of Newton's second law and linear continuum elastodynamics. *Proceedings of the Royal Society A, 463*, 855–880.
Mishuris, G. S., Movchan, A. B., & Slepyan, L. I. (2009). Localised knife waves in a structured interface. *Journal of the Mechanics and Physics of Solids, 57*, 1958–1979.
Mishuris, G. S., & Slepyan, L. I. (2014). Brittle fracture in a periodic structure with internal potential energy. *Proceedings of the Royal Society A, 470*, 20130821.
Mishuris, S., Movchan, A. B., & Slepyan, L. I. (2019). Waves in elastic bodies with discrete and continuous dynamic microstructure. *Philosophical Transactions of the Royal Society A, 378*, 20190313.
Mishuris, G. S., Movchan, A. B., & Slepyan, L. I. (2020). Localized waves at a line of dynamic inhomogeneities. *Journal of the Mechanics and Physics of Solids, 138*, 1–16.
Movchan, A. B., Brun, M., Slepyan, L. I., & Glaccu, G. F. (2015). Dynamic multi-structure in modelling a transition flexural wave. *Mathematika, 61*, 444–456.

Munk, W. H. (1974). Sound channel in an exponentially stratified ocean. *Journal of the Acoustical Society of America, 55*, 220.

Needleman, A., & Rosakis, A. J. (1999). The effect of bond strength on dynamic interfacial crack growth. *Journal of the Mechanics and Physics of Solids, 47*(3), 569–596.

Ostrovsky, A. A., & Potapov, A. I. (1988). *The modulated waves in linear media with dispersion*. Gor'ky University.

Ozisik, M., & Akbarov, S. D. (2003). Rayleigh-wave propagation in a half-plane covered with a prestressed layer under complete and incomplete interfacial contact. *Mechanics of Composite Materials,39*, 177–182.

Peets, T., Tamm, K., & Engelbrecht, J. (2021). Modeling of complex signals in nerves. *Mathematics and Computers in Simulation, 182*, 381–392.

Peskin, C. S. (2010). *Wave momentum*. NYU: Courant Institute of Mathematical Sciences.

Pethiyagoda, R., McCue, S. W., & Moroney, T. J. (2015). Wake angle for surface gravity waves on a finite depth fluid. *Physics of Fluids,27*, Article 061701.

Pohozaev, S. (1965). Eigenfunctions of the equation $\Delta u + \lambda f(u) = 0$. *Soviet Mathematics - Doklady, 6*, 1408–1411.

Porter, R. (2019). An extended linear shallow-water equation. *Journal of Fluid Mechanics, 876*, 413–427.

Prigogine, I. (1980). *From being to becoming*. Freeman and Company: W.H.

Rabaud, M., & Moisy, F. (2013). Ship wakes: Kelvin or mach angle? *Physical Review Letters, 110*, Article 214503.

Radi, E., & Loret, B. (2008). Mode I intersonic crack propagation in poroelastic media. *Mechanics of Materials, 40*, 524–548.

Rakhmatulin, K. A. (1947). Propagation of a wave of unloading. *Prikladnaya Matematika i Mekhanika, 11*(3), 379–400.

Rayleigh, L. (1877). *The Theory of Sound* (Vol. 1). London: Macmillan.

Rayleigh, L. (1885). On waves propagated along the plane surface of an elastic solid. *Proceedings of the London Mathematical Society, s1–17*(1), 4–11.

Rayleigh, L. (1902). On the pressure of vibrations. *Philosophical Magazine, 3*, 338–346.

Rayleigh, L. (1905). On the momentum and pressure of gaseous vibrations. *Philosophical Magazine, 10*, 364–374.

Rice, J. R., & Levy, N. (1972). The part-through surface crack in an elastic plate. *Journal of Applied Mechanics,39*(1), 185–194.

Roger, D. (2020). *Viscoelastic waves and rays in layered media* (2nd ed.). Cambridge University Press.

Russell, J. S. (1845a). *Report on waves*. British Association for the Advancement of Science.

Russell, J. S. (1845b). *Report on waves: Made to the meetings of the british association in 1842–43*. London.

Samudrala, O., Huang, Y., Rosakis, A. J., & Geubelle, P. H. (2002). Subsonic and intersonic mode II crack propagation with a rate-dependent cohesive zone. *Journal of the Mechanics and Physics of Solids, 50*, 1231–1268.

Saraikin, V. A., & Slepyan, L. I. (1979). Plane problem of the dynamics of a crack in an elastic solid. *Mechanics of Solids, 14*(4), 46–62.

Scott, R. H., Roche, P. A., & Knight, G. (2005). The use of the Hilbert-Huang transform for the analysis of structural vibration data. *Proceedings of the Institution of Civil Engineers - Structures and Buildings,158*(1), 47–56.

Slepyan, L. I. (1967). The strain wave in a bar with vibration-isolated masses. *Mechanics of Solids, 2*, 57–64.

Slepyan, L. I. (1972). *Transient elastic waves*. Leningrad: Sudostroenie.

Slepyan, L. I. (1981). Dynamics of a crack in a lattice. *Soviet Physics - Doklady, 26*, 538–540.

Slepyan, L. I. (1982). The relation between the solutions of mixed dynamic problems for a continuous elastic medium and a lattice. *Soviet physics - Doklady,27*, 771–772.

Slepyan, L. I. (2016). Mechanical wave momentum from the first principles. *Wave Motion, 68*, 283–290.
Slepyan, L. I., & Brun, M. (2012). Driving forces in moving-contact problems of dynamic elasticity: Indentation, wedging and free sliding. *Journal of the Mechanics and Physics of Solids,60.*
Slepyan, L. I., & Troyankina, L. V. (1969). The failure of brittle bar under longitudinal impact. *Mechanics of Solids, 4*, 57–64.
Slepyan, L. I., & Troyankina, L. V. (1983). On the theory of crack grows under cyclic loads. *Mechanics of Solids, 18*, 101–124.
Slepyan, L. I., & Troyankina, L. V. (1984). Fracture wave in a chain structure. *Journal of Applied Mechanics and Technical Physics,25*(6), 921–927.
Slepyan, L., Krylov, V., & Parnes, R. (1995). Solitary waves in an inextensible, flexible, helicoidal fiber. *Physical Review Letters, 74*(14), 2725–2728.
Slepyan, L. I., Movchan, A. B., & Ayzenberg-Stepanenko, M. V. (1995). Spontaneous crack propagation in a lattice model. *Journal of the Mechanics and Physics of Solids, 43*(8), 1497–1521.
Slepyan, L., Krylov, V., & Rosenau, P. (1998). Solitary waves in a flexible, arbitrary elastic helix. *Journal of Engineering Mechanics, 124*(9), 966–970.
Slepyan, L. I. (2001). Feeding and dissipative waves in fracture and phase transition I. *Journal of the Mechanics and Physics of Solids, 49*(3), 25–67.
Slepyan, L. I. (2001). Feeding and dissipative waves in fracture and phase transition II. *Journal of the Mechanics and Physics of Solids, 49*(3), 69–106.
Slepyan, L. I. (2001). Feeding and dissipative waves in fracture and phase transition III. *Journal of the Mechanics and Physics of Solids, 49*(12), 2839–2875.
Slepyan, L. I. (2002). *Models and phenomena in fracture mechanics*. Berlin: Springer.
Slepyan, L. I., & Ayzenberg-Stepanenko, M. V. (2004). Localized transition waves in bistable-bond lattices. *Journal of the Mechanics and Physics of Solids, 52*(7), 1447–1479.
Slepyan, L., Cherkaev, A., & Cherkaev, E. (2005). Transition waves in bistable structures II. *Journal of the Mechanics and Physics of Solids, 53*(2), 407–436.
Slepyan, L. I., Mishuris, G. S., & Movchan, A. B. (2010). Crack in a lattice waveguide. *International Journal of Fracture, 162*, 91–106.
Slepyan, L. I. (2010). Dynamic crack growth under Rayleigh wave. *Journal of the Mechanics and Physics of Solids, 58*, 636–655.
Slepyan, L. I. (2010). Wave radiation in lattice fracture. *Acoustical Physics, 56*(6), 962–971.
Slepyan, L. I. (2015a). On the energy partition in oscillations and waves. *Proceedings of the Royal Society A, 471*, 20140838.
Slepyan, L. I., Ayzenberg-Stepanenko, M. V., & Mishuris, G. S. (2015). Forerunning mode transition in a continuous waveguide. *Journal of the Mechanics and Physics of Solids, 78*, 32–45.
Slepyan, L. I. (2018). Structural discontinuity as generalized strain. *International Journal of Engineering Sciences, 130*, 199–214.
Slepyan, L. I. (2019). About some projects with attendant circumstances. *Philosophical Transactions of the Royal Society A, 377*, 20190097.
Slepyan, L. I. (2022). Phase shift in forced oscillations and waves. *European Journal of Mechanics A Solids,96*, Article 104762.
Slepyan, L. I. (2022). *Forced waves in a uniform waveguide.* https://doi.org/10.1016/j.ijengsci.2022.103628
Sobolev, S. L. (1935). Le problème de Cauchy dans l'espace des distributions. *Doklady Akademii Nauk SSSR,3*(7), 291–294.
Sobolev, S. L. (1937). Some questions of the propagation theory of oscillations. ONTI.
Sommerfeld, A. (1904) "Zur Elektronentheorie (3 Teile)," Nachrichten der Kgl. Gesellschaft der Wissenschaften zu Gõttingen, math.-naturwiss. Klasse, 99–130, 363–439; 201–235 (1905).
Sommerfeld, A. (1949). *Partial differential equations in physics*. New York: Academic Press.
Stokes, G. G. (1847). On the theory of oscillatory waves. *Transactions of the Cambridge Philosophical Society, 8*, 441–455.

Verboomen, J., Van Hertem, D., Schavemaker, P. H., Kling, W. L., & Belmans, R. (2005). Phase shifting transformers: principles and applications. In *International Conference on Future Power Systems 2005*. IEEE.

Whitham, G. B. (1974). *Linear and nonlinear waves*. NY: Wiley.

Wiener, N., & Hopf, E. (1931). Über eine Klasse singulärer Integralgleichungen. *Sitzungsberichte der Preussischen Akademie der Wissenschaften, Physikalisch-mathematische Klasse,* 696–706.

Yoffe, E. H. (1951). The Moving Griffith Crack. *Philosophical Magazine, 42*, 739–750.

Zel'dovich, Ya. B., & Raizer, Yu. P. (2002). *Physics of Shock Waves and High-Temperature Hydrodynamic Phenomena. Mineola*. NY: Dover.

Zilbergleit, A. S., & Nuller, B. M. (1977). Generalized orthogonality of homogeneous solutions in dynamic problems of elasticity theory. *Doklady Akademii Nauk SSSR, 234*(2), 333–335.

Zwillinger, D. (1997). *Handbook of differential equations* (3rd ed.). Academic.

The manufacturer's authorised representative in the EU is Springer Nature Customer Service Centre GmbH, Europaplatz 3, 69115 Heidelberg, Germany. If you have any concerns regarding our products, please contact ProductSafety@springernature.com

Printed and bound by CPI Group (UK) Ltd, Croydon, CR0 4YY
17/09/2026
02218437-0010